INTRODUCTION TO SUPERSYMMETRY AND SUPERGRAVITY

INTRODUCTION TO SUPERSYMMETRY AND SUPERGRAVITY

Extended Second Edition

Peter West

King's College
University of London
Strand
London

World Scientific
Singapore • New Jersey • London • Hong Kong

Published by

World Scientific Publishing Co. Pte. Ltd.

P O Box 128, Farrer Road, Singapore 9128

USA office: 687 Hartwell Street, Teaneck, NJ 07666

UK office: 73 Lynton Mead, Totteridge, London N20 8DH

ISBN 981-02-0098-6
      981-02-0099-4 (pbk)

Printed in Singapore by Singapore National Printers Ltd.

Preface to the Second Edition

"When swept out of its normal channel, life scatters into innumerable streams. It is difficult to foresee which it will take in its treacherous and winding course. Where today it flows in shallows, like a rivulet over sand banks, so shallow that the shoals are visible, tomorrow it will flow quickly and full."

Mikhail Sholokhov

"Men willingly believe what they wish"

Caesar

The first edition of this book was completed in 1986, however, much of the material was written long before. It focused on the development of four-dimensional supersymmetric models including supergravity with emphasis on their ultraviolet properties. Already in 1983, our understanding of the finiteness of rigid supersymmetric theories had led to the realization that supersymmetry was most unlikely to solve the celebrated inconsistency of quantum mechanics and gravity. This, and the fact that many aspects of supersymmetric theories had been worked out, lead to a search for new ideas. It was inevitable that string theory, which had been extensively developed in the late 1960's and early 1970's would be revived from its dormant state. We recall that supersymmetry was discovered independently in two ways, one of which was within the superstring which contained it as a symmetry. Also during the dormant stage, theoreticians had developed BRST symmetry, conformal models, the vertex operator representation of Lie algebras, the use of the gauge group E_8 for grand unified models, even within the context of ten-dimensional supersymmetric theories and gained further understanding of anomalies. All these enabled the solution of some of the problems which the original pioneers of string theory had encountered.

The superstring possesses two-dimensional supersymmetry on its world sheet. Consequently, it requires a knowledge of two-dimensional supersymmetry algebras and the models which realize them. The more interesting string theories also possess space-time supersymmetry and it is thought that any contact they have with reality is via a four-dimensional supersymmetric theory. Hence, the study of string theory involves a knowledge of supersymmetry in general.

Although there are a number of arguments to suggest that supersymmetry may have a place in Nature, there is at present no experimental evidence for supersymmetry. Hopefully, the next generation of accelerators will find such evidence. However, even if this is not the case it would seem that supersymmetry has entered the culture of both mathematics and theoretical physics and will be the subject of study for the foreseeable future both by itself and within the context of superstring theories. Supersymmetry has introduced into theoretical physics, almost at a subconscious level, a number of ideas which are likely to be important. This process has become further apparent from the development of string theory; our understanding of which has been revealed to be far from complete and probably requiring the introduction of new physical concepts. One likely casualty is our notion of space-time, our perception of which has been subtlety changing as supersymmetry and string theory have evolved.

Taking into account the above developments, I have taken this opportunity to add to the second edition five and a half new chapters on two-dimensional supersymmetry, extending the short discussion of this subject which already existed in the first edition. After an introduction, the topics discussed are two-dimensional supersymmetry algebras, their irreducible representations, rigid and local (i.e., supergravity) theories of supersymmetry both in x-space and superspace. These theories include the actions of the superstring and the heterotic string. We also give a discussion of the superconformal algebras in two dimensions and the final chapter contains an account of Green's functions and operator product expansions for $N = 2$ superconformal models.

These chapters should equip the reader with the supersymmetry necessary for the study of superstring theory and superconformal models. However, another virtue is that two-dimensional theories, having fewer component fields, are in general much simpler than their four-dimensional analogues and so illustrate more clearly many of the principles of supersymmetric theories given in the earlier chapters. I have also made a number of minor changes in the original text and supplied a number of problems in Appendix C. I have, however, resisted any temptation to include any string theory other than that contained in the first edition. The reader may find much preliminary material for a book, by this author, on string theory in the reviews of Refs. 245 and 258.

I would like to thank Paul Howe and George Papadopoulos for many discussions on two-dimensional supersymmetry and Myrna Guarisco for typing the manuscript.

Geneva
June, 1989

Preface

This book has evolved out of a number of courses that I have given on supersymmetry and supergravity. While giving these lectures I became convinced of the need for a book which contained a pedagogical introduction to most aspects of supersymmetric theories. Although the content of this book has been to some extent constrained by those areas that were my research activities at the time I wrote my lecture notes, most major areas relevant for an introduction are covered, as well as some more advanced topics. Some of the latter are concerned with the quantum properties of supersymmetric theories and the construction of supergravity theories. The final chapter contains a discussion of free gauge covariant string field theory.

Supersymmetric theories have had an important influence on the theoretical physics community. It has encouraged the quest for a single unified theory of physics and has lead to a wider understanding of what can constitute the space-time we live in. On a more general level, it has made more acceptable the study of ideas which are at first sight rather distantly related to experimental data. Some effort has been made to present a step-by-step and necessarily technical derivation of the results. However, by studying the subject itself, it is hoped that the reader will also come to appreciate more fully the concepts that may be abstracted from supersymmetric theories.

I would like to express my gratitude to King's College, the California Institute of Technology and CERN where this manuscript was written and typed. I also wish to thank my collaborators for the insights they have shared with me.

London
June, 1986

Peter West

Contents

Preface to the Second Edition   v

Preface   vii

1. Introduction   1
2. The Supersymmetry Algebra   5
3. Alternative Approaches to the Supersymmetry Algebra   15
4. Immediate Consequences of the Supersymmetry Algebra   17
5. The Wess-Zumino Model   19
6. $N = 1$ Supersymmetric Gauge Theory: Super QED   25
7. $N = 1$ Yang-Mills Theory and the Noether Technique   27
8. The Irreducible Representations of Supersymmetry   32
9. Simple Supergravity: Linearized $N = 1$ Supergravity   42
10. Invariance of Simple Supergravity   50
11. Tensor Calculus of Rigid Supersymmetry   53
     11.1 Supermultiplets   53
     11.2 Combination of Supermultiplets   56
     11.3 Action Formulas   58
12. Theories of Extended Rigid Supersymmetry   64
     12.1 $N = 2$ Yang-Mills   66
     12.2 $N = 2$ Matter   70
     12.3 The General $N = 2$ Rigid Theory   72
     12.4 The $N = 4$ Yang-Mills Theory   74
13. The Local Tensor Calculus and the Coupling of Supergravity
     to Matter   75
14. Superspace   85
     14.1 An Elementary Account of $N = 1$ Superspace   85
     14.2 $N = 1$ Superspace   87
     14.3. $N = 2$ Superspace   108
15. Superspace Formulations of Rigid Supersymmetric Theories   113
     15.1 $N = 1$ Superspace Theories: The Wess-Zumino Model   113
     15.2 $N = 1$ Yang-Mills Theory   115

15.3 A Geometrical Approach to $N = 1$ Supersymmetric
      Yang-Mills Theory   119
15.4 $N = 2$ Superspace Theories   126
16. Superspace Formulation of $N = 1$ Supergravity   132
    16.1 Geometry   132
    16.2 The Superspace Constraints   136
    16.3 Analysis of the Superspace Constraints   143
    16.4 Superspace Supergravity from x-Space Supergravity   149
17. $N = 1$ Super-Feynman Rules   157
    17.1 General Formalism   158
    17.2 The Wess-Zumino Multiplet   160
    17.3 Super Yang-Mills Theory   166
    17.4 Applications of $N = 1$ Super-Feynman Rules   169
    17.5 Divergence in Super-Feynman Graphs   174
    17.6 One-Loop Infinities in a General $N = 1$ Supersymmetric
          Theory   177
    17.7 The Background-Field Method   180
    17.8 The Superspace Background-Field Method   182
18. Ultra-violet Properties of the Extended Rigid Supersymmetric
    Theories   186
    18.1 The Anomalies Argument   187
    18.2 The Non-Renormalization Argument   190
    18.3 Finite $N = 2$ Supersymmetric Rigid Theories   193
    18.4 Explicit Breaking and Finiteness   195
19. Spontaneous Breaking of Supersymmetry and Realistic Models   204
    19.1 Tree-Level Breaking of Supersymmetry   205
    19.2 Quantum Breaking of Supersymmetry   208
    19.3 The Gauge Hierarchy Problem   209
    19.4 Comments on the Construction of Realistic Models   213
20. Currents in Supersymmetric Theories   216
    20.1 General Considerations   216
    20.2 Currents in the Wess-Zumino Model   226
    20.3 Currents in $N = 1$ Super Yang-Mills Theory   228
    20.4 Quantum Generated Anomalies   228
    20.5 Currents and Supergravity Formulations   230
21. Introduction to Two-Dimensional Supersymmetric Models and
    Superstring Actions   233
    21.1 2-Dimensional Models of Rigid Supersymmetry   233
    21.2 Coupling of 2-Dimensional Matter to Supergravity   239

22. Two-Dimensional Supersymmetry Algebras 248
    22.1 Conventions in Two-Dimensional Minkowski and Euclidean
        Spaces 248
    22.2 Superalgebras in Two-Dimensions 259
    22.3 Irreducible Representations of Two-Dimensional
        Supersymmetry 266
23. Two-Dimensional Superspace and the Construction of Models 270
    23.1 Minkowski Superspaces 270
    23.2 Euclidean Superspaces 293
24. Superspace Formulations of Two-Dimensional Supergravities 297
    24.1 Geometrical Framework 297
    24.2 (1, 0) Supergravity 301
    24.3 (1, 1) Supergravity 312
25. The Superconformal Group 318
    25.1 The Conformal Group in Arbitrary Dimensions 319
    25.2 The Two-Dimensional Conformal Group 323
    25.3 The (1, 1) Superconformal Group 327
    25.4 The (2, 2) Superconformal Group 334
26. Green Functions and Operator Product Expansions in (2, 2)
    Superconformal Models 341
    26.1 Two and Three Point Green Functions 341
    26.2 Chiral Correlators in (2, 2) Superconformal Models 351
    26.3 Super Operator Product Expansions 356
27. Gauge Covariant Formulation of Strings 363
    27.1 The Point Particle 363
    27.2 The Bosonic String 365
    27.3 Oscillator Formalism 369
    27.4 The Gauge Covariant Theory at Low Levels 371
    27.5 The Finite Set 376
    27.6 The Infinite Set 378
    27.7 The Master Set 379
    27.8 The On-Shell Spectrum of the Master Set 386
Appendix A: An Explanation of our Choice of Conventions 389
Appendix B: List of Reviews and Books 401
Appendix C: Problems 403
References 411
Subject Index 421

Chapter 1

Introduction

"Thus fortune on our first endeavour smiles"

Virgil

This book has grown out of a number of lectures given at various summer schools and is intended to be an introduction to Supersymmetry and Supergravity. It is pedagogical in the sense that complete proofs are almost always given. After more than ten years of intensive work on supersymmetric theories, it is impossible to cover the vast literature in a single volume. Rather than discuss many topics in an incomplete way, we have decided to restrict our attention to those areas which are essential for a beginner to learn, are important for the development of the subject as a whole and can be covered in a reasonably short space in a pedagogical manner. This included almost all of rigid supersymmetry and $N = 1$ supergravity. Excluded were the extended supergravity theories, superstring theories, Kaluza-Klein dimensional reduction and an extensive discussion of the phenomenological implications of supersymmetry. The reader is referred to Appendix B for some reviews on these important subjects.

In view of the introductory nature of this review, we have not attempted to give the most systematic treatment, or give the complete mathematical background, for fear that the reader should become lost in the details. Often, only those features which are necessary for a step-by-step derivation of the desired result are given. In this sense, this review is rather low-brow, but hopefully easily understandable. It is to be expected that the reader should feel that he is on the tip of the iceberg of some deeper framework. This is particularly true, for example, in Chapter 14 on superspace.

A substantial part of the review has been written at the same time as the book of Ref. 6. This is a more scholarly work in which the conceptual, as well as mathematical theory behind supersymmetry is given. From this base the entire theory of supersymmetry is derived in detail in a complete manner. The resulting book is somewhat lengthy, but is systematic and complete in those topics that it covers. Care has been taken to avoid too much overlap, and it is hoped that these two works will complement each other. In the later chapters, however, there is some similarity and I would like to thank Peter van Nieuwenhuizen for reading these chapters and making many helpful suggestions.

*　　*　　*

Supersymmetry was discovered by Golfand and Liktman.[1] A theory invariant under a non-linear realization of supersymmetry was given by Akulov and Volkov.[2] In a separate development,[7] supersymmetry was introduced as a two-dimensional symmetry of the world sheet within the context of string theories. However, supersymmetry only became widely known when this two-dimensional symmetry was generalized to four dimensions and used to construct the Wess-Zumino model.[3]

To this day, there is no firm evidence that supersymmetry is realized in Nature. Neither is there any completely compelling reason to believe that supersymmetry is required to resolve any of the paradoxes of our present theories of physics. However, it is possible that supersymmetry may be required to explain the new phenomena found already, or in the near future, in particle accelerators. On the theoretical side, there are also some reasons to hope that supersymmetry is required. In Nature, there are at least two vastly different energy scales: the weak scale (100 GeV) and the Planck scale (10^{19} GeV). There are also some reasons to believe that there should be one or more intermediate scales. Although the origin of these vastly different scales is unknown, it is considered to be natural to have a theory in which phenomena at the lowest scale are not polluted by much larger effects arising from the higher scales. Some supersymmetric theories are natural in this sense, and it is a consequence of this argument that the superpartners of the observed particles ought to have masses around the weak scale and hence should be seen in the near future (see Chapter 19). This particular property of supersymmetric theories is a consequence of the fact that the spin-zero states are related by supersymmetry to states of spin $\frac{1}{2}$.

The most uncertain aspect of the standard model of weak and electromagnetic interactions is the spin-zero sector. In fact, many of the 19 free parameters of this model arise due to the undetermined interactions of the spin-zero fields with themselves and the spin-$\frac{1}{2}$ fields. It is natural to hope that some of these free parameters are fixed in a supersymmetric theory. In fact, this has not been achieved within the context of supersymmetric models with only one supersymmetry, but it is likely to be the case should one succeed in constructing a realistic model with more than one supersymmetry.

On a different mass scale, it has been hoped that supersymmetric theories would provide a consistent theory of gravity and quantum mechanics and, at the same time, unify gravity with all the other forces of nature. The most promising candidates that could achieve this long-awaited result are superstring theories. In this context it is, however, worth remembering that gravity may not be a fundamental force, but may arise due to a dynamical mechanism.

It is striking that supersymmetry seems as if it might provide answers to so many of the outstanding unanswered questions. What is remarkable is that, were supersymmetry to answer these questions, then it would relate phenomena at the Planck scale to those at the weak scale.

Although supersymmetry was not universally embraced at its conception, in more recent years it has become the major research interest of a substantial number of the theoretical physics community. Such a devotion of resources to a subject which has no firm contact with the observed world is not an uncommon feature of human behaviour, but it is a new phenomenon for theoretical physics. This is partly a consequence of the hopes for super-symmetry outlined above, as well as the very rich structure of supersymmetric theories that have led to a succession of developments which in turn fuelled further interest. Some of these developments include the construction of rigid extended supersymmetry theories and supergravity theories; superconformal invariance, that is, the finiteness of a class of the former theories; the further development of superstring theories; the construction of realistic supersymmet-ric models and the use of supersymmetry to simplify proofs of mathematical theorems.

This book begins, of course, with the supersymmetry algebra. It is shown that this algebra is a natural consequence, within the framework of quantum field theory, of demanding either a Fermi-Bose symmetry or that physics should realize the Poincaré group and an internal symmetry group in a non-trivial manner (see Chapters 2, 3, 4 and 5).

The irreducible representations of this supersymmetry algebra (Chapter 8) describe the possible on-shell states of supersymmetric theories. It is explained how one may systematically construct all supersymmetric theories from a knowledge of only their on-shell states. One simply finds the supersymmetry transformations that rotate the fields subject to their field equations (i.e., the on-shell states) into each other, and then finds an invariant action constructed from unconstrained fields. This technique is illustrated for $N = 1$ rigid super-symmetric theories in Chapters 5, 6 and 7, and the theories of rigid super-symmetry are constructed in Chapter 12. The word "rigid" means that the parameter of supersymmetry transformations is independent of space-time.

The $N = 1$ supergravity theory is constructed in Chapter 9, by first finding the linearized theory which is invariant under only rigid supersymmetry. Then, using the Noether method, we find the complete locally supersymmetric theory. The invariance of this result is established in Chapter 11.

Having obtained these supersymmetric theories, we can then couple them together. This is most easily achieved using the supersymmetric tensor cal-culus. As the name suggests, by analogy with general relativity, one takes multiplets of supersymmetry, that is, sets of x-space fields that transform into themselves under supersymmetry, and finds rules for combining them to form new supermultiplets. This, with a knowledge of how to construct invariants, allows one to find the most general coupling. In Chapters 12 and 13 this is performed for rigid and local supersymmetry respectively.

This supersymmetry tensor calculus is given in x-space and allows one to keep supersymmetry manifest at every step of the construction. It cannot, however, be used to keep supersymmetry manifest during quantum calculations. This is achieved at the classical and quantum level by using superspace (Chapter 14). Here, the supersymmetry arises as transformations on an 8-dimensional coset space called superspace. Four of the co-ordinates of superspace commute, while the remaining four anticommute. This construction is a generalization of the realization of the Poincaré group on Minkowski space. Supermultiplets are then given by fields, called superfields, on superspace. The formulation of the theories of rigid supersymmetry and $N = 1$ supergravity are given in Chapters 15 and 16.

In Chapter 17 we discuss how to calculate quantum effects in superspace, and in Chapter 18 the ultraviolet properties of supersymmetric theories are derived. These include the existence of a large class of finite quantum field theories. In Chapter 19, a brief discussion of the theoretical aspects of the construction of realistic models is given. Chapter 20 discusses the supermultiplet structure to which the currents in supersymmetric theories belong. This enables one to find formulations of supergravity, and also has implications for the quantum effects in supersymmetric theories.

Finally, with the new interest in string theories in mind we discuss in Chapter 21, the two-dimensional supersymmetric theories that underlie superstring theories, while, in Chapter 22, we give an introduction to the gauge covariant formulation of string theories.

Chapter 2

The Supersymmetry Algebra

In the 1960's, with the growing awareness of the significance of internal symmetries such as SU(2) and larger groups, physicists attempted to find a symmetry which would combine in a non-trivial way the space-time Poincaré group with an internal symmetry group. After much effort it was shown that such an attempt was impossible within the context of a Lie group. Coleman and Mandula[4] showed on very general assumptions that any Lie group which contained the Poincaré group P, whose generators P_a and J_{ab} satisfy the relations

$$[P_a, P_b] = 0$$

$$[P_a, J_{bc}] = (\eta_{ab}P_c - \eta_{ac}P_b)$$

$$[J_{ab}, J_{cd}] = -(\eta_{ac}J_{bd} + \eta_{bd}J_{ac} - \eta_{ad}J_{bc} - \eta_{bc}J_{ad}) \qquad (2.1)$$

and an internal symmetry group G with generators T_s such that

$$[T_r, T_s] = f_{rst}T_t \qquad (2.2)$$

must be a direct product of P and G; or in other words

$$[P_a, T_s] = 0 = [J_{ab}, T_s] \qquad (2.3)$$

They also showed that G must be of the form of a semisimple group with additional $U(1)$ groups.

It is worthwhile to make some remarks concerning the status of this no-go theorem. Clearly there are Lie groups that contain the Poincaré group and internal symmetry groups in a non-trivial manner; however the theorem states that these groups lead to trivial physics. Consider, for example, two-body scattering; once we have imposed conservation of angular momentum and momentum the scattering angle is the only unknown quantity. If there were a Lie group that had a non-trivial mixing with the Poincaré group then there would be further space-time associated generators. The resulting conservation laws will further constrain, for example, two-body scattering, and so the scattering angle can only take on discrete values. However, the scattering process is expected to be analytic in the scattering angle, θ, and hence we must conclude that the process does not depend on θ at all.

Essentially the theorem shows that if one used a Lie group that contained an internal group which mixed in a non-trivial manner with the Poincaré group then the S-matrix for all processes would be zero. The theorem assumes among other things, that the S-matrix exists and is non-trivial, the vacuum is nondegenerate and that there are no massless particles. It is important to realize that the theorem only applies to symmetries that act on S-matrix elements and not on all the other many symmetries that occur in quantum field theory. Indeed it is not uncommon to find examples of the latter symmetries. Of course, no-go theorems are only as strong as the assumptions required to prove them.

In a remarkable paper Gelfand and Likhtman[1] showed that provided one generalized the concept of a Lie group one could indeed find a symmetry that included the Poincaré group and an internal symmetry group in a non-trivial way. In this section we will discuss this approach to the supersymmetry group; having adopted a more general notion of a group, we will show that one is led, with the aid of the Coleman-Mandula theorem, and a few assumptions, to the known supersymmetry group.

Since the structure of a Lie group, at least in some local region of the identity, is determined entirely by its Lie algebra it is necessary to adopt a more general notion than a Lie algebra. The vital step in discovering the supersymmetry algebra is to introduce generators, $Q_\alpha{}^i$, which satisfy anti-commutation relations, i.e.

$$\{Q_\alpha{}^i, Q_\beta{}^j\} = Q_\alpha{}^i Q_\beta{}^j + Q_\beta{}^j Q_\alpha{}^i$$

$$= \text{some other generator} \qquad (2.4)$$

The significance of the i and α indices will become apparent shortly. Let us therefore assume that the supersymmetry group involves generators P_a, J_{ab}, T_s and possibly some other generators which satisfy commutation relations, as well as the generators $Q_\alpha{}^i (i = 1, 2, \ldots, N)$. We will call the former generators which satisfy Eqs. (2.1), (2.2) and (2.3) to be even and those satisfying Eq. (2.4) to be odd generators.

Having let the genie out of the bottle we promptly replace the stopper and demand that the supersymmetry algebra have a Z_2 graded structure. This simply means that the even and odd generators must satisfy the rules:

$$[\text{even}, \text{even}] = \text{even}$$

$$\{\text{odd}, \text{odd}\} = \text{even}$$

$$[\text{even}, \text{odd}] = \text{odd} \qquad (2.5)$$

We must still have the relations

$$[P_a, T_s] = 0 = [J_{ab}, T_s] \tag{2.6}$$

since the even (bosonic) subgroup must obey the Coleman-Mandula theorem.

Let us now investigate the commutator between J_{ab} and $Q_\alpha{}^i$. As a result of Eq. (2.5) it must be of the form

$$[Q_\alpha{}^i, J_{ab}] = (b_{ab})_\alpha{}^\beta Q_\beta{}^i \tag{2.7}$$

since by definition the $Q_\alpha{}^i$ are the only odd generators. We take the α indices to be those rotated by J_{ab}. As in a Lie algebra we have some generalized Jacobi identities. If we denote an even generator by B and an odd generator by F we find that

$$[[B_1, B_2], B_3] + [[B_3, B_1], B_2] + [[B_2, B_3], B_1] = 0$$

$$[[B_1, B_2], F_3] + [[F_3, B_1], B_2] + [[B_2, F_3], B_1] = 0$$

$$\{[B_1, F_2], F_3\} + \{[B_1, F_3], F_2\} + [\{F_2, F_3\}, B_1] = 0$$

$$[\{F_1, F_2\}, F_3] + [\{F_1, F_3\}, F_2] + [\{F_2, F_3\}, F_1] = 0 \tag{2.8}$$

The reader may verify, by expanding each bracket that these relations are indeed identically true.

The identity

$$[[J_{ab}, J_{cd}], Q_\alpha{}^i] + [[Q_\alpha{}^i, J_{ab}], J_{cd}] + [[J_{cd}, Q_\alpha{}^i], J_{ab}] = 0 \tag{2.9}$$

upon use of Eq. (2.7) implies the result

$$[b_{ab}, b_{cd}]_\alpha{}^\beta = -\eta_{ac}(b_{bd})_\alpha{}^\beta - \eta_{bd}(b_{ac})_\alpha{}^\beta$$
$$+ \eta_{ad}(b_{bc})_\alpha{}^\beta + \eta_{bc}(b_{ad})_\alpha{}^\beta \tag{2.10}$$

This means that the $(b_{cd})_\alpha{}^\beta$ form a representation of the Lorentz algebra or in otherwords the $Q_\alpha{}^i$ carry a representation of the Lorentz group. We will select $Q_\alpha{}^i$ to be in the $(0, \frac{1}{2}) \oplus (\frac{1}{2}, 0)$ representation of the Lorentz group, i.e.

$$[Q_\alpha{}^i, J_{ab}] = \frac{1}{2}(\sigma_{ab})_\alpha{}^\beta Q_\beta{}^i \tag{2.11}$$

We can choose $Q_\alpha{}^i$ to be a Majorana spinor, i.e.

$$Q_\alpha{}^i = C_{\alpha\beta} \bar{Q}^{\beta i} \tag{2.12}$$

where $C_{\alpha\beta} = -C_{\beta\alpha}$ is the charge conjugation matrix (see Appendix A). This does not represent a loss of generality since, if the algebra admits complex conjugation as an involution we can always redefine the supercharges so as to satisfy (2.12) (see Note 1 at the end of this chapter).

The above calculation reflects the more general result that the $Q_\alpha{}^i$ must belong to a realization of the even (bosonic) subalgebra of the supersymmetry group. This is a simple consequence of demanding that the algebra be Z_2 graded. The commutator of any even generator B_1, with $Q_\alpha{}^i$ is of the form

$$[Q_\alpha{}^i, B_1] = (h_1)_{\alpha j}{}^{i\beta} Q_\beta{}^j \tag{2.13}$$

The generalized Jacobi identity

$$[[Q_\alpha{}^i, B_1], B_2] + [[B_1, B_2], Q_\alpha{}^i] + [[B_2, Q_\alpha{}^i], B_1] = 0 \tag{2.14}$$

implies that

$$[h_1, h_2]_{\alpha j}{}^{i\beta} Q_\beta{}^j = [Q_\alpha{}^i, [B_1, B_2]] \tag{2.15}$$

or in other words the matrices h represent the Lie algebra of the even generators.

The above remarks imply that

$$[Q_\alpha{}^i, T_r] = (l_r)^i{}_j Q_\alpha{}^j + (t_r)^i{}_j (i\gamma_5)_\alpha{}^\beta Q_\beta{}^j \tag{2.16}$$

where $(l_r)^i{}_j + i\gamma_5(t_r)^i{}_j$ represent the Lie algebra of the internal symmetry group. This results from the fact that $\delta_\beta{}^\alpha$ and $(\gamma_5)_\alpha{}^\beta$ are the only invariant tensors which are scalar and pseudoscalar.

The remaining odd-even commutator is $[Q_\alpha{}^i, P_a]$. A possibility that is allowed by the generalized Jacobi identities that involve the internal symmetry group and the Lorentz group is

$$[Q_\alpha{}^i, P_a] = c(\gamma_a)_\alpha{}^\beta Q_\beta{}^i \tag{2.17}$$

However, the $[[Q_\alpha{}^i, P_a], P_b] + \cdots$ identity implies that the constant $c = 0$, i.e.

$$[Q_\alpha{}^i, P_a] = 0 \tag{2.18}$$

More generally we could have considered $(c\gamma_a + d\gamma_a\gamma_5)Q$, on the right-hand side of (2.17), however, then the above Jacobi identity and the Majorana condition imply that $c = d = 0$. (See Note 2 at the end of this chapter.) Let us finally consider the $\{Q_\alpha{}^i, Q_\beta{}^j\}$ anticommutator. This object must be composed of even generators and must be symmetric under interchange of $\alpha \leftrightarrow \beta$ and $i \leftrightarrow j$. The even generators are those of the Poincaré group, the internal symmetry group and other even generators which, from the Coleman-Mandula theorem, commute with the Poincaré group, i.e., they are scalar and pseudo-scalar. Hence the most general possibility is of the form.

$$\{Q_\alpha{}^i, Q_\beta{}^j\} = r(\gamma^a C)_{\alpha\beta} P_a \delta^{ij} + s(\sigma^{ab} C)_{\alpha\beta} J_{ab} \delta^{ij} + C_{\alpha\beta} U^{ij} + (\gamma_5 C)_{\alpha\beta} V^{ij} \tag{2.19}$$

We have not included a $(\gamma^b \gamma_5 C)_{\alpha\beta} L_b{}^{ij}$ term as the (Q, Q, J_{ab}) Jacobi identity implies that $L_b{}^{ij}$ mixes nontrivially with the Poincaré group and so is excluded by the no-go theorem.

The fact that we have only used numerically invariant tensors under the Poincaré group is a consequence of the generalized Jacobi identities between two odd and one even generators.

To illustrate the argument more clearly, let us temporarily specialize to the case $N = 1$ where there is only one supercharge, Q_α. Equation (2.19) then reads

$$\{Q_\alpha, Q_\beta\} = r(\gamma^a C)_{\alpha\beta} P_a + s(\sigma^{ab} C)_{\alpha\beta} J_{ab}.$$

Using the Jacobi identity,

$$\{[P_a, Q_\alpha], Q_\beta\} + \{[P_a, Q_\beta], Q_\alpha\} + [\{Q_\alpha, Q_\beta\}, P_a] = 0,$$

we find that

$$0 = s(\sigma^{cd} C)_{\alpha\beta} [J_{cd}, P_a] = s(\sigma^{cd} C)_{\alpha\beta} (-\eta_{ac} P_d + \eta_{ad} P_c),$$

and, consequently, $s = 0$. We are free to scale the generator P_a in order to bring $r = 2$.

Let us now consider the commutator of the generator of the internal group and the supercharge. For only one supercharge, Eq. (2.16) reduces to

$$[Q_\alpha, T_r] = l_r Q_\alpha + i(\gamma_5)_\alpha^\beta t_r Q_\beta.$$

Taking the adjoint of this equation, multiplying by $(i\gamma^0)$ and using the definition of the Dirac conjugate given in Appendix A, we find that

$$[\bar{Q}^\alpha, T_r] = l_r^* \bar{Q}^\alpha + \bar{Q}^\beta (it_r^*)(\gamma_5)_\beta^{\ \alpha}.$$

Multiplying by $C_{\gamma\alpha}$ and using Eq. (2.12), we arrive at the equation

$$[Q_\alpha, T_r] = l_r^* Q_\alpha + it_r^* (\gamma_5)_\alpha^{\ \beta} Q_\beta.$$

Comparing this equation with the one we started from, we therefore conclude that

$$l_r^* = l_r \qquad t_r^* = t_r.$$

The Jacobi identity

$$[\{Q_\alpha, Q_\beta\}, T_r] + [[T_r, Q_\alpha], Q_\beta] + [[T_r, Q_\beta], Q_\alpha] = 0$$

results in the equation

$$[0 + (l_r \delta_\alpha^{\ \gamma} + it_r (\gamma_5)_\alpha^{\ \gamma}) 2(\gamma_a C)_{\gamma\beta} P_a] + \alpha \leftrightarrow \beta = 0$$

$$= 2P_a \{l_r (\gamma_a C)_{\alpha\beta} + it_r (\gamma_5 \gamma_a C)_{\alpha\beta}\} + (\alpha \leftrightarrow \beta) = 0.$$

Since $(\gamma_a C)_{\alpha\beta}$ and $(\gamma_5 \gamma_a C)_{\alpha\beta}$ are symmetric and antisymmetric in $\alpha\beta$ respectively, we conclude that $l_r = 0$ but t_r has no constant placed on it. Consequently, we find that we have only one internal generator R and we may scale it such that

$$[Q_\alpha, R] = i(\gamma_5)_\alpha^{\ \beta} Q_\beta.$$

The $N = 1$ supersymmetry algebra is summarized in Eq. (2.27).

Let us now return to the extend supersymmetry algebra. The even generators $U^{ij} = -U^{ji}$ and $V^{ij} = -V^{ji}$ are called central charges[5] and are often also denoted by Z. It is a consequence of the generalized Jacobi identities $((Q, Q, Q)$ and $(Q, Q, Z))$ that they commute with all other generators including themselves, i.e.,

$$[U^{ij}, \text{anything}] = 0 = [V^{ij}, \text{anything}] \qquad (2.20)$$

We note that the Coleman-Mandula theorem allowed a semi-simple group plus $U(1)$ factors. Their role in supersymmetric theories will emerge in later sections.

In general, we should write, on the right-hand side of (2.19), $(\gamma^a C)_{\alpha\beta}\omega^{ij}P_a + \cdots$, where ω^{ij} is an arbitrary real symmetric matrix. However, one can show that it is possible to redefine (rotate and rescale) the supercharges, whilst preserving the Majorana condition, in such a way as to bring ω^{ij} to the form $\omega^{ij} = r\delta^{ij}$ (see Note 3 at the end of this chapter). The $[P_a, \{Q_\alpha{}^i, Q_\beta{}^j\}] + \cdots = 0$ identity implies that $s = 0$ and we can normalize P_a by setting $r = 2$ yielding the final result

$$\{Q_\alpha{}^i, Q_\beta{}^j\} = 2(\gamma_a C)_{\alpha\beta}\delta^{ij}P^a + C_{\alpha\beta}U^{ij} + (\gamma_5 C)_{\alpha\beta}V^{ij} \qquad (2.21)$$

In any case r and s have different dimensions and so it would require the introduction of a dimensional parameter in order that they were both nonzero.

Had we chosen another irreducible Lorentz representation for $Q_\alpha{}^i$ other than $(j + \frac{1}{2}, j) \oplus (j, j + \frac{1}{2})$ we would not have been able to put P_a, i.e., a $(\frac{1}{2}, \frac{1}{2})$ representation, on the right-hand side of Eq. (2.21). The simplest choice is $(0, \frac{1}{2}) \oplus (\frac{1}{2}, 0)$. In fact this is the only possible choice (see Note 4).

Finally, we must discuss the constraints placed on the internal symmetry group by the generalized Jacobi identity. This discussion is complicated by the particular way the Majorana constraint of Eq. (2.12) is written. A two-component version of this constraint is

$$\bar{Q}_{\dot{A}i} = (Q_A{}^i)^*; \; A, \dot{A} = 1, 2 \qquad (2.22)$$

(see Appendix A for two-component notation). Equations (2.19) and (2.16) then become

$$\{Q_A{}^i, \bar{Q}_{\dot{B}j}\} = -2i(\sigma^a)_{A\dot{B}}\delta^i{}_j P_a$$

$$\{Q_A{}^i, Q_B{}^j\} = \varepsilon_{AB}(U^{ij} + iV^{ij})$$

$$[Q_A{}^i, J_{ab}] = +\tfrac{1}{2}(\sigma_{ab})_A{}^B Q_B{}^i \qquad (2.23)$$

and

$$[Q_A{}^i, T_r] = (l_r + it_r)^i{}_j Q_A{}^j \qquad (2.24)$$

Taking the complex conjugate of the last equation and using the Majorana condition we find that

$$[Q_{\dot{A}i}, T_r] = Q_{\dot{A}k}(U_r^\dagger)^k{}_i \tag{2.25}$$

where $(U_r)^i{}_j = (l_r + it_r)^i{}_j$. The $(Q, \bar{Q}, T)$ Jacobi identity then implies that $\delta^i{}_j$ be an invariant tensor of G, i.e.,

$$U_r + U_r^\dagger = 0 \tag{2.26}$$

Hence U_r is an antihermitian matrix and so represents the generators of the unitary group $U(N)$. However, taking account of the central charge terms in the (Q, Q, T) Jacobi identity one finds that there is for every central charge an invariant antisymmetric tensor of the internal group and so the possible internal symmetry group is further reduced. If there is only one central charge, the internal group is $Sp(N)$ while if there are no central charges it is $U(N)$.

To summarize, once we have adopted the rule that the algebra be Z_2 graded and contain the Poincaré group and an internal symmetry group then the generalized Jacobi identities place very strong constraints on any possible algebra. In fact, once one makes the further assumption that $Q_\alpha{}^i$ are spinors under the Lorentz group then the algebra is determined to be of the form of equations (2.1), (2.6), (2.11), (2.16), (2.18) and (2.21).

The simplest algebra is for $N = 1$ and takes the form

$$\{Q_\alpha, Q_\beta\} = 2(\gamma_a C)_{\alpha\beta} P^a$$

$$[Q_\alpha, P_a] = 0$$

$$[Q_\alpha, J_{cd}] = \tfrac{1}{2}(\sigma_{cd})_\alpha{}^\beta Q_\beta$$

$$[Q_\alpha, R] = i(\gamma_5)_\alpha{}^\beta Q_\beta \tag{2.27}$$

as well as the commutation relations of the Poincaré group. We note that there are no central charges (i.e., $U^{11} = V^{11} = 0$), and the internal symmetry group becomes just a chiral rotation with generator R.

We now wish to prove three of the statements made above. This is done here rather than in the above text, in order that the main line of argument should not become obscured by technical points. These points are best clarified in two-component notation.

Note 1: Suppose we have an algebra that admits a complex conjugation as an involution; for the supercharges this means that

$$(Q_A{}^i)^* = b_i{}^j Q_{\dot{A}j}; \quad (Q_{\dot{A}j})^* = d^j{}_k Q_A{}^k$$

There is no mixing of the Lorentz indices since $(Q_A{}^i)^*$ transforms like $Q_{\dot{A}i}$, namely in the $(0, \tfrac{1}{2})$ representation of the Lorentz group, and not like $Q_A{}^i$ which is in the $(\tfrac{1}{2}, 0)$ representation. The lowering of the i index under * is at this point purely a notational device. Two successive * operations yield the unit operation and this implies that

$$(b_i^j)^* d_k^j = \delta^i_k \tag{2.28}$$

and in particular that b_i^j is an invertible matrix. We now make the redefinitions

$$Q'_A{}^i = Q_A{}^i$$
$$Q'_{Ai} = b_i^j Q_{Aj} \tag{2.29}$$

Taking the complex conjugate of Q'_{Ai}, we find

$$(Q'_A{}^i)^* = (Q_A{}^i)^* = b_i^j Q_{Aj} = Q'_{Aj}$$

while

$$(Q'_{Ai})^* = (b_i^j)^*(Q_{Aj})^* = (b_i^j)^* d^j_k Q_A{}^k = Q_A{}^i \tag{2.30}$$

using Eq. (2.28).

Thus the $Q'_A{}^i$ satisfy the Majorana condition, as required. If the Q's do not initially satisfy the Majorana condition, we may simply redefine them so that they do.

Note 2:   Suppose the $[Q_A, P_a]$ commutator were of the form

$$[Q_A, P_a] = e(\sigma_a)_{A\dot{B}} Q^{\dot{B}} \tag{2.31}$$

where e is a complex number and for simplicity we have suppressed the i index. Taking the complex conjugate (see Appendix A), we find that

$$[Q_{\dot{A}}, P_a] = -e^*(\sigma_a)_{B\dot{A}} Q^B \tag{2.32}$$

Consideration of the $\big[[Q_A, P_a], P_b\big] + \cdots = 0$ Jacobi identity yields the result

$$-|e|^2(\sigma_a)_{A\dot{B}}(\sigma^b)^{C\dot{B}} - (a \leftrightarrow b) = 0 \tag{2.33}$$

Consequently $e = 0$ and we recover the result

$$[Q_A, P_a] = 0 \tag{2.34}$$

Note 3:   The most general form of the $Q^{Ai}, Q^{\dot{B}}_j$ anticommutator is

$$\{Q^{Ai}, Q^{\dot{B}}_j\} = -2iU^i_j(\sigma^m)^{A\dot{B}} P_m + \text{terms involving other Dirac matrices} \tag{2.35}$$

Taking the complex conjugate of this equation and comparing it with itself, we find that U is a Hermitian matrix

$$(U^i_j)^* = U^j_i \tag{2.36}$$

We now make a field redefinition of the supercharge

$$Q'^{Ai} = B^i_j Q^{Aj} \tag{2.37}$$

and its complex conjugate

$$Q'^{\dot{A}}_i = (B^i_j)^* Q^{\dot{A}}_j \tag{2.38}$$

Upon making this redefinition in Eq. (2.35), the U matrix becomes replaced by

$$U'^i_j = B^i_k U^k_l(B^j_l)^* \quad \text{or} \quad U' = BUB^\dagger \tag{2.39}$$

Since U is a Hermitian matrix, we may diagonalize it in the form $c_i \delta^i_j$ using a unitarity matrix B. We note that this preserves the Majorana condition on Q^{Ai}. Finally, we may scale $Q^i \to (1/\sqrt{c^i})Q^i$ to bring U to the form $U = d_i\delta^i{}_j$, where $d_i = \pm 1$. In fact, taking $A = B = 1$ and $i = j = k$, we realize that the right-hand side of Eq. (2.35) is a positive definite operator and since the energy $-iP_0$ is assumed positive definite, we can only find $d_i = +1$. The final result is

$$\{Q^{Ai}, Q^{\dot{B}}{}_j\} = -2i\delta^i{}_j(\sigma^m)^{A\dot{B}}P_m \tag{2.40}$$

Note 4: Let us suppose that the supercharge Q contains an irreducible representation of the Lorentz group other than $(0,\frac{1}{2}) \oplus (\frac{1}{2},0)$, say, the representation $Q_{A_1 \ldots A_n, \dot{B}_1 \ldots \dot{B}_m}$ where the A and B indices are understood to be separately symmetrized and $n + m$ is odd in order that Q is odd and $n + m > 1$. By projecting the $\{Q, Q^\dagger\}$ anticommutator we may find the anti-commutator involving $Q_{A_1 \ldots A_n, \dot{B}_1 \ldots \dot{B}_m}$ and its hermitian conjugate. Let us consider in particular the anti-commutator involving $Q = Q_{11\ldots 1, \dot{1}\dot{1}\ldots \dot{1}}$, this must result in an object of spin $n + m > 1$. However, by the Coleman Mandula no-go theorem no such generator can occur in the algebra and so the anticommutator must vanish, i.e., $QQ^\dagger + Q^\dagger Q = 0$.

Assuming the space on which Q acts as a positive definite norm, one such example being the space of on-shell states, we must conclude that Q vanishes. However if $Q_{11\ldots 1, \dot{1}\dot{1}\ldots \dot{1}}$ vanishes, so must $Q_{A_1 \ldots A_n; \dot{B}_1 \ldots \dot{B}_m}$ by its Lorentz properties, and we are left only with the $(0,\frac{1}{2}) \oplus (\frac{1}{2},0)$ representation.

Although the above discussion started with the Poincaré group, one could equally well have started with the conformal or (anti)de Sitter groups and obtained the superconformal and super (anti)de Sitter algebras. For completeness, we now list these algebras. The superconformal algebra which has the generators P_n, J_{mn}, D, K_n, $Q^{\alpha i}$ and $S^{\alpha i}$ is given by the Lorentz group plus:

$$[J_{mn}, P_k] = \eta_{nk}P_m - \eta_{mk}P_n$$

$$[J_{mn}, K_k] = \eta_{nk}K_m - \eta_{mk}K_n$$

$$[D, P_K] = -P_K \qquad [D, K_K] = +K_K$$

$$[P_m, K_n] = -2J_{mn} + 2\eta_{mn}D \qquad [K_n, K_m] = 0 \qquad [P_n, P_m] = 0$$

$$[Q_\alpha{}^i, J_{mn}] = \tfrac{1}{2}(\gamma_{mn})_\alpha{}^\beta Q_\beta{}^i \qquad [S_\alpha{}^i, J_{mn}] = \tfrac{1}{2}(\gamma_{mn})_\alpha{}^\beta S^{\beta i}$$

$$\{Q_\alpha{}^i, Q_\beta{}^j\} = -2(\gamma^n C)^{-1}{}_{\alpha\beta}P_n\delta^{ij}$$

$$\{S_\alpha{}^i, S_\beta{}^j\} = +2(\gamma^n C^{-1})_{\alpha\beta}K_n\delta^{ij}$$

$$[Q_\alpha{}^i, D] = \tfrac{1}{2}Q_\alpha{}^i \qquad [S_\alpha{}^i, D] = -\tfrac{1}{2}S_\alpha{}^i$$

$$[Q_\alpha{}^i, K_n] = -(\gamma_n)_\alpha{}^\beta S_\beta{}^i \qquad [S_\alpha{}^i, P_n] = (\gamma_n)_\alpha{}^\beta Q_\beta{}^i$$

$$[Q_\alpha{}^i, T_r] = (\delta_\alpha{}^\beta(\tau_{r_1})^i{}_j + (\gamma_5)_\alpha{}^\beta(\tau_{r_2})^i{}_j)Q_\beta{}^j$$

$$[S_\alpha{}^i, T_r] = (\delta_\alpha{}^\beta(\tau_{r_1})^i{}_j - (\gamma^5)_\alpha{}^\beta(\tau_{r_2})^i{}_j)Q_\beta{}^j$$

$$[Q_\alpha{}^i, A] = -i(\gamma_5)_\alpha{}^\beta Q_\beta{}^i\left(\frac{4-N}{4N}\right)$$

$$[S_\alpha{}^i, A] = \frac{4-N}{4N}i(\gamma_5)_\alpha{}^\beta S_\beta{}^i$$

$$\{Q_\alpha{}^i, S_\beta{}^j\} = -2(C^{-1})_{\alpha\beta}D\delta^{ij} + (\gamma^{mn}C^{-1})_{\alpha\beta}J_{mn}\delta^{ij} + 4i(\gamma_5 C^{-1})_{\alpha\beta}A\delta^{ij}$$
$$\qquad -2(\tau_{r_1})^{ij}(C^{-1})_{\alpha\beta} + ((\tau_{r_2})^{ij}(\gamma_5 C^{-1})_{\alpha\beta})T_r \tag{2.41}$$

The T_r and A generate $U(N)$ and $\tau_1 + \gamma_5 \tau_2$ are in the fundamental representation of $SU(N)$.

The case of $N = 4$ is singular and one can have either

$$[Q_\alpha{}^i, A] = 0 \qquad \text{or} \qquad [Q_\alpha{}^i, A] = -i(\gamma_5)_\alpha{}^\beta Q_\beta{}^i$$

and similarly for $S_\alpha{}^i$ and A. One may verity that both possibilities are allowed by the $N = 4$ Jacobi identities and so form acceptable superalgebras.

The anti-de Sitter superalgebra has generators M_{mn}, $T_{ij} = -T_{ji}$ and $Q^{\alpha i}$, and is given by

$$[M_{mn}, M_{pq}] = \eta_{np} M_{mq} + 3 \text{ terms}$$

$$[M_{mn}, T_{ij}] = 0 \qquad [Q_\alpha{}^i, M_{mn}] = \tfrac{1}{2}(\gamma_{mn} C^{-1})_\alpha{}^\beta Q_\beta{}^i$$

$$[Q_\alpha{}^i, T^{jk}] = -2i(\delta^{ij} Q_\alpha{}^k - \delta^{ik} Q_\alpha{}^j)$$

$$\{Q_\alpha{}^i, Q_\beta{}^j\} = \delta^{ij}(\gamma_{mn} C^{-1})_{\alpha\beta} i M_{mn} + (C^{-1})_{\alpha\beta} T^{ij}$$

$$[T^{ij}, T^{kl}] = -2i(\delta^{jk} T^{il} + 3 \text{ terms}) \qquad\qquad (2.42)$$

Chapter 3

Alternative Approaches to the Supersymmetry Algebra

Although the supersymmetry algebra was presented in Chapter 2 from the viewpoint of allowing the Poincaré group and an internal symmetry group to coexist in a non-trivial manner, it can also be obtained, with hindsight, from at least two other approaches.

Perhaps the most intuitive is that of demanding that there exist a Fermi-Bose symmetry. Adopting this approach let us assume that the symmetry possesses a representation which contains at least one scalar A, only one spinor χ_α as well as other bosonic fields. Then we must have a transformation between A and χ_α.

The transformation of the scalar, A into the fermion, χ_α must, if it is linear, be of the form

$$\delta A = \bar{\varepsilon}^\alpha \chi_\alpha \qquad (3.1)$$

On dimensional grounds $\bar{\varepsilon}^\alpha$ must have dimension one-half. It must also be an anticommuting parameter, as A and χ_α obey Bose-Einstein and Fermi-Dirac statistics respectively. The transformation for χ_α, if we assume it to be linear and take account of dimensions and Lorentz invariance and parity must be of the form

$$\delta\chi_\alpha = (\gamma^a)_\alpha{}^\beta \partial_a A \varepsilon_\beta + \text{terms involving the other fields} \qquad (3.2)$$

Of course to have a symmetry it is not sufficient just to write down a set of transformations; one must also ensure that these transformations obey a closed algebra. Equations (3.1) and (3.2) imply that

$$[\delta_1, \delta_2]A = \bar{\varepsilon}_2 \gamma^a \varepsilon_1 \partial_a A - (1 \leftrightarrow 2) + \text{other terms} \qquad (3.3)$$

Consequently, we find that demanding a Fermi-Bose symmetry which is linear must, as a result of the different dimensions of bosonic and fermionic fields, involve space-time translations as well as particles of different spin. The reader may verify that the commutator of two supersymmetries on χ_α when the first term in $\delta\chi_\alpha$ alone is taken is not a simple translation. This reflects the fact that spin-zero fields other than A are required. This approach to supersymmetry may also be carried to completion and the reader is referred to Chapters 5 and 6 for the construction of the Wess-Zumino model and the $N = 1$ Yang-Mills respectively.

Yet another approach to supersymmetry is to write down a general field theory and demand that the theory have the improved ultraviolet behaviour which is the well-known hallmark of supersymmetry theories. For example, consider the vacuum energy which is given for a particle of mass m and spin j by

$$\frac{1}{2}(-1)^{2j}(2j+1)\int d^3k\sqrt{k^2+m_j^2}$$

$$=\frac{1}{2}(-1)^{2j}(2j+1)\int d^3k\sqrt{k^2}\left(1+\frac{1}{2}\frac{m_j^2}{k^2}-\left(\frac{m_j^2}{k^2}\right)^2+\cdots\right)$$

Demanding that the quartic, quadratic and logarithmic divergence be absent implies that

$$\sum_j(-1)^{2j}(2j+1)=0 \qquad \sum_j(-1)^{2j}(2j+1)m_j^2=0 \qquad \text{and}$$

$$\sum_j(-1)^{2j}(2j+1)m_j^4=0$$

The first condition states that there are equal numbers of fermionic and bosonic degrees of freedom while the last conditions are also satisfied if the particles have the same mass. This fact was noticed long ago by Pauli.

Taking the simplest such possibility, i.e., one Majorana fermion and two spin-0 fields, one can write down the most general renormalizable coupling. Demanding that this theory which initially has ten free parameters have only a wave function renormalization places restrictions on the coupling constants and masses of this theory. In fact, one finds the Wess-Zumino model given in Chapter 5 which one would discover admitted supersymmetry. One could also demand that that one had a theory of spin 1/2 and spin 1 and only one infinite wave-function renormalization and then one would presumably find the $N=1$ Yang-Mills theory of Chapter 6.

Chapter 4

Immediate Consequences of the Supersymmetry Algebra

The supercharge $Q_\alpha{}^i$ is in the spin-$\frac{1}{2}$ representation of the Lorentz group and so its action on a state of spin j will result in a state of spin $j \pm \frac{1}{2}$. Hence, supersymmetry is a symmetry which mixes particles of different spin, that is mixes fermions and bosons. The Z_2 graded structure of the supersymmetry is now seen to be a necessary consequence of the fact that bosons and fermions obey Bose-Einstein and Fermi-Dirac statistics respectively. That is, in the classical limit a boson A and fermion χ_α obey the relations

$$[A, A] = 0 \qquad [A, \chi_\alpha] = 0$$

$$\{\chi_\alpha, \chi_\beta\} \equiv \chi_\alpha \chi_\beta + \chi_\beta \chi_\alpha = 0 \tag{4.1}$$

In other words, the anticommuting nature of the supersymmetry parameter ε_α is necessary in order to be compatible with the above relations.

The relation $[P_a, Q_\alpha{}^i] = 0$ implies that

$$[P_a{}^2, Q_\alpha{}^i] = 0 \tag{4.2}$$

Hence $P_a{}^2$ is a Casimir operator of the supersymmetry algebra and so particles in any irreducible representation of supersymmetry will have the same mass. One does not have to look at the particle tables for long to realize that this fact is not observed, even approximately, in nature. One of the major obstacles in applying supersymmetry to nature is to find ways of breaking supersymmetry whilst preserving its desirable features and predictive power.

Another consequence of the supersymmetry algebra which has far reaching consequences is that the energy, P_0, in supersymmetric theories is positive. To see this let us first multiply Eq. (2.21) by $C^{\delta\beta}$ to obtain

$$\{Q_\alpha{}^i, \bar{Q}^{\beta j}\} = -2\delta^{ij}(\gamma^a P_a)_\alpha{}^\beta - \delta_\alpha{}^\beta U^{ij} - (\gamma_5)_\alpha{}^\beta V^{ij} \tag{4.3}$$

Taking $i = j$, multiplying by γ^0 and performing a trace over Dirac and i indices we find that

$$i \sum_k \text{Tr}[\{Q_\alpha{}^k, \bar{Q}^{\beta k}\}\gamma^0] = \sum_{\alpha,k} Q_\alpha{}^k (Q_\alpha{}^k)^\dagger + \text{h.c.} = -2iN \text{ Tr} \gamma^0 \gamma^a P_a = 8iN P_0 \geqslant 0$$

Therefore

$$iP_0 \geqslant 0 \tag{4.4}$$

Finally, let us establish the following very useful theorem.

Theorem: In any representation of supersymmetry in which P_a is a one-to-one operator there are equal numbers of fermion and boson degrees of freedom.

Proof: Let us divide the representation of supersymmetry into two sets, the fermionic set and the bosonic set and apply the operator $\{Q_\alpha{}^i, Q_\beta{}^j\}$ to the bosonic set. The action of the first $Q_\alpha{}^i$ will be to map the bosons into the fermionic set and the second $Q_\beta{}^j$ will map the resulting fermions back into the bosonic set (Fig. 4.1).

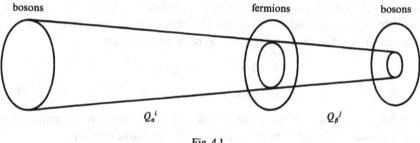

Fig. 4.1.

Now if P_a is a one-to-one onto operator then $\{Q_\alpha{}^i, Q_\beta{}^i\} = 2(\gamma_a C)_{\alpha\beta} P^a$ must be a one-to-one, onto operator which implies that each $Q_\alpha{}^i$ must map in a one-to-one fashion and so we must have equal numbers of bosonic and fermionic degrees of freedom. (Strictly speaking we should consider the case $\alpha = \beta = 1$ and $i = j = 1$, say.)

This theorem applies to on-shell states and off-shell field representations in which $P_a = \partial_a$ is indeed a one-to-one operator.

Chapter 5

The Wess-Zumino Model

The first four-dimensional model in which supersymmetry was linearly realized was found by Wess and Zumino,[3] by studying two-dimensional dual models.[7] In this chapter we rediscover supersymmetry along the lines given in Chapter 4 and discuss the Wess-Zumino model which is the simplest model of $N = 1$ supersymmetry.

Let us assume that the simplest model possesses one fermion χ_α which is a Majorana spinor, i.e.,

$$\chi_\alpha = C_{\alpha\beta}\bar{\chi}^\beta \tag{5.1}$$

On shell, that is, when

$$\partial\!\!\!/\chi = \text{interaction} \tag{5.2}$$

χ_α has two degrees of freedom or two helicity states. Applying our rule concerning equal numbers of fermionic and bosonic degrees of freedom of the previous chapter to the on-shell states we find that we must add two bosonic degrees of freedom to χ_α in order to form a realization of supersymmetry. These could either be two spin-zero particles or one massless vector particle which also has two helicity states on-shell. We will consider the former possibility in this section and the latter possibility, which is the $N = 1$ Yang-Mills theory, in the next chapter.

In Chapter 8 we will show that these considerations are indeed correct. An irreducible representation of $N = 1$ supersymmetry can be carried either by one parity even spin-zero state, one parity odd spin-zero state and one Majorana spin-$\frac{1}{2}$, or by one massless spin-one and one Majorana spin-$\frac{1}{2}$.

Taking the former possibility we have a Majorana spinor χ_α and two spin-zero states which we will assume to be represented by a scalar field A and pseudoscalar field B. For simplicity we will begin by constructing the free theory; the fields A, B, χ_α are then subject to

$$\partial^2 A = \partial^2 B = \partial\!\!\!/\chi = 0 \tag{5.3}$$

We now wish to construct the supersymmetry transformations that are carried by this irreducible realization of supersymmetry. Since $\bar{\varepsilon}^\alpha Q_\alpha$ is dimensionless and Q_α has mass dimension $+\frac{1}{2}$, the parameter $\bar{\varepsilon}^\alpha$ must have dimension

$-\frac{1}{2}$. On grounds of linearity, dimension, Lorentz invariance and parity we may write down the following set of transformations:

$$\delta A = \bar{\varepsilon} Q A = \bar{\varepsilon} \chi \qquad \delta B = i\bar{\varepsilon}\gamma_5 \chi$$

$$\delta\chi = \partial\!\!\!/(\alpha A + \beta i\gamma_5 B)\varepsilon \tag{5.4}$$

where α and β are undetermined parameters.

The variation of A is straightforward; however, the appearance of a derivative in $\delta\chi$ is the only way to match dimensions once the transformations are assumed to be linear. The reader will find no trouble verifying that these transformations do leave the set of field equations of Eq. (5.3) intact.

We can now test whether the $N = 1$ supersymmetry algebra of Chapter 2 is represented by these transformations. The commutator of two supersymmetries on A is given by

$$[\delta_1, \delta_2]A = [\bar{\varepsilon}_1 Q, \bar{\varepsilon}_2 Q]A \tag{5.5}$$

which, using Eq. (2.27), becomes

$$[\delta_1, \delta_2]A = 2\bar{\varepsilon}_2\gamma^a\varepsilon_1 P_a A$$

$$= 2\bar{\varepsilon}_2\gamma^a\varepsilon_1 \partial_a A$$

since $\qquad\qquad\qquad P_a = \partial_a \tag{5.6}$

On the other hand the transformation laws of Eq. (5.4) imply that

$$[\delta_1, \delta_2]A = \bar{\varepsilon}_2\partial\!\!\!/(\alpha A + i\gamma_5\beta B)\varepsilon_1 - (1 \leftrightarrow 2)$$

$$= 2\alpha\bar{\varepsilon}_2\partial\!\!\!/\varepsilon_1 A \tag{5.7}$$

The term involving B drops out because of the properties of Majorana spinors (see Appendix A). Provided $\alpha = +1$ this is indeed the 4-translation required by the algebra. We therefore set $\alpha = +1$. The calculation for B is similar and yields $\beta = +1$. For the field χ_α the commutator of two supersymmetries gives the result

$$[\delta_{\varepsilon_1}, \delta_{\varepsilon_2}]\chi = \partial\!\!\!/[\bar{\varepsilon}_1\chi + i\gamma_5\bar{\varepsilon}_1 i\gamma_5\chi]\varepsilon_2 - (1 \leftrightarrow 2)$$

$$= -\tfrac{1}{4}\bar{\varepsilon}_1\gamma^R\varepsilon_2\partial\!\!\!/[\gamma_R + i\gamma_5\gamma_R i\gamma_5]\chi - (1 \leftrightarrow 2)$$

$$= +\tfrac{1}{2}\bar{\varepsilon}_2\gamma^a\varepsilon_1 2\partial\!\!\!/\gamma_a\chi$$

$$= 2\bar{\varepsilon}_2\partial\!\!\!/\varepsilon_1\chi - \bar{\varepsilon}_2\gamma^a\varepsilon_1\gamma_a\partial\!\!\!/\chi \tag{5.8}$$

The above calculation makes use of a Fierz rearrangement (see Appendix A) as well as the properties of Majorana spinors. However, χ_α is subject to its equation of motion, i.e., $\partial\!\!\!/\chi = 0$, implying the final result

$$[\delta_1, \delta_2]\chi = 2\bar{\varepsilon}_2 \slashed{\partial}\varepsilon_1 \chi \tag{5.9}$$

which is the consequence dictated by the supersymmetry algebra. The reader will have no difficulty verifying that the fields A, B and χ_α and the transformations

$$\delta A = \bar{\varepsilon}\chi, \qquad \delta B = i\bar{\varepsilon}\gamma_5\chi$$

$$\delta\chi = \slashed{\partial}(A + i\gamma_5 B)\varepsilon \tag{5.10}$$

form a representation of the whole of the supersymmetry algebra *provided A, B* and χ_α are on-shell (i.e., $\partial^2 A = \partial^2 B = \slashed{\partial}\chi = 0$).

We now wish to consider the fields A, B and χ_α when they are no longer subject to their field equations. The Lagrangian from which the above field equations follow is

$$L = -\tfrac{1}{2}(\partial_\mu A)^2 - \tfrac{1}{2}(\partial_\mu B)^2 - \tfrac{1}{2}\bar{\chi}\slashed{\partial}\chi \tag{5.11}$$

It is easy to prove that the action $\int d^4 x L$ is indeed invariant under the transformation of Eq. (5.10). This invariance is achieved *without the use of the field equations*. The trouble with this formulation is that the fields A, B and χ_α do not form a realization of the supersymmetry algebra when they are no longer subject to their field equations, as the last term in Eq. (5.8) demonstrates. It will prove useful to introduce the following terminology. We shall refer to an irreducible representation of supersymmetry carried by fields which are subject to their equations of motion as an *on-shell representation*. We shall also refer to a Lagrangian as being *algebraically on-shell* when it is formed from fields which carry an on-shell representation, that is, do not carry a representation of supersymmetry off-shell, and the Lagrangian is invariant under these on-shell transformations. The Lagrangian of Eq. (5.11) is then an algebraically on-shell Lagrangian.

That A, B and χ_α cannot carry a representation of supersymmetry off-shell can be seen without any calculation, since these fields do not satisfy the rule of equal numbers of fermions and bosons which was given earlier. Off-shell, A and B have two degrees of freedom, but χ_α has four degrees of freedom. Clearly, the representations of supersymmetry must change radically when enlarged from on-shell to off-shell.

A possible way out of this dilemma would be to add two bosonic fields F and G which would restore the fermion-boson balance. However, these additional fields would have to occur in the Lagrangian so as to give rise to no on-shell states. As such, they must occur in the Lagrangian in the form $+\tfrac{1}{2}F^2 + \tfrac{1}{2}G^2$ assuming the free action to be only bilinear in the fields and consequently be of mass dimension two. On dimensional grounds their supersymmetry transformations must be of the form

$$\delta F = \bar{\varepsilon}\partial\!\!\!/\chi \qquad \delta G = i\bar{\varepsilon}\gamma_5\partial\!\!\!/\chi \tag{5.12}$$

where we have tacitly assumed that F and G are scalar and pseudoscalar respectively. The fields F and G cannot occur in δA on dimensional grounds, but can occur in $\delta\chi_\alpha$ in the form

$$\delta\chi = [(\mu F + i\tau\gamma_5 G) + \partial\!\!\!/(A + i\gamma_5 B)]\varepsilon \tag{5.13}$$

where μ and τ are undetermined parameters.

We note that we can only modify transformation laws in such a way that on-shell (i.e., when $F = G = \partial\!\!\!/\chi = \partial^2 A = \partial^2 B = 0$) we regain the on-shell transformation laws of Eq. (5.10).

We must now test if these new transformations do form a realization of the supersymmetry algebra. In fact, straightforward calculation shows they do, provided $\mu = \tau = +1$. This representation of supersymmetry involving the fields A, B, χ_α, F and G was found by Wess and Zumino[3] and we now summarize their result:

$$\delta A = \bar{\varepsilon}\chi \qquad \delta B = i\bar{\varepsilon}\gamma_5\chi$$

$$\delta\chi = [F + i\gamma_5 G + \partial\!\!\!/(A + i\gamma_5 B)]\varepsilon$$

$$\delta F = \bar{\varepsilon}\partial\!\!\!/\chi \qquad \delta G = i\bar{\varepsilon}\gamma_5\partial\!\!\!/\chi \tag{5.14}$$

The action which is invariant under these transformations, is given by the Lagrangian

$$A = \int d^4x \left\{ -\frac{1}{2}(\partial_\mu A)^2 - \frac{1}{2}(\partial_\mu B)^2 - \frac{1}{2}\bar{\chi}\partial\!\!\!/\chi + \frac{1}{2}F^2 + \frac{1}{2}G^2 \right\} \tag{5.15}$$

As expected the F and G fields occur as squares without derivatives and so lead to no on-shell states.

The above construction of the Wess-Zumino model is typical of that for a general free supersymmetric theory. We begin with the on-shell states, given for any model in Chapter 8, and construct the on-shell transformation laws. We can then find the Lagrangian which is invariant without use of the equations of motion, but contains no auxiliary fields. One then tries to find a set of auxiliary fields that give an off-shell algebra. Once this is done one can find a corresponding off-shell action. How one finds the nonlinear theory from the free theory is discussed in the later chapters.

The first of these two steps is always possible; however, there is no sure way of finding auxiliary fields that are required in all models, except with a few rare exceptions. This fact is easily seen to be a consequence of our rule for equal numbers of fermi and bose degrees of freedom in any representation of supersymmetry. It is only spin 0's, when represented by scalars, that have the same number of field components off-shell as they have on-shell states. For example, a Majorana spin-$\frac{1}{2}$ when represented by a spinor χ_α has a jump of 2 degrees of freedom between on and off-shell and a massless spin-1 boson when represented by a vector A_μ has a jump of 1 degree of freedom. In the latter case it is important to subtract the one gauge degree of freedom from A_μ thus leaving 3 field components off-shell (see next chapter). Since the increase in the number of degrees of freedom from an on-shell state to the off-shell field representing it changes by different amounts for fermions and bosons, the fermion-boson balance which holds on-shell will not hold off-shell if we only introduce the fields that describe the on-shell states. The discrepancy must be made up by fields, like F and G, that lead to no on-shell states. These latter type of fields are called *auxiliary fields*. The whole problem of finding representations of supersymmetry amounts to finding the auxiliary fields.

Unfortunately, it is not at all easy to find the auxiliary fields. Although the fermi-bose counting rule gives a guide to the number of auxiliary fields it does not actually tell you what they are, or how they transform. In fact, the auxiliary fields are only known for almost all $N = 1$ and 2 supersymmetry theories and for a very few $N = 4$ theories and not for the higher N theories. In particular, they are not known for the $N = 8$ supergravity theory.

Theories for which the auxiliary fields are not known can still be described by a Lagrangian in the same way as the Wess-Zumino theory can be described without the use of F and G, namely, by the so called algebraically on-shell Lagrangian formulation, which for the Wess-Zumino theory was given in Eq. (5.11). Such 'algebraically on-shell Lagrangians' are not too difficult to find at least at the linearized level. As explained in Chapter 8 we can easily find the relevant on-shell states of the theory. The algebraically on-shell Lagrangian then consists of writing down the known kinetic terms for each spin.

Of course, we are really interested in the interacting theories. The form of the interactions is however often governed by symmetry principles such as gauge invariance in the above example or general coordinate invariance in the case of gravity theories. When the form of the interactions is dictated by a local symmetry there is a straightforward, although maybe very lengthy way of finding the nonlinear theory from the linear theory. This method, called Noether coupling, is described in Chapter 7. In one guise or another this technique has been used to construct nonlinear "algebraically on-shell Lagrangians" for all supersymmetric theories.

The reader will now ask himself whether algebraically on-shell Lagrangians may be good enough. Do we really need the auxiliary fields? This question will be addressed in the next chapter, but the following example is a warning against over-estimating the importance of a Lagrangian that is invariant under a set of transformations that mix fermi-bose fields, but do not obey any particular algebra.

Consider the Lagrangian

$$L = -\tfrac{1}{2}(\partial_\mu A)^2 - \tfrac{1}{2}\bar{\chi}\partial\!\!\!/\chi \tag{5.16}$$

whose corresponding action is invariant under the transformations

$$\delta A = \bar{\varepsilon}\chi \qquad \delta\chi = \partial\!\!\!/A\varepsilon \tag{5.17}$$

However, this theory has nothing to do with supersymmetry. The algebra of transformations of Eq. (5.17) does not close on or off-shell without generating transformations which, although invariances of the free theory, can never be generalized to be invariances of an interacting theory. In fact, the on-shell states do not even have the correct fermi-bose balance required to form an irreducible representation of supersymmetry. This example illustrates the fact that the "algebraically on-shell Lagrangians" rely for their validity, as supersymmetric theories, on their on-shell algebra.

As a final remark in this section it is worth pointing out that the problem of finding the representations of any group is a mathematical question not dependent on any dynamical considerations for its resolution. Thus the questions of which are physical fields and which are auxiliary fields is a model-dependent statement.

Chapter 6

$N = 1$ Supersymmetric Gauge Theory: Super QED[8]

For simplicity we will first consider an Abelian gauge group. The construction of this model follows the same pattern as the Wess-Zumino model except for the added complication of gauge invariance. The first step is to find the irreducible representation which is carried by the on-shell states which consist of one Majorana spinor, λ_α and one massless vector A_a. The fields λ_α and A_a are subject to their free field equations, i.e.

$$\not{\partial}\lambda = \partial^a f_{ab} = 0 \qquad (6.1)$$

where

$$f_{ab} = \partial_a A_b - \partial_b A_a$$

The field A_a has the usual gauge transformation

$$\delta A_a = \partial_a \Lambda \qquad (6.2)$$

while

$$\delta\lambda = 0 \qquad (6.3)$$

As a result of the Majorana condition λ can carry no charge.

On dimensional grounds and Lorentz invariance the most general transformation between A_μ and λ_α is of the form

$$\delta A_a = \bar{\varepsilon}\gamma_a\lambda \qquad (6.4)$$

$$\delta\lambda = (\alpha\sigma^{ab}f_{ab} + \beta\partial_a A^a)\varepsilon \qquad (6.5)$$

where α and β are constants. In this case we require that the supersymmetry (Eqs. (6.4) and (6.5)) and gauge transformations form a closed algebra. The commutator of a supersymmetry transformation and a gauge transformation on λ is

$$[\delta_\Lambda, \delta_\varepsilon]\lambda = \delta_\Lambda(\alpha\sigma^{ab}f_{ab} + \beta\partial_a A^a)\varepsilon$$

$$= \beta\partial_a\partial^a\Lambda\varepsilon \qquad (6.6)$$

Since the right-hand side is not one of these transformations we must take $\beta = 0$.

The commutator of two supersymmetries on A_a is given by

$$[\delta_{\varepsilon_1}, \delta_{\varepsilon_2}] A_\mu = \bar{\varepsilon}_2 \gamma_\mu (\alpha \sigma^{cd} f_{cd} \varepsilon_1) - (1 \leftrightarrow 2)$$

$$= -4\bar{\varepsilon}_2 \not{\partial} \varepsilon_1 \alpha A_\mu + 2\partial_\mu (2\alpha \bar{\varepsilon}_2 \not{A} \varepsilon_1) \tag{6.7}$$

In order to get a translation of the magnitude predicted by the supersymmetry algebra we must choose $\alpha = -\frac{1}{2}$; however, we note that in this case we not only have a translation, but also a gauge transformation. This is a feature which occurs in all supersymmetries that involve gauge invariance and it originates from a choice of gauge (see Chapter 11).

The reader may verify the supersymmetry closure on λ subject to $\not{\partial}\lambda = 0$ and so the transformations

$$\delta A_a = \bar{\varepsilon} \gamma_a \lambda$$

$$\delta \lambda = -\tfrac{1}{2} \sigma^{cd} f_{cd} \varepsilon \tag{6.8}$$

form an irreducible realization of supersymmetry and leave invariant the field equations which A_a and λ are subject to.

An action which is invariant under these transformations, without the use of the field equations is given by

$$A = \int d^4 x \left\{ -\frac{1}{4} (f_{ab})^2 - \frac{1}{2} \bar{\lambda} \not{\partial} \lambda \right\} \tag{6.9}$$

We now wish to find a representation of supersymmetry that is off-shell, in other words closes not subject to the field equations. Again our fermi-bose counting rule tells us that we must add fields. However, we note from Eq. (6.7) that the commutator of two supersymmetries contains a gauge transformation as well as a translation. Consequently, we can only safely apply the theorem of Chapter 3 to the number of gauge invariant states, for on these states $\{Q_\alpha, Q_\beta\} = 2(\gamma_\mu C)_{\alpha\beta} P^\mu$. The field A_a has four components minus one gauge degree of freedom and so has 3 off-shell states. The field λ_α on the other hand has four degrees of freedom. The simplest possibility is that we add one auxiliary field, D, which has dimension two. The most general transformations, assuming D to be a pseudoscalar, which reduce on-shell to those of Eq. (6.8) are given by

$$\delta A_a = \bar{\varepsilon} \gamma_a \lambda$$

$$\delta \lambda = (-\tfrac{1}{2} \sigma^{cd} f_{cd} + ia\gamma_5 D)\varepsilon$$

$$\delta D = i\bar{\varepsilon} \gamma_5 \not{\partial} \lambda \tag{6.10}$$

Closure confirms the choice of one auxiliary field, D which is a pseudoscalar and requires $a = +1$.

The invariant action is given by

$$A = \int d^4 x \left\{ -\frac{1}{4} (f_{cd})^2 - \frac{1}{2} \bar{\lambda} \not{\partial} \lambda + \frac{1}{2} D^2 \right\} \tag{6.11}$$

Chapter 7

$N = 1$ Yang-Mills Theory and the Noether Technique

Any theory whose nonlinear form is determined by a gauge principle can be constructed by a Noether procedure.[9] Because of the importance of the Noether technique in constructing theories of supergravity we will take this opportunity to illustrate the technique within the framework of the simpler theory of supersymmetric Yang-Mills theory.[10]

Let us begin by considering the construction of the Yang-Mills theory itself from the linearized (free) theory. At the free level the theory is invariant under two distinct transformations: rigid and local Abelian transformations. Rigid transformations belong to a group S with generators R_i which satisfy

$$[R_i, R_j] = s_{ijk} R_k \tag{7.1}$$

The structure constants s_{ijk} may be chosen to be totally antisymmetric. Under these rigid transformations the vector fields $A_a{}^i$ transform as

$$\delta A_a{}^i = s^{ijk} T_j A_a{}^k \tag{7.2}$$

where T_j are the infinitesimal group parameters. The other type of transformations are local Abelian transformations:

$$\delta A_a{}^i = \partial_a \Lambda^i \tag{7.3}$$

Clearly both these transformations form a closed algebra.[†] The linearized theory which is invariant under the transformations of Eqs. (7.2) and (7.3) is

$$\mathring{A} = \int d^4x \left\{ -\frac{1}{4} f_{ab}^i f_{ab}^i \right\}$$

where

$$f_{ab}^i = \partial_a A_b{}^i - \partial_b A_a{}^i \tag{7.4}$$

The nonlinear theory is found in a series of steps, the first of which is to make the rigid transformations local, i.e., $T_j = T_j(x)$. Now, $\mathring{A}$ is no longer invariant under

$$\delta A_a{}^i = s^{ijk} T_j(x) A_{ak} \tag{7.5}$$

[†] In particular one finds $[\delta_A, \delta_T] A_a^i = \partial_a(s^{ijk} T_j \Lambda_k)$.

but its variation may be written in the form

$$\delta \mathring{A} = \int d^4x \{ (\partial_a T_k(x)) j^a{}_k \} \tag{7.6}$$

where

$$j^a{}_k = -s_{kij} A_b{}^i f^{ab}_j \tag{7.7}$$

Consider, however the action, A_1 given by

$$A_1 = \mathring{A} - \frac{1}{2} g \int d^4x (A_{ai} j^{ai}) \tag{7.8}$$

where g is the gauge coupling constant; it is invariant *to order g^0 provided* we combine the local transformation $T_i(x)$ with the local transformation $\Lambda_i(x)$ with the identification $\Lambda_i(x) = (1/g) T_i(x)$. That is, the initially separate local and rigid transformations of the linearized theory become knitted together into a single local transformation given by

$$\delta A_a{}^i = \frac{1}{g} \partial_a T^i(x) + s^{ijk} T_j(x) A_a{}^k(x) \tag{7.9}$$

The first term in the transformation of $\delta A_a{}^i$ yields in the last term in A_1 just that term which cancels the unwanted variation of $\mathring{A}$.

We now continue with this process of amending the Lagrangian and transformations order-by-order in g until we obtain an invariant Lagrangian.

The variation of A_1 under the transformation of Eq. (7.9) is of order g and is given by

$$\delta A_1 = \int d^4x \{ -g(A_a{}^i A_b{}^j s_{kij})(A_b^l \partial_a T^m s_{klm}) \} \tag{7.10}$$

An action invariant to order g is

$$A_2 = A_1 + \int d^4x \frac{g^2}{4} (A_a{}^i A_b{}^j s_{kij})(A^{bl} A^{am} s_{klm})$$

$$= -\frac{1}{4} (F_{ab}{}^i)^2 \tag{7.11}$$

where

$$F_{ab}{}^i = \partial_a A_b{}^i - \partial_b A_a{}^i - g s^{ijk} A_a{}^j A_b{}^k \tag{7.12}$$

In fact the action A_2 is invariant under the transformations of Eq. (7.9) to all orders in g and so represents the final answer, and is, of course, the well-known action of Yang-Mills theory. The commutator of two transformations on $A_a{}^i$ is

$$[\delta_{T_1}, \delta_{T_2}]A_a{}^i = s^{ijk}T_{2j}\left(\frac{1}{g}\partial_a T_{1k} + s_{klm}T_1{}^l A_a{}^m\right) - (1 \leftrightarrow 2)$$

$$= \frac{1}{g}\partial_a T_{12}{}^i + s^{ijk}T_{12j}A_{ak} \qquad (7.13)$$

where

$$T_{12i} = s^{ijk}T_{2j}T_{1k}$$

and so the transformations form a closed algebra.

For supergravity and other local theories the procedure is similar, although somewhat more complicated. The essential steps are to first make the rigid transformations local and find invariant Lagrangians order by order in the appropriate gauge coupling constant. This is achieved in general not only by adding terms to the action, but also adding terms to the transformation laws of the field. If the latter process occurs one must also check the closure order by order in the gauge coupling constant.

Although one can use a Noether procedure which relies on the existence of an action, one can also use a Noether method which uses the transformation laws alone. This works, in the Yang-Mills case, as follows: upon making the rigid transformations local as in Eq. (7.5) one finds that the algebra no longer closes, i.e.,

$$[\delta_\Lambda, \delta_T]A_a{}^i = \partial_a\left(s^{ijk}(T_j\Lambda_k)\right) - s^{ijk}(\partial_a T_j)\Lambda_k \qquad (7.14)$$

The cure for this is to regard the two transformations as simultaneous and knit them together as explained above. Using the new transformation for $A_a{}^i$ of Eq. (7.9) we then test the closure to order g^0. In fact, in this case the closure works to all orders in g and the process stops here; in general however, one must close the algebra order by order in the coupling constant modifying the transformation laws and the closure relations for the algebra. Having the full transformations it is then easy to find the full action when that exists.

Let us return to the supersymmetric Yang-Mills theory. The linearized theory was given in the preceeding chapter on super QED. The fields $A_a{}^i$, λ^i, D^i have the rigid transformations T_i and local transformations $\Lambda^i(x)$ given by

$$\delta A_a{}^i = s^{ijk}T_j A_{ak} \qquad \delta\lambda^i = s^{ijk}T_j\lambda_k$$
$$\delta D^i = s^{ijk}T^j D^k \qquad (7.15)$$

and

$$\delta A_a{}^i = \partial_a\Lambda^i \qquad \delta D^i = 0, \delta\lambda^i = 0 \qquad (7.16)$$

The supersymmetry transformations are

$$\delta A_a{}^i = \bar{\varepsilon}\gamma_a \lambda^i$$

$$\delta \lambda^i = -\tfrac{1}{2}\sigma^{cd} f_{cd}^i \varepsilon + i\gamma_5 D^i \varepsilon$$

$$\delta D^i = i\bar{\varepsilon}\gamma_5 \partial\!\!\!/\lambda^i \tag{7.17}$$

These transformations form a closed algebra, and leave invariant the following linearized Lagrangian

$$L = -\tfrac{1}{4}(f_{cd}^i)^2 - \tfrac{1}{2}\bar{\lambda}^i \partial\!\!\!/\lambda_i + \tfrac{1}{2}D_i{}^2 \tag{7.18}$$

Let us use the Noether method on the algebra to find the nonlinear theory. Making the rigid transformation on $A_a{}^i$ local we must, as in the Yang-Mills case, knit the rigid and the local transformations together (i.e., $\Lambda^i(x) = (1/g)T^i(x)$) to gain closure of gauge transformations on $A_\mu{}^i$. Closure of supersymmetry and gauge transformations implies that the rigid transformations on λ^i and D^i also become local. This particular closure also requires that all the supersymmetry transformations are modified to involve covariant quantities. For example, we find that on D^i

$$[\delta_T, \delta_\varepsilon]D^i = i\bar{\varepsilon}\gamma_5 \gamma^a s^{ijk}(\partial_a T_j)\lambda_k \tag{7.19}$$

and as a result we must replace $\partial_a \lambda^i$ by $\mathcal{D}_a \lambda^i = \partial_a \lambda^i - gs^{ijk}A_{aj}\lambda_k$ in the δD^i of Eq. (7.17) and then the commutator $[\delta_A, \delta_\varepsilon]$ is zero to all orders in g. The algebra then takes the form

$$\delta A_a{}^i = \bar{\varepsilon}\gamma_a \lambda^i$$

$$\delta \lambda^i = (-\tfrac{1}{2}\sigma^{cd} F_{cd}{}^i + i\gamma_5 D^i)\varepsilon$$

$$\delta D^i = i\bar{\varepsilon}\gamma_5 \mathcal{D}\!\!\!/\lambda^i \tag{7.20}$$

where

$$F_{ab}{}^i = \partial_a A_b{}^i - \partial_b A_a{}^i - gs^{ijk}A_a{}^j A_b{}^k$$

We must now verify that the above supersymmetry transformations close. For other supersymmetric gauge theories one must add further terms to the supersymmetry transformations in order to regain closure. However, in this case gauge covariance and dimensional analysis ensure that there are no possible terms that one can add to these supersymmetry transformations and so the transformations of Eqs. (7.20) must be the complete laws for the full theory. The reader may verify that there are no inconsistencies by showing that the algebra does indeed close.

The action invariant under these transformations is

$$A = \int d^4 x \left\{ -\frac{1}{4}(F_{ab}{}^i)^2 - \frac{1}{2}\bar{\lambda}^i \mathcal{D}\!\!\!/\lambda_i + \frac{1}{2}D_i{}^2 \right\} \tag{7.21}$$

One could also have used the Noether procedure on the action. Gauge invariance implies that the action is that given in Eq. (7.21). Demanding that this gauge invariant action be supersymmetric requires us to modify the supersymmetry transformations to those of Eq. (7.20).

Chapter 8
The Irreducible Representations of Supersymmetry[11]

In this chapter we wish to find the irreducible representations of supersymmetry, or, put another way, we want to know what is the possible particle content of supersymmetric theories. As is well known the irreducible representations of the Poincaré group are found by the Wigner method of induced representations.[12] This method consists of finding a representation of a subgroup of the Poincaré group and boosting it up to a representation of the full group. In practice, one adopts the following recipe: We choose a given momentum q^μ which satisfies $P_\mu{}^2 = 0$ or $P_\mu{}^2 = -m^2$ depending which case we are considering. We find the subgroup H which leaves q^μ intact and find a representation of H on the $|q^\mu\rangle$ states. We then induce this representation to the whole of the Poincaré group denoted P, in the usual way. In this construction there is a one-to-one correspondence between points of P/H and four-momentum which satisfies $P_\mu{}^2 = 0$ or $P_\mu{}^2 = -m^2$. One can show that the result is independent of the choice of momentum q^μ one starts with.

In what follows we will not discuss the irreducible representations in general, but only that part applicable to the rest frame, i.e., the representations of H in the states at rest. We can do this safely in the knowledge that once the representation of H on the rest-frame states is known then the representation of P is uniquely given and that every irreducible representation of the Poincaré group can be obtained by considering every irreducible representation of H.

In terms of physics the procedure has a simple interpretation, namely, the properties of a particle are determined entirely by its behaviour in a given frame (i.e., for given q^μ). The general behaviour is obtained from the given q^μ by boosting either the observer or the frame with momentum q^μ to one with arbitrary momentum.

The procedure outlined above for the Poincaré group can be generalized to any group of the form $S \otimes T$ where the symbol $\otimes$ denotes the semi-direct product of the groups S and T where T is Abelian. It also applies to the supersymmetry group and we shall take it for granted that the above recipe is the correct procedure and does in fact yield all irreducible representations of the supersymmetry group.

Let us first consider the massless case $q_\mu q^\mu = 0$, for which we choose the standard momentum $q^\mu{}_s = (m, 0, 0, m)$ for our "rest frame". We must now find H whose group elements leave $q^\mu{}_s = (m, 0, 0, m)$ intact. Clearly this contains $Q_\alpha{}^i$, P_μ and T_s, since these generators all commute with P_μ and so rotate the states with $q^\mu{}_s$ into themselves. We neglect central charges for the massless case.

Under the Lorentz group the action of the generator $\frac{1}{2}\Lambda^{\mu\nu}J_{\mu\nu}$ creates an infinitesimal transformation $q^\mu \rightarrow \Lambda^\mu{}_\nu q^\nu + q^\mu$. Hence $q^\mu{}_s$ is left invariant provided the parameters obey the relations

$$\Lambda_{30} = 0, \qquad \Lambda_{10} + \Lambda_{13} = 0, \qquad \Lambda_{20} + \Lambda_{23} = 0 \tag{8.1}$$

Thus the Lorentz generators in H are

$$T_1 = J_{10} + J_{13}, \qquad T_2 = J_{20} + J_{23}, \qquad J = J_{12} \tag{8.2}$$

These generators form the algebra

$$[T_1, J] = -T_2$$

$$[T_2, J] = +T_1$$

$$[T_1, T_2] = 0 \tag{8.3}$$

The reader will recognise this to be the Lie algebra of E_2, the group of translations and rotations on a two-dimensional plane.

Now the only unitary representations of E_2 which are finite dimensional have T_1 and T_2 trivially realized, i.e.

$$T_1|q^\mu{}_s\rangle = T_2|q^\mu{}_s\rangle = 0 \tag{8.4}$$

This results from the theorem that all nontrivial unitary representations of noncompact groups are infinite dimensional. We will assume we require finite-dimensional representations of H.

Hence, for the Poincaré group, in the case of massless particles, finding representations of H results in finding representations of E_2 and consequently for the generator J alone. We choose our states so that

$$J|\lambda\rangle = i\lambda|\lambda\rangle \tag{8.5}$$

(Our generators are antihermitian.) In fact, J is the helicity operator and we select λ to be integer or half-integer (i.e., $J = \mathbf{q} \times \mathbf{J}/|\mathbf{q}|$ evaluated at $q = (0, 0, 0, m)$ where $J_i = \varepsilon_{ijk}J_{jk}$; $i, j = 1, 2, 3$).

Let us now consider the action of the supercharges $Q_\alpha{}^i$ on the rest-frame states, $|q^\mu{}_s\rangle$. The calculation is easiest when performed using the two-component formulation of the supersymmetry algebra of Eq. (2.23). On rest-frame states we find that

$$\{Q^{Ai}, Q^{\dot{B}}{}_j\} = -2\delta^i_j(\sigma_\mu)^{A\dot{B}} q^\mu{}_s,$$

$$= -2\delta^i_j(\sigma_0 + \sigma_3)^{A\dot{B}} m = +4m\delta^i_j \begin{pmatrix} 0 & 0 \\ 0 & 1 \end{pmatrix}^{A\dot{B}} \tag{8.6}$$

In particular these imply the relations

$$\{Q^{1i}, Q^i{}_j\} = 0$$

$$\{Q^{2i}, Q^2{}_j\} = 4m\delta^i_j$$

$$\{Q^i{}_i, Q^{2j}\} = \{Q^i{}_i, Q^{2j}\} = 0 \tag{8.7}$$

The first relation implies that

$$\langle q^\mu{}_s | (Q^{1i}(Q^{1i})^* + (Q^{1i})^* Q^{1i}) | q^\mu{}_s \rangle = 0 \tag{8.8}$$

Demanding that the norm on physical states be positive definite yields

$$Q_2{}^i | q^\mu{}_s \rangle = Q_{\dot{2}i} | q^\mu{}_s \rangle = 0. \tag{8.9}$$

Hence, all generators in H have zero action on rest-frame states except J, T_s, P_μ, $Q_1{}^i$ and Q_{1i}. Using Eq. (2.23) we find that

$$[Q_1{}^i, J] = \frac{1}{2}(\sigma_{12})_1{}^1 Q_1{}^i$$

$$= -\frac{i}{2} Q_1{}^i \tag{8.10}$$

Similarly, we find that complex conjugation implies

$$[(Q_1{}^i)^*, J] = +\frac{i}{2}(Q_1{}^i)^* \tag{8.11}$$

The relations between the remaining generators summarized in Eqs. (8.7), (8.10), (8.11) and (2.24) can be summarized by the statement that $Q_1{}^i$ and $(Q_1{}^i)^*$ form a Clifford algebra, act as raising and lowering operators for the helicity operator J and transform under the N and $\bar{N}$ representation of $SU(N)$.

We find the representations of this algebra in the usual way; we choose a state of given helicity, say λ, and let it be the vacuum state for the operator $(Q_1{}^i)$, i.e.

$$(Q_1{}^i)|\lambda\rangle = 0$$

$$J|\lambda\rangle = i\lambda|\lambda\rangle \tag{8.12}$$

The states of this representation are then

$$|\lambda\rangle = |\lambda\rangle$$

$$|\lambda - \tfrac{1}{2}, i\rangle = (Q_1{}^i)^* |\lambda\rangle$$

$$|\lambda - 1, [ij]\rangle = (Q_1{}^i)^* (Q_1{}^j)^* |\lambda\rangle \qquad (8.13)$$

etc. These states have the helicities indicated and belong to the $[ijk\ldots]$ anti-symmetric representation of $SU(N)$. The series will terminate after the helicity $\lambda - (N/2)$, as the next state will be an object antisymmetric in $N + 1$ indices[†]. The states have helicities from λ to $\lambda - (N/2)$, there being $N!/(m!(N-m)!)$ states with helicity $\lambda - (m/2)$.

To obtain a set of states which represent particles of both helicities we must add to the above set, the representation with helicities from $-\lambda$ to $-\lambda + (N/2)$. The exception is the so-called CPT self-conjugate sets of states which automatically contain both helicity states.

The representations of the full supersymmetry group are obtained by boosting the above states in accordance with the Wigner method of induced representations.

Hence the massless irreducible representation of $N = 1$ supersymmetry comprises only the two states

$$|\lambda\rangle$$

$$|\lambda - \tfrac{1}{2}\rangle = (Q_1)^* |\lambda\rangle \qquad (8.14)$$

with helicities λ and $\lambda - \tfrac{1}{2}$ and since

$$Q_1^* Q_1^* |\lambda\rangle = 0 \qquad (8.15)$$

there are no more states.

To obtain a CPT invariant theory we must add states of the opposite helicities, i.e., $-\lambda$ and $-\lambda + \tfrac{1}{2}$. For example, if $\lambda = \tfrac{1}{2}$ we get on-shell helicity states 0 and $\tfrac{1}{2}$ and their CPT conjugates with helicities $-\tfrac{1}{2}$, 0, giving a theory with two spin 0's and one Majorana spin-$\tfrac{1}{2}$. Alternatively, if $\lambda = 2$ then we get on-shell helicity states $3/2$ and 2 and their CPT self conjugates with helicity $-3/2$ and -2; this results in a theory with one spin 2 and one spin 3/2 particles. These on-shell states are those of the Wess-Zumino model and $N = 1$ supergravity respectively. Later in this discussion we will give a complete account of these theories.

For $N = 4$ with $\lambda = 1$ we get the massless states

$$|1\rangle, |\tfrac{1}{2}, i\rangle, |0, [ij]\rangle, |-\tfrac{1}{2}, [ijk]\rangle, |-1, [ijkl]\rangle \qquad (8.16)$$

This is a CPT self-conjugate theory with one spin-1, four spin-$\tfrac{1}{2}$ and six spin-0 particles.

[†] Since there are only N labels this object vanishes identically.

Table 8.1 below gives the multiplicity for massless irreducible representations which have maximal helicity 1 or less.

Table 8.1   Multiplicities for massless irreducible representations with maximal helicity 1 or less.

Spin \ N	1	1	2	2	4
Spin 1	—	1	1	—	1
Spin $\frac{1}{2}$	1	1	2	2	4
Spin 0	2	—	2	4	6

We see that as N increases, the multiplicities of each spin and the number of different types of spin increases. The simplest theories are those for $N = 1$. The one in the first column is the Wess-Zumino model and the one in the second column is the $N = 1$ supersymmetric Yang-Mills theory. The latter contains one spin 1 and one spin $\frac{1}{2}$, consistent with the formula for the lowest helicity $\lambda - (N/2)$, which in this case gives $1 - \frac{1}{2} = \frac{1}{2}$. The $N = 4$ multiplet is *CPT* self conjugate, since in this case we have $\lambda - (N/2) = 1 - 4/2 = -1$. The Table stops at N equal to 4 since when N is greater than 4 we must have particles of spin greater than 1. (Clearly, $N > 4$ implies that $\lambda - (N/2) = 1 - (N/2) < -1$.) This leads us to the well-known statement that the $N = 4$ supersymmetric theory is the maximally extended Yang-Mills theory.

The content for massless on-shell representations with a maximum helicity 2 is given in Table 8.2.

Table 8.2   Multiplicity for massless on-shell representations with maximal helicity 2.

Spin \ N	1	2	3	4	5	6	7	8
Spin 2	1	1	1	1	1	1	1	1
Spin $\frac{3}{2}$	1	2	3	4	5	6	8	8
Spin 1		1	3	6	10	16	28	28
Spin $\frac{1}{2}$			1	4	11	26	56	56
Spin 0				2	5	30	70	70

The $N = 1$ supergravity theory contains only one spin-2 graviton and one spin-3/2 gravitino. It is often referred to as simple supergravity theory. For the $N = 8$ supergravity theory, $\lambda - (N/2) = 2 - \frac{8}{2} = -2$. Consequently it is *CPT* self conjugate and contains all particles from spin 2 to spin 0. Clearly, for theories in which N is greater than 8, particles of higher than spin 2 will occur. Thus, the $N = 8$ theory is the maximally extended supergravity theory.

It is claimed that this theory is in fact the largest possible supergravity theory. This contention rests on the widely-held belief that it is impossible to consistently couple particles of spin $\frac{5}{2}$ to other particles.

We now consider the massive irreducible representations of supersymmetry. We take our rest-frame momentum to be

$$q^{\mu}{}_s = (m, 0, 0, 0) \tag{8.17}$$

The corresponding little group is then generated by

$$P_m, \; Q^{\alpha i}, \; T^r, \; Z_1{}^{ij}, \; Z_2{}^{ij}, \; J_m \equiv \tfrac{1}{2}\varepsilon_{mnr}J^{nr} \tag{8.18}$$

where $m, n, r = 1, 2, 3$ for the present discussion. The J_m generate the group $SU(2)$. Let us first consider the case where the central charges are trivially realized.

When acting on the rest-frame states the supercharges obey the algebra

$$\{Q^{Ai}, (Q^{Bj})^*\} = 2\delta^A{}_B \delta^i{}_j m$$

$$\{Q^{Ai}, Q^{Bj}\} = 0 \tag{8.19}$$

The action of the T^r is that of $U(N)$ while the $SU(2)$ rotation generators satisfy

$$[J_m, J_n] = \varepsilon_{mnr} J_r$$

$$[Q^{Ai}, J_m] = i(\sigma_m)^A{}_B Q^{Bi} \tag{8.20}$$

where (σ_m) are the Pauli matrices. We note that as far as $SU(2)$ is concerned the dotted spinor Q^{Ai} behaves like the undotted spinor Q_{Ai}.

We observe that unlike the massless case none of the supercharges are trivially realized and so the Clifford algebra they form has $4N$ elements, that is, twice as many as those for the massless case. The unique irreducible representation of the Clifford algebra is found in the usual way. We define a Clifford vacuum

$$Q_A{}^i|q^{\mu}{}_s\rangle = 0, \qquad \forall = A, i \tag{8.21}$$

and the representation is carried by the states

$$|q^{\mu}{}_s\rangle, (Q_A{}^i)^*|q^{\mu}{}_s\rangle, (Q_A{}^i)^*(Q_B{}^j)^*|q^{\mu}{}_s\rangle, \dots \tag{8.22}$$

Due to the anticommuting nature of the $(Q_A{}^j)^*$ this series terminates when one applies $(2N + 1)Q^*$'s.

The structure of the above representation is not particularly apparent since it is not clear how many particles of a given spin it contains.

The properties of the Clifford algebra are more easily displayed by defining the real generators

$$\Gamma^i_{2A-1} = \frac{1}{2m}(Q^{Ai} + (Q^{Ai})^*)$$

$$\Gamma^i_{2A} = \frac{i}{2m}(Q^{Ai} - (Q^{Ai})^*) \tag{8.23}$$

where the

$$\Gamma^i_p = (\Gamma^i_1, \Gamma^i_2, \Gamma^i_3, \Gamma^i_4) \tag{8.24}$$

are hermitian. The Clifford algebra of Eq. (8.19) now becomes

$$\{\Gamma^i_p, \Gamma^j_q\} = \delta^{ij}\delta_{pq} \tag{8.25}$$

The $4N$ elements of the Clifford algebra carry the group $SO(4N)$ in the standard manner; the $4N(4N-1)/2$ generators of $SO(4N)$ being

$$O^{ij}_{mn} = \tfrac{1}{2}[\Gamma^i_m, \Gamma^j_n] \tag{8.26}$$

As there are an even number of elements in the basis of the Clifford algebra, we may define a "parity" (γ_5) operator

$$\Gamma_{4N+1} = \prod_{p=1}^{4} \prod_{i=1}^{N} \Gamma^i_p \tag{8.27}$$

which obeys the relations

$$(\Gamma_{4N+1})^2 = +1$$

$$\{\Gamma_{4N+1}, \Gamma^i_p\} = 0 \tag{8.28}$$

Indeed, the irreducible representation of Eq. (8.22) is of dimension 2^{2N} and transforms according to an irreducible representation of $SO(4N)$ of dimension 2^{2N-1} with $\Gamma_{4N+1} = -1$ and another of dimension 2^{2N-1} with $\Gamma_{4N+1} = +1$. Now any linear transformation of the Q's, Q^*'s (for example $\delta Q = rQ$) can be represented by a generator formed from the commutator of the Q's and Q^*'s (for example, $r[Q, Q^*]$). In particular the $SU(2)$ rotation generators are given by

$$s_k = -\frac{i}{4m}(\sigma_k)^A{}_B[Q^{jB}, (Q^{jA})^*] \tag{8.29}$$

One may easily verify that

$$[Q^{jA}, s_k] = i(\sigma_k)^A{}_B Q^{Bj} \tag{8.30}$$

The states of a given spin will be classified by that subgroup of $SO(4N)$ which commutes with the appropriate $SU(2)$ rotation subgroup of $SO(4N)$. This will be the group generated by all generators bilinear in Q, Q^* that have their two-component index contracted, i.e.,

$$\Lambda^i_j = \frac{i}{2m}[Q^{Ai},(Q^{Aj})^*]$$

$$k^{ij} = \frac{i}{2m}[Q^{Ai},Q_A{}^j] \tag{8.31}$$

and $(k^{ij})^\dagger = k_{ij}$. It is easy to verify that the Λ^i_j, k^{ij} and k_{ij} generate the group $USp(2N)$ and so the states of a given spin are labeled by representations of $USp(2N)$. That the group is $USp(2N)$ is most easily seen by defining

$$Q_A{}^a = \begin{cases} Q_A{}^i\delta_i{}^a & a = 1,\dots,N \\ \varepsilon_{AB}(Q^{Bi})^* & a = N+1,\dots,2N \end{cases} \tag{8.32}$$

for then the generators Λ^i_j, k^{ij} and k_{ij} are given by

$$s^{ab} = \frac{i}{2m}[Q^{Aa},Q_A{}^b]. \tag{8.33}$$

Using the fact that

$$\{Q_A{}^a,Q_B{}^b\} = \varepsilon_{AB}\Omega^{ab} \tag{8.34}$$

where

$$\Omega^{ab} = \begin{pmatrix} 0 & 1 \\ -1 & 0 \end{pmatrix}$$

we can verify that

$$[s^{ab},s^{cd}] = \Omega^{ac}s^{cd} + \cdots \tag{8.35}$$

which is the algebra of $USp(2N)$.

The particle content of a massive irreducible representation is given by the following.

Theorem[21]: If our Clifford vacuum is a scalar under the $SU(2)$ spin group and the internal symmetry group, then the irreducible massive representation of supersymmetry has the following content

$$2^{2N} = \left[\frac{N}{2},(0)\right] + \left[\frac{N-1}{2},(1)\right] + \cdots + \left[\frac{N-\kappa}{2},(\kappa)\right]$$

$$+ \cdots + [0,(N)] \tag{8.36}$$

where the first entry in the bracket denotes the spin and the last entry, say (k) denotes which kth fold antisymmetric traceless irreducible representation of $USp(2N)$ that this spin belongs to.

Table 8.3   Some massive representations (without central charges) labelled in terms of the $USp(2N)$ representations.

Spin \ N	1			2			3		4
Spin 2			1		1			1	1
Spin $\frac{3}{2}$		1	2		1	4	1	6	8
Spin 1	1	2	1	1	4	$5+1$	6	$14+1$	27
Spin $\frac{1}{2}$	1	2	1	4	$5+1$	4	14	$14'+6$	48
Spin 0	2	1		5	4	1	$14'$	14	42

Table 8.4   Some massive representations with one central charge ($|Z| = m$). All states are complex.

Spin \ N	2		4		6		8
Spin 2						1	1
Spin $\frac{3}{2}$				1	1	6	8
Spin 1		1	1	4	6	$14+1$	27
Spin $\frac{1}{2}$	1	2	4	$5+1$	14	$14'+6$	48
Spin 0	2	1	5	4	$14'$	14	42

Consider an example with two supercharges. The classifying group is $USp(4)$ and the 2^4 states are one spin 1, four spin 1/2, and five spin 0 corresponding to the **1**-, **4**- and **5**-dimensional representations of $USp(4)$. For more examples see Table 8.3.

Should the Clifford vacuum carry spin and belong to a non-trivial representation of the internal group $U(N)$, then the irreducible representation is found by taking the tensor product of the vacuum and the representation given in the above theorem.

The introduction of non-trivial central charges changes the massive irreducible representation in an important way. Clearly, they change the algebra of the Q's on the rest-frame states and it is this Clifford algebra which determines the number of states in the corresponding irreducible representation of supersymmetry. In the massless case the spread of helicities is from 0 to $N/2$ ($N/4$ in the *CPT* self-conjugate case) and has dimension $2^{N+1}(2^N)$ corresponding to N non-trivially realized elements in the Clifford vacuum. In fact, in the massless case one can easily show that non-trivial central charges lead to negative mass states. In the massive case without central charges the spins are from 0 to $N/2$ and the dimension is 2^{2N} corresponding to a $4N$-

dimensional Clifford algebra. This conclusion is the same for central charges except for the special case when the central charge has the same value as the mass. In this case the Clifford algebra is reduced from having $4N$ elements to only $2N$ elements in the same way as the massless case. The dimension and spread of spin of the corresponding irreducible representations are divided by two and become the same as for the massless case. (For examples see Table 8.4.) Central charge irreducible representations often occur during spontaneous symmetry breaking.

For an account of irreducible representations of supersymmetry that possess central charges see the reviews in Ref. 21. The account of the massive irreducible representations of supersymmetry given here is along similar lines to the review by Ferrara and Savoy given in Ref. 21.

Chapter 9
Simple Supergravity: Linearized $N = 1$ Supergravity

We now wish to construct supergravity along similar lines to the method used to find the Wess-Zumino model and $N = 1$ super QED. In this section we will start with the on-shell states and construct the linearized theory, without and then with auxiliary fields. Because of the complexity of supergravity it is particularly useful to examine the much simpler linearized theory which displays many of the features of the full nonlinear theory.

The irreducible representations of supersymmetry which include a spin-2 graviton contain either a spin-3/2 or a spin-5/2 fermion. The spin-5/2 particle would seem to have considerable problems in coupling to other fields and so we will choose the spin-3/2 particle.

As in the Yang-Mills case, the linearized theory possesses rigid supersymmetry and local Abelian gauge invariances. The latter invariances are required, in order that the fields do describe the massless on-shell states alone without involving ghosts.[†]

These on-shell states are represented by a symmetric second rank tensor field, $h_{\mu\nu}$ ($h_{\mu\nu} = h_{\nu\mu}$) and a Majorana vector spinor, $\psi_{\mu\alpha}$. For these fields to represent a spin-2 particle and a spin-3/2 particle they must possess the infinitesimal gauge transformations

$$\delta h_{\mu\nu} = \partial_\mu \xi_\nu(x) + \partial_\nu \xi_\mu(x)$$

$$\partial \psi_{\mu\alpha} = \partial_\mu \eta_\alpha(x) \tag{9.1}$$

The unique ghost free gauge-invariant, free field equations are

$$E_{\mu\nu} = 0, \qquad R^\mu = 0 \tag{9.2}$$

where $E_{\mu\nu} = R^L_{\ \mu\nu} - \frac{1}{2}\eta_{\mu\nu} R^L$, $R^{Lab}_{\ \ \mu\nu}$ being the linearized Riemann tensor given by

$$R^{Lab}_{\ \ \mu\nu} = -\partial_a \partial_\mu h_{b\nu} + \partial_b \partial_\mu h_{a\nu} + \partial_a \partial_\nu h_{b\mu} - \partial_b \partial_\nu h_{a\mu}$$

and

[†]We recall that a rigid symmetry is one whose parameters are space-time independent while a local symmetry has space-time dependent parameters.

$$R^\mu = \varepsilon^{\mu\nu\rho\kappa} i\gamma_5 \gamma_\nu \partial_\rho \psi_\kappa \qquad R^{Lb}{}_\mu = R^{Lab}{}_{\mu\nu} \delta^\mu_a$$

$$R^L = R^{La}{}_\mu \delta^\mu_a \tag{9.3}$$

For an explanation of this point see van Nieuwenhuizen.[13]

We must now search for the supersymmetry transformations that form an invariance of these field equations and represent the supersymmetry algebra on-shell. On dimensional grounds the most general transformation is

$$\delta h_{\mu\nu} = \tfrac{1}{2}(\bar\varepsilon\gamma_\mu\psi_\nu + \bar\varepsilon\gamma_\nu\psi_\mu) + \delta_1 \eta_{\mu\nu}\bar\varepsilon\gamma^\kappa\psi_\kappa$$

$$\delta\psi_\mu = +\delta_2 \sigma^{ab}\partial_a h_{b\mu}\varepsilon + \delta_3 \partial_\nu h^\nu{}_\mu \varepsilon \tag{9.4}$$

The parameters δ_1, δ_2 and δ_3 are parameters whose values will be determined by the demand that the set of transformations which comprise the supersymmetry transformations of Eq. (9.4) and the gauge transformations of Eq. (9.1) should form a closing algebra when the field equations of Eq. (9.2) hold. At the linearized level the supersymmetry transformations are linear rigid transformations, that is, they are *first order* in the fields $h_{\mu\nu}$ and $\psi_{\mu\alpha}$ and parameterized by *constant* parameters ε^α.

Carrying out the commutator of a Rarita-Schwinger gauge transformation, $\eta_\alpha(x)$ of Eq. (9.1) and a supersymmetry transformation, ε of Eq. (9.4) on $h_{\mu\nu}$, we get:

$$[\delta_\eta, \delta_\varepsilon] h_{\mu\nu} = \tfrac{1}{2}(\bar\varepsilon\gamma_\mu \partial_\nu \eta + \bar\varepsilon\gamma_\nu \partial_\mu \eta) + \delta_1 \eta_{\mu\nu} \bar\varepsilon \slashed{\partial} \eta \tag{9.5}$$

This is a gauge transformation with parameter $\tfrac{1}{2}\bar\varepsilon\gamma_\mu\eta$ on $h_{\mu\nu}$ provided $\delta_1 = 0$. Similarly, calculating the commutator of a gauge transformation of $h_{\mu\nu}$ and a supersymmetry transformation on $h_{\mu\nu}$ automatically yields the correct result zero. However, carrying out the commutator of a supersymmetry transformation and an Einstein gauge transformation on $\psi_{\mu\alpha}$ yields

$$[\delta_{\xi_\mu}, \delta_\varepsilon] \psi_\mu = +\delta_2 \sigma^{ab} \partial_a(\partial_\mu \xi_b)\varepsilon + \delta_3 \partial_\nu \partial^\nu \xi_\mu \varepsilon + \delta_3 \partial_\nu \partial_\mu \xi^\mu \varepsilon \tag{9.6}$$

which is a Rarita-Schwinger gauge transformation on ψ_μ provided $\delta_3 = 0$. Hence we take $\delta_1 = \delta_3 = 0$.

We must test the commutator of two supersymmetries. On $h_{\mu\nu}$ we find the commutator of two supersymmetries to give

$$[\delta_{\varepsilon_1}, \delta_{\varepsilon_2}] h_{\mu\nu} = +\tfrac{1}{2}\{\bar\varepsilon_2 \gamma_\mu \delta_2 \sigma^{ab} \partial_a h_{b\nu}\varepsilon_2 + (\mu \leftrightarrow \nu)\} - (1 \leftrightarrow 2)$$

$$= \delta_2 \{\bar\varepsilon_2 \gamma^b \varepsilon_1 \partial_\mu h_{b\nu} - \bar\varepsilon_2 \gamma^a \varepsilon_1 \partial_a h_{\mu\nu} - (\mu \leftrightarrow \nu)\} \tag{9.7}$$

This is a gauge transformation on $h_{\mu\nu}$ with parameter $\delta_2 \bar\varepsilon_2 \gamma^b \varepsilon_1 h_{b\nu}$ as well as a space-time translation. The latter coincides with that dictated by the supersymmetry group provided $\delta_2 = -1$ which is the value we now adopt.

It is important to stress that linearized supergravity differs from the Wess-Zumino model in that one must take into account the gauge transformations of Eqs. (9.1) as well as the rigid supersymmetry transformations of Eq. (9.4) in order to obtain a closed algebra. The resulting algebra is the $N = 1$ super-symmetry algebra when supplemented by gauge transformations. This algebra reduces to the $N = 1$ supersymmetry algebra only on gauge-invariant states.

For the commutator of two supersymmetries on ψ_μ we find

$$[\delta_{\varepsilon_1}, \delta_{\varepsilon_2}]\psi_\mu = -\sigma^{ab}\partial_a\varepsilon_2 \frac{1}{2}(\varepsilon_1\gamma_b\psi_\mu + \bar{\varepsilon}_1\gamma_\mu\psi_b) - (1 \leftrightarrow 2)$$

$$= +\frac{1}{2\cdot 4}\bar{\varepsilon}_1\gamma_R\varepsilon_2\sigma^{ab}\partial_a\gamma^R(\gamma_b\psi_\mu + \gamma_\mu\psi_b) - (1 \leftrightarrow 2)$$

$$= \frac{1}{8}\bar{\varepsilon}_1\gamma_R\varepsilon_2\sigma^{ab}\gamma^R\left(\gamma_b\psi_{a\mu} + \frac{1}{2}\gamma_\mu\psi_{ab}\right)$$

$$+ \partial_\mu\left(\frac{1}{8}\bar{\varepsilon}_1\gamma_R\varepsilon_2\sigma^{ab}\gamma^R\gamma_b\psi_a\right) - (1 \leftrightarrow 2) \tag{9.8}$$

where $\psi_{\mu\nu} = \partial_\mu\psi_\nu - \partial_\nu\psi_\mu$. Using the different forms of the Rarita-Schwinger equation of motion, given by

$$R^\mu = 0$$

$$\Leftrightarrow \gamma^\mu\psi_{\mu\nu} = 0$$

$$\Leftrightarrow \psi_{\mu\nu} + \tfrac{1}{2}i\gamma_5\varepsilon_{\mu\nu\rho\kappa}\psi^{\rho\kappa} = 0 \tag{9.9}$$

we find the final result to be

$$[\delta_{\varepsilon_1}, \delta_{\varepsilon_2}]\psi_\mu = 2\bar{\varepsilon}_2\gamma^c\varepsilon_1\partial_c\psi_\mu + \partial_\mu(-\bar{\varepsilon}_2\gamma^c\varepsilon_1\gamma\psi_c$$

$$+ \tfrac{1}{8}\bar{\varepsilon}_1\gamma_R\varepsilon_2\sigma^{ab}\gamma^R\gamma_b\psi_a) - (1 \leftrightarrow 2) \tag{9.10}$$

This is the required result: a translation and a gauge transformation on ψ_μ.

The reader can verify that the transformations of Eq. (9.4) with the values of $\delta_1 = \delta_3 = 0$, $\delta_2 = -2$ do indeed leave the equations of motion of $h_{\mu\nu}$ and $\psi_{\mu\alpha}$ invariant.

Having obtained an irreducible representation of supersymmetry carried by the fields $h_{\mu\nu}$ and $\psi_{\mu\alpha}$ when subject to their field equations we can now find the algebraically on-shell Lagrangian. The action (Freedman, van Niuwen-huizen & Ferrara [14], Deser & Zumino [15]) from which the field equations of Eq. (9.2) follow, is

$$A = \int d^4x\left\{-\frac{1}{2}h^{\mu\nu}E_{\mu\nu} - \frac{1}{2}\bar{\psi}_\mu R^\mu\right\} \tag{9.11}$$

It is invariant under the transformations of Eq. (9.4) provided we adopt the values for the parameters δ_1, δ_2 and δ_3 found above. This invariance holds without use of the field equations, as it did in the Wess-Zumino case.

We now wish to find a linearized formulation which is built from fields which carry a representation of supersymmetry without imposing any restrictions, (i.e., equations of motion) namely, we find the auxiliary fields. As a guide to their number we can apply our fermi-bose counting rule which, since the algebra contains gauge transformations, applies only to the gauge invariant states. On-shell $h_{\mu\nu}$ has two helicities and so does $\psi_{\mu\alpha}$; however off-shell $h_{\mu\nu}$ contributes $(5 \times 4)/2 = 10$ degrees of freedom minus 4 gauge degrees of freedom giving 6 bosonic degrees of freedom. On the other hand, off-shell $\psi_{\mu\alpha}$ contributes $4 \times 4 = 16$ degrees of freedom minus 4 gauge degrees of freedom, giving 12 fermionic degrees of freedom. Hence, the auxiliary fields must contribute 6 bosonic degrees of freedom. If there are n auxiliary fermions there must be $4n + 6$ bosonic auxiliary fields.

Let us assume that a minimal formulation exists, that is, there are no auxiliary spinors. Let us also assume that the bosonic auxiliary fields occur in the Lagrangian as squares without derivatives (like F and G) and so are of dimension two. Hence we have 6 bosonic auxiliary fields; it only remains to find their Lorentz character and transformations. We will assume that they consist of a scalar M, a pseudoscalar N and a pseudovector b_μ, rather than an antisymmetric tensor or 6 spin-0 fields. We will give the motivating arguments for this later.

Another possibility is the two fields A_μ and $a_{\kappa\lambda}$ which possess the gauge transformations $\delta A_\mu = \partial_\mu \Lambda$; $\delta a_{\kappa\lambda} = \partial_\kappa \Lambda_\lambda - \partial_\lambda \Lambda_\kappa$. A contribution $\varepsilon_{\mu\nu\rho\kappa} A^\mu \partial^\nu a^{\rho\kappa}$ to the action would not lead to propagating degrees of freedom.

The transformations of the fields $h_{\mu\nu}$, $\psi_{\mu\alpha}$, M, N and b_μ must reduce on-shell to the on-shell transformations found above. This restriction, dimensional arguments and the fact that if the auxiliary fields are to vanish on-shell they must vary into field equations gives the transformations to be[16,17]

$$\delta h_{\mu\nu} = \tfrac{1}{2}(\bar{\varepsilon}\gamma_\mu\psi_\nu + \bar{\varepsilon}\gamma_\nu\psi_\mu)$$

$$\delta\psi_{\mu\alpha} = -\sigma^{ab}\partial_a h_{b\mu}\varepsilon - \tfrac{1}{3}\gamma_\mu(M + i\gamma_5 N)\varepsilon + b_\mu i\gamma_5\varepsilon + \delta_6\gamma_\mu\slashed{b}i\gamma_5\varepsilon$$

$$\delta M = \delta_4\bar{\varepsilon}\gamma \cdot R$$

$$\delta N = \delta_5 i\bar{\varepsilon}\gamma_5\gamma \cdot R$$

$$\delta b_\mu = +\delta_7 i\bar{\varepsilon}\gamma_5 R_\mu + \delta_8 i\bar{\varepsilon}\gamma_5\gamma_\mu\gamma \cdot R \qquad (9.12)$$

The parameters δ_4, δ_5, δ_6, δ_7 and δ_8 are determined by the restriction that the above transformations of Eq. (9.12) and the gauge transformations of Eq.

Introduction to Supersymmetry and Supergravity

(9.1) should form a closed algebra. For example, the commutator of two supersymmetries on M gives

$$[\delta_{\varepsilon_1}, \delta_{\varepsilon_2}]M = \delta_4\{-\bar{\varepsilon}_2\gamma^\mu\varepsilon_1\partial_\mu M + 16\bar{\varepsilon}_2\sigma^{\mu\nu}i\gamma_5\varepsilon_1(1 + 3\delta_6)\partial_\mu b_\nu\} \quad (9.13)$$

which is the required result provided $\delta_4 = -\frac{1}{2}$ and $\delta_6 = -\frac{1}{3}$. Carrying out the commutator of two supersymmetries on all fields we find a closing algebra provided

$$\delta_4 = -\tfrac{1}{2}, \quad \delta_5 = -\tfrac{1}{2}, \quad \delta_6 = -\tfrac{1}{3}, \quad \delta_7 = \tfrac{3}{2} \quad \text{and} \quad \delta_8 = -\tfrac{1}{2} \quad (9.14)$$

We henceforth adopt these values for the parameters. An action which is constructed from the fields $h_{\mu\nu}$, $\psi_{\mu\alpha}$, M, N and b_μ and is invariant under the transformations of Eq. (9.12) with the above values of the parameters is

$$A = \int d^4x \left\{ -\frac{1}{2}h_{\mu\nu}E^{\mu\nu} - \frac{1}{2}\bar{\psi}_\mu R^\mu - \frac{1}{3}(M^2 + N^2 - b_\mu^2) \right\} \quad (9.15)$$

This is the action of linearized $N = 1$ supergravity and upon elimination of the auxiliary field M, N and b_μ it reduces to the algebraically on-shell Lagrangian of Eq. (9.11).

The Nonlinear Theory

The full nonlinear theory of supergravity can be found from the linearized theory discussed above by applying the Noether technique of Chapter 7. Just as in the case of Yang-Mills the reader will observe that the linearized theory possesses the local Abelian invariances of Eq. (9.1) as well as the rigid (i.e., constant parameter) supersymmetry transformations of Eq. (9.4).

We proceed just as in the case of the Yang-Mills theory and make the parameter of rigid transformations space-time dependent, i.e., set $\varepsilon = \varepsilon(x)$ in Eq. (9.4). The linearized action of Eq. (9.15) is then no longer invariant, but its variation must be of the form

$$\delta A_0 = \int d^4x \partial_\mu \bar{\varepsilon}^\alpha j^\mu{}_\alpha \quad (9.16)$$

since it is invariant when $\bar{\varepsilon}^\alpha$ is a constant. The object $j^\mu{}_\alpha$ is proportional to $\psi_{\mu\alpha}$ and linear in the bosonic fields $h_{\mu\nu}$, M or N and b_μ. As such, on dimensional grounds, it must be of the form

$$j_{\mu\alpha} \propto \partial_\tau h_{\rho\mu}\psi_{\nu\beta} + \cdots$$

Consider now the action, A, given by

$$A_1 = A_0 - \frac{\kappa}{4}\int d^4x \bar{\psi}^\mu j_\mu \quad (9.17)$$

where κ is the gravitational constant. The action A is invariant to order κ^0 *provided* we combine the now local supersymmetry transformations of Eq. (9.12) with a local Abelian Rarita-Schwinger gauge transformation of Eq. (9.1) with parameter $\eta(x) = (2/\kappa)\varepsilon(x)$. That is, we make a transformation

$$\delta\psi_\mu = \frac{2}{\kappa}\partial_\mu\varepsilon(x) - \partial_a h_{b\mu}\sigma^{ab}\varepsilon(x) - \frac{1}{3}\gamma_\mu(M + i\gamma_5 N)\varepsilon(x)$$

$$+ i\gamma_5\left(b_\mu - \frac{1}{3}\gamma_\mu\not{b}\right)\varepsilon(x) \tag{9.18}$$

the remaining fields transforming as before except that ε is now space-time dependent.

As in the Yang-Mills case the two invariances of the linearized action become knitted together to form one transformation, the role of gauge coupling being played by the gravitational constant, κ. The addition of the term $(-\kappa/4)\bar{\psi}^\mu j_\mu$ to A_0 does the required job; its variation is

$$-\frac{\kappa}{4}\cdot 2\cdot\left(\frac{2}{\kappa}\right)(\partial_\mu\bar{\varepsilon})j^\mu + \text{terms of order } \kappa^1 \tag{9.19}$$

The order κ^1 terms do not concern us at the moment. (Note $j_{\mu\alpha}$ is linear in $\psi_{\mu\alpha}$ and so we get a factor of 2 from $\delta\psi_{\mu\alpha}$).

In fact, one can carry the Noether procedure out in the context of pure gravity where one finds at the linearized level the rigid translation

$$\delta h_{\mu\nu} = \zeta^\lambda\partial_\lambda h_{\mu\nu} \tag{9.20}$$

and the local gauge transformation

$$\delta h_{\mu\nu} = \partial_\mu\xi_\nu + \partial_\nu\xi_\mu \tag{9.21}$$

These become knitted together at the first stage of the Noether procedure to give

$$\delta h_{\mu\nu} = \frac{1}{\kappa}\partial_\mu\zeta_\nu + \frac{1}{\kappa}\partial_\nu\zeta_\mu + \zeta^\nu\partial_\lambda h_{\mu\nu} \tag{9.22}$$

since $\xi_\nu = (1/\kappa)\zeta_\nu$. This variation of $h_{\mu\nu}$ contains the first few terms of an Einstein general coordinate transformation of the vierbein which is given in terms of $h_{\mu\nu}$ by

$$e_\mu{}^a = \eta_\mu{}^a + \kappa h_\mu{}^a \tag{9.23}$$

We proceed in a similar way to the Yang-Mills case. We obtain order by order in κ an invariant Lagrangian by adding terms to the Lagrangian and in this case also adding terms to the transformations of the fields. For example, if we added a term to $\delta\psi_\mu$ say, $\delta\bar{\psi}_\mu = \cdots + \bar{\varepsilon}X_\mu\kappa$ then from the linearized action we receive a contribution $-\kappa\bar{\varepsilon}X_\mu R^\mu$ upon variation of ψ_μ. It is necessary at each step (order of κ) to check that the transformations of the fields form a closed algebra. In fact, any ambiguities that arise in the procedure are resolved by demanding that the algebra closes.

The final set of transformations[16,17] is

$$\delta e_\mu{}^a = \kappa\bar{\varepsilon}\gamma^a\psi_\mu$$

$$\delta\psi_\mu = 2\kappa^{-1}D_\mu(w(e,\psi))\varepsilon + i\gamma_5\left(b_\mu - \frac{1}{3}\gamma_\mu\slashed{b}\right)\varepsilon - \frac{1}{3}\gamma_\mu(M + i\gamma_5 N)\varepsilon$$

$$\delta M = -\frac{1}{2}e^{-1}\bar{\varepsilon}\gamma_\mu R^\mu - \frac{\kappa}{2}i\bar{\varepsilon}\gamma_5\psi_\nu b^\nu - \kappa\bar{\varepsilon}\gamma^\nu\psi_\nu M + \frac{\kappa}{2}\bar{\varepsilon}(M + i\gamma_5 N)\gamma^\mu\psi_\mu$$

$$\delta N = -\frac{e^{-1}}{2}i\bar{\varepsilon}\gamma_5\gamma_\mu R^\mu + \frac{\kappa}{2}\bar{\varepsilon}\psi_\nu b^\nu - \kappa\bar{\varepsilon}\gamma^\nu\psi_\nu N - \frac{\kappa}{2}i\bar{\varepsilon}\gamma_5(M + i\gamma_5 N)\gamma^\mu\psi_\mu$$

$$\delta b_\mu = \frac{3i}{2}e^{-1}\bar{\varepsilon}\gamma_5\left(g_{\mu\nu} - \frac{1}{3}\gamma_\mu\gamma_\nu\right)R^\nu + \kappa\bar{\varepsilon}\gamma^\nu b_\nu\psi_\mu - \frac{\kappa}{2}\bar{\varepsilon}\gamma^\nu\psi_\nu b_\mu$$

$$- \frac{\kappa}{2}i\bar{\psi}_\mu\gamma_5(M + i\gamma_5 N)\varepsilon - \frac{i\kappa}{4}\varepsilon_\mu{}^{bcd}b_b\bar{\varepsilon}\gamma_5\gamma_c\psi_d \tag{9.24}$$

where

$$R^\mu = \varepsilon^{\mu\nu\rho\kappa}i\gamma_5\gamma_\nu D_\rho(w(e,\psi))\psi_\kappa$$

$$D_\mu(w(e,\psi)) = \partial_\mu + w_{\mu ab}\frac{\sigma^{ab}}{4}$$

and

$$w_{\mu ab} = \frac{1}{2}e^\nu{}_a(\partial_\mu e_{b\nu} - \partial_\nu e_{b\mu}) - \frac{1}{2}e_b{}^\nu(\partial_\mu e_{a\nu} - \partial_\nu e_{a\mu})$$

$$- \frac{1}{2}e_a{}^\rho e_b{}^\sigma(\partial_\rho e_{\sigma c} - \partial_\sigma e_{\rho c})e_\mu{}^c$$

$$+ \frac{\kappa^2}{4}(\bar{\psi}_\mu\gamma_a\psi_b + \bar{\psi}_a\gamma_\mu\psi_b - \bar{\psi}_\mu\gamma_b\psi_a) \tag{9.25}$$

They form a closed algebra, the commutator of two supersymmetries on any field being

$$[\delta_{\varepsilon_1}, \delta_{\varepsilon_2}] = \delta_{\text{supersymmetry}}(-\kappa\xi^\nu\psi_\nu) + \delta_{\text{general coordinate}}(2\xi_\mu)$$

$$+ \delta_{\text{Local Lorentz}}\left(-\frac{2\kappa}{3}\varepsilon_{ab\lambda\rho}b^\lambda\xi^\rho\right.$$

$$\left. -\frac{2\kappa}{3}\bar{\varepsilon}_2\sigma_{ab}(M + i\gamma_5 N)\varepsilon_1 + 2\xi^d w_d{}^{ab}\right) \qquad (9.26)$$

where

$$\xi_\mu = \bar{\varepsilon}_2\gamma_\mu\varepsilon_1$$

The transformations of Eq. (9.24) leave invariant the action

$$A = \int d^4x \left\{\frac{e}{2\kappa^2}R - \frac{1}{2}\bar{\psi}_\mu R^\mu - \frac{1}{3}e(M^2 + N^2 - b_\mu b^\mu)\right\} \qquad (9.27)$$

where

$$R = R_{\mu\nu}{}^{ab}e_a{}^\mu e_b{}^\nu$$

and

$$R_{\mu\nu}{}^{ab}\frac{\sigma_{ab}}{4} = [D_\mu, D_\nu]$$

The auxiliary fields M, N and b_μ may be eliminated to obtain the nonlinear algebraically on-shell Lagrangian which was the form in which supergravity was originally found in Refs. 14 and 15.

As discussed in Chapter 7 one could also build up the nonlocal theory by working with the algebra of field transformations alone.

Chapter 10

Invariance of Simple Supergravity

At first sight the presence of terms quartic in $\psi_{\mu\alpha}$ in the supergravity action would seem to indicate that it may be very lengthy to prove that the action of Eq. (9.27) is actually invariant under the transformations of Eq. (9.24). These difficult terms result from expressing w_μ^{ab} in terms of e_μ^a and ψ_μ^α. A rather straightforward proof of invariance[18,19] can be achieved by judicious handling of the e_μ^a and ψ_μ^α terms in w_μ^{ab}. Rather than simply regard the action as a function of e_μ^a, ψ_μ^α, M, N and b_μ, it is better to regard it as a function of e_μ^a, ψ_μ^α, M, N, b_μ and w_μ^{ab} where w_μ^{ab} itself is the function of e_μ^a and ψ_μ^α of Eq. (9.25), i.e., $A(e_\mu^a, \psi_\mu^\alpha, M, N, b_\mu, w_\mu^{ab}(e, \psi))$. The variation of the action is then given by

$$\delta A = \int d^4x \left\{ \frac{\delta A}{\delta e_\mu^a} \delta e_\mu^a + \frac{\delta A}{\delta \psi_\mu^\alpha} \delta \psi_\mu^\alpha + \frac{\delta A}{\delta M} \delta M \right.$$
$$\left. + \frac{\delta A}{\delta N} \delta N + \frac{\delta A}{\delta b_\mu} \delta b_\mu + \frac{\delta A}{\delta w_\mu^{ab}} \delta w_\mu^{ab} \right\} \tag{10.1}$$

where δw_μ^{ab} is the variation of w_μ^{ab} obtained by varying e_μ^a and ψ_μ^α in Eq. (9.25). The advantage of this is that the last term vanishes since

$$\frac{\delta A}{\delta w_{\mu ab}} \bigg|_{w_{\mu ab} = w_\mu^{ab}(e, \psi)} = 0 \tag{10.2}$$

This is a consequence of the fact that the functorial form of w_μ^{ab} given by Eq. (9.25) is just that found from the equation of motion of w_μ^{ab} in the first-order formalism, i.e.,

$$\frac{\delta A}{\delta w_{\mu ab}} = 0 \tag{10.3}$$

The net result of this is that

$$\delta A = \int d^4x \left\{ \frac{\delta A}{\delta e_\mu^a} \delta e_\mu^a + \frac{\delta A}{\delta \psi_\mu^\alpha} \delta \psi_\mu^\alpha + \frac{\delta A}{\delta N} \delta N + \frac{\delta A}{\delta M} \delta M + \frac{\delta A}{\delta b_\mu} \delta b_\mu \right\} \tag{10.4}$$

or equivalently we do not vary w_μ^{ab} in the action. This is sometimes called the

1.5 order formalism, but it is really more of a trick than a formalism. Adopting this trick we will now show the invariance of simple supergravity.

Let us first consider terms in the variation that do not involve auxiliary fields. The variation of the Einstein part is given by

$$\delta \int \frac{e}{2\kappa^2}(e_a{}^\mu e_b{}^\nu R_{\mu\nu}{}^{ab})\,d^4x = \int d^4x \left\{ \frac{1}{\kappa}\{\bar{\varepsilon}\gamma^\mu\psi_a\}\left\{-R_\mu{}^a + \frac{1}{2}e_\mu{}^a R\right\}\right\} \quad (10.5)$$

The variations of the Rarita-Schwinger part of the Lagrangian gives the following three terms

$$\delta \int \left(-\frac{i}{2}\bar{\psi}_\mu\gamma_5 e_\nu{}^a\gamma_a D_\rho\psi_\kappa \varepsilon^{\mu\nu\rho\kappa}\right) d^4x = \int d^4x \left\{ -\frac{i}{\kappa}\bar{\varepsilon}\vec{D}_\mu\gamma_5\gamma_\nu D_\rho\psi_\kappa\varepsilon^{\mu\nu\rho\kappa}\right.$$

$$-\frac{i}{\kappa}\bar{\psi}_\mu\gamma_5\gamma_\nu\vec{D}_\rho D_\kappa\varepsilon\varepsilon^{\mu\nu\rho\kappa}$$

$$\left.-\frac{\kappa}{2}i\bar{\varepsilon}\gamma^a\psi_\nu\bar{\psi}_\mu\gamma_5\gamma_a D_\rho\psi_\kappa\varepsilon^{\mu\nu\rho\kappa}\right\} \quad (10.6)$$

The second term of the above equation, upon using Eq. (9.27) takes the form

$$\frac{-i}{2\cdot4\kappa}\bar{\psi}_\mu\gamma_5\gamma_\nu R_{\rho\kappa}{}^{cd}\sigma_{cd}\varepsilon\varepsilon^{\mu\nu\rho\kappa} = \frac{-i}{2\cdot4\kappa}\bar{\varepsilon}\sigma_{cd}\gamma_\nu\gamma_5\psi_\mu R_{\rho\kappa}{}^{cd}\varepsilon^{\mu\nu\rho\kappa} \quad (10.7)$$

Integrating the first term of Eq. (10.6) by parts and neglecting surface terms we find that it equals

$$+\frac{i}{\kappa}\bar{\varepsilon}\gamma_5[D_\mu,\gamma_\nu]D_\rho\psi_\kappa\varepsilon^{\mu\nu\rho\kappa}$$

$$+\frac{i}{\kappa}\bar{\varepsilon}\gamma_5\gamma_\nu D_\mu D_\rho\psi_\kappa\varepsilon^{\mu\nu\rho\kappa} \quad (10.8)$$

The second of these terms becomes

$$+\frac{i}{2\cdot4\kappa}\bar{\varepsilon}\gamma_5\gamma_\nu R_{\rho\kappa}{}^{cd}\sigma_{cd}\psi_\mu\varepsilon^{\mu\nu\rho\kappa} \quad (10.9)$$

Collecting the terms of Eq. (10.9) with that of Eq. (10.7) gives the result

$$\frac{+i}{2\cdot4\kappa}\bar{\varepsilon}\gamma_5(\gamma_\nu\sigma_{cd}+\sigma_{cd}\gamma_\nu)\psi_\mu R_{\rho\kappa}{}^{cd}\varepsilon^{\mu\nu\rho\kappa} = \frac{1}{4\kappa}\bar{\varepsilon}\gamma_f\psi_\mu\varepsilon_{fvcd}\varepsilon^{\mu\nu\rho\kappa}R_{\rho\kappa}{}^{cd}e$$

$$= -\frac{1}{2\kappa}\bar{\varepsilon}\gamma^a\psi_\mu\{e_a{}^\mu R - 2R_a{}^\mu\}e \quad (10.10)$$

The term of Eq. (10.10) exactly cancels the variation of the Einstein action given in Eq. (10.5).

Consequently, we are just left with the first term of Eq. (10.8), i.e.,

$$+\frac{i}{\kappa}\bar{\varepsilon}\gamma_5[D_\mu,\gamma_\nu]D_\rho\psi_\kappa\varepsilon^{\mu\nu\rho\kappa} \qquad (10.11)$$

and the last term of Eq. (10.6), i.e.,

$$-\frac{\kappa}{2}i\bar{\varepsilon}\gamma^a\psi_\nu\bar{\psi}_\mu\gamma_5\gamma_a D_\rho\psi_\kappa\varepsilon^{\mu\nu\rho\kappa} \qquad (10.12)$$

Performing a Fierz transformation on the latter term it becomes

$$-\frac{\kappa}{2\cdot4}i\bar{\varepsilon}\gamma^a\gamma_R\gamma_a\gamma_5 D_\rho\psi_\kappa\varepsilon^{\mu\nu\rho\kappa}\bar{\psi}_\mu\gamma_R\psi_\nu = +\frac{\kappa}{4}i\bar{\varepsilon}\gamma_c\gamma_5 D_\rho\psi_\kappa\varepsilon^{\mu\nu\rho\kappa}\bar{\psi}_\mu\gamma^c\psi_\nu \quad (10.13)$$

The term of Eq. (10.11) is best evaluated by going to inertial coordinates, that is, we set $\partial_\mu e_\nu{}^a = 0$; it becomes

$$+\frac{i}{4\kappa}\bar{\varepsilon}\gamma_5[\sigma^{cd},\gamma_\nu]w_{\mu cd}D_\rho\psi_\kappa\varepsilon^{\mu\nu\rho\kappa} = \frac{i}{\kappa}\bar{\varepsilon}\gamma_5\gamma^c D_\rho\psi_\kappa w_{\mu c\nu}\varepsilon^{\mu\nu\rho\kappa}$$

$$= +\frac{\kappa}{4}i\bar{\varepsilon}\gamma_5\gamma^c D_\rho\psi_\kappa\bar{\psi}_\mu\gamma_c\psi_\nu\varepsilon^{\mu\nu\rho\kappa} \qquad (10.14)$$

This term cancels with that of Eq. (10.13).

It only remains to show that the variations that contain auxiliary fields cancel. The variation of the auxiliary-field terms of the action give

$$\delta\int d^4x\left\{-\frac{1}{3}(M^2 + N^2 - b_\mu^2)\right\}e$$

$$= \int d^4x\bar{\varepsilon}\left\{(M + i\gamma_5 N)\frac{1}{3}\gamma_\mu + \left(b_\mu - \frac{1}{3}\not{b}\gamma_\mu\right)i\gamma_5\right\}R^\mu \qquad (10.15)$$

The only other such terms come from the auxiliary field terms in the gravitino variations in the Rarita-Schwinger term; these are

$$= \int\left\{-\left(\bar{\varepsilon}\left[\frac{1}{3}(M + i\gamma_5 N)\gamma_\mu + i\left(b_\mu - \frac{1}{3}\not{b}\gamma_\mu\right)\gamma_5\right]R^\mu\right\}d^4x \qquad (10.16)$$

This term cancels that of Eq. (10.15) and so we have shown the invariance of the supergravity action to all orders in κ.

Chapter 11

Tensor Calculus of Rigid Supersymmetry

Given any symmetry group there are rules for taking any two representations and obtaining a third. In conjunction with the rules for building invariants these rules enable the systematic construction of actions. This is true not only for internal groups, but also for space-time groups such as the Poincaré group or the conformal group. For the case of the Poincaré group, we work with four vectors and higher rank tensors. The rules for combining tensors to obtain tensors of a higher rank are obvious. Invariants are obtained by contraction with the Minkowski metric $\eta_{\mu\nu}$.

A similar analysis can be given for the supersymmetry group, and comes under the name of tensor calculus. In the rigid case it was worked out by Wess and Zumino.[20] The rules of tensor calculus are equivalent to the superspace manipulations of Chapter 14. Although the superspace formulation is in some ways more elegant and certainly more useful at the quantum level, it is the tensor calculus which is most useful for constructing classical actions, in particular, the couplings of supergravity to matter. The reason for this is that tensor calculus works with component fields grouping them into sets (supermultiplets) which have well-defined transformation rules under supersymmetry. As such the effect of any manoeuvre on the supermultiplets can easily be evaluated in terms of expressions involving component fields.

11.1 Supermultiplets

The basic supermultiplet called a general scalar multiplet, has component-field content

$$\mathbb{C} = (C; \zeta_\alpha; H, K, A_\mu; \lambda_\alpha; D) \tag{11.1}$$

This supermultiplet is defined to have the following supersymmetry and chiral transformation with parameters ε_α and α respectively:

$$\delta C = i\bar{\varepsilon}\gamma_5\zeta \tag{11.2a}$$

$$\delta\zeta = (A - i\gamma_5\slashed{\partial}C + H + i\gamma_5 K)\varepsilon - \alpha i\gamma_5\zeta \tag{11.2b}$$

$$\delta H = \bar{\varepsilon}\lambda + \bar{\varepsilon}\slashed{\partial}\zeta + 2\alpha\kappa \tag{11.2c}$$

$$\delta K = i\bar{\varepsilon}\gamma_5\lambda + i\bar{\varepsilon}\gamma_5\slashed{\partial}\zeta - 2\alpha H \tag{11.2d}$$

$$\delta A_a = \bar{\varepsilon}\gamma_a\lambda + \bar{\varepsilon}\partial_a\zeta \tag{11.2e}$$

$$\delta\lambda = (-\tfrac{1}{2}\sigma^{cd}f_{cd} + i\gamma_5 D)\varepsilon + i\alpha\gamma_5\lambda \tag{11.2f}$$

$$\delta D = i\bar{\varepsilon}\gamma_5\partial\!\!\!/\lambda \tag{11.2g}$$

where

$$f_{cd} = \partial_c A_d - \partial_d A_c$$

The above transformations can be derived in the following way: we start with the pseudoscalar C, invariant under chiral transformations, and write down the most general form of δC, which is (11.2a). We then write the most general form for $\delta\zeta$ which preserves the Majorana condition, namely

$$\delta\zeta = (H + i\gamma_5 K + A\!\!\!/ + i\gamma^\mu\gamma^5 U_\mu + \sigma^{\mu\nu}t_{\mu\nu}) - i\alpha\gamma_5\zeta' \tag{11.3}$$

and now enforce the supersymmetry on $(\delta_1, \delta_2)C$. We find as a consequence that $U_\mu = \partial_\mu C, t_{\mu\nu} = 0$ and $\zeta' = \zeta$, so that we are left with Eq. (2b) as the most general $\delta\zeta$ compatible with the algebra. Next we make an *ansatz* for $\delta H, \delta K$ and δA_μ and enforce the algebra on ζ, and so on until we have derived transformation laws for the entire multiplet (11.2).

The same technique can be used to enforce the more complicated algebra of supergravity on the multiplet.

In fact, the general scalar multiplet is a 'reducible' representation of supersymmetry. A word is necessary here about the use of the word 'reducible'. As we saw in Chapter 8 an irreducible representation of supersymmetry has its component fields on-shell and so would be of no use in the construction of actions. By 'irreducible' in this context, and often when discussing superspace, we mean a multiplet which has no subset of fields with the properties that

a) It's fields transform under supersymmetry amongst themselves.

b) None of its fields satisfy their equations of motion, (i.e., are on-shell) although some of its fields may satisfy other differential equations. To signify this unconventional use of this word we will put it in inverted commas.

One 'irreducible' sub-super multiplet is called the *curl multiplet* and is given by

$$\lambda = (\lambda_\alpha, f_{cd}, D) \tag{11.4}$$

In this supermultiplet the antisymmetric tensor f_{cd} is subject to the constraint

$$\varepsilon^{abcd}\partial_b f_{cd} = 0 \tag{11.5}$$

and is of the form

$$f_{cd} = \partial_c A_d - \partial_d A_c \tag{11.6}$$

This is the supermultiplet used to construct $N = 1$ super Yang-Mills and is discussed in Chapter 6. We could also write it in the form

$$\mathbb{V} = (A_\mu, \lambda_\alpha, D) \tag{11.7}$$

where A_μ is subject to the gauge transformation

$$A_\mu \to A_\mu + \partial_\mu \Lambda \tag{11.8}$$

Another sub-supermultiplet of $\mathbb{C}$ is the *chiral multiplet* and it is obtained by setting the supermultiplet λ to zero. This implies that

$$\lambda_\alpha = A_\mu - \partial_\mu v = D = 0$$

and leaves a supermultiplet, S of the form

$$S = (v, C; \zeta; H, K) \tag{11.9a}$$

or we may rewrite it as a scalar multiplet, i.e.,

$$\delta S = (C; \zeta; M, N, \partial_\mu v; 0, 0) \tag{11.9b}$$

Clearly this is the multiplet used to construct the Wess-Zumino Lagrangian in Chapter 5.

Another chiral submultiplet which has a different chiral weight to the one above is formed from

$$\mathbb{H} = (H, K; \lambda + \partial\!\!\!/\zeta; \partial_a A^a, D + \partial^2 c) \tag{11.10}$$

The starting components, $A + iB$ of a chiral multiplet are complex and so in general it can admit an arbitrary chiral weight n. The corresponding chiral transformations of a chiral multiplet with components

$$\mathbb{A} = (A, B, \chi_\alpha, F, G) \tag{11.11}$$

are

$$\delta_c A = \alpha n B \qquad \delta_c B = -\alpha n A$$

$$\delta_c \chi = \alpha(n-1) i \gamma_5 \chi$$

$$\delta_c F = \alpha(2-n) G \qquad \delta G = -\alpha(2-n) F \tag{11.12}$$

The supersymmetry transformations are the same regardless of the chiral weight.

$$\delta A = \bar{\varepsilon}\chi \qquad \delta B = i\bar{\varepsilon}\gamma_5\chi$$

$$\delta \chi = (F + i\gamma_5 G + \partial\!\!\!/(A + i\gamma_5 B))\varepsilon$$

$$\delta F = \bar{\varepsilon}\partial\!\!\!/\chi \qquad \delta G = i\bar{\varepsilon}\gamma_5\partial\!\!\!/\chi$$

In the chiral multiplets given above, that of Eq. (11.9) has chiral weight zero whereas that of Eq. (11.10) has chiral weight 2.

Finally we can obtain another submultiplet by setting the supermultiplet $\mathbb{H}$ to zero. The remaining supermultiplet is called the *linear multiplet* and has the component field

$$\mathbb{L} = (C, \zeta, v_\mu) \tag{11.13}$$

where v_μ is subject to the constraint

$$\partial^\mu v_\mu = 0 \tag{11.14}$$

In fact the most general irreducible supermultiplets can be obtained from a general supermultiplet of Eq. (11.1) which possesses Lorentz indices. In the case of one vector index we have

$$\mathbb{C}_\mu = (C_\mu, \zeta_{\mu\alpha}, H_\mu, K_\mu, A_{\mu\nu}, \lambda_{\mu\alpha}, D_\mu) \tag{11.15}$$

The supersymmetry transformations are the same as for $\mathbb{C}$ except for the presence of the extra μ index which is carried along in a trivial manner, i.e.,

$$\delta C_\mu = i\bar{\varepsilon}\gamma_5\zeta_\mu \tag{11.16}$$

11.2   Combination of Supermultiplets

We now wish to combine two supermultiplets to obtain a third one. This operation is analogous to the combination of two vectors to form a second-rank tensor. The combinations are as follows:

(a) We can combine two general supermultiplets in a symmetric way to find a third one,

$$\mathbb{C}_1 \cdot \mathbb{C}_2 = \mathbb{C}_3 \tag{11.17}$$

with components

$$C_3 = C_1 C_2$$

$$\zeta_3 = C_2\zeta_1 + C_1\zeta_2$$

$$H_3 = H_1 C_2 + H_2 C_1 - \frac{i}{2}\bar{\zeta}_1\gamma_5\zeta_2$$

$$K_3 = K_1 C_2 + K_2 C_1 - \frac{1}{2}\bar{\zeta}_1\zeta_2$$

$$A_{\mu 3} = A_{\mu 1} C_2 + A_{\mu 2} C_1 + \frac{i}{2}\bar{\zeta}_1\gamma_\mu\gamma_5\zeta_2$$

$$\lambda_3 = \lambda_1 C_2 + \left[+\frac{1}{2}(K_1 + i\gamma_5 H_1) + \frac{i}{2}A_1\gamma_5 - \frac{1}{2}(\partial\!\!\!/C_1) \right]\zeta_2 + (1 \leftrightarrow 2)$$

$$D_3 = C_2 D_1 + C_1 D_2 + H_1 H_2 + K_1 K_2 - \partial^\mu C_1 \partial_\mu C_2 - A^\mu{}_1 A_{\mu 2} - \bar{\lambda}_1\zeta_2$$

$$\qquad - \bar{\lambda}_2\zeta_1 - \bar{\zeta}_1\partial\!\!\!/\zeta_2 - \bar{\zeta}_2\partial\!\!\!/\zeta_1 \tag{11.18}$$

This result is obtained by varying $C_1 C_2$, which is defined to be C_3, under supersymmetry to obtain the next component

$$\delta(C_1 C_2) = i\bar{\varepsilon}\gamma_5(C_1\zeta_2 + C_2\zeta_1)$$

The operation is repeated on $C_1\zeta_2 + C_2\zeta_1$ to find the next component, and so on.

(b) We can combine two chiral multiplets in a symmetric way to form a third one of chiral weight $n_3 = n_1 + n_2$:

$$\mathbb{A}_1 \cdot \mathbb{A}_2 = \mathbb{A}_3 \tag{11.19}$$

with components

$$A_3 \equiv A_1 A_2 - B_1 B_2$$

$$B_3 = A_1 B_2 + B_1 A_2$$

$$\chi_3 = (A_1 - i\gamma_5 B_1)\zeta_2 + (A_2 - i\gamma_5 B_1)\zeta_2$$

$$F_3 = A_1 F_2 + B_1 G_2 + A_2 F_1 + B_2 G_1 - \bar{\chi}_1\chi_2$$

$$G_3 = A_1 G_2 - B_1 F_2 + A_2 G_1 - B_2 F_1 + i\bar{\chi}_1\gamma_5\chi_2 \tag{11.20}$$

Multiplication of $\mathbb{A}$ with the two constant multiplets

$$\mathbb{1}_+ = (1, 0; 0; 0, 0) \quad \text{and} \quad \mathbb{1}_- = (0, 1; 0; 0, 0) \tag{11.21}$$

gives $\mathbb{A} \cdot \mathbb{1}_+ = \mathbb{A}$ and $\mathbb{A} \cdot \mathbb{1}_- = (-B, A; -i\gamma_5\chi; G, -F)$, the "parity flipped" chiral multiplet.

(c) We can combine two chiral multiplets of the same chiral weight, $n_1 = n_2$, in a symmetric way to form a general multiplet

$$\mathbb{A}_1 \times \mathbb{A}_2 = \mathbb{C} \tag{11.22}$$

with components

$$C \equiv A_1 A_2 + B_1 B_2$$

$$\zeta = (B_1 - i\gamma_5 A_1)\chi_2 + (B_2 - i\gamma_5 A_2)\chi_1$$

$$H = +F_1 B_2 + F_2 B_1 + A_1 G_2 + A_2 G_1$$

$$K = -F_1 A_2 - F_2 A_1 + B_1 G_2 + B_2 G_1$$

$$A_\mu = B_1 \overleftrightarrow{\partial}_\mu A_2 + B_2 \overleftrightarrow{\partial}_\mu A_1 - i\bar{\chi}_1\gamma_5\gamma_\mu\chi_2$$

$$\lambda = +(G_1 + i\gamma_5 F_1)\chi_2 + (G_2 + i\gamma_5 F_2)\chi_1 - \partial_a(B_1 + i\gamma_5 A_1)\gamma^a\chi_2$$

$$\qquad -\partial_a(B_2 + i\gamma_5 A_2)\gamma^a\chi_1$$

$$D = 2F_1 F_2 + 2G_1 G_2 - 2\partial_\mu A_1 \partial^\mu A_2 - 2\partial_\mu B_1 \partial^\mu B_2 - \bar{\chi}_1\slashed{\partial}\chi_2 - \bar{\chi}_2\slashed{\partial}\chi_1 \tag{11.23}$$

(d) The particular combination

$$\mathbb{A}_1 \wedge \mathbb{A}_2 \equiv (\mathbb{1}_- \cdot \mathbb{A}_1) \times \mathbb{A}_2 \qquad (11.24)$$

is antisymmetric in the fields of $\mathbb{A}_1$ and $\mathbb{A}_2$ and has as lowest component $C = A_1 B_2 - A_2 B_1$.

One can of course take any combination of component fields and form a third supermultiplet by systematic variation. However, the most useful combinations are those given above.

A further element of tensor calculus is based on taking a component field and forming a multiplet from it. A particularly useful multiplet is obtained by taking as first components the F and G fields of a chiral multiplet. Since

$$\delta F = \bar{\varepsilon} \partial \chi \qquad (11.25)$$

we obtain the chiral supermultiplet $T\mathbb{A}$ given by

$$T\mathbb{A} = (F, G, \partial \chi, \partial^2 A, \partial^2 B) \qquad (11.26)$$

We note that

$$T T \mathbb{A} = \partial^2 \mathbb{A}$$

The multiplet $T\mathbb{A}$ is sometimes called the *kinetic multiplet*.

11.3 Action Formulas

Having found multiplets of supersymmetry and rules for combining them, it only remains to construct invariant actions. We do this separately for the general and the chiral multiplets.

Given a general supermultiplet, the integral over its highest component, denoted by

$$\int d^4 x [\mathbb{C}]_D \equiv \int d^4 x D \qquad (11.27)$$

is invariant under supersymmetry transformations, since D varies into a divergence and therefore $\delta \int d^4 x [\mathbb{C}]_D = 0$.

Given a chiral multiplet, we can construct an invariant by integrating over its F-component:

$$\int d^4 x [\mathbb{A}]_F \equiv \int d^4 x F \qquad (11.28)$$

since F varies into a divergence. Note, however, that $[\mathbb{A}]_F$ will only be chirally invariant if the chiral weight of $\mathbb{A}$ is 2.

Example

a) The model of Wess and Zumino[3] is based on a single chiral multiplet. As a Lagrangian, we take the most general "invariant" which can be constructed without use of coupling constants with negative dimensions,

$$L_{W-Z} = -\frac{1}{4}[A \times A]_D - \frac{m}{2}[A \cdot A]_F - \frac{\lambda}{3}[A \cdot A \cdot A]_F \qquad (11.29)$$

A term $[A \cdot TA]_F$ gives nothing new since it differs only by a divergence from $-\frac{1}{2}[A \times A]_D$, and a term $[A]_F$ can always be shifted away by combinations of shifts $A \to A + \alpha 1_+ + \beta 1_-$ and constant chiral transformations.

The reader may use the rules above to evaluate the action and obtain the result

$$L_{W-Z} = \{-\tfrac{1}{2}(\partial_a A)^2 - \tfrac{1}{2}(\partial_a B)^2 - \tfrac{1}{2}\bar{\chi}\partial\chi + \tfrac{1}{2}F^2 + \tfrac{1}{2}G^2\}$$
$$- m\{AF + GB - \tfrac{1}{2}\bar{\chi}\chi\} - \lambda\{(A^2 - B^2)F + 2GAB - \bar{\chi}(A - i\gamma_5 B)\chi\}$$
$$(11.30)$$

b) Consider the linear multiplet of Eq. (11.13); an invariant action is given by

$$\int d^4x \frac{1}{2}[L \cdot L]_D \qquad (11.31)$$

Evaluating this expression using the above equations we find that it equals

$$\int d^4x \left\{ -\frac{1}{2}(\partial_\mu C)^2 - \frac{1}{2}\bar{\chi}\partial\chi - v_\mu{}^2 \right\} \qquad (11.32)$$

This is an alternative description [47] of the supersymmetric states $(0^+, 0^-, \tfrac{1}{2})$ which are also represented by the Wess-Zumino Lagrangian.

c) Super QED[20]

Consider two chiral multiplets S_1 and S_2 which have gauge transformations given by

$$\delta S_1 = gSS_2 \qquad \delta S_2 = -gSS_1 \qquad (11.33)$$

where S is a chiral multiplet and is the parameter of the gauge transformations. The gauge field resides in the general supermultiplet V which has the gauge transformation

$$\delta V = \partial S \qquad (11.34)$$

Here ∂S is a general superfield of Eq. (11.9b).

Clearly we can gauge away the C, ζ, H and K components of V leaving the Abelian gauge invariance of A_μ. However, having done this we must consider how we can remain in the gauge

$$C = \zeta = H = K = 0 \qquad (11.35)$$

The supersymmetry variation of ζ in this gauge is

$$\delta\zeta = A\varepsilon \tag{11.36}$$

The gauge variation of ζ is given by

$$\delta\zeta = \chi \tag{11.37}$$

and so we can maintain the choice $\zeta = 0$ by making a compensating gauge transformation with $\chi = -A\varepsilon$. A similar procedure is required for H and K. The net effect of this compensation on the remaining fields A_μ, λ and D is that the commutator of two supersymmetries gives not only a space-time translation, but also a gauge transformation as discussed in Chapter 7.

Let us now define the two general scalar multiplets

$$V_{\mathrm{I}} = \tfrac{1}{2}(S_1 \times S_1 + S_2 \times S_2)$$

$$V_{\mathrm{II}} = (S_1 \wedge S_2) \tag{11.38}$$

Under a gauge transformation the above quantities transform as

$$\delta V_{\mathrm{I}} = -2g V_{\mathrm{II}} \cdot \partial S \qquad \delta V_{\mathrm{II}} = -2g V_{\mathrm{I}} \cdot \partial S \tag{11.39}$$

This result is a consequence of the following identities which are easily established

$$(S \cdot S_1) \times S_2 - (S \cdot S_2) \times S_1 = 2(S_1 \wedge S_2) \cdot \partial S$$

$$(S \cdot S_1) \times S_1 = -(S_1 \times S_1) \cdot \partial S \tag{11.40}$$

A general scalar supermultiplet which is invariant under gauge transformations is given by

$$\tfrac{1}{4}[V_{\mathrm{I}} \cdot (e^{2gV} + e^{-2gV}) + V_{\mathrm{II}} \cdot (e^{2gV} - e^{-2gV})] \tag{11.41}$$

Consequently an action invariant under supersymmetry and gauge transformations is

$$A^{\mathrm{S.QED}} = \int d^4x \frac{1}{4}[V_{\mathrm{I}} \cdot (e^{2gV} + e^{-2gV}) + V_{\mathrm{II}} \cdot (e^{2gV} - e^{-2gV})]_D \tag{11.42}$$

Using the Wess-Zumino gauge choice and the tensor calculus explained above we can evaluate $A^{\mathrm{S.QED}}$ to find

$$A^{\mathrm{S.QED}} = \int d^4x \left\{ \left[-\frac{1}{2}(\mathcal{D}_a A_1)^2 - \frac{1}{2}(\mathcal{D}_a B_1)^2 - \frac{1}{2}\bar{\chi}_1 \slashed{\mathcal{D}} \chi_1 + (1 \leftrightarrow 2) \right] \right.$$

$$+ \left[-\frac{1}{4}f_{\mu\nu}^{~2} - \frac{1}{2}\bar{\lambda}\slashed{\partial}\lambda + \frac{1}{2}D^2 \right] + \left[+gD(A_1 B_2 - A_2 B_1) \right.$$

$$\left. - g\bar{\lambda}[(A_1 + i\gamma_5 B_1)\chi_2 - (A_2 + i\gamma_5 B_2)\chi_1]] \right\} \tag{11.43}$$

where

$$\mathscr{D}_a A_1 = \partial_a A_1 - g A_a A_2$$

$$\mathscr{D}_a A_2 = \partial_a A_2 + g A_a A_1$$

and similarly for B_1 and B_2.

Also

$$\mathscr{D}_a \chi_1 = \partial_a \chi_1 + g A_a \chi_2$$

Using the tensor calculus on the Noether method described in Chapter 7 one can find the most general renormalizable $N = 1$ supersymmetric theory. The invariant action is of the form

$$A = A^{\mathrm{YM}} + A^{\mathrm{matter}} + A^{\mathrm{mass}} + A^{\mathrm{int}} + A^{\mathrm{linear}}$$

where

$$A^{\mathrm{YM}} = \int d^4 x \left\{ -\frac{1}{4}(F_{\mu\nu}{}^s)^2 - \bar{\lambda}^s \mathscr{D} \lambda_s + \frac{1}{2} D_s{}^2 \right\}$$

$$A^{\mathrm{matter}} = \int d^4 x \left\{ -\frac{1}{2}(\mathscr{D}_\mu z)^2 - \frac{1}{2} \bar{\chi}_{\mathrm{L}} \mathscr{D} \chi_{\mathrm{L}} + |f|^2 \right.$$

$$\left. + g(T^s)^a{}_b (\chi_{\mathrm{L}a} z^b \lambda_s - \bar{\lambda}_s z_a{}^* \chi_{\mathrm{L}}{}^b + D_s z_a{}^* z^b) \right\}$$

$$A^{\mathrm{mass}} = -\int d^4 x\, m_{ab} \left\{ f^a z^b - \frac{1}{4} \chi_{\mathrm{L}}{}^{ca} \chi_{\mathrm{L}}{}^b + \mathrm{h.c.} \right\}$$

$$A^{\mathrm{int}} = -\int d^4 x\, d_{abe} \left\{ z^a z^b f^e - \frac{1}{2} \chi_{\mathrm{L}}{}^{ca} z^b \chi_{\mathrm{L}}{}^e + \mathrm{h.c.} \right\}$$

$$A^{\mathrm{linear}} = \int d^4 x (D_s \xi^s + \mu_a f^a + \mathrm{h.c.}) \tag{11.44}$$

In the above

$$F_{\mu\nu}{}^s = \partial_\mu A_\nu{}^s - \partial_\nu A_\mu{}^s - g f^{srt} A_\mu{}^r A_\nu{}^t$$

$$\mathscr{D}_\mu \lambda^s = \partial_\mu \lambda^s - g f^{srt} A_{r\mu} \lambda^t$$

$$\mathscr{D}_\mu z^a = \partial_\mu z^a - g(A_\mu)^a{}_b z^b$$

$$\mathscr{D}_\mu \chi_{\mathrm{L}} = \partial_\mu \chi_{\mathrm{L}} - g(A_\mu)^a{}_b \chi_{\mathrm{L}}{}^b$$

$$\chi_{\mathrm{L}} = \tfrac{1}{2}(1 + \gamma_5)\chi; \qquad \chi_{\mathrm{L}}{}^{ca} = C^{-1} \chi_{\mathrm{L}}{}^a$$

and $(A_\mu)^a{}_b = A_\mu{}^s (T_s)^a{}_b$ where $(T_s)^a{}_b$ are the representation of the generators of the gauge G to which the chiral matter belongs. The above action will be gauge invariant if

$$m_{ab}(T^s)^a{}_c + m_{ab}(T^s)^b{}_c = 0$$

$$d_{abe}(T^s)^e{}_c + d_{aec}(T^s)^e{}_b + d_{ebc}(T^s)^e{}_a = 0$$

$$\mu_a(T_s)^a{}_b = 0$$

$$\zeta^s f_{rst} = 0 \qquad\qquad (11.45)$$

The latter two equations imply that the linear terms belong to a chiral singlet and a $U(1)$ gauge multiplet respectively.

The transformations that leave the above action invariant are

$$\delta A^s = \bar{\varepsilon}\gamma_\mu \lambda^s$$

$$\delta\lambda^s = (-\tfrac{1}{2}\sigma_{\mu\nu}F^{s\mu\nu} + i\gamma_5 D^s)\varepsilon$$

$$\delta D^s = i\bar{\varepsilon}\gamma_5 \slashed{\mathcal{D}}\lambda^s$$

and

$$\delta z^a = \bar{\varepsilon}\chi_{\mathrm{L}}{}^a$$

$$\delta\chi_{\mathrm{L}}{}^a = 2f^a \varepsilon_{\mathrm{L}} + 2\slashed{\mathcal{D}}z^a \varepsilon_{\mathrm{R}}$$

$$\delta f^a = \bar{\varepsilon}\slashed{\mathcal{D}}\chi_{\mathrm{L}}{}^a \qquad\qquad (11.46)$$

The fact that the chiral matter transformation laws involve the Yang-Mills fields is due to the choice of the Wess-Zumino gauge and the corresponding compensating transformation. The relation of the above fields to the usual description of the Wess-Zumino multiplet is

$$z = \tfrac{1}{2}(A - iB) \qquad \chi_{\mathrm{L}} = \tfrac{1}{2}(1 + \gamma_5)\chi \qquad f = F + iG \qquad (11.47)$$

The reader who wishes to verify the above invariance may wish to use the relations

$$\mathcal{D}_\mu z_a{}^* = \partial_\mu z_a{}^* + g z_b{}^*(A_\mu)^b{}_a$$

$$\delta\bar{\chi}_{\mathrm{L}a} = \bar{\varepsilon}_{\mathrm{L}}2f_a{}^* - \bar{\varepsilon}_{\mathrm{R}}2\slashed{\mathcal{D}}z_a{}^* \qquad \delta f_a{}^* = -\bar{\chi}_{\mathrm{L}a}\slashed{\mathcal{D}}\varepsilon \qquad (11.48)$$

The equations of motion for the auxiliary fields are

$$f_a{}^* = + m_{ab}z^b + d_{abc}z^b z^c + \mu_a$$

$$D^s = -g(T^s)^a{}_b z_a{}^* z^b + \zeta^s \qquad\qquad (11.49)$$

and as such the classical potential is of the form

$$V = |f_a|^2 + \tfrac{1}{2}(D^s)^2 \qquad\qquad (11.50)$$

We note that in terms of the superpotential

$$W(\varphi_a) = \mu_a \varphi^a + \frac{1}{2}m_{ab}\varphi^a \varphi^b + \frac{d_{abc}}{3}\varphi^a \varphi^b \varphi^c \qquad\qquad (11.51)$$

the auxiliary fields are given by

$$f_a^* = \frac{\partial W}{\partial z^a} \tag{11.52}$$

while the coupling to $\chi_L{}^{ca}\chi_L{}^b$ is of the form

$$\frac{1}{2}\frac{\partial^2 W}{\partial z_a^* z_b^*} \tag{11.53}$$

These expressions in terms of the superpotential are a simple consequence of expanding the superspace action in terms of component fields.

Chapter 12

Theories of Extended Rigid Supersymmetry

In this chapter we wish to construct theories of rigid supersymmetry in x-space, their superspace formulations being given later. The possible on-shell states of supersymmetric massless theories with spins one and less are given by the irreducible representations of supersymmetry discussed in Chapter 8 and are listed in the table below.

Table 12.1   Multiplicities for massless irreducible representations with maximal helicity 1 or less.

Spin \ N	1	1	2	2	4
Spin 1	—	1	—	1	1
Spin $\frac{1}{2}$	1	1	2	2	4
Spin 0	2	—	4	2	6

As noted previously, theories with greater than four supercharges must have spins greater than one. In the context of the revised supersymmetric no-go theorem, they are the most symmetric theories with spins not exceeding one. They are consistent in the sense that they are renormalizable and unitary. The inclusion of a spin-3/2 field requires a spin-2 field in order to propagate causally, but it now seems very unlikely that supergravity theories are renormalizable. It is thought, however, that some string theories are renormalizable and unitary.

The first theory listed under $N = 2$ has 2 spin-$\frac{1}{2}$ Majorana states which are singlets under $SU(2)$ and 4 spin-0 states which are a complex doublet under $SU(2)$. We call this multiplet the hypermultiplet or $N = 2$ matter. The second theory listed under $N = 2$ has 1 spin-1, 2 Majorana spin-1/2, and 2 spin-0 states. These particles are $SU(2)$ singlets, doublets and singlets respectively. This multiplet is called the $N = 2$ Yang-Mills multiplet.

The unique $N = 4$ multiplet has 1 spin-1, 4 Majorana spin-$\frac{1}{2}$ and 6 spin-0 in the singlet, vector and self-dual antisymmetric tensor representation of $SU(4)$.

The above Yang-Mills multiplets must be put in the adjoint representation of the gauge group while that of the $N = 2$ matter multiplet can belong to any real representation of the gauge group. An important property of these extended theories is that they are vector-like, that is, the fermions of both handedness belong to the same representation of the gauge group. This is an inevitable consequence of the fact that the supercharges commute with the gauge group. The hypermultiplet, when considered in the "rest frame" $(m, 0, 0, m)$ has as its Clifford vacuum the helicity $+\frac{1}{2}$ state. This state is related to the helicity $-\frac{1}{2}$ state by the action of two supercharges. Consequently both the $+\frac{1}{2}$ and $-\frac{1}{2}$ states must belong to the same representation of the gauge group. For the super Yang-Mills models, all fields are in the same representation as the gauge fields, namely the adjoint representation. This means that if these models contain the particles of the standard model (Glashow-Salam-Weinberg theory) then for every chiral fermion in the standard model they will contain a fermion of the opposite chirality, but in the same representation (the so-called mirror particles). Of course, to find a realistic model one must find some way of splitting the observed fermions from their mirror particles in mass. The inclusion of mirror particles is quite a general feature of all extended theories and can only be avoided by using nonlinear realizations, which is, in some sense, cheating, or by including higher-spin fields which transform non-trivially under the gauge group.

For the construction of $N = 2$ models we will be required to write down Majorana conditions on the supercharges, supersymmetry parameter, etc. The usual Majorana property on a spinor is

$$\bar{\lambda}_{\alpha i} = \lambda^{\beta i} C_{\beta \alpha}$$

The maximal internal symmetry transformation compatible with such a condition is $U(N)$ which is realized as follows:

$$\delta \lambda^{\alpha i} = \left((S_1')^i{}_j \delta^\alpha{}_\beta + (S'')^i{}_j (\gamma_5)^\alpha{}_\beta \right) \lambda^{\beta j}$$

where S' is real and antisymmetric (i.e., $\Lambda_2 i_2 \sigma_2$ for $N = 2$) while S'' is complex and symmetric (i.e., $S'' = i\sigma_2(\Lambda_3 \sigma_1, \Lambda_0 i\sigma_2, \Lambda_1 \sigma_3)$ for $N = 2$). The proof of this fact is simple, and is essentially the same as that given for the algebra in Chapter 1.

For the $N = 2$ case there exists a numerically invariant second rank tensor $\varepsilon_{ij} = -\varepsilon_{ji}$ and it is often preferable to use an alternative Majorana condition which is obtained from the one above by a field redefinition. (We recall that any reality condition can be brought to the above form by a field redefinition.) Let us make the redefinition

$$\lambda'^{\alpha j} = e \varepsilon^{ji} \left(\frac{1 + \gamma_5}{2} \right)^\alpha{}_\beta \lambda^{\beta i} + d \left(\frac{1 - \gamma_5}{2} \right)^\alpha{}_\beta \lambda^{\beta j}$$

where

$$\varepsilon^{12} = -\varepsilon^{21} = +1$$

Choosing $+ie^* = d$ we find the Majorana condition becomes[22,67]

$$\lambda'^{\alpha i} = (i\gamma_5 C)^{\alpha\beta}\varepsilon^{ij}\overline{\lambda}'_{\beta j}$$

where we have adopted the convention that complex conjugation lowers an i or a j index. This condition is sometimes called the symplectic Majorana condition.

The symmetry transformation for λ' is deduced from its definition and the transformation of $\lambda^{\alpha i}$ and is given by

$$\delta\lambda'^{\alpha i} = ((C_1)_j{}^i\delta^{\alpha}{}_{\beta} + (\gamma_5)^{\alpha}{}_{\beta}(C_2)^i{}_j)\lambda'^{\beta j}$$

where

$$C_2 = \mathbb{1}\Lambda_0, \qquad C_1 = i(\Lambda_1\sigma_1, \Lambda_2\sigma_2, \Lambda_3\sigma_3)$$

The advantage of this type of Majorana constraint becomes clear. The $SU(2)$ (generated by C_1) is realized without a γ_5 and since ε_{ij} is a numerically invariant tensor of $SU(2)$ it is obviously allowed by the new Majorana constraint. The $U(1)$, however, is realized with a γ_5 transformation. The situation with the original Majorana condition is the opposite and so although both constraints allow a $U(2)$ symmetry, less of this symmetry is realized in an obvious way with the original Majorana condition. Clearly, in two-component formalism these distinctions are irrelevant. The reader may verify the consistency of the new Majorana condition under complex conjugation.

In the discussion of $N = 2$ theories which follows we will adopt the Majorana constraint

$$\varepsilon_{\alpha i} = (i\gamma_5 C)_{\alpha\beta}\varepsilon_{ij}\overline{\varepsilon}^{\beta j}$$

Clearly, having a parameter $\varepsilon_{\alpha i}$ without a Majorana constraint would be equivalent to having four supersymmetries rather than two.

We now construct the rigid theories of $N = 2$ supersymmetry in turn beginning with $N = 2$ Yang-Mills theory. We will follow the procedure for the construction of the Wess-Zumino model given in Chapter 5; that is, begin with the on-shell states, find an on-shell supersymmetry linking them and construct an invariant "on-shell action." We then find an off-shell description, i.e., the auxiliary fields and a corresponding invariant action.

12.1 $N = 2$ Yang-Mills[23]

I. The on-shell states, which belong to representations of (rigid) $SU(2)$, may be represented by the fields A, B, A_μ and $\lambda_{\alpha i}$. The fields A, B, and A_μ are real and singlets under $SU(2)$ whereas the spin-1/2 is a doublet. To remain with only one doublet of spin 1/2, as required by the on-shell states, we must impose a Majorana condition. This condition is of the form

$$\lambda_{\alpha i} = (i\gamma_5 C)_{\alpha\beta} \varepsilon_{ij} \overline{\lambda}^{\beta j} \tag{12.1}$$

The symbol $\varepsilon_{ij} = -\varepsilon_{ji} = \varepsilon^{ij}$ can be used to raise and lower $SU(2)$ indices as follows:

$$\lambda_\alpha^{\ i} = \varepsilon^{ij} \lambda_{\alpha j} \quad \text{and} \quad \lambda_\alpha^{\ i} \varepsilon_{ij} = \lambda_{\alpha j} \tag{12.2}$$

We leave it as an exercise for the reader to deduce the $U(1)$ weights of A, B, $\lambda_{\alpha i}$ and A_μ. Since these fields represent the on-shell states, they obey, neglecting interaction terms, their equations of motion,

$$\partial^2 A = \partial^2 B = (\partial\!\!\!/\lambda_i)\alpha = \partial^\nu f_{\mu\nu} = 0 \tag{12.3}$$

where

$$f_{\mu\nu} = \partial_\mu A_\nu - \partial_\nu A_\mu \tag{12.4}$$

We first examine the linearized or free theory as this is a consistent theory which is much simpler in structure than the interacting theory.

II. We now wish to find the supersymmetry transformations between the on-shell states. On grounds of dimension and linearity these must be of the form

$$\delta A = i\overline{\varepsilon}^i \lambda_i, \qquad \delta B = \overline{\varepsilon}^i \gamma_5 \lambda_i$$

$$\delta A_\mu = +\overline{\varepsilon}^i \gamma_\mu \lambda_i$$

$$\delta \lambda_i = -\tfrac{1}{2} c_1 f^{\mu\nu} \sigma_{\mu\nu} \varepsilon_i - ic_2 \partial\!\!\!/(A - i\gamma_5 B)\varepsilon_i + c_3 \partial_\mu A^\mu \varepsilon_i \tag{12.5}$$

where c_1, c_2 and c_3 are constants. The linearized theory should be invariant under local Abelian transformations

$$\delta A^s_{\ \mu} = \partial_\mu \Lambda^s, \qquad \delta A^s = \delta B^s = \delta \lambda^s_{\ \alpha i} = 0 \tag{12.6}$$

and rigid G_1 transformations with parameter T^r (i.e., T^r are independent of x^μ).

$$\delta A^r = f^{rst} T^s A^t, \qquad \delta A_\mu^{\ r} = f^{rst} T^s A_\mu^{\ t}, \qquad \text{etc.} \tag{12.7}$$

where f^{rst} are the structure constants of the group G_1 which will become the gauge group of the interacting theory. All fields are in the adjoint representation of G_1 as they must transform the same way as A_μ. This is a consequence of the fact that supersymmetry and the gauge group G_1 commute. In the above supersymmetry transformations and in what follows the group indices have not been written, but are understood to be present.

The supersymmetry transformations, however, must form a closed algebra. In particular, carrying out the commutator of supersummetry and local transformations on $\lambda_{\alpha i}$ yields

$$[\delta_\Lambda, \delta_\varepsilon]\lambda_{\alpha i} = c_3(\partial_\mu \partial^\mu \Lambda)\varepsilon_{\alpha i} \tag{12.8}$$

This is not a recognizable symmetry of the theory and so it implies that $c_3 = 0$. The commutator of two supersymmetries on A yields

$$[\delta_1, \delta_2]A = 2c_2 \bar{\varepsilon}_2{}^i \gamma^\mu \varepsilon_{1i} \partial_\mu A \tag{12.9}$$

and so to agree with the supersymmetry algebra we choose $c_2 = +1$. The terms containing B and $F_{\mu\nu}$ vanish due to subtracting the $(1 \leftrightarrow 2)$ exchange and the Majorana properties of $\varepsilon_{\alpha i}$. On A_μ we find that the commutator of two supersymmetries yields a transformation of the correct magnitude provided $c_1 = +1$, but one also finds a gauge transformation. This phenomenon also occurs in the $N = 1$ Yang-Mills theory when working only with the fields $(A_\mu, \lambda_\alpha, D)$. It results from working in the Wess-Zumino gauge[13] and is a consequence of the compensating transformations required to maintain this gauge choice under supersymmetry. The situation in $N = 2$ Yang-Mills theory is similar. We find that the algebra closes on $\lambda_{\alpha i}$ provided we use the field equations. The above supersymmetry transformations are an "on-shell algebra."

III. We must build an invariant action which can only be of the form

$$A = \int d^4x \left[-\frac{1}{4}F^2{}_{\mu\nu} - \frac{1}{2}(\partial_\mu A)^2 - \frac{1}{2}(\partial_\mu B)^2 - \frac{1}{2}\bar{\lambda}^i \not{\partial}\lambda_i \right] \tag{12.10}$$

This action is invariant without the use of field equations and as such is an "on-shell action".

It is appropriate at this point to comment on the drawbacks of formulating theories using "on-shell actions." Although one can calculate with such actions it is rather difficult to use them to establish general results. For example, one may wish to find all possible interacting theories. However, adding interaction terms modifies the field equations and so requires new on-shell supersymmetry transformations. In effect, one must construct each new interaction term from scratch. This is not such a serious problem for the theories of extended rigid supersymmetry as the renormalizable couplings are few in number. However, it is a very serious difficulty when trying to find the most general coupling of matter to supergravity.

Another difficulty concerns the quantization of supersymmetric theories. It is much simpler to handle the symmetries of a theory that is being quantized if the symmetries can be manifestly realized. Clearly this is not the case when working with the "on-shell actions."

IV. In view of the above discussion we now wish to find a supermultiplet that closes without the use of the classical field equations. That is, we must find the auxiliary fields. A useful guide to the number of auxiliary fields is provided by demanding that their addition gives an off-shell multiplet with equal numbers of bosonic and fermionic degrees of freedom. The spin-1/2 $\lambda_{\alpha i}$ has 8 degrees of freedom while A, B and A_μ provide only $1 + 1 + 3 = 5$, respectively. The vector A_μ provides only 3 degrees of freedom as the fermi-bose rule is proved using the algebra relation $\{Q, Q\} \sim P_\mu$. As noted previously, this only holds for gauge invariant quantities and hence we must in effect subtract the gauge degree of freedom of A_μ. As such, we require 3 auxiliary bosonic degrees of freedom. One possibility is that we have an $SU(2)$ triplet of auxiliary fields $X^{ij} = X^{ji}$ which are of dimension two. Let us assume that this is the case. We must now construct the supersymmetry transformations among the fields

$$A, B, A_\mu, \lambda_{\alpha i} \text{ and } X^{ij} \qquad (12.11)$$

where

$$(X^{ij})^* \equiv X_{ij} = X^{kl}\varepsilon_{ki}\varepsilon_{lj}$$

Demanding linearity, matching dimensions and using the fact that when $X^{ij} = 0$ the on-shell algebra reduces to that given earlier, we find that

$$\delta A, \delta B \text{ and } \delta A_\mu$$

must be the same as before while

$$\delta\lambda_i = -\tfrac{1}{2}F^{\mu\nu}\sigma_{\mu\nu}\varepsilon_i - i\slashed{\partial}(A - i\gamma_5 B)\varepsilon_i - i(\underline{\tau})_i{}^j\varepsilon_j\underline{X}$$

$$\delta\underline{X} = +ic_4\bar{\varepsilon}^i(\underline{\tau})_i{}^j\slashed{\partial}\lambda_j \qquad (12.12)$$

where $X_i{}^j = (\underline{\tau})_i{}^j\underline{X}$, the $(\underline{\tau})_i{}^j$ are the Pauli matrices and the constant c_4 is determined to be $c_4 = +1$ by demanding closure.

The corresponding invariant action is

$$A = \int d^4x \left\{ -\frac{1}{4}F^2{}_{\mu\nu} - \frac{1}{2}(\partial_\mu A)^2 - \frac{1}{2}(\partial_\mu B)^2 - \frac{1}{2}\bar{\lambda}^i\slashed{\partial}\lambda_i + \frac{1}{2}X^2 \right\} \qquad (12.13)$$

Having constructed the linearized theory we now wish to find the corresponding nonlinear theory—that is, the theory invariant under the now local gauge group G_1. Consequently, we must make the parameter T^r space-time dependent, i.e., $T^r(x)$. To recover gauge invariance we must introduce covariant derivatives and knit together the linearized local Abelian transformations with the now local G_1 transformations by identifying

$$\Lambda^r(x) = \frac{1}{g}T^r(x) \qquad (12.14)$$

For a detailed discussion of the Noether technique, see Chapter 9. The resulting gauge-invariant action is

$$A^1 = \text{Tr} \int d^4x \left\{ -\frac{1}{4}F^2_{\mu\nu} - \frac{1}{2}\bar{\lambda}^i \not{D}\lambda_i - \frac{1}{2}(D_\mu A)^2 - \frac{1}{2}(D_\mu B)^2 + \frac{1}{2}X^2 \right\} \quad (12.15)$$

where

$$F_{\mu\nu} = \partial_\mu A_\nu - \partial_\nu A_\mu - g[A_\mu, A_\nu]$$

$$D_\mu A = \partial_\mu A - g[A_\mu, A] \quad \text{etc.} \quad (12.16)$$

In the above, $A_\mu = A_\mu^s T_s$, where T_s are the generators of G_1 in the adjoint representation.

The action A^1 is, however, no longer supersymmetric. We can recover supersymmetry by adding gauge-invariant terms to A^1 and to the super-symmetry transformation laws. Such additions to the action can only be terms quartic in the spin-0 fields A and B and Yukawa type terms.

The final result for the transformation laws is

$$\delta A = i\bar{\varepsilon}^i \lambda_i \qquad \delta B = \bar{\varepsilon}^i \gamma_5 \lambda_i \qquad \delta A_\mu = \bar{\varepsilon}^i \gamma_\mu \lambda_i$$

$$\delta \lambda_i = -\frac{1}{2}F^{\mu\nu}\sigma_{\mu\nu}\varepsilon_i - i\gamma^\mu D_\mu(A - i\gamma_5 B)\varepsilon_i + ig[A, B]\gamma_5 \varepsilon_i - i(\underline{\tau})_i{}^j \varepsilon_j \underline{X}$$

$$\delta \underline{X} = i\bar{\varepsilon}^i(\underline{\tau})_i{}^j \{\gamma^\mu D_\mu \lambda_j + g[A - i\gamma_5 B, \lambda_j]\} \quad (12.17)$$

and the supersymmetric action is

$$A^{N=2\,\text{YM}} = \text{Tr} \int d^4x \left\{ -\frac{1}{4}F^2_{\mu\nu} - \frac{1}{2}(D_\mu A)^2 - \frac{1}{2}(D_\mu B)^2 + \frac{1}{2}\underline{X}^2 \right.$$

$$\left. -\frac{1}{2}\bar{\lambda}^i \gamma^\mu D_\mu \lambda_i - \frac{ig}{2}\bar{\lambda}^i[A - i\gamma_5 B, \lambda_i] - \frac{g^2}{2}([A, B])^2 \right\} \quad (12.18)$$

The above description of $N = 2$ Yang-Mills theory does not involve a central charge in the sense that the fields carry an off-shell realization of the super-symmetry algebra of Chapter 2 in which the central charge is trivially realized. For an alternative description of $N = 2$ Yang-Mills theory where one of the spin-zero fields is represented by a conserved vector, the reader is referred to Ref. 24.

12.2 $N = 2$ Matter[25]

We now repeat in outline the above procedure for $N = 2$ matter.

I. The on-shell states are represented by the fields A^{ia} and ψ^a as well as their complex conjugates $(A^{ia})^* = A_{ia}$ and ψ_a. The spin-0 fields are a complex doublet under $SU(2)$, while ψ^a is a singlet. The fields A^{ia} and ψ^a can be in any representation R of a group G_2, while their complex conjugates are of course in the representation $\bar{R}$, the index a being the group index.

II. The transformation laws are determined by linearity, matching dimensions, and closure to be of the form

$$\delta A^{ia} = +\bar{\varepsilon}^i \psi^a$$

$$\delta \psi^a = +2\gamma^\mu \partial_\mu A^{ia} \varepsilon_i \qquad (12.19)$$

We note that carrying out the commutator of two supersymmetries on A^{ia} yields the correct result, namely,

$$[\delta_1, \delta_2] A^{ia} = +2\bar{\varepsilon}_2{}^i \gamma^\mu \partial_\mu A^{ja} \varepsilon_{1j} - (1 \leftrightarrow 2)$$

$$= 2\bar{\varepsilon}_2{}^k \partial\!\!\!/ \varepsilon_{1k} A^{ia} \qquad (12.20)$$

Here, we have used the identity

$$\delta_i^k \delta_j^l = \tfrac{1}{2} \delta_i^l \delta_j^k + \tfrac{1}{2} (\tau)_i{}^l (\tau)_j{}^k \qquad (12.21)$$

and the fact that $\varepsilon^{ki}(\tau)_i{}^l$ is symmetric in k and l while $\bar{\varepsilon}^k \gamma_\mu \varepsilon^l$ is antisymmetric in k and l due to the $i\gamma_5$ matrix in the Majorana condition. The reader may verify that the transformation closes on ψ provided $\partial\!\!\!/ \psi = 0$.

III. The on-shell action is

$$A = \int d^4 x \left\{ -|\partial_\mu A^{ia}|^2 - \frac{1}{2} \bar{\psi}_a \partial\!\!\!/ \psi^a \right\} \qquad (12.22)$$

IV. To find an off-shell formulation we require a contribution of four auxiliary bosonic degrees of freedom. There are, however, several ways to achieve this. We now review the different possible off-shell formulations.

(a) We can add a complex doublet of dimension 2 auxiliary fields, F^{ia}. The supersymmetry transformations[15] are

$$\delta A^{ia} = +\bar{\varepsilon}^i \psi^a \qquad (12.23)$$

$$\delta \psi^a = +2\gamma^\mu \partial_\mu A^{ia} \varepsilon_i - 2\gamma_5 F^{ia} \varepsilon_i$$

$$\delta F^{ia} = +\bar{\varepsilon}^i \gamma_5 \gamma^\mu \partial_\mu \psi^a$$

In this case, the algebra closes without using the equations of motion, but with an off-shell central charge, i.e.,

$$[\delta_1, \delta_2] A^i = \bar{\varepsilon}_2{}^i 2 \partial\!\!\!/ A^j \varepsilon_{1j} - \bar{\varepsilon}_2{}^i 2 i\gamma_5 \varepsilon_{1j} F^j - (1 \leftrightarrow 2)$$

$$= 2\bar{\varepsilon}_2{}^j \partial\!\!\!/ \varepsilon_{1j} A^i + 2i \bar{\varepsilon}_2{}^j \gamma_5 \varepsilon_{1j} F^i$$

the transformation of the multiplet under the central charge being

$$\delta A^i = w F^i \qquad \delta \psi = iw\gamma_5 \partial\!\!\!/ \psi \qquad \delta F^i = w \partial^2 A^i$$

The reader may verify that the commutation of two supersymmetries and a central charge commute; for example

$$[\delta_\varepsilon, \delta_w] A^i = w i \bar\varepsilon^i \gamma_5 \partial\!\!\!/\psi - \bar\varepsilon^i w i \gamma_5 \partial\!\!\!/\psi = 0$$

An off-shell central charge[24] is one that vanishes when the equations of motion are used, in this case the equations of motion are

$$\delta_w A^{ia} = \delta_w \psi^a = \delta_w F^{ia} = 0$$

The corresponding invariant action is

$$A = \int d^4x \left\{ -|\partial_\mu A^{ia}|^2 - \frac{1}{2}\bar\psi_a \gamma^\mu \partial_\mu \psi^a + |F^{ia}|^2 \right\} \tag{12.24}$$

Clearly the $N = 2$ hypermultiplet is composed of two $N = 1$ Wess-Zumino multiplets. One could also consider making the $N = 2$ matter from one Wess-Zumino multiplet $(A, B; \chi_\alpha; F, G)$ and one linear multiplet (C, ζ_α, V_μ) with $\partial^\mu V_\mu = 0$. Although there is no guarantee that this can be made supersymmetric, in fact it can and it is our next possible formulation.

(b) In this formulation[26] the on-shell states belong to the different representations of $SU(2)$ discussed in Chapter 2. If we start with a doublet, helicity-1/2 Clifford vacuum, $|1/2, i\rangle$, the on-shell states are an $SU(2)$ doublet of spin-1/2 Majorana fermions and the four spin-0's consisting of a triplet and a singlet. We now realize the spin-1/2 by the $SU(2)$ Majorana field $\lambda_{\alpha i}$, the triplet of spin 0 by $L^{ij} = L^{ji}$, but the singlet spin 0 by V_μ, which is subject to $\partial_\mu V^\mu = 0$. The off-shell theory now requires only two auxiliary fermions which we take to be the fields S and P. The resulting action is of the form

$$A = \int d^4x \left\{ -\frac{1}{2}(\partial_\mu L^{ij})^2 - \frac{1}{2}\bar\lambda^i \partial\!\!\!/\lambda_i + \frac{1}{2}V_\mu^2 + \frac{1}{2}S^2 + \frac{1}{2}P^2 \right\} \tag{12.25}$$

The transformations may be deduced as before (for example, $\delta L^{ij} = \bar\varepsilon^i \lambda^j + \bar\varepsilon^j \lambda^i$) and close without an off-shell central charge.

(c) There exists another off-shell formulation which is an extension of (b). This consists of relaxing the constraint $\partial^\mu V_\mu = 0$ and imposing it by a supersymmetric set of Lagrange-multipliers. This formulation is rather complicated,[50] but important for superspace quantization.

(d) There also exists an extension of the type (a) off-shell formulation given above that does not involve an off-shell central charge.[27]

There do not appear to be any renormalizable self interactions for $N = 2$ matter.[28] (We note that in formulation (a) the Yukawa term is $\bar\psi A_i \psi$ which is not $SU(2)$ invariant.)

12.3   The General $N = 2$ Rigid Theory

We are now in a position to couple the $N = 2$ matter to $N = 2$ Yang-Mills. Normally, when one couples two multiplets together one introduces additional coupling constants. However, in this case the coupling constants are the gauge coupling constants. This follows from the fact that the coupling of A_μ to the $N = 2$ matter fields is determined by the gauge coupling constant g. By $N = 2$ supersymmetry, however, the coupling of any component of the $N = 2$ Yang-Mills multiplet to the $N = 2$ hypermultiplet is determined by the gauge coupling constant. These theories have the important property of having only one coupling constant—they are truly grand unified theories. The coupling between the multiplets can be easily found by starting from the linearized theory whose action is the sum of the linearized action of each theory. We now gauge covariantize in accordance with the fact that the $N = 2$ matter is in the group representation $R(\bar{R})$ and recover supersymmetry by adding terms to the action and transformation laws. The $N = 2$ super-Yang-Mills fields have the same transformations as before and were given in Eq. (12.17). The $N = 2$ matter transformation law is

$$\delta A^{ia} = \bar{\varepsilon}^i \psi^a$$

$$\delta\psi^a = +2\gamma^\mu D_\mu A^{ia}\varepsilon_i - 2\gamma_5 F^{ia}\varepsilon_i - 2ig(T_s)^a{}_b A^{ib}(A_s - i\gamma_5 B_s)\varepsilon_i$$

$$\delta F^{ia} = \bar{\varepsilon}^i\gamma_5\gamma^\mu D_\mu\psi^a - 2g\bar{\varepsilon}^i\gamma_5\lambda_{js}A^{jb}(T_s)^a{}_b + g\bar{\varepsilon}^i(i\gamma_5 A_s - B_s)(T^s)^a{}_b\psi^b \quad (12.26)$$

where

$$D_\mu A^{ia} = \partial_\mu A^{ia} - g(A_\mu)^a{}_b A^{ib}$$

$$D_\mu A^*_{ia} = \partial_\mu A^*_{ia} + g(A_\mu)^b{}_a A^*_{ib} \quad (12.27)$$

while the invariant action is

$$A = A^{\text{YM}} + A^{\text{matter}} + A^{\text{interaction}} + A^{\text{mass}} \quad (12.28)$$

where

$$A^{\text{YM}} = \text{Tr} \int d^4x \left\{ -\frac{1}{4}F^2_{\mu\nu} - \frac{1}{2}(D_\mu A)^2 - \frac{1}{2}(D_\mu B)^2 - \frac{1}{2}\bar{\lambda}^i\gamma^\mu D_\mu\lambda_i \right.$$

$$\left. + \frac{1}{2}X^2 - \frac{ig}{2}\bar{\lambda}^i[A - i\gamma_5 B, \lambda_i] - \frac{g^2}{2}([A, B])^2 \right\}$$

$$A^{\text{matter}} = \int d^4x \left\{ -D_\mu A^{ia}D^\mu A^*_{ia} - \frac{1}{2}\bar{\psi}_a\gamma^\mu D_\mu\psi^a + |F^{ia}|^2 \right\} \quad (12.29)$$

$$A^{\text{interaction}} = \int d^4x \left\{ -g\bar{\lambda}^{is}(T_s)^a{}_b A^*_{ia}\psi^b \right.$$

$$+ g\bar{\psi}_a A^{ib}(T_s)^a{}_b\lambda_{is} + \frac{ig}{2}(T_s)^a{}_b\bar{\psi}_a(B_s - i\gamma_5 A_s)\gamma_5\psi^b$$

$$\left. + g^2 A^*_{ia}A^{ic}(T_s T_t)^a{}_c(A_t A_s + B_t B_s) + 4gX^{ij}A^*_{ic}A^c_j(T_s)^c{}_d \right\}$$

$$A^{\text{mass}} = m \int d^4x \left\{ iF^{ia} A^*_{ia} - iF^*_{ia} A^{ia} + \frac{1}{2} \bar{\psi}_a \gamma_5 \psi^a + 2ig B_s (T^s)^a{}_b A^*_{ia} A^{bi} \right\}$$

$$(12.30)$$

The above action for $m = 0$ is clearly $U(2)$ and dilatation invariant, but as expected it is also $N = 2$ superconformally invariant.

One further term that can be added is a term linear in X^{ij} for those theories where X^{ij} is a gauge singlet. This occurs when X^{ij} comes from a $U(1)$ $N = 2$ Yang-Mills multiplet. This multiplet can couple to $N = 2$ matter multiplets that have non-trivial $U(1)$ weights, but this term breaks $U(2)$.

12.4   The $N = 4$ Yang-Mills Theory[29]

The $N = 4$ Yang-Mills multiplet has as its on-shell states one spin-1, four spin-$\frac{1}{2}$ and six spin-0 particles. It is clearly composed of an $N = 2$ Yang-Mills multiplet and one $N = 2$ matter multiplet in the adjoint representation. Since in the absence of the mass term this coupling is unique, $N = 4$ Yang-Mills must be given by Eq. (12.25) for $m = 0$ with the $N = 2$ matter being in the adjoint representation. It only remains to cast this action in a manifestly $SU(4)$ or $O(4)$ invariant form and find the $N = 4$ on-shell supersymmetry transformations. This model does not possess $U(4)$ symmetry due to its being a CPT self-conjugate multiplet and the fact that the $U(1)$ factor commutes, for $N = 4$, with the supercharge.

In the $O(4)$ formulation the fields are A_μ, $\lambda_{\alpha i} A_{ij}$ and B_{ij} where $\lambda_{\alpha i}$ is an $O(4)$ Majorana spinor, i.e.

$$\lambda_{\alpha i} = C_{\alpha \beta} \bar{\lambda}^\beta{}_i$$

and A_{ij} and B_{ij} belongs to the triplet representation of $O(4)$. That is, they are real antisymmetric self-dual tensors of rank two

$$A_{ij} = -\tfrac{1}{2} \varepsilon_{ijkl} A_{kl}, \qquad B_{ij} = +\tfrac{1}{2} \varepsilon_{ijkl} B_{kl}$$

In the $SU(4)$ formulation the fields are A_μ, $\lambda'_{\alpha i}$ and ϕ_{ij}. Here $\lambda'_{\alpha i}$ is a chiral spinor in the fundamental representation of $SU(4)$ while ϕ_{ij} are in the second rank antisymmetric self-dual **6** of $SU(4)$.

$$\phi_{ij} = -\phi_{ji}$$

$$(\phi_{ij})^* \equiv \phi^{ij} = \tfrac{1}{2} \varepsilon^{ijkl} \phi_{kl}$$

We leave it as an exercise to the reader to deduce the on-shell $N = 4$ transformation laws and action. This theory is $N = 4$ superconformally invariant and has no known off-shell formulation that does not involve constrained fields.[30] An off-shell formulation has been given in Ref. 24.

Chapter 13

The Local Tensor Calculus and the Coupling of Supergravity to Matter

The most popular method of constructing realistic models of supersymmetry is within the context of supergravity (see Chapter 19). In order to build these models it is important to know the most general coupling of supergravity to matter. In certain quarters there is an impression that this coupling was handed down on the Tables of Moses. In fact, it was constructed using the local tensor calculus, which is the generalization to supergravity of the rigid tensor calculus of Chapter 11. Various matter coupling had previously been constructed using the Noether method.[31]

The local tensor calculus which was developed in Refs. 32, 33, 34 and 35 has, as its building blocks, off-shell representations. In particular, it requires an off-shell representation of supergravity. In common with the rest of this book we will use the old minimal formulation with the fields $e_\mu{}^n$, $\psi_\mu{}^\alpha$, M, N and b_μ. Their transformations are given, in Chapter 9, and the algebra of minimal supergravity is

$$[\delta_{\varepsilon_1}, \delta_{\varepsilon_2}] = \delta_{GC}(2\xi^\mu) + \delta_{SS}(-\kappa\xi^\nu\psi_\nu) + \delta_{LL}\left(-\frac{2\kappa}{3}\varepsilon_{ab\lambda\rho}b^\lambda\xi^\rho\right.$$

$$\left. + 2\xi^d w_d{}^{ab} - \frac{2\kappa}{3}\bar{\varepsilon}_2\sigma^{ab}(M + i\gamma_5 N)\varepsilon_1\right) \tag{13.1}$$

where $\xi^\mu = \bar{\varepsilon}_2\gamma^\mu\varepsilon_1$. The matter multiplets must also have transformations which lead to this closed algebra.

The first problem is to find the local equivalent of the general multiplet $(C, \zeta_\alpha; H, K, A_\mu; \lambda_\alpha, D)$. There are two ways to find this result. We could use the Noether technique to couple this general multiplet to supergravity. This can be achieved either by coupling massive QED to supergravity or we could generalize the action $\int d^4x D$, of rigid supersymmetry, to supergravity. We will return to this last possibility later. The most pedagogical method to find the local general multiplet is to Noether couple up the algebra of transformations. We begin with the rigid multiplet whose transformation laws are given in Eq. (11.2) of Chapter 11. We then replace ε_α by $\varepsilon_\alpha(x)$ in the these transformation laws. The algebra no longer closes; for example

$$[\delta_1, \delta_2]\zeta_\alpha = 2\bar{\varepsilon}_1 \not{\partial}\varepsilon_2 \zeta + [(\partial_\mu i\bar{\varepsilon}_1)\gamma_5 \zeta i\gamma^\mu \gamma_5 \varepsilon_2 - (1 \leftrightarrow 2)] \tag{13.2}$$

This last term can be eliminated by supercovariantizing the derivative of C that occurs in $\delta\zeta_\alpha$, i.e.,

$$\delta\zeta = (\not{A} + H + i\gamma_5 \kappa + i\gamma^\mu \gamma_5 \mathscr{D}_\mu C)\varepsilon \tag{13.3}$$

where

$$\mathscr{D}_\mu C = \partial_\mu C - \frac{i\kappa}{2}\bar{\psi}_\mu \gamma_5 \zeta \tag{13.4}$$

Since

$$\delta(\mathscr{D}_\mu C) = i\bar{\varepsilon}\gamma_5(\partial_\mu \zeta) + O(\kappa) \tag{13.5}$$

we find that the algebra closes on ζ_α up to terms of order κ^0. The story for the closure on the other fields is the same. The whole algebra will close up to terms of order κ^0 provided we replace all the space-time derivatives with supercovariant derivatives. If a field F has supervariation $\delta F = \varepsilon^\alpha f_\alpha$ then the supercovariant derivative of F is

$$\mathscr{D}_\mu F = \partial_\mu F - \frac{\kappa}{2}\bar{\psi}_\mu^\alpha f_\alpha \tag{13.6}$$

The algebra is now

$$[\delta_{\varepsilon_1}, \delta_{\varepsilon_2}] = 2\bar{\varepsilon}_1(x)\gamma^\mu \varepsilon_2(x)\partial_\mu + O(\kappa) \tag{13.7}$$

and it agrees with the supergravity algebra in the limit $\kappa \to 0$.

We now close the algebra order by order in κ adding terms to the transformations of the fields of the general multiplet and modifying the closure of the algebra so that at each order in κ it agrees with the algebra of Eq. (13.1). Since the transformation laws involve ψ_μ^α and its supervariation introduces the fields M, N and b_μ we expect to find the auxiliary fields in the transformation laws of the matter fields.

The final result of this lengthy calculation is

$$\delta C = i\bar{\varepsilon}\gamma_5 \zeta$$

$$\delta\zeta = (\not{A} - i\gamma_5 \not{\mathscr{D}}C + H + i\gamma_5 K)\varepsilon$$

$$\delta H = \bar{\varepsilon}\lambda + \bar{\varepsilon}\not{\mathscr{D}}\zeta - \kappa\bar{\eta}\zeta$$

$$\delta K = i\bar{\varepsilon}\gamma_5 \lambda + i\bar{\varepsilon}\gamma_5 \not{\mathscr{D}}\zeta + i\kappa\bar{\eta}\gamma_5 \zeta$$

$$\delta A_a = \bar{\varepsilon}\gamma_a \lambda + \bar{\varepsilon}\mathscr{D}_a \zeta + \frac{\kappa}{2}\bar{\eta}\gamma_a \zeta$$

$$\delta\lambda = (-\tfrac{1}{2}\sigma^{ab}\hat{F}_{ab} + i\gamma_5 D)\varepsilon$$

$$\delta D = i\bar{\varepsilon}\gamma_5 \not{\mathscr{D}}\lambda$$

where

$$\mathcal{D}_a C = \partial_a C - \frac{i\kappa}{2}\bar{\psi}_a\gamma_5\zeta$$

$$\mathcal{D}_a \zeta = D_a\zeta - \frac{\kappa}{2}(\slashed{A} - i\gamma_5\slashed{\mathcal{D}}C + H + i\gamma_5\kappa)\psi_a - \frac{i\kappa}{2}b_a\gamma_5\zeta$$

$$\hat{F}_{ab} = \left[D_aA_b - \frac{\kappa}{2}(\bar{\psi}_a\gamma_b\lambda + \bar{\psi}_a\hat{D}_b\zeta - \bar{\zeta}\hat{D}_a\psi_b)\right] - (a\leftrightarrow b)$$

$$\hat{D}_a\psi_b = D_a\psi_b + \frac{i\kappa}{2}\gamma_5 b_a\psi_b$$

$$\hat{D}_a\lambda = D_a\lambda + \frac{i\kappa}{2}\gamma_5 b_a\lambda - \frac{\kappa}{2}\left(-\frac{\sigma^{cd}}{2}\hat{F}_{cd} + i\gamma_5 D\right)\psi_a$$

$$\eta = -\frac{1}{3}(M + i\gamma_5 N + i\slashed{b}\gamma_5)\varepsilon \tag{13.8}$$

These transformation close on the algebra of Eq. (13.1).

The occurrence of b_μ in the derivatives reflects the fact that the general multiplet is a multiplet of superconformal gravity.

As in the rigid case the general supermultiplet is not "irreducible." However, in the local case the decomposition is a little more complicated. Let us first consider the Maxwell multiplet. The field A_α does not transform into only λ_α, D and a gradient term. However, the field

$$A'_a = A_a + \frac{\kappa}{2}\bar{\psi}_a\zeta \tag{13.9}$$

transforms as

$$\delta A'_\mu = \bar{\varepsilon}\gamma_\mu\lambda + \partial_\mu(\bar{\varepsilon}\zeta) \tag{13.10}$$

that is, it transforms into λ_α plus a gradient term. We may write the field strength as

$$\hat{F}_{ab} = e_a{}^\mu e_b{}^\nu\left(\partial_\mu A'_\nu - \frac{\kappa}{2}\bar{\psi}_\mu\gamma_\nu\lambda - (\mu\leftrightarrow\nu)\right) \tag{13.11}$$

Clearly, the multiplet

$$\lambda_\alpha = (\lambda_\alpha; \hat{F}_{ab}; D) \tag{13.12}$$

forms an irreducible representation of local supersymmetry; the role of the gauge field being played by A'_μ. The gauge multiplet

$$Y = (A'_\mu; \lambda; D) \tag{13.13}$$

forms a representation of the local algebra plus gauge transformations ($\delta A_\mu = \partial_\mu \Lambda$).

The chiral supermultiplet is obtained by setting the Maxwell multiplet to zero. We find $\lambda = D = 0$ while $A'_\mu = \partial_\mu A$. The remaining chiral supermultiplet is

$$(A, B; \chi_\alpha; F, G) \tag{13.14}$$

where $B = C$, $\chi_\alpha = \zeta_\alpha$, $F = H$, $G = K$ and it has the transformations

$$\delta A = \bar{\varepsilon}\chi; \qquad \delta B = i\bar{\varepsilon}\gamma_5\chi$$

$$\delta \chi = [F + i\gamma_5 G + \hat{\slashed{D}}(A + i\gamma_5 B)]\varepsilon$$

$$\delta F = \bar{\varepsilon}\hat{\slashed{D}}\chi - \kappa\bar{\eta}\chi$$

$$\delta G = i\bar{\varepsilon}\gamma_5\hat{\slashed{D}}\chi + i\kappa\bar{\eta}\gamma_5\chi \tag{13.15}$$

where

$$\hat{D}_a A = \partial_a A - \frac{\kappa}{2}\bar{\psi}_a\chi, \qquad \hat{D}_a B = \partial_a B - \frac{i\kappa}{2}\bar{\psi}_a\gamma_5\chi$$

$$\hat{D}_a\chi = \left(D_a - \frac{i\kappa}{2}b_a\gamma_5\right)\chi - \frac{\kappa}{2}(\hat{\slashed{D}}(A + i\gamma_5 B) + F + i\gamma_5 G)\psi_a$$

The analogue of the rigid multiplet H of Eq. (11.10) cannot have as its first components H and K as these fields do not rotate into the same spinor under supersymmetry. The solution to this dilemma is to add terms proportional to $-\frac{2}{3}CN$ and $+\frac{2}{3}CM$ to H and K respectively. These fields both vary into the spinor

$$\psi = \lambda + \hat{\slashed{D}}\zeta + \frac{i\kappa}{3}\slashed{b}\gamma_5\zeta + \frac{\kappa}{3}(M - i\gamma_5 N)\zeta + \frac{2}{3}i\gamma_5 C\xi \tag{13.16}$$

For the definition of ξ see below. The supersymmetry variation of this spinor yields the last two components of this new multiplet. The final result is

$$\mathbb{H} \equiv \mathbb{H}(V) = \left(H - \frac{2}{3}CN, K + \frac{2}{3}CM, \psi, P, Q\right) \tag{13.17}$$

where

$$P = \partial_a A^a - \frac{\kappa^2}{2}\bar{\psi}_\mu\gamma\cdot\psi A^\mu - \frac{\kappa}{2}\bar{\psi}_a(\gamma^a\lambda + \hat{D}^a\zeta) - \frac{2}{3}gC$$

$$+ \frac{2\kappa}{3}b^a\hat{D}_a C + \kappa(MH + \kappa N) - \frac{1}{6}\kappa^2\bar{\psi}_a(M + i\gamma_5 N)\gamma^a\zeta$$

$$+ \frac{\kappa}{4}\bar{\zeta}\gamma\cdot R + \frac{\kappa^2}{12}i\bar{\psi}\cdot\gamma\slashed{b}\gamma_5\zeta + \frac{i\kappa^2}{12}b^a\bar{\psi}_a\gamma_5\zeta$$

$$Q = D + \partial_a \left(\partial^a C - \frac{i\kappa}{2} \bar{\psi}_a \gamma_5 \zeta - \frac{\kappa^2}{2} \bar{\psi}_a \gamma \cdot \psi \hat{D}^a C \right.$$

$$- \frac{i\kappa}{2} \bar{\psi}_a \gamma_5 \hat{D}_a \zeta + \frac{2}{3} Cf - \frac{2}{3} \kappa b_a A^a + \frac{\kappa}{3}(HN - KM)$$

$$- \frac{i\kappa^2}{6} \bar{\psi}_a (M + i\gamma_5 N) \gamma_5 \gamma^a \zeta - \frac{i\kappa}{12} \bar{\zeta} \gamma_5 \gamma \cdot R$$

$$\left. + \frac{\kappa}{12} \bar{\psi}_a \not{b} \gamma^a \zeta + \frac{\kappa^2}{12} \bar{\psi}_a \zeta b^a \right) \tag{13.18}$$

The symbols ξ, f and g form part of a chiral multiplet that is constructed entirely from supergravity fields. It is called the Einstein multiplet and is given by

$$U = (-M, -N, \xi, f, g) \tag{13.19}$$

where

$$\xi = \frac{1}{2}(e\gamma \cdot R + \gamma^a (M + i\gamma_5 N)\psi_a + i\gamma_5 b^a \psi_a)$$

$$f = -\frac{1}{2}R + \frac{1}{4}\bar{\psi}_\mu \gamma_\nu \gamma^\mu R^\nu - \frac{1}{4}\bar{\psi}^\mu (M + i\gamma_5 N)\psi_\mu$$

$$+ \frac{i}{4}\bar{\psi}_\mu b^\mu \gamma_5 \gamma \cdot \psi - \frac{2}{3}(M^2 + N^2 + \frac{1}{2}b_a{}^2)$$

$$g = -e^{-1}\partial_\mu (eb^\mu) + \frac{i}{2}\bar{\psi}_\mu \gamma_5 R^\mu - \frac{i}{2}\bar{\psi}_\mu \sigma^{\mu\nu}\gamma_5 R_\nu$$

$$- \frac{i}{4}\bar{\psi}_\mu \gamma_5 (M + i\gamma_5 N)\psi^\mu + \frac{1}{4}\bar{\psi}_\mu b^\mu \gamma \cdot \psi \tag{13.20}$$

The components of this multiplet are obtained by successive variations of its first components M and N.

The local linear multiplet is obtained by setting the supermultiplet $\mathbb{M} = 0$, it is given by

$$\mathbb{L} = (C, \zeta, A_a) \tag{13.21}$$

where A_a is subject to the constraint $P = 0$.

Given a scalar multiplet $S = (A, B, \chi_a, F, G)$ we can form a "general multiplet"

$$\hat{D}S = (B, \chi_a, F, G, \hat{D}_a A, 0, 0) \tag{13.22}$$

Substituting $\hat{D}S$ into $H(\hat{D}S)$ of Eq. (13.17) we obtain the local analogue of

$\mathbb{T}S$, i.e.,

$$\mathbb{T}(S) = \mathbb{H}(\hat{D}S) \tag{13.23}$$

A useful chiral supermultiplet can be constructed out of the gauge multiplet λ_α.

$$W^2 = \lambda^\alpha \cdot \lambda_\alpha - (\bar{\lambda}\lambda, i\bar{\lambda}\gamma_5\lambda, 2(-\tfrac{1}{2}\sigma^{ab}F_{ab} + i\gamma_5 D)\lambda, (-F^{\mu\nu}F_{\mu\nu} - 2\bar{\lambda}\slashed{\mathcal{D}}\lambda + 2D^2),$$

$$(\tfrac{1}{2}\varepsilon^{\mu\nu\rho\kappa}F_{\mu\nu}F_{\rho\kappa} + i\slashed{\mathcal{D}}^\mu(\bar{\lambda}\gamma_\mu\gamma_5\lambda)))$$

We now wish to combine two supermultiplets into a third supermultiplet. The construction proceeds exactly for the rigid case; we select an object bilinear in component fields and vary it successively to find the other components of the composite supermultiplet. The result for the combination of chiral supermultiplets $\mathbb{A}_1$ and $\mathbb{A}_2$ denoted $\mathbb{A}_1 \cdot \mathbb{A}_2$ is identical to the rigid case of Eq. (11.20). The results for $\mathbb{C}_1 \cdot \mathbb{C}_2$, $\mathbb{A}_1 \times \mathbb{A}_2$ and $\mathbb{A}_1 \wedge \mathbb{A}_2$ where $\mathbb{C}_1$ and $\mathbb{C}_2$ are general multiplets are also the same as in the rigid case Eqs. (11.18), (11.22) and (11.24) provided one replaces all space-time derivatives with the covariant derivatives i.e., we make the substitutions

$$\partial_\mu C \to \hat{D}_\mu C \qquad \partial_\mu A \to \hat{D}_\mu A \qquad \partial_\mu B \to \hat{D}_\mu B$$

$$\partial_\mu \chi \to \hat{D}_\mu \chi \qquad \partial_\mu \zeta \to \hat{D}_\mu \zeta \tag{13.24}$$

The derivatives of the other component fields do not occur in these formulae.

The last part of the tensor calculus is to find invariant density formulae that generalize the rigid density formulae

$$[S]_F = \int d^4x F \quad \text{and} \quad [V]_D = \int d^4x D \tag{13.25}$$

These are found by using the Noether method. Consider the variation of $\int d^4x D$

$$\delta\left(\int d^4x D\right) = \int d^4x (-\partial_\mu \bar{\varepsilon}) i\gamma_5 \gamma^\mu \lambda + O(\kappa). \tag{13.26}$$

An invariant action to order κ^0 is given by

$$\int d^4x \left\{ D + \frac{i\kappa}{2} \bar{\psi}_\mu \gamma_5 \gamma^\mu \lambda + O(\kappa) \right\} \tag{13.27}$$

As explained previously, we now add terms to the action gaining invariance order by order in κ; the local transformations of $D, \lambda, \dots$ being already known in Eq. (13.8). The F density formula is obtained in the same way and the final results are

$$[V]_D = \int d^4xe \left\{ D - \frac{i\kappa}{2}\bar{\psi}\cdot\gamma\gamma_5\lambda - \frac{2}{3}\kappa NH + \frac{2}{3}\kappa MK \right.$$

$$- \frac{2}{3}\kappa A_a\left(b^a + \frac{3}{8}e^{-1}\kappa\varepsilon^{\mu\rho a\tau}\bar{\psi}_\rho\gamma_\tau\psi_\mu\right)$$

$$\left. - \frac{\kappa}{3}\bar{\zeta}\left(i\gamma_5\gamma\cdot R + \frac{3}{8}\kappa\varepsilon^{\mu\rho a\tau}\psi_a\bar{\psi}_\rho\gamma_\tau\psi_\mu\right) - \frac{2}{3}\kappa^2 cL_{S-G} \right\} \quad (13.28)$$

where L_{S-G} is Lagrangian for $N = 1$ supergravity and

$$[S]_F = \int d^4xe \left\{ F + \frac{1}{2}\bar{\psi}\cdot\gamma\chi + \frac{1}{4}\bar{\psi}_\mu\gamma^{\mu\nu}(A + i\gamma_5 B)\psi_\nu - (NB + MA) \right\} \quad (13.29)$$

This completes the construction of the local tensor calculus.

It is now straightforward to construct any coupling to supergravity. We begin by noting that the supergravity action A^{S-G} is given by

$$A^{S-G} = -\frac{3}{2\kappa^2}[1]_D = \frac{1}{3}[U]_F \quad (13.30)$$

The cosmological constant is

$$m[1]_F = -em\{M - \tfrac{1}{4}\bar{\psi}_\mu\sigma^{\mu\nu}\psi_\nu\}$$

The kinetic action A^C for a chiral matter supermultiplet S can be found in two ways:

$$A^C = \tfrac{1}{4}[S \times S]_D = \tfrac{1}{4}[S \cdot \mathbb{T}(S)]_F \quad (13.31)$$

The results in component fields is given after a short calculation by

$$A^C = \int d^4xe \left\{ -\frac{1}{2}\hat{D}_a A\hat{D}_a A - \frac{1}{2}\hat{D}_a B\hat{D}_a B - \frac{1}{2}\bar{\chi}\mathcal{D}\chi \right.$$

$$+ \frac{1}{2}F^2 + \frac{1}{2}G^2 + \frac{i\kappa}{2}\bar{\psi}_\mu\gamma_5\gamma^\mu(G + i\gamma_5 F - \hat{D}_\alpha(B + i\gamma_5 A)\gamma^\alpha)\chi$$

$$- \frac{2}{3}\kappa N(FB + GA) + \frac{2}{3}e\kappa(GB - FA)M$$

$$- \frac{2\kappa}{3}\left(b^a + \frac{3}{8}\kappa e^{-1}\varepsilon^{\mu\rho a\tau}\bar{\psi}_\rho\gamma_\tau\psi_\mu\right)\left((B\hat{D}_a A - A\hat{D}_a B) - \frac{i}{2}\bar{\chi}\gamma_5\gamma_a\chi\right)$$

$$\left. - \frac{\kappa}{3}\bar{\chi}(B - i\gamma_5 A)\left(i\gamma_5\gamma\cdot R + \frac{3}{8}\kappa^2\varepsilon^{\mu\rho a\tau}\psi_a\bar{\psi}_\rho\gamma_\tau\psi_\mu\right) - \frac{\kappa^2}{3}(A^2 + B^2)L_{S-G} \right\}$$

$$(13.32)$$

We observe the presence of an $(A^2 + B^2)R$ term and in this sense the coupling is non-minimal. The usual interaction term is given by

$$A^I = \left[\frac{m}{2} S \cdot S + \frac{\lambda}{3!} S \cdot S \cdot S\right]_F \tag{13.33}$$

and we leave the evaluation of its components to the reader.

The most general coupling of a scalar multiplet is given by

$$[f(S) \times S]_D + [g(S)]_F \tag{13.34}$$

where f and g are arbitrary functions S. We may rewrite the first term as $[f(S) \times T(S)]_F$. It is lengthy, but straightforward to evaluate this in terms of component fields. The results was given in Ref. 39. Although it looks as if the action depends on two function f and g, in fact its dependence is on one function which is a combination of f and g. This results from the ability to rescale the supergravity fields. For a specific choice of the function $f(S)$ one can, even before scaling, find a theory that has no non-minimal $(A^2 + B^2)R$ term.

The Maxwell action is given by

$$[W^2]_F = \int d^4x e \left\{-\frac{1}{4} F_{\mu\nu} F^{\mu\nu} - \frac{1}{2} \bar{\lambda} \not{D} \lambda \right.$$
$$\left. + \frac{1}{2} D^2 - \frac{\kappa}{4} \bar{\psi}_\tau \sigma^{\mu\nu} \gamma^\tau \lambda F_{\mu\nu} - \frac{i\kappa}{4} \bar{\lambda} \gamma^a \gamma_5 \psi_a D \right\} \tag{13.35}$$

Again one may consider more complicated actions; for example, $[h(S)W^2]_F$ where h is an arbitrary function of S.

We now consider the coupling of Maxwell fields to chiral multiplets. The gauge fields are contained in the general multiplet V which now has a gauge transformation

$$\delta V = \hat{D}\Lambda \tag{13.36}$$

where Λ is a chiral supermultiplet. The field A_a does not transform as a gauge field; however, the field $A'_a = A_a + (\kappa/2)\hat{\psi}_a \zeta$ does transform under the above gauge transformation by a pure gradient. This is in agreement with out previous discussion of the Maxwell sub-supermultiplet. Given two scalar multiplets S_1 and S_2 we specify their gauge transformations to be

$$\delta S_1 = g\Lambda S_1, \qquad \delta S_2 = -g\Lambda S_2 \tag{13.37}$$

The gauge invariant coupling of S_1 and S_2 to supergravity is given by

$$\left[\frac{1}{8}(S_1 \times S_1 + S_2 \times S_2)(e^{2gV} + e^{-2gV}) + (S_1 \wedge S_2)(e^{2gV} - e^{-2gV})\right]_D \quad (13.38)$$

The evaluation in terms of components can be found in Ref. 35.

We now consider the local extension of the rigid Fayet-Iliopoulous term ($\int d^4 x D$). Since the local D-density formula involves all the component fields of V the expression $[V]_D$ is not gauge invariant. In fact, the local generalization[35] is

$$[e^V]_D \quad (13.39)$$

This still does not look gauge invariant, however, it can be shown that the supergravity fields can undergo compensating transformations. The transformation that compensates for the gauge transformation of A_μ is a local chiral transformation. It can be shown that a local Fayet-Iliopoulous term can only be added to action which are locally R invariant.[36,37] The most general coupling involving Yang-Mills fields, chiral matter multiplets and supergravity was given in Refs. 38–40.

Finally we discuss the possible on-shell counterterms of $N = 1$ supergravity. In this case $M = N = b_\mu = 0$ and so the Einstein multiplet vanishes. The only nonvanishing supermultiplet W_{ABC} has as its first component $t_{(ABC)}$ where

$$t_{ABC} \approx (\sigma^\mu)_A{}^{\dot{D}}(\sigma^\nu)_{B\dot{D}}(D_\mu \psi_{\nu c} - D_\nu \psi_{\mu c}) \quad (13.40)$$

(See Chapter 15.) Since $t_{(ABC)}$ does not involve the gravitino equation of motion this supermultiplet does not vanish on-shell. The next component of W_{ABC} involves the Weyl tensor. The supermultiplet W_{ABC} is chiral and so is

$$W_1{}^2 = W_{ABC} W^{ABC}$$

In fact, the superconformal action is given by

$$[W_1{}^2]_F \quad (13.41)$$

The on-shell counterterm[34,41] at three loops is

$$[W^2 \times W^2]_D \quad (13.42)$$

while that at the $(4k - 1)$th loop is

$$[(W^2 \times W^2)^4]_D \quad (13.43)$$

In fact, there exist counterterms at every odd-loop order above two loops. These are easily constructed, i.e.,

$$[(W^2)^n \times W^2]_F \quad (13.44)$$

One may wonder, whether the local tensor calculus for other auxiliary-field formulations yield different results, i.e., different couplings of supermatter to supergravity. Clearly, one can only compare the couplings after the elimination of the respective auxiliary fields. In fact, the tensor calculus[42] of the new minimal formulation only allows coupling of R invariant matter. It has been shown that these new minimal matter couplings are contained in the R-invariant subset of the matter couplings of old minimal supergravity couplings.[43]

The above discussion has presented the local tensor calculus based on the local Poincaré superalgebra. An alternative method[44] is to use a tensor calculus based on the local superconformal algebra and use appropriate compensating multiplets to construct non-conformally invariant couplings. This method has the advantage that the superconformal algebra has more symmetry and so, for example, the appearance of fields in the covariant derivatives is more tightly constrained by straightforward gauge principles. It is this approach that has been pursued in $N = 2$ supergravity.[45]

Chapter 14

Superspace[46]

In this chapter, we wish to find a method of formulating supersymmetric theories, that keeps the supersymmetry manifest at every step of the calculation. This is particularly useful for the quantization of supersymmetric theories.

14.1 An Elementary Account of $N = 1$ Superspace

In the following section we will give a group theoretical derivation of superspace and its properties. Here however, we give an elementary account of $N = 1$ superspace for the reader who wishes to use superspace, but does not wish to work through all the concepts required to derive superspace as a coset space.

Superspace is an 8-dimensional manifold parametrized by coordinates x^μ, θ^α which obey the relations

$$x^\mu x^\nu - x^\nu x^\mu = 0$$

$$x^\mu \theta_\alpha - \theta_\alpha x^\mu = 0$$

$$\theta_\alpha \theta_\beta + \theta_\beta \theta_\alpha = 0$$

that is the θ_α's are anticommuting coordinates which are Majorana spinors $\theta_\alpha = C_{\alpha\beta}\bar{\theta}^\beta$. The supersymmetry transformations and translations are realized on this manifold by the transformations

$$x'^\mu = x^\mu - \bar{\varepsilon}\gamma^\mu \theta + a^\mu$$

$$\theta'_\alpha = \theta_\alpha + \varepsilon_\alpha$$

where ε_α is an anticommuting Majorana spinor. The Lorentz transformations and translations are given by

$$x'^\mu = x^\mu - \omega^\mu{}_\nu x^\nu + a^\mu$$

$$\theta'_\alpha = \theta_\alpha - \omega_{\mu\nu}\tfrac{1}{4}(\gamma^{\mu\nu})_\alpha{}^\beta \theta_\beta$$

We now check that the supersymmetry transformations do represent the supersymmetry algebra and yield a translation of the appropriate magnitude. Carrying out two supertransformations with parameters ε_2 and ε_1 in succession yields

$$x_{12} = x_1 - \bar{\varepsilon}_1 \gamma^\mu (\theta + \varepsilon_2)$$

$$\theta_{12\alpha} = \theta_\alpha + \varepsilon_{1\alpha} + \varepsilon_{2\alpha}$$

Taking the term with 1 and 2 interchanged we find

$$x_{12} - x_{21} = -2\bar{\varepsilon}_1 \gamma^\mu \varepsilon_2$$

$$\theta_{12} - \theta_{21} = 0$$

as expected.

Superfields are functions on superspace, $\varphi(x^\mu, \theta_\alpha)$ and the component content can be found by Taylor expanding in θ_α

$$\varphi(x^\mu, \theta_\alpha) = C(x) + i\bar{\theta}\gamma_5 \zeta(x) - \frac{1}{2}\bar{\theta}\theta\kappa(x) + \frac{i}{2}\bar{\theta}\gamma_5\theta H(x) - \frac{i}{2}\bar{\theta}\gamma_\nu\gamma_5\theta A^\nu(x)$$

$$- \bar{\theta}\theta\bar{\theta}i\gamma_5\left(\lambda(x) + \frac{i}{2}\displaystyle{\not}\partial\zeta(x)\right) + \frac{1}{4}(\bar{\theta}\theta)^2\left(D(x) - \frac{1}{2}\partial^2 C(x)\right)$$

There is no $\theta_\alpha \theta_\beta \theta_\gamma \theta_\delta \theta_\varepsilon$, as it is a 5-index object antisymmetric in all its indices. The terms bilinear in θ_α are $\frac{1}{2}\theta_\alpha\theta_\beta N^{\alpha\beta}$ and we can rewrite this by expanding $N^{\alpha\beta}$ in the complete set

$$N^{\alpha\beta} = (C^{-1})^{\alpha\beta}K - (\gamma^5 C^{-1})^{\alpha\beta}H + (\gamma^\mu \gamma^5 C^{-1})^{\alpha\beta}A_\mu$$

Hence $\varphi(x^\mu, \theta^\alpha)$ contains the component fields $(C, \zeta_\alpha; H, K, A_\mu; \lambda_\alpha; D)$.

Multiplying two superfields φ_1 and φ_2 together we obtain a third superfield

$$\phi_3 = \phi_1\phi_2 = C_1 C_2 + \bar{\theta}(C_1\zeta_{2\alpha} + C_2\zeta_{1\alpha}) + \cdots$$

which has the component fields $(C_1 C_2, C_1\zeta_{2\alpha} + C_2\zeta_{1\alpha}, \ldots)$.

We define a scalar superfield to transform under supersymmetry as

$$\phi'(x', \theta') = \phi(x, \theta)$$

and so for a small variation

$$\delta\phi = \phi(x', \theta') - \phi(x, \theta) = \phi(x^\mu - \bar{\varepsilon}\gamma^\mu\theta, \theta + \varepsilon) - \phi(x, \theta) = +\bar{\varepsilon}Q\phi$$

Consequently the supercharge Q_α is given by

$$Q_\alpha = \frac{\partial}{\partial\theta^\alpha} - (\gamma_\mu\theta)_\alpha \partial^\mu$$

The transformations of the component fields are discussed in the next section, as are the representations of the other generators of the super Poincaré algebra.

In order to construct actions it is desirable to first find covariant derivatives although the operator $\partial_m = P_m$ commutes with supersymmetry as $[Q_\alpha, P_m] = 0$, the spinorial derivative $\partial/\partial\bar{\theta}^\alpha$ does not transform covariantly

$$\frac{\partial}{\partial\bar{\theta}^\alpha} = \frac{\partial\bar{\theta}'^\beta}{\partial\bar{\theta}^\alpha}\frac{\partial}{\partial\bar{\theta}^\beta} + \frac{\partial x'^\nu}{\partial\bar{\theta}^\alpha}\frac{\partial}{\partial x'^\nu} = \frac{\partial}{\partial\bar{\theta}'^\alpha} + (\gamma^\mu\varepsilon)_\alpha\frac{\partial}{\partial x'^\mu}$$

The derivative, given by

$$D_\alpha = \frac{\partial}{\partial\bar{\theta}^\alpha} + (\partial\!\!\!/\theta)_\alpha$$

however, does transform covariantly. This can also be checked by confirming its commutator with the supercharge Q_α vanishes i.e.:

$$\{Q_\alpha, D_\beta\} = 0$$

The reader may now jump to Eq. (14.40) of the next section if he wishes.

14.2 $N = 1$ Superspace

General Formalism

All the properties of superspace may be systematically derived by viewing it as a coset space. We now briefly review this formalism. Given any group G with a subgroup H, we can define an equivalence relation between any two elements g_1 and g_2 of G:

$$g_1 \sim g_2 \quad \text{iff} \quad g_2^{-1}g_1 \in H$$

This relation is such that if we put equivalent elements of G into the same set then G is divided into disjoint sets. The collection of such sets is the coset space denoted by G/H. We may label a given coset space by taking any group element that belongs to it. The action of the group G on the coset is given by acting on the chosen representative by the natural action of G. That is, if g is any element of G its action on the coset, whose chosen representative is g_1, is the coset that contains the element gg_1. It is easy to verify that this action is independent of the representative element used.

The coset manifold G/H may be parametrized by coordinates ξ^π ($\pi = 1\ldots$, $dimG - dimH$) and so the representative elements may be written $L(\xi^\pi)$. A useful parametrization is to use the exponential map. Any group element may be written in the form

$$g = \exp(\xi^n K_n)\exp(w^i H_i)$$

where H_i are the generators of H and K_n are the remaining generators. A useful coset representative is obtained by taking $w^i = 0$.

Given the coset space G/H we may wish to place on it geometric objects such as vierbeins and spin-connections. It is desirable that these objects reflect the symmetries of the coset manifold. Invariant vierbiens $e_\pi{}^n$ and spin connections $\omega_\pi{}^i$ are, for reductive coset spaces given by the formula

$$L^{-1}(\xi)\partial_\pi L(\xi) = e_\pi{}^n K_n + w_\pi{}^i H_i$$

The covariant derivative is given by

$$D_N = e_N{}^\pi(\partial_\pi + w_\pi{}^i T_i)$$

where T_i are the generators of H appropriate to the field being acted on. For a detailed discussion of these results see Ref. 6.

We will begin for simplicity with the $N = 1$ super Poincaré group, denoted SP, and construct the coset space of SP/L; where L is the Lorentz group.

The general element of the SP group can be written in the form

$$g_0 = \exp\{a^\mu P_\mu + \varepsilon^A Q_A + \varepsilon^{\dot A} Q_{\dot A}\}\exp\{\tfrac{1}{2}w^{mn}J_{mn}\} \tag{14.1}$$

The coset space SP/L, called $N = 1$ superspace, is parameterized by coordinates $(x^\mu, \theta^A, \theta^{\dot A}) = z^\pi$ where the point z^π corresponds to the group element

$$e^{z^\pi K_\pi} = \exp\{x^\mu P_\mu + \theta^A Q_A + \theta^{\dot A} Q_{\dot A}\} \tag{14.2}$$

where $K_\pi = (P_\mu, Q_A, Q_{\dot A})$. We have chosen our coset representative to be the one with $w^{mn} = 0$.

Under the action of the group element g_0 of Eq. (14.1) the point z^π is shifted to the point z'^π where z'^π is given by the equation

$$g_0 e^{z^\pi K_\pi} = e^{z'^\pi K_\pi} e^{1/2 w'^{mn}J_{mn}} \tag{14.3}$$

To lowest order in a^μ w^{mn} and ε_α we find that

$$e^{z'^\pi K_\pi} = \exp\{z^\pi K_\pi + a^\mu P_\mu + \varepsilon^A Q_A + \varepsilon^{\dot A} Q_{\dot A} + \tfrac{1}{2}[a^\mu P_\mu + \varepsilon^A Q_A$$
$$+ \varepsilon^{\dot A} Q_{\dot A} + w^{mn}J_{mn}, z^\pi K_\pi] + \cdots\}\exp\{\tfrac{1}{2}w^{mn}J_{mn}\} \tag{14.4}$$

In the above we have used the relation

$$e^{\varepsilon A}e^B = \exp\left\{B + \varepsilon A + \varepsilon \sum_{n=1}^{\infty}\frac{1}{(n+1)!}\underbrace{[\cdots[A,B]\cdots,B]}_{\leftarrow n\,\text{times}\,\rightarrow} + O\left(\varepsilon^2\right)\right\}$$

and so

$$x'^\mu = x^\mu + a^\mu + i\varepsilon^A(\sigma^\mu)_{A\dot A}\theta^{\dot A} - i\theta^A(\sigma^\mu)_{A\dot A}\varepsilon^{\dot A} + w^{\mu\nu}x_\nu$$

$$\theta'^A = \theta^A + \varepsilon^A + \tfrac{1}{4}(\sigma^{mn})^A{}_B\theta^B w_{mn}$$

$$\theta'^{\dot A} = \theta^{\dot A} + \varepsilon^{\dot A} - \tfrac{1}{4}\theta^{\dot B}(\bar\sigma^{mn})_{\dot B}{}^{\dot A}w_{mn} \tag{14.5}$$

For notational simplicity we will write curved indices $\theta^{\underline{A}}$ as θ^A in most of the following equations. Since Q_A and P_μ commute these transformations represent the supersymmetry transformation to all orders in ε^A. Carrying out the commutator of two supersymmetries we do indeed find that these transformations realize the $N = 1$ super-Poincaré group. For example, in the case of two supersymmetry transformations

$$g_1 g_2 = \exp\{\varepsilon_1{}^A Q_A + \varepsilon_1{}^{\dot{A}} Q_{\dot{A}}\} \exp\{\varepsilon_2{}^B Q_B + \varepsilon_2{}^{\dot{B}} Q_{\dot{B}}\} \tag{14.6}$$

$$= \exp\{(\varepsilon_1{}^A + \varepsilon_2{}^A) Q_A + (\varepsilon_1{}^{\dot{A}} + \varepsilon_2{}^{\dot{A}}) Q_{\dot{A}}$$

$$- \tfrac{1}{2} \varepsilon_1{}^{\dot{A}} \varepsilon_2{}^B \{Q_{\dot{A}}, Q_B\} - \tfrac{1}{2} \varepsilon_1{}^A \varepsilon_2{}^{\dot{B}} \{Q_A, Q_{\dot{B}}\} + \cdots \} \tag{14.7}$$

Alternatively carrying out the two successive transformations on $e^{z \cdot K}$, namely $g_1(g_2 e^{z \cdot K}) = e^{z_{12} \cdot K}$ we find that

$$x_{12}{}^m = x_2{}^m + i\varepsilon_1{}^A (\sigma^m)_{A\dot{A}} (\theta^{\dot{A}} + \varepsilon_2{}^{\dot{A}}) - i(\theta^A + \varepsilon_2{}^A)(\sigma^m)_{A\dot{A}} \varepsilon_1{}^{\dot{A}}$$

$$\theta_{12}{}^A = \theta^A + \varepsilon_1{}^A + \varepsilon_2{}^A \qquad \theta_{12}{}^{\dot{A}} = \theta^{\dot{A}} + \varepsilon_1{}^{\dot{A}} + \varepsilon_2{}^{\dot{A}} \tag{14.8}$$

Comparing terms of the form $\varepsilon_1 \varepsilon_2$ and subtracting the term with 1 and 2 interchanged we find that

$$\{Q_{\dot{A}}, Q_B\} = -2i(\sigma^m)_{\dot{A}B} P_m \tag{14.9}$$

We now consider functions (superfields) defined on $N = 1$ superspace. A scalar superfield is defined by the equation

$$\phi'(z') = \phi(z) \tag{14.10}$$

In order to agree with the usual method of evaluating commutators in the supersymmetry community, which we will discuss shortly, we adopt the passive interpretation of the group action, namely,

$$U(g)\phi(z) = \phi(\tau(g)z) \tag{14.11}$$

where $\tau(g)z = z'$. For the passive interpretation of transformations we find that

$$\delta\phi = \phi(z + \delta z) - \phi(z) = +\delta g^N f_N{}^\pi \partial_\pi \phi(z)$$

$$= +\delta g^N X_N \phi(z) \tag{14.12}$$

where for small group element δg^N the point $z^\pi \to z'^\pi = z^\pi + \delta g^N f_N{}^\pi$. Consequently, the infinitesimal operators (Killing vectors) corresponding to the supersymmetric transformation on superspace are

$$X_N = f_N{}^\pi \partial_\pi = (l_m, l_A, l_{\dot{A}}, l_{mn}) \tag{14.13}$$

The letter l stands for the concrete realization of the generators $P_m, Q_A, Q_{\dot{A}}, J_{mn}$ in terms of differential operators. Examining Eq. (14.5) we find that

$$l_m = \partial_m = \delta_m{}^\mu \partial_\mu$$

$$l_A = \frac{\partial}{\partial \theta^A} + i(\sigma^m)_{A\dot{A}} \theta^{\dot{A}} \partial_m$$

$$l_{\dot{A}} = \frac{\partial}{\partial \theta^{\dot{A}}} + i(\sigma^m)_{A\dot{A}} \theta^A \partial_m \tag{14.14}$$

$$l_{mn} = -(x_m \partial_n - x_n \partial_m) - \frac{1}{2} \theta^B (\sigma^{mn})_B{}^A \frac{\partial}{\partial \theta^A} + \frac{1}{2} \theta^{\dot{B}} (\bar{\sigma}^{mn})_{\dot{B}}{}^{\dot{A}} \frac{\partial}{\partial \theta^{\dot{A}}} \tag{14.15}$$

We adopt the following rules for differentiation with respect to anti-commuting coordinates (see Appendix).

$$\frac{\partial}{\partial \theta^A} \theta^B = \delta_A{}^B \qquad \frac{\partial}{\partial \theta^{\dot{A}}} \theta^{\dot{B}} = \delta_{\dot{A}}{}^{\dot{B}} \tag{14.16}$$

The derivative $\partial/\partial \theta^A$ must anticommute with θ's; for example,

$$\frac{\partial}{\partial \theta^A} (\theta^B \theta^C) = \delta_A{}^B \theta^C - \theta^B \delta_A{}^C \tag{14.17}$$

The properties of $\partial/\partial \theta^A$ under complex conjugation are given in the appendix on conventions.

These operators obey the algebra

$$\{l_A, l_B\} = 0 \qquad \{l_A, l_{\dot{A}}\} = 2i(\sigma^m)_{A\dot{A}} l_m$$

$$[l_m, l_A] = 0 \qquad [l_A, l_{mn}] = +\tfrac{1}{2}(\sigma_{mn})_A{}^B l_B \tag{14.18}$$

These relations are consistent with demanding that the infinitesimal operators represent the algebra. In carrying out the commutator of two transformations on a superfield $\phi(z)$ we may take into account the fact that we should slip δ_1 past the operation of the first δ_2. The action of one supersymmetry is by definition

$$\delta \phi = (\varepsilon^A l_A + \varepsilon^{\dot{A}} l_{\dot{A}}) \phi$$

while the action of two of them is as follows:

$$\begin{aligned}
[\delta_1, \delta_2] \phi &= \delta_1 [(\varepsilon_2{}^A l_A + \varepsilon_2{}^{\dot{A}} l_{\dot{A}}) \phi] - (1 \leftrightarrow 2) \\
&= [\varepsilon_2{}^A l_A + \varepsilon_2{}^{\dot{A}} l_{\dot{A}}, \varepsilon_1{}^B l_B + \varepsilon_1{}^{\dot{B}} l_{\dot{B}}] \phi \\
&= -\varepsilon_2{}^A \varepsilon_1{}^{\dot{B}} \{l_A, l_{\dot{B}}\} \phi - \varepsilon_2{}^{\dot{A}} \varepsilon_1{}^B \{l_{\dot{A}}, l_B\} \phi \\
&= (-\varepsilon_2{}^A \varepsilon_1{}^{\dot{B}} 2i(\sigma^m)_{A\dot{B}} \partial_m - (1 \leftrightarrow 2)) \phi \\
&= -2i\varepsilon_2{}^A \varepsilon_1{}^{\dot{B}} (\sigma^m)_{A\dot{B}} \partial_m \phi + \text{h.c.} \tag{14.19}
\end{aligned}$$

where use has been made of Eq. (14.18). Alternatively we may use the super-symmetry algebra,

$$[\varepsilon_1{}^A Q_A + \varepsilon_1{}^{\dot{A}} Q_{\dot{A}}, \varepsilon_2{}^B Q_B + \varepsilon_2{}^{\dot{B}} Q_{\dot{B}}]\phi = -\varepsilon_1{}^A \varepsilon_2{}^{\dot{B}}\{Q_A, Q_{\dot{B}}\}\phi - (1 \leftrightarrow 2)$$

$$= +\varepsilon_2{}^A \varepsilon_1{}^{\dot{B}}(-2i(\sigma^m)_{A\dot{B}})P_m\phi + \text{h.c.} \quad (14.20)$$

These results are the same, since for small a^m

$$\delta_{a_m}\phi = (U(e^{a^m P_m}) - 1)\phi = a^m l_m\phi = a^m \partial_m\phi \quad (14.21)$$

We may, consider more general representations than scalar superfields, by giving the superfield indices corresponding to the representation of the Lorentz group to which it belongs. For a superfield ϕ_p

$$U(g)\phi_p(z) = D_p{}^q(e^{-w^{mn}J_{mn}1/2})\phi_q(\tau(g)z) \quad (14.22)$$

where g is given in Eq. (14.1) and $D_p{}^q$ is the matrix appropriate to the representation parameterized by q. Because of the fact that the generators of $N = 1$ supersymmetry themselves form a representation of the Lorentz group, the subgroup rotation in Eq. (14.22) is not dependent on z. A spinor superfield has the following transformation

$$U(g)\phi^A(z) = [\exp(-\tfrac{1}{4}w^{mn}\sigma_{mn})]^A{}_B\phi^B(\tau(g)z) \quad (14.23)$$

The infinitesimal generators action on ϕ_p are unchanged except for l_{mn} which acquires an extra term due to Lorentz rotation; for example

$$l_{mn}\phi^C = \left\{-(x_m\partial_n - x_n\partial_m) - \frac{1}{2}\theta^B(\sigma^{mn})_B{}^A\frac{\partial}{\partial\theta^A}\right.$$

$$\left. + \frac{1}{2}\theta^{\dot{B}}(\bar{\sigma}^{mn})_{\dot{B}}{}^{\dot{A}}\frac{\partial}{\partial\theta^{\dot{A}}}\right\}\phi^C - \frac{1}{2}(\sigma_{mn})^C{}_B\phi^B \quad (14.24)$$

Given a superfield $\phi(x^\mu, \theta^A, \theta^{\dot{A}})$ we may expand it in a Taylor series in θ^A and $\theta^{\dot{A}}$, the coefficients in this expansion being functions of x^μ

$$\phi(x^\mu, \theta^A, \theta^{\dot{A}}) = C(x) + \theta^A\chi_A(x) + \theta^{\dot{A}}\chi_{\dot{A}}(x) + \tfrac{1}{2}\theta^A\theta_A f(x)$$

$$+ \tfrac{1}{2}\theta^{\dot{A}}\theta_{\dot{A}}k(x) + \theta^A\theta^{\dot{B}}A_{A\dot{B}}(x) + \tfrac{1}{2}\theta^A\theta_A\theta^{\dot{B}}\lambda_{\dot{B}}(x)$$

$$+ \tfrac{1}{2}\theta^{\dot{A}}\theta_{\dot{A}}\theta^B\lambda_B(x) + \tfrac{1}{4}\theta^A\theta_A\theta^{\dot{B}}\theta_{\dot{B}}D(x) \quad (14.25)$$

The series terminates after $\theta^A\theta_A\theta^{\dot{B}}\theta_{\dot{B}}$ as $\theta^A\theta^B\theta^C = 0$.

If this superfield is a scalar superfield then we may find the effects of a supertransformation upon the component fields. Recalling Eq. (14.12)

$$\delta\phi = \delta C + \theta^A\delta\chi_A + \cdots \quad (14.26)$$

$$= (\varepsilon^A Q_A + \varepsilon^{\dot A} Q_{\dot A})\phi$$

$$= \phi(x^\mu + i\varepsilon^A(\sigma^\mu)_{A\dot A}\theta^{\dot A} - i\theta^A(\sigma^\mu)_{A\dot A}\varepsilon^{\dot A}, \theta^A + \varepsilon^A, \theta^{\dot A} + \varepsilon^{\dot A})$$

$$\quad - \phi(x^\mu, \theta^A, \theta^{\dot A})$$

$$= \varepsilon^A \chi_A + \varepsilon^{\dot A}\chi_{\dot A} + i\varepsilon^A(\sigma^\mu)_{A\dot A}\theta^{\dot A}\partial_\mu C - i\theta^A(\sigma^\mu)_{A\dot A}\varepsilon^{\dot A}\partial_\mu C$$

$$\quad + \theta^A \varepsilon_A f + \theta^{\dot A}\varepsilon_{\dot A}k + \varepsilon^A\theta^{\dot B}A_{A\dot B} + \theta^A \varepsilon^{\dot B}A_{A\dot B} + \cdots \tag{14.27}$$

Comparing coefficients of θ, θ^2, etc. we find that

$$\delta C = \varepsilon^A \chi_A + \varepsilon^{\dot A}\chi_{\dot A}$$

$$\delta \chi_A = +f\varepsilon_A + A_{A\dot B}\varepsilon^{\dot B} - i(\sigma^\mu)_{A\dot A}\varepsilon^{\dot A}\partial_\mu c$$

$$\delta \chi_{\dot A} = +k\varepsilon_{\dot A} - A_{A\dot B}\varepsilon^A - i(\sigma^\mu)_{A\dot A}\varepsilon^A \partial_\mu c$$

$$\delta f = +\varepsilon^{\dot A}\lambda_{\dot A} - i\varepsilon^{\dot B}(\sigma^\mu)_{B\dot B}\partial_\mu \chi^B$$

$$\delta k = +\varepsilon^B \lambda_B - i\varepsilon^B(\sigma^\mu)_{B\dot B}\partial_\mu \chi^{\dot B}$$

$$\delta A_{A\dot B} = \varepsilon_{\dot B}\lambda_A - i\varepsilon^B(\sigma^\mu)_{B\dot B}\partial_\mu \chi^{\dot B} + i\varepsilon^{\dot C}(\sigma^\mu)_{A\dot C}\partial_\mu \chi_{\dot B} - \varepsilon_A \lambda_{\dot B}$$

$$\delta \lambda_A = -i(\sigma^\mu)_{A\dot A}\varepsilon^{\dot A}\partial_\mu k + D\varepsilon_A + i(\sigma^\mu)_C{}^{\dot B}\partial_\mu A_{A\dot B}\varepsilon^C$$

$$\delta \lambda_{\dot A} = -i\partial_\mu f(\sigma^\mu)_{A\dot A}\varepsilon^{\dot A} + D\varepsilon_{\dot A} - i(\sigma^\mu)^C{}_{\dot C}\partial_\mu A_{C\dot A}\varepsilon^{\dot C}$$

$$\delta D = -i\varepsilon^B(\sigma^\mu)_{B\dot B}\partial_\mu \lambda^{\dot B} - i\varepsilon^{\dot A}(\sigma^\mu)_{C\dot A}\partial_\mu \lambda^C \tag{14.28}$$

It is useful to also have these transformation rules for a scalar supermultiplet in 4-component form; using the Appendix on conventions and making some field redefinition so as to be more in agreement with the literature we find that

$$\delta C = i\bar\varepsilon \gamma_5 \zeta$$

$$\delta \zeta = (H + i\gamma_5 K - i\gamma_5 \partial c + \slashed A)\varepsilon$$

$$\delta H = \bar\varepsilon \lambda' + \bar\varepsilon \slashed\partial \zeta$$

$$\delta K = i\bar\varepsilon \gamma_5 \lambda' + i\bar\varepsilon \gamma_5 \slashed\partial \zeta$$

$$\delta A_\mu = \bar\varepsilon \gamma_\mu \lambda' + \bar\varepsilon \partial_\mu \zeta$$

$$\delta \lambda' = (-\sigma^{\mu\nu}\partial_\mu A_\nu + i\gamma_5 D')\varepsilon$$

$$\delta D' = i\bar\varepsilon \gamma_5 \slashed\partial \lambda' \tag{14.29}$$

where

$$\zeta = -i\gamma_5 \chi, \; H = \frac{i}{2}(-f + k), \; K = -\frac{1}{2}(f + k)$$

$$A^{A\dot B} = (\sigma^\mu)^{A\dot B}A_\mu, \; \lambda' = -2i\gamma_5 \lambda + \frac{1}{2}\slashed\partial \chi$$

$$D' = 2D + \frac{1}{2}\partial^2 c \qquad (14.30)$$

As a final check we may verify closure on the x-component fields

$$[\delta_1, \delta_2]C = -i\varepsilon_2{}^A(\sigma^\mu)_{A\dot{A}}\varepsilon_1{}^{\dot{A}}\partial_\mu C - i\varepsilon_2{}^{\dot{A}}(\sigma^\mu)_{A\dot{A}}\varepsilon_1{}^A\partial_\mu C - (1 \leftrightarrow 2)$$

$$= -2i\varepsilon_2{}^A(\sigma^\mu)_{A\dot{A}}\varepsilon_1{}^{\dot{A}}\partial_\mu C + \text{h.c.} \qquad (14.31)$$

which agrees with Eqs. (14.20) and (14.21) at $\theta = 0$.

We could, of course, have carried out the active interpretation, but then the commutation would have to be interpreted differently. For example, in this case, we would find that

$$\delta C = -(\varepsilon^A\chi_A + \varepsilon^{\dot{A}}\chi_{\dot{A}})$$

and the commutator must be carried out without slipping δ_1 past the action of δ_2. In this approach one first transforms the multiplet, i.e.

$$(C, \chi, \ldots) \rightarrow (C - \varepsilon_2{}^A\chi_A - \varepsilon_2{}^{\dot{A}}\chi_{\dot{A}}, \ldots)$$

and then substitutes this new multiplet into the transformation rule under δ_1, i.e.

$$[\delta_1, \delta_2]C = \{-\varepsilon_1{}^A(+i(\not{\partial})_{A\dot{A}}C \cdot \varepsilon_2{}^{\dot{A}} - f\varepsilon_{2A} - A_{A\dot{B}}\varepsilon_2{}^{\dot{B}})$$

$$- \varepsilon_1{}^{\dot{A}}(+i(\not{\partial})_{A\dot{A}}\varepsilon_2{}^A(-k\varepsilon_{2\dot{A}} + A_{B\dot{A}}\varepsilon_2{}^B))\} - (1 \leftrightarrow 2)$$

$$= -2i\varepsilon_1{}^A(\not{\partial})_{A\dot{A}}\varepsilon_2{}^{\dot{A}}C + \text{h.c.}$$

$$= +2i\varepsilon_1{}^A\sigma^\mu{}_{A\dot{A}}\varepsilon_2{}^{\dot{A}}\partial_\mu C + \text{h.c.}$$

$$= -2i\varepsilon_2{}^A\sigma^\mu{}_{A\dot{A}}\varepsilon_1{}^{\dot{A}}\partial_\mu C + \text{h.c.}$$

This result is indeed in agreement with (14.20).

Vielbeins, Connections and Covariant Derivatives

The vielbeins and connections can be found from the usual formula

$$1 + \delta z^\pi e_\pi{}^r K_r + \delta z^\pi w_\pi{}^{mn}\tfrac{1}{2}J_{mn}$$

$$= \exp\{-x^\mu P_\mu - \theta^A Q_A - \theta^{\dot{A}}Q_{\dot{A}}\}$$

$$\cdot \exp\{(x^\mu + \delta x^\mu)P_\mu + (\theta^A + \delta\theta^A)Q_A + (\theta^{\dot{A}} + \delta\theta^{\dot{A}})Q_{\dot{A}}\} \qquad (14.32)$$

$$= 1 + \delta x^\mu P_\mu + \delta\theta^A Q_A + \delta\theta^{\dot{A}}Q_{\dot{A}}$$

$$+ \delta\theta^{\dot{A}}\theta^A i(\sigma^\mu)_{A\dot{A}}P_\mu + \delta\theta^A(i\sigma^\mu)_{A\dot{A}}\theta^{\dot{A}}P_\mu \qquad (14.33)$$

This result is easily found by using the formula

$$e^{-A}de^A = \left(\frac{e^{-A} - 1}{-A}\right) \wedge dA = dA - \frac{1}{2}A \wedge dA + \cdots$$

where $1 \wedge dA = dA$, $A \wedge dA = [A, dA]$, $A^2 \wedge dA = [A, [A, dA]]$, etc. Since $[Q, P] = 0$ the series terminates after the second term. Hence we find that

$$e_\mu{}^m = \delta_\mu{}^m \qquad e_\mu{}^A = e_\mu{}^{\dot A} = 0$$

$$e_A{}^m = \theta^A i(\sigma^m)_{A\dot A} \qquad e_{\dot A}{}^m = i(\sigma^m)_{A\dot A}\theta^{\dot A}$$

$$e_A{}^B = \delta_A{}^B \qquad e_{\dot A}{}^{\dot B} = \delta_{\dot A}{}^{\dot B} \quad \text{and} \quad e_A{}^{\dot B} = e_{\dot B}{}^A = 0 \qquad (14.34)$$

and the $w_\pi{}^{mn}$ connection vanishes. In matrix notation this result becomes

$$E_\pi{}^M = \begin{pmatrix} \delta_\mu{}^m & 0 & 0 \\ i(\sigma^m)_{A\dot A}\theta^{\dot A} & \delta_A{}^B & 0 \\ i(\sigma^m)_{A\dot A}\theta^A & 0 & \delta_{\dot A}{}^{\dot B} \end{pmatrix} \qquad (14.35)$$

The inverse super vielbein is defined by

$$E_A{}^M E_M{}^\pi = \delta_A{}^\pi \quad \text{or} \quad E_M{}^A E_A{}^N = \delta_M{}^N \qquad (14.36)$$

and takes the form

$$E_M{}^A = \begin{pmatrix} \delta_m{}^\mu & 0 & 0 \\ -i(\sigma^\mu)_{A\dot A}\theta^{\dot A} & \delta_A{}^B & 0 \\ -i(\sigma^\mu)_{A\dot A}\theta^A & 0 & \delta_{\dot A}{}^{\dot B} \end{pmatrix} \qquad (14.37)$$

Using this vielbein and spin connection we can construct a covariant derivative. The covariant derivative is given by

$$D_M = E_M{}^A(\partial_A + \tfrac{1}{2}w_A{}^{mn}J_{mn})$$

$$= (D_m, D_A, D_{\dot A}) \qquad (14.38)$$

where

$$D_m = \partial_m$$

$$D_A = \frac{\partial}{\partial\theta^A} - i(\sigma^m)_{A\dot A}\theta^{\dot A}\partial_m$$

$$D_{\dot A} = \frac{\partial}{\partial\theta^{\dot A}} - i(\sigma^m)_{A\dot A}\theta^A\partial_m \qquad (14.39)$$

The covariant derivatives obey the relations

$$[\partial_m, D_A] = 0 = [\partial_m, D_{\dot A}] = 0$$

$$\{D_A, D_{\dot B}\} = -2i(\sigma^m)_{A\dot B}\partial_m$$

$$\{D_A, D_B\} = \{D_{\dot A}, D_{\dot B}\} = 0 \qquad (14.40)$$

This algebra is very similar to the algebra for l_A, $l_{\dot A}$ and of Eq. (14.18) except for an overall sign.

The reader may check that the covariant derivative constructed above does the job it advertised. An explicit calculation shows that

$$[D_A, l_n] = \{D_A, Q_B\} = \{D_A, Q_{\dot{B}}\} = 0 \qquad [D_A, l_{mn}] = 0 \qquad (14.41)$$

The torsion and curvature of $N = 1$ superspace corresponding to the vielbein and spin connection of Eq. (14.35) are found from the equation

$$[D_M, D_N] = T_{MN}{}^R D_R + \tfrac{1}{2} R_{MN}{}^{mn} J_{mn} \qquad (14.42)$$

This implies that

$$R_{MN}{}^{rs} = 0 \qquad (14.43)$$

and all the torsions vanish except

$$T_{A\dot{A}}{}^m = -2i(\sigma^m)_{A\dot{A}} \qquad (14.44)$$

Consequently $N = 1$ superspace is a manifold without curvature, but with torsion and as such it is not a Riemannian manifold.[†] It follows that any theory of local supersymmetry which is based on such a Riemannian super manifold cannot contain rigid superspace as a limiting case.

An important property of covariant derivatives is that they do not form an irreducible representation of the supersymmetry group. For example, each of the superfields

$$D_A \phi, D_{\dot{A}} \phi, \partial_m \phi \qquad (14.45)$$

transform into themselves and can therefore be used to place covariant constraints on ϕ, as we shall see shortly.

The covariant derivatives can be used to derive the transformation laws of the x-space component fields. Given any superfield, ϕ the supersymmetry transformation of its lowest component $C = \phi|_{\theta=0}$ is given by

$$\delta\phi|_{\theta=0} = \delta C = [(\varepsilon^A l_A + \varepsilon^{\dot{A}} l_{\dot{A}})\phi]_{\theta=0}$$
$$= [(\varepsilon^A D_A + \varepsilon^{\dot{A}} D_{\dot{A}})\phi]|_{\theta=0} \qquad (14.46)$$

This equation holds as the actions of D_A and Q_A coincide at $\theta = 0$. The higher components of the super field may become the lowest components of another superfield by using the covariant derivative. Consider the superfields

$$\phi, D_A\phi, D_{\dot{A}}\phi, -\tfrac{1}{2}D^A D_A\phi, -\tfrac{1}{2}D^{\dot{A}} D_{\dot{A}}\phi, \tfrac{1}{2}[D_{\dot{B}}, D_A]\phi + \cdots \qquad (14.47)$$

Their order $\theta = 0$ components are

$$C, \chi_A, \chi_{\dot{A}}, f, k, A_{A\dot{B}}, \ldots \qquad (14.48)$$

respectively. The supersymmetry transformations of these component fields can be found by manipulating the covariant derivative and using only the

[†] equiped with its usual torsion free connection.

rules of Eq. (14.40):

$$\delta C = (\varepsilon^A D_A + \varepsilon^{\dot{A}} D_{\dot{A}})\phi|_{\theta=0}$$

$$= \varepsilon^A \chi_A + \varepsilon^{\dot{A}} \chi_{\dot{A}} \tag{14.49}$$

$$\delta \chi_A = [(\varepsilon^B D_B + \varepsilon^{\dot{B}} D_{\dot{B}}) D_A \phi]_{\theta=0}$$

$$= \left[-\tfrac{1}{2}\varepsilon_A D^C D_C \phi + \tfrac{1}{2}\varepsilon^{\dot{B}}\{D_{\dot{B}}, D_A\}\phi + \tfrac{1}{2}\varepsilon^{\dot{B}}[D_{\dot{B}}, D_A]\phi \right]_{\theta=0} \tag{14.50}$$

$$= f\varepsilon_A - i(\sigma^m)_{A\dot{B}}\varepsilon^{\dot{B}}\partial_m c + A_{A\dot{B}}\varepsilon^{\dot{B}} \tag{14.51}$$

etc. In deriving these results we have used the fact that

$$D^C \theta_A|_{\theta=0} = \varepsilon^{CB} D_B \theta_C|_{\theta=0} = -\delta^C{}_A \tag{14.52}$$

The scalar superfield $\phi(z)$ contained fields of spin 0, spin $\tfrac{1}{2}$ and spin 1. However, the simplest model of supersymmetry, the Wess-Zumino model, did not contain a spin-1 field. The resolution of this difficulty is that one can find submultiplets of the scalar supermultiplet which transform into themselves under supersymmetry and yet are not subject to their equations of motion. These submultiplets can be found by imposing conditions on ϕ using the covariant derivative. The lowest dimensional covariant constraint is

$$D_{\dot{A}}\phi = 0 \tag{14.53}$$

We note that the canonical (= naive) dimension of the covariant derivatures and coordinates is given by

$$\dim[D_A] = \dim[d\theta] = \tfrac{1}{2} \qquad \dim[\theta] = -\tfrac{1}{2}$$

while

$$\dim[\partial_\mu] = \dim[x_\mu^{-1}] = 1$$

The x-space component fields and their transformation laws can be found by using the method given above. Let

$$z = \phi|_{\theta=0} \qquad \chi_A = D_A \phi|_{\theta=0}$$

$$f = -\tfrac{1}{2}D^A D_A \phi|_{\theta=0} \tag{14.54}$$

Clearly, there are no more x-space components after f. The use of $D_{\dot{B}}$ gives fields which are space-time derivatives of those given already. The transformation laws are

$$\delta z = [(\varepsilon^A D_A + \varepsilon^{\dot{A}} D_{\dot{A}})\phi]_{\theta=0} = \varepsilon^A \chi_A \tag{14.55}$$

$$\delta \chi_A = [(\varepsilon^B D_B + \varepsilon^{\dot{B}} D_{\dot{B}}) D_A \phi]_{\theta=0}$$

$$= f\varepsilon_A - 2i(\sigma^\mu)_{A\dot{B}}\varepsilon^{\dot{B}}\partial_\mu z \tag{14.56}$$

$$\delta f = [(\varepsilon^B D_B + \varepsilon^{\dot{B}} D_{\dot{B}})(-\tfrac{1}{2} D^A D_A \phi)]_{\theta=0}$$

$$= -2i\varepsilon^{\dot{B}} (\sigma^\mu)_{A\dot{B}} \partial_\mu \chi^A \qquad (14.57)$$

we have used the fact that $D_A D_B D_C = 0$.

In 4-component notation these transformation rules read

$$\delta A = \bar{\varepsilon}\chi \qquad \delta B = i\bar{\varepsilon}\gamma_5\chi$$

$$\delta\chi = [F + i\gamma_5 G + \partial\!\!\!/(A + i\gamma_5 B)]\varepsilon \qquad (14.58)$$

$$\delta F = \bar{\varepsilon}\partial\!\!\!/\chi \qquad \delta G = i\bar{\varepsilon}\gamma_5\partial\!\!\!/\chi \qquad (15.59)$$

where

$$f = F + iG \qquad z = \frac{-1}{2}(A + iB) \qquad \chi^\alpha = \begin{pmatrix} \chi^A \\ (\chi_A)^* \end{pmatrix} \qquad (14.60)$$

The reader will recognise that this is the representation which is used to construct the Wess-Zumino model and we will refer to it as the chiral or Wess-Zumino multiplet.

The field content remaining after imposing $D_{\dot{A}}\phi = 0$ can also be found by making a convenient choice of basis.

Consider the new coordinates[68]

$$y^\mu = x^\mu - i\theta^A(\sigma^\mu)_{A\dot{B}}\theta^{\dot{B}}$$

$$\zeta^A = \theta^A \qquad \zeta^{\dot{A}} = \theta^{\dot{A}} \qquad (14.61)$$

then

$$\frac{\partial}{\partial\theta^{\dot{A}}} = \frac{\partial\zeta^{\dot{B}}}{\partial\theta^{\dot{A}}}\frac{\partial}{\partial\zeta^{\dot{B}}} + \frac{\partial y^\mu}{\partial\theta^{\dot{A}}}\frac{\partial}{\partial y^\mu} = \frac{\partial}{\partial\zeta^{\dot{A}}} + \zeta^A(\sigma^\mu)_{A\dot{A}}\frac{\partial}{\partial y^\mu}$$

$$\frac{\partial}{\partial\theta^A} = \frac{\partial}{\partial\zeta^A} - i(\sigma^\mu)_{A\dot{B}}\zeta^{\dot{B}}\frac{\partial}{\partial y^\mu}$$

$$\frac{\partial}{\partial y^\mu} = \frac{\partial}{\partial x^\mu} \qquad (14.62)$$

Using the Appendix the reader may verify that y^μ is not a real coordinate. The constraint $D_{\dot{A}}\phi = 0$ in these new coordinates becomes

$$\frac{\partial}{\partial\zeta^{\dot{B}}}\phi(y, \zeta^A, \zeta^{\dot{B}}) = 0 \qquad (14.63)$$

which only tells us that ϕ is independent of $\zeta^{\dot{A}}$ and must therefore be of the form

$$\phi = z(y) + \zeta^A \chi_A(y) + \tfrac{1}{2}\zeta^A \zeta_A f(y)$$

$$= \exp\{-\theta^A(\sigma_\mu)_{A\dot{B}}\theta^{\dot{B}}\partial^\mu\}\{z(x) + \theta^A \chi_A(x) + \tfrac{1}{2}\theta^A \theta_A f(x)\} \qquad (14.64)$$

The last step is obtained by replacing y as in Eq. (14.61) and Taylor expanding, i.e., $f(x + a) = (\exp a^\mu \partial_\mu)f(x)$.

Imposing the additional constraint

$$D_A \phi = 0 \qquad (14.65)$$

in addition to that of Eq. (14.55) (this latter equation actually follows by complex conjugation if ϕ is real) leads to the result

$$\{D_A, D_{\dot{B}}\}\phi = -2i(\sigma^\mu)_{A\dot{B}}\partial_\mu\phi = 0$$

or

$$\partial_\mu\phi = 0 \qquad (14.66)$$

This equation implies that all the component fields of ϕ are constants. Examining their supersymmetry transformation laws implies that the only consistent superfield is given by

$$C = \text{constant} \qquad \chi_\alpha = f = k = A_\mu = \lambda_\alpha = D = 0 \qquad (14.67)$$

The constraint of lowest dimension after $D_{\dot{A}}\phi = 0$ we may impose on ϕ is

$$D^A D_A \phi = 0 \qquad (14.68)$$

However in this case, we may also demand that ϕ be real and so $D^{\dot{A}}D_{\dot{A}}\phi = 0$. The resulting field content is

$$\phi|_{\theta=0} = C \qquad D_A\phi|_{\theta=0} = \chi_A \qquad D_{\dot{A}}\phi|_{\theta=0} = \chi_{\dot{A}}$$

$$\tfrac{1}{2}[D_A, D_{\dot{B}}]\phi|_{\theta=0} = A_{A\dot{B}} = A_\mu(\sigma^\mu)_{A\dot{B}} \qquad (14.69)$$

Clearly higher spinorial derivatives will only give x-space component fields which are space-time derivatives of those given above. The transformation rules are

$$\delta C = \varepsilon^A \chi_A + \varepsilon^{\dot{A}}\chi_{\dot{A}}$$

$$\delta\chi_A = -i(\sigma^\mu)_{A\dot{B}}\varepsilon^{\dot{B}}\partial_\mu C + A_{A\dot{B}}\varepsilon^{\dot{B}}$$

$$\delta A_{A\dot{B}} = i\varepsilon^{\dot{C}}(\sigma_\mu)_{A\dot{C}}\partial^\mu\chi_{\dot{B}} + i\varepsilon_{\dot{B}}(\sigma^\mu)_{A\dot{C}}\partial_\mu\chi^{\dot{C}}$$

$$\qquad - i\varepsilon_A(\sigma^\mu)_{C\dot{B}}\partial_\mu\chi^C - i\varepsilon^B(\sigma^\mu)_{B\dot{B}}\partial^\mu\chi_A \qquad (14.70)$$

Applying $D^A D^{\dot{B}}$ to Eq. (14.68) we find that

$$\partial^\mu A_\mu = 0 \qquad (14.71)$$

In 4-component notation and in terms of the fields of Eq. (14.30) these transformation laws read

$$\delta C = i\bar{\varepsilon}\gamma_5 \zeta$$

$$\delta \zeta = (-i\gamma_5 \partial\!\!\!/C + A\!\!\!/)\varepsilon$$

$$\delta A_\mu = -\bar{\varepsilon}\sigma_{\mu\nu}\partial^\nu\zeta \qquad (14.72)$$

This multiplet is called the linear multiplet, and yields in fact an alternative formulation to the Wess-Zumino model.[47]

The other dimension-one constraint

$$\tfrac{1}{2}[D_A, D_{\dot{B}}]\phi = 0 \qquad (14.73)$$

cannot be used to define a new superfield. It implies setting $A_\mu = 0$ and in the transformation laws of Eq. (14.29) requires that the superfield be a constant superfield ($C = $ constant, the rest zero).

Let us now see whether one could have imposed a dimension-3/2 constraint on ϕ of Eq. (14.25), i.e.,

$$D_{\dot{B}}D^A D_A\phi = 0 \qquad (14.74)$$

This constraint sets λ' in Eq. (14.30) equal to zero. The corresponding field content is given by Eq. (14.29) and is

$$(C; \zeta; H, K, A_\mu; \lambda' = 0, D' = 0) \qquad (14.75)$$

while $\partial_\mu A_\nu - \partial_\nu A_\mu = 0$, that is, $A_\mu = \partial_\mu A$. This is, of course, the Wess-Zumino multiplet.

Finally, one may consider a multiplet that begins with λ'. Thus, instead of imposing constraints, one consider a particular superfield

$$W_A = D_A \bar{D}^2 \phi \quad \phi \text{ real} \qquad (14.76)$$

where

$$\bar{D}^2 = D^{\dot{A}}D_{\dot{A}}$$

This multiplet only contains,

$$(f_{\mu\nu} = \partial_\mu A_\nu - \partial_\nu A_\mu; \lambda'; D') \qquad (14.77)$$

and does not contain the multiplet of Eq. (14.75). We call this the Yang-Mills multiplet.

The above discussion can also be carried out for superfields with Lorentz indices, but these indices make little change in the discussion, since they rotate by quantities which are not themselves functions of superspace.

Super Integration and δ functions

Before constructing an invariant integration we must decide how to integrate with anticommuting variables. Given only one anticommuting variable θ we define[52]

$$\int d\theta = 0 \quad \text{and} \quad \int d\theta\, \theta = 1 \tag{14.78}$$

The boundary of integration is not given, but it is assumed not to contribute. This definition is the only one possible that is invariant under shifts $\theta \to \theta + \varepsilon$. Therefore integration is identical to differentiation. Given any function $f(\theta) = a + \theta b$, we have that

$$\int d\theta\, f(\theta) = b = \frac{\partial}{\partial\theta} f(\theta) \tag{14.79}$$

This rather impoverished definition of integration does not seem to lead to topologically interesting results.

Integration of many anticommuting variables is a straightforward generalization of the above. That is, for two variables θ_1, θ_2

$$\int d\theta_1\, d\theta_2\, \theta_2 \theta_1 = 1$$

while

$$\int d\theta_1\, d\theta_2\, \theta_1 = 0 \quad \text{etc.} \tag{14.80}$$

We now construct the invariant integration on the coset space SP/L by using the super determinant of the super vielbien. Using Eq. (14.35) and the definition of the super determinant we find that

$$\det E_\pi{}^M = 1 \tag{14.81}$$

Hence the volume element is

$$d^8 z = d^4 x\, d^2\theta\, d^2\bar\theta \tag{14.82}$$

where

$$\int d^2\theta\, \theta^2 = 1 \quad \text{and} \quad \int d^2\bar\theta\, \bar\theta^2 = 1 \tag{14.83}$$

or alternatively

$$\int d^2\theta = -\frac{1}{4}\partial^2 \equiv -\frac{1}{4}\partial^A \partial_A$$

and   (14.84)

$$\int d^2\bar{\theta} = -\frac{1}{4}\bar{\partial}^2 \equiv -\frac{1}{4}\partial^{\dot{A}}\partial_{\dot{A}}$$

where

$$\partial_A = \frac{\partial}{\partial\theta^A} \quad \text{and} \quad \partial_{\dot{A}} = \frac{\partial}{\partial\theta^{\dot{A}}}$$

Under the space-time integral we may replace ∂_A and $\partial_{\dot{A}}$ by D_A and $D_{\dot{A}}$ respectively. Hence, using the above definition integration we find the results

$$\int d^4x\, d^4\theta\, D_A\phi = \int d^4x\, d^4\theta\, Q_A\phi = \int d^4x\, d^4\theta\, \partial_A\phi = 0 \qquad (14.85)$$

and similarly for $D_{\dot{A}}$. Integration by parts goes as follows

$$\int d^4x\, d^4\theta (D_A\phi_1)\phi_2 = -\int d^4x\, d^4\theta\, \phi_1 D_A\phi_2 \qquad (14.86)$$

This gives a way of seeing how to construct supersymmetric invariant expressions.

Given a scalar superfield, an invariant is given by

$$\int d^4x\, d^2\theta\, d^2\bar{\theta}\, \phi = \int d^4x \frac{D^2\bar{D}^2}{16}\phi = \frac{1}{4}\int D(x)\, d^4x \qquad (14.87)$$

Another way of seeing the invariance of this integral is to note that $D(x)$ is the highest dimension field and as the supersymmetry parameter ε has dimension $-\frac{1}{2}$, the supervariation of $D(x)$, which is linear, must be a space-time divergence.

For a chiral superfield ($D_{\dot{A}}\phi = 0$), $\int d^4x\, d^4\theta\, \phi = 0$; however integrating over a subspace of superspace we can find an invariant which is given by

$$\left\{\int d^4x\, d^2\theta\, \phi + \text{h.c.}\right\} = \int d^4x \frac{1}{2}(f + f^*) \qquad (14.88)$$

The chiral integral may be written as a full superspace integral by using the relationship

$$\bar{D}^2 D^2\phi = +16\partial^2\phi \qquad (14.89)$$

Then

$$\int d^4x\, d^2\theta\, \phi = \int d^4x\left[-\frac{1}{4}D^2\phi\right] = \int d^4x\left(-\frac{1}{4}D^2\right)\left(+\frac{\bar{D}^2 D^2}{16\partial^2}\phi\right)$$

$$= -\int d^4x\, d^4\theta\left(\frac{D^2\phi}{4\partial^2}\right) \qquad (14.90)$$

An alternative way of deriving this result is to recognize that the constraint $D_{\dot{A}}\phi$ is solved by

$$\phi = -\tfrac{1}{4}\bar{D}^2 U \tag{14.91}$$

Then, we find that

$$\int d^4x\, d^2\theta\, \phi = \int d^4x\, d^4\theta\, U \tag{14.92}$$

In analogy with usual x-space we may define a δ function in θ-space. Returning to the case of a one-dimensional anticommuting variable θ we define $\delta(\theta - \theta')$ by

$$\int d\theta\, f(\theta)\delta(\theta - \theta') = f(\theta') \tag{14.93}$$

for all functions f. In particular, we have that

$$\int d\theta\, \delta(\theta - \theta') = 1 = \frac{\partial}{\partial \theta}\delta(\theta - \theta'). \tag{14.94}$$

As a result of the anticommuting nature of θ the δ function is none other than

$$\delta(\theta - \theta') = \theta - \theta' \tag{14.95}$$

For $N = 1$ superspace we define the $\delta^4(\theta_1 - \theta_2) \equiv \delta_{12}$ to be such that

$$\int d\theta_1\, f(\theta_1)\delta(\theta_1 - \theta_2) = f(\theta_2) \tag{14.96}$$

for all functions f. In particular this implies that

$$\int d^4\theta_1\, \delta^4(\theta_1 - \theta_2) = 1 = \frac{\partial_1^2 \bar{\partial}_1^2}{16}\delta^4(\theta_1 - \theta_2) \tag{14.97}$$

Under a space-time integral we may use the formula

$$D_1^2 \bar{D}_1^2 \delta_{12} = +16 \tag{14.98}$$

The δ function can be represented by

$$\delta_{12} = 4(\theta_1 - \theta_2)^2(\bar{\theta}_1 - \bar{\theta}_2)^2 \tag{14.99}$$

Some properties of the δ function which are extremely useful for quantum calculations using super Feynman rules are given by

$$\delta_{12}\delta_{12} = \delta_{12}D_A\delta_{12} = \delta_{12}D_{\dot{A}}\delta_{12}$$
$$= \delta_{12}\bar{D}^2\delta_{21} = \delta_{12}D_{\dot{A}}D^2\delta_{12} = \delta_{12}D^2\delta_{12}$$
$$= \delta_{12}\bar{D}^2 D_A\delta_{12} = 0 \tag{14.100}$$

while

$$\delta_{12} D^2 \bar{D}^2 \delta_{21} = \delta_{12} \bar{D}^2 D^2 \delta_{21}$$

$$= \delta_{12} D^A \bar{D}^2 D_A \delta_{12} = \delta_{12} \qquad (14.101)$$

These results can easily be established using the explicit form of the δ function given in Eq. (14.99).

We can now define functional differentiation with respect to superfields. For a scalar superfield ϕ we define

$$\frac{\delta\phi(x',\theta')}{\delta\phi(x,\theta)} = \delta^4(x - x')\delta^4(\theta - \theta') = \delta^8(z - z') \qquad (14.102)$$

and hence

$$\frac{\delta}{\delta\phi(x,\theta)} \int f(\phi) \, d^8z = \frac{\partial f}{\partial \phi}(\phi(x,\theta)). \qquad (14.103)$$

It will also be useful to describe the functional derivative with respect to chiral superfields. This derivative should be such that it maintains the defining condition $D_{\dot{A}}\phi = 0$, that is,

$$D_{\dot{A}} \frac{\delta\varphi(x',\theta')}{\delta\varphi(x,\theta)} = 0 \qquad (14.104)$$

We therefore adopt the definition

$$\frac{\delta\varphi(x',\theta)}{\delta\varphi(x,\theta)} = -\frac{1}{4} \bar{D}^2 \delta^4(x - x')\delta^4(\theta - \theta') \qquad (14.105)$$

As a result we find that

$$\frac{\delta}{\delta\varphi(x,\theta)} \int d^4x' \, d^2\theta' \, f(\varphi(x',\theta'))$$

$$= \int d^4x' \, d^2\theta' \, f'(\varphi) \frac{\delta\varphi(x',\theta')}{\delta\varphi(x,\theta)}$$

$$= \int d^4x' \, d^2\theta' \, f'(\varphi) \left(-\frac{D^2}{4} \delta^4(x - x')\delta^4(\theta - \theta') \right)$$

$$= \int d^4x \, d^2\theta' \left(-\frac{D^2}{4} \right) f'(\varphi)\delta^8(z - z')$$

$$= f'(\varphi(x,\theta)) \qquad (14.106)$$

Extended Superspace: General Formalism

We now construct the coset space, Extended Super Poincaré group/Lorentz group or extended superspace. We recall the extended supersymmetry algebra of Chapter 2; we have the commutators of the Poincaré group plus

$$\{Q^{Ai}, Q^{Bj}\} = 2i\varepsilon^{AB}(\Omega^e)^{ij}Z_e$$

$$\{Q^{\dot{A}}_{i}, Q^{\dot{B}}_{j}\} = 2i\varepsilon_{\dot{A}\dot{B}}(\Omega^e)_{ij}Z_e$$

$$\{Q^{Ai}, Q^{\dot{B}}_{j}\} = -2i(\sigma^\mu)^{A\dot{B}}P_\mu\delta^i_j \qquad (14.107)$$

$$[Q^{Ai}, J_{\mu\nu}] = \tfrac{1}{2}(\sigma_{\mu\nu})^A_{\ B}Q^{Bi} \qquad (14.108)$$

$$[Q^{Ai}, P_\mu] = 0 \qquad (14.109)$$

$$[Q^{Ai}, T^r] = (t^r)^i_{\ j}Q^{Aj} \qquad (14.110)$$

where $(\Omega^e)^{ij} = -(\Omega^e)^{ji}$; $i, j = 1, 2, \ldots, N$.

$$(\Omega^e)_{ij} = [(\Omega^e)^{ij}]^* \qquad (14.111)$$

and the Z_e are choosen to be real. The reality property of Ω^e follows from the Majorana property of Q. The general group element is of the form

$$g = \exp\{a^\mu P_\mu + \varepsilon^A_{\ i}Q_A^{\ i} + \varepsilon^{\dot{A}i}Q_{\dot{A}i} + w^e Z_e + \tfrac{1}{2}w^{mn}J_{mn}\} \qquad (14.112)$$

and the coset elements are labelled by the group elements,

$$e^{z^\pi K_\pi} = \exp\{x^\mu P_\mu + \theta^A_{\ i}Q_A^{\ i} + \theta^{\dot{A}i}Q_{\dot{A}i} + z^e Z_e\} \qquad (14.113)$$

Thus a coset space is parametrized by

$$z^\pi = (x^\mu, \theta^A_{\ i}, \theta^{\dot{A}i}, z^e).$$

Under the action of the group element g the coset point goes to the point z'^π where

$$ge^{z^\pi K_\pi} = e^{z'^\pi K_\pi} \qquad (14.114)$$

Evaluating the above equation we find that under a supersymmetry transformation

$$x'^\mu = x^\mu + i\varepsilon^A_{\ j}(\sigma^\mu)_{A\dot{B}}\theta^{\dot{B}j} - i\theta^A_{\ j}(\sigma^\mu)_{A\dot{B}}\varepsilon^{\dot{B}j}$$

$$\theta'^A_{\ j} = \theta^A_{\ j} + \varepsilon^A_{\ j}; \qquad \theta'^{\dot{A}j} = \theta^{\dot{A}j} + \varepsilon^{\dot{A}j}$$

$$z'^e = z^e + i\varepsilon^A_{\ i}\theta_{Aj}(\Omega^e)^{ij} + i\varepsilon^{\dot{A}i}\theta_{\dot{A}}^{\ j}(\Omega^e)_{ij} \qquad (14.115)$$

while under a central change $z^e \to z'^e = z^e + w^e$ while x and θ are inert.

Scalar superfields on extended superspace are of the form

$$\phi(z') \equiv \phi(z) = \phi(x^\mu, \theta^A_{\ j}, \theta^{\dot{A}j}, z^e) \qquad (14.116)$$

and the action of the supersymmetry group is defined as

$$U(g)\phi(z) = \phi(z') \qquad (14.117)$$

The infinitesimal generators representing the supersymmetry algebra are defined by

$$\delta g^N f_N{}^A \partial_A \phi = \delta g^N X_N \phi \tag{14.118}$$

where for small δg^N

$$z'^\pi \equiv z^\pi + \delta g^N f_N{}^\pi \tag{14.119}$$

The infinitesimal operators are given by

$$l_\mu = \partial_\mu$$

$$l_A{}^i = \frac{\partial}{\partial \theta^A{}_i} + i(\sigma^\mu)_{A\dot{B}} \theta^{\dot{B}j} \partial_\mu + i(\Omega^e)^{ij} \theta_{Aj} \frac{\partial}{\partial z^e}$$

$$l_{\dot{A}i} = \frac{\partial}{\partial \theta^{\dot{A}i}} + i(\sigma^\mu)_{A\dot{A}} \theta^A \partial_\mu + i\theta_{\dot{A}}{}^j (\Omega^e)_{ij} \frac{\partial}{\partial z^e}$$

$$l^e = \frac{\partial}{\partial w^e}$$

$$l_{\mu\nu} = -(x_m \partial_n - x_n \partial_m) - \frac{1}{2} \theta^B (\sigma^{mn})_B{}^A \frac{\partial}{\partial \theta^A} + \frac{1}{2} \theta^{\dot{B}} (\bar{\sigma}^{mn})_{\dot{B}}{}^{\dot{A}} \frac{\partial}{\partial \theta^{\dot{A}}}$$

$$l_s = -\theta^A{}_i (t_r)^i{}_j \frac{\partial}{\partial \theta^A{}_j} - \theta^{\dot{A}i} [(t_r)^i{}_j]^* \frac{\partial}{\partial \theta^{\dot{A}j}} \tag{14.120}$$

Expanding an arbitrary superfield in a Taylor expansion in z^e, $\theta^A{}_i$, $\theta^{\dot{B}j}$ leads to coefficients which are functions of x^μ alone; however there are an infinite number of such x-component fields corresponding to the fact that the z^e are bosonic coordinates and their expansion will not in general terminate. The expansion in θ will terminate at $(\theta^A{}_i)^4 (\theta^{\dot{B}j})^4$. The expansion reads

$$\phi(x^\mu, \theta^A{}_i, \theta^{\dot{B}j}, z^e) = \phi^{(0)}(x^\mu, \theta^A{}_i, \theta^{\dot{B}j}) + \phi^{(1)}_{(e)}(x^\mu, \theta^A{}_i, \theta^{\dot{B}j}) z^{(e)} + \cdots$$

where

$$\phi^{(0)}(x^\mu, \theta^A{}_i, \theta^{\dot{B}j}) \equiv \phi(x^\mu, \theta^A{}_i, \theta^{\dot{B}}{}_j, 0)$$

$$= C^{(0)}(x^\mu) + \theta^A{}_i \chi_A^{(0)i} + \cdots$$

$$\phi^{(1)}_{(e)}(x^\mu, \theta^A{}_i, \theta^{\dot{B}}) \equiv \frac{d\phi}{dz^e}(x^\mu, \theta^A{}_i, \theta^{\dot{B}}{}_j, z^2)|_{z^e = 0}$$

$$= C^{(1)}_{(e)}(x^\mu) + \theta^A{}_i \chi^{(1)i}_{(e)A} + \cdots \quad \text{etc.} \tag{14.121}$$

To investigate how to avoid infinite sets of x^μ-component fields we will arm ourselves with the covariant derivatives.

The supervierbein and spin connection are defined by the equation

$$e^{-z \cdot T} e^{(z+\delta z) \cdot T} = \delta z^{\pi} (e_{\pi}{}^{m} P_{m} + e_{\pi i}{}^{A} Q_{A}{}^{i} + e_{\pi}{}^{\dot{B}j} Q_{\dot{B}j} + e_{\pi}{}^{e} z_{e})$$
$$+ \delta z^{\pi} \tfrac{1}{2} \Omega_{\pi}{}^{mn} J_{mn}. \tag{14.122}$$

A straightforward calculation yields the results

$$e_{\mu}{}^{m} = \delta_{\mu}{}^{m} \qquad e_{\mu}{}^{A}{}_{i} = 0 \qquad e_{\mu}{}^{\dot{B}j} = 0 \qquad e_{\mu}{}^{e} = 0$$

$$e_{\underline{A}}{}^{iB}{}_{j} = \delta_{A}{}^{B} \delta_{j}{}^{i} \qquad e_{\underline{A}}{}^{i\dot{B}j} = 0$$

$$e_{\underline{A}}{}^{im} = +i(\sigma^{m})_{A\dot{B}} \theta^{\dot{B}i} \qquad e_{\underline{A}}{}^{ie} = +i(\Omega^{e})^{ij} \theta_{Aj}$$

$$e_{\underline{\dot{A}i}}{}^{B}{}_{j} = 0 \qquad e_{\underline{\dot{A}i}}{}^{\dot{B}j} = \delta_{\dot{A}}{}^{\dot{B}} \delta_{i}{}^{j}$$

$$e_{\underline{\dot{A}i}}{}^{m} = i(\sigma^{m})_{A\dot{A}} \theta^{A}{}_{i} \qquad e_{\underline{\dot{A}i}}{}^{e} = +i(\Omega^{e})_{ij} \theta_{\dot{A}}{}^{j}$$

$$e_{\underline{e}}{}^{m} = e_{\underline{e}}{}^{A}{}_{j} = e_{\underline{e}}{}^{\dot{B}j} = 0 \qquad e_{\underline{e}}{}^{e'} = \delta_{e}{}^{e'} \tag{14.123}$$

and

$$\Omega_{\pi}{}^{mn} = 0 \tag{14.124}$$

The inverse vierbein has the following non-zero components

$$e_{m}{}^{\mu} = \delta_{m}{}^{\mu}$$

$$e_{A}{}^{i\mu} = -i(\sigma^{\mu})_{A\dot{B}} \theta^{\dot{B}i}, \qquad e_{A}{}^{iB}{}_{\underline{j}} = \delta_{A}{}^{B} \delta_{j}{}^{i}$$

$$e_{\dot{B}i}{}^{\mu} = -i(\sigma^{\mu})_{A\dot{B}} \theta^{A}{}_{i}; \qquad e_{\dot{A}i}{}^{\dot{B}j} = \delta_{\dot{A}}{}^{\dot{B}} \delta_{i}{}^{j}$$

$$e_{A}{}^{ie} = -i(\Omega^{e})^{ij} \theta_{Aj}; \qquad e_{\dot{B}i}{}^{e} = -i(\Omega^{e})_{ij} \theta_{\dot{B}}$$

$$e_{e}{}^{e'} = \delta_{e}{}^{e'}. \tag{14.125}$$

The covariant derivative, is therefore given by

$$D_{M} = \left(E_{M}{}^{\pi} (\partial_{\pi} + \tfrac{1}{2} \Omega_{\pi}{}^{mn} J_{mn})\right) \tag{14.126}$$

which equals

$$D_{M} = (D_{m}, D_{A}{}^{i}, D_{\dot{B}j}, D_{e})$$

where

$$D_{m} = \partial_{m}$$

$$D_{A}{}^{i} = \frac{\partial}{\partial \theta^{A}{}_{i}} - i(\sigma^{\mu})_{A\dot{B}} \theta^{\dot{B}i} \partial_{\mu} - i(\Omega^{e})^{ij} \theta_{Aj} \frac{\partial}{\partial z^{e}}$$

$$D_{\dot{A}i} = \frac{\partial}{\partial \theta^{\dot{A}i}} - i(\sigma^{\mu})_{B\dot{A}} \theta^{B}{}_{i} \partial_{\mu} - i(\Omega^{e})_{ij} \theta_{\dot{A}}{}^{j} \frac{\partial}{\partial z^{e}}.$$

$$D_{e} = \frac{\partial}{\partial z^{e}}. \tag{14.127}$$

The covariant derivative commutes with the supercharges $Q_A{}^i$ and $Q_{\dot{B}}{}^j$ and all the generators of the supersymmetry group; they also obey the relations

$$\{D_A{}^i, D_B{}^j\} = +2i(\Omega^e)^{ij}\varepsilon_{AB}D_e$$

$$\{D_{\dot{A}i}, D_{\dot{B}j}\} = +2i(\Omega^e)_{ij}\varepsilon_{\dot{A}\dot{B}}D_e$$

$$\{D_A{}^i, D_{\dot{B}j}\} = -2i(\sigma^\mu)_{A\dot{B}}\delta^i_j\partial_\mu$$

$$[\partial_\mu, D_A{}^i] = [\partial_\mu, D_{\dot{B}j}] = 0 = [\partial_\mu, D_z]$$

$$[D_z, \text{anything}] = 0 \tag{14.128}$$

We may now use the covariant derivatives to place covariant constraints on the general superfield. A constraint one may impose is

$$D_e\phi = \frac{\partial}{\partial z^e}\phi = 0 \tag{14.129}$$

This ensures that the superfield does not depend on z^e and so does not contain an infinite number of x-space component fields.

The lowest dimensional constraint one may impose is

$$D_A{}^i\phi = 0 \tag{14.130}$$

This constraint implies that

$$\{D_{\dot{A}i}, D_{\dot{B}j}\}\phi = +2i\varepsilon_{\dot{A}\dot{B}}(\Omega^e)_{ij}D_e\phi = 0 \tag{14.131}$$

or, in other words, $D_e\phi = 0$. In this case

$$\frac{\partial}{\partial z^e}\phi = 0 \tag{14.132}$$

and so ϕ has no z^e dependence. The x-space component content of ϕ can be best extracted by using covariant derivatives; the $\theta = 0$ components of the superfields

$$\phi, D_A{}^i\phi, D_A{}^iD_B{}^j\phi, D_A{}^iD_B{}^jD_C{}^k\phi, D_A{}^iD_B{}^jD_C{}^kD_D{}^l\phi, \ldots \tag{14.133}$$

by

$$z, \chi_A{}^i, C_{AB}{}^{ij}, \lambda_{ABC}{}^{ijk}, d_{ABCD}{}^{ijkl}, \ldots \tag{14.134}$$

Their transformation properties may be extracted by using the rules of the D algebra. The above tensors have various symmetry properties such as $C_{AB}{}^{ij} = -C_{BA}{}^{ji}$ which follow from the algebra of the covariant derivatives.

In general one may consider superfields with Lorentz and internal indices by

$$\phi_{ij\ldots;AB\ldots;\dot{A}\dot{B}\ldots} \tag{14.135}$$

Their transformation properties are obvious.

In constructing the coset space of extended superspace one could have placed the central charge generators in the isotropy group. This, however, gives the same result due to the Abelian nature of these generators. If they were in the isotropy group, then the superfields could carry indices corresponding to the central charge generators. The equivalence of these two formulations is obtained by taking the Fourier transform, i.e.

$$\phi(\ldots, z) = \int \phi_k(\ldots) e^{ikz}\, dk. \tag{14.136}$$

14.3 $N = 2$ Superspace

In this section we wish to describe some of the important $N = 2$ superfields. At first sight it may seem that one could have two central charges corresponding to

$$(\Omega^e)^{ij} Z_e = \varepsilon^{ij} Z^{(1)} + i\varepsilon^{ij} Z^{(2)} \tag{14.137}$$

However, one of the central charges, can be eliminated by a chiral rotation. In this case $(\Omega^e)^{ij} = -i\varepsilon^{ij}$ where ε^{ij} has the same properties for raising and lowering indices as ε^{AB}. As a result, there is only one central charge and we find that

$$\{D_A{}^i, D_B{}^j\} = 2\varepsilon_{AB}\varepsilon^{ij} D_z \tag{14.138}$$

where for the case of one central change we denote the central charge derivative by D_z. This transformation would not be available in nonchirally invariant theories and in this case two central charges may be possible. We will begin by considering a chiral multiplet W;

$$D_{\dot Ai} W = 0, \tag{14.139}$$

which, as discussed above, must have no central charge dependence

$$D_z W = 0 \tag{14.140}$$

The x-space component fields are given by the $\theta = 0$ components of the following superfields

$$W, D_A{}^i W, D^{(ij)} W, D_{(AB)} W, D_A{}^i D_B{}^j D_C{}^k W; D_A{}^i D_B{}^j D_C{}^k D_D{}^l W. \tag{14.141}$$

denoted by

$$z, \chi_A{}^i, C^{ij}, F_{AB}, t^{ijk}_{ABC}, t^{ijkl}_{ABCD} \tag{14.142}$$

where

$$D^{ij} = D^{Ai} D_A{}^j \text{ and } D_{AB} = D_A{}^i D_{Bi} \tag{14.143}$$

In the above use has been made in the third and fourth terms of the vanishing of the central charge derivative

$$D^i{}_i W = D_z W = 0 \tag{14.144}$$

In the fifth term we find that

$$t^{Ai}{}_{AiC}{}^k = 0; \qquad t^{ijk}_{ABC} = -t^{jik}_{BAC} \tag{14.145}$$

and similarly for

$$t^{ijkl}_{ABCD} \tag{14.146}$$

Using the fact that

$$t^{ijk}_{(ABC)} = t^{(ijk)}_{ABC} \tag{14.147}$$

we may deduce that

$$t^{ijk}_{ABC} = \varepsilon_{AB} s_C{}^j \varepsilon^{ik} + \varepsilon_{AB} \varepsilon^{jk} s_C{}^i + \varepsilon^{ij} \varepsilon_{AC} \lambda_B{}^k + \varepsilon^{ij} \varepsilon_{BC} \lambda_A{}^k. \tag{14.148}$$

Application of Eq. (14.145) implies that

$$s_C{}^j = -\lambda_B{}^k \tag{14.149}$$

and taking the trace on $\varepsilon_{ji} \varepsilon^{CB}$ yields

$$t_A{}^{iB}{}_{iB}{}^k = +6\lambda_A{}^k \tag{14.150}$$

Similarly one may deduce that the only non zero component of t^{ijkl}_{ABCD} is

$$d = D^{ij} D_{ij} W \tag{14.151}$$

The supersymmetry transformations of the component fields of Eq. (14.42) are given by taking the $\theta = 0$ components in the usual way to find the results.

$$\delta z = (\varepsilon^A{}_i D_A{}^i + \varepsilon^{\dot{A}i} D_{\dot{A}i}) W$$

$$= \varepsilon^A{}_i \chi_A{}^i$$

$$\delta \chi_A{}^i = (\varepsilon^B{}_j D_B{}^j + \varepsilon^{\dot{B}j} D_{\dot{B}j}) D_A{}^i W$$

$$= -\frac{1}{2} F_{AB} \varepsilon^{Bi} - 2i(\partial)_{A\dot{B}} z \varepsilon^{\dot{B}i} - \frac{1}{2} C^{ij} \varepsilon_{Aj}$$

$$\delta C^{jk} = (\varepsilon^B{}_i D_B{}^i + \varepsilon^{\dot{B}i} D_{\dot{B}i}) D^{jk} W$$

$$= 2(\varepsilon^{Bj} \lambda_B{}^k + \varepsilon^{Bk} \lambda_B{}^j) - 2i\varepsilon^{\dot{B}j}(\partial)^A{}_{\dot{B}} \chi_A{}^k$$

$$- 2i\varepsilon^{\dot{B}k}(\partial)^A{}_{\dot{B}} \chi_A{}^j$$

$$\delta F_{AB} = (\varepsilon_i{}^c D_c{}^i + \varepsilon^{\dot{B}i} D_{\dot{B}i})(D^k{}_A D_{Bk}) W$$

$$= -2i\varepsilon^{\dot{B}i}(\partial)_{A\dot{B}} \chi_{Bi} - 2i\varepsilon^{\dot{B}i}(\partial)_{B\dot{B}} \chi_{Ai}$$

$$+ 2(\varepsilon_{Ai} \lambda^i{}_B + \varepsilon_{Bi} \lambda^i{}_A)$$

$$\delta\lambda_A{}^k = +\frac{1}{6}(\varepsilon^c{}_j D_c{}^j + \varepsilon^{\dot{c}j}D_{\dot{c}j})D_A{}^i D_i{}^k W$$

$$= \frac{1}{6\cdot4}\varepsilon_A{}^k d + \frac{i}{2}(\not{\partial})^B{}_{\dot{c}}\varepsilon^{\dot{c}k}F_{AB} - \frac{i}{2}(\not{\partial})_{\dot{c}A}\varepsilon^{\dot{c}j}c_j{}^k$$

$$\delta d = (\varepsilon_k{}^A D_A{}^k + \varepsilon^{\dot{A}k}D_{\dot{A}k})D^{ij}D_{ij}W$$

$$= -6\cdot8\varepsilon^{\dot{A}k}i(\partial)_{\dot{A}}{}^B\lambda_{Bk} \tag{14.152}$$

In fact, this multiplet is reducible, we may set C^{jk} to be real, i.e.

$$C^{jk} = (C^{jk})^* = \varepsilon^{jj'}C^*_{j'k'}\varepsilon^{kk'} \tag{14.153}$$

The above supersymmetry transformations then imply that

$$\lambda_B{}^k + i(\not{\partial})_B{}^{\dot{A}}\chi_{\dot{A}}{}^k = 0$$

$$d - 6\cdot4\cdot4\partial^2 z = 0$$

$$\varepsilon^{\mu\nu\rho\kappa}\partial^\nu F_{\rho\kappa} = 0 \tag{14.154}$$

where

$$F_{AB} = (\sigma^{\mu\nu})_{AB}\mathscr{F}_{\mu\nu}; \qquad F_{\mu\nu} = \tfrac{1}{2}(\mathscr{F}_{\mu\nu} + \mathscr{F}^*_{\mu\nu})$$

As a result of imposing this constraint we are left with the following fields[58]

$$z, \chi_A{}^i, C^{jk}(\text{real}), F_{\mu\nu} \tag{14.155}$$

where $F_{\mu\nu} = \partial_\mu A_\nu - \partial_\nu A_\mu$. These are the fields of the $N = 2$ Yang-Mills theory.

In terms of superspace the constraint of Eq. (14.153) is given by

$$D^{jk}W = \bar{D}^{jk}\overline{W} \tag{14.156}$$

One could also impose the constraint that C^{jk} be imaginary. This change can be achieved with the substitution $W \to iW$ and so it leads to a multiplet which is simply a field redefinition of the multiplet of Eq. (14.155). Examination of the multiplet shows that C^{ij} is the lowest dimensional field on which a constraint can be imposed without setting the multiplet on-shell.

An interesting example[49] of a multiplet which does possess a central charge is the hypermultiplet. This multiplet is described by a superfield ϕ_i ($i = 1, 2$) which satisfies the constraints

$$D_A{}^i\phi_j = \tfrac{1}{2}\delta_j{}^i D_A{}^k\phi_k$$

$$D_{\dot{A}i}\phi_j + D_{\dot{A}j}\phi_i = 0 \tag{14.157}$$

Using the symbol ε^{ij} these constraints may be expressed in the form

$$D_{Ai}\phi_j + D_{Aj}\phi_i = 0 \text{ or } D_{Ai}\phi_j = -\tfrac{1}{2}\varepsilon_{ij}D_A{}^k\phi_k$$

$$D_{Ai}\phi_j = -\tfrac{1}{2}\varepsilon_{ij}D_A{}^k\phi_k. \tag{14.158}$$

We will first perform some D-algebra manipulation to discover which are the independent components and then give the supersymmetry transformations.

Consider applying $D_A{}^i$ on $D_B{}^k\phi_k$:

$$D_A{}^iD_B{}^k\phi_k = 2\varepsilon^{ik}\varepsilon_{AB}D_z\phi_k - D_B{}^kD_A{}^i\phi_k$$

$$= 2\varepsilon_{AB}D_z\phi^i - \tfrac{1}{2}D_B{}^iD_A{}^k\phi_k \tag{14.159}$$

As a result we see that

$$D_{(A}{}^iD_{B)}{}^k\phi_k = 0$$

$$D^{ik}\phi_k = -8D_z\phi^i = -2D^k{}_k\phi^i \tag{14.160}$$

where

$$D^{ik} = D^{Ai}D_A{}^k \tag{14.161}$$

Similarly, we gain the results

$$D_{(\dot{A}}{}^iD_{\dot{B})}{}^k\phi_k = 0$$

$$\bar{D}^{ik}\phi_k = -8iD_z\phi^i = -2\bar{D}^k{}_k\phi^i \tag{14.162}$$

Let us now apply $D_A{}^i$ to $D_{\dot{B}}{}^k\phi_k$

$$D_A{}^iD_{\dot{B}}{}^k\phi_k = -2i(\sigma^\mu)_{A\dot{B}}\partial_\mu\phi^i - D_{\dot{B}}{}^kD_A{}^i\phi_k$$

$$= -2i(\sigma^\mu)_{A\dot{B}}\partial_\mu\phi^i - \tfrac{1}{2}D_{\dot{B}}{}^iD_A{}^k\phi_k$$

$$= -i(\partial\!\!\!/)_{A\dot{B}}\phi^i + \tfrac{1}{4}D_A{}^iD_{\dot{B}}{}^k\phi_k \tag{14.163}$$

As a result we find that

$$D_A{}^iD_{\dot{B}}{}^k\phi_k = -\tfrac{4}{3}i(\partial\!\!\!/)_{A\dot{B}}\phi^i \tag{14.164}$$

Similarly we arrive at the result

$$D_A{}^iD_{\dot{B}}{}^k\phi_k = \tfrac{4}{3}i(\partial\!\!\!/)_{A\dot{B}}\phi^i \tag{14.165}$$

It finally remains to discover the covariant derivatives of $D_z\phi^i$, starting with $D_A{}^k$ and using the fact that D_z commutes with $D_A{}^k$ and $D_{\dot{A}}{}^k$. We obtain

$$D_A{}^kD_z\phi_k = D_zD_A{}^k\phi_k$$

$$= +D_A{}^k(\tfrac{1}{4}\bar{D}^j{}_j\phi_i)$$

$$= -i(\partial\!\!\!/)_A{}^{\dot{B}}D_{\dot{B}}{}^k\phi_k. \tag{14.166}$$

and similarly

$$D_{\dot{B}}{}^k D_z \phi_k = D_z D_{\dot{B}}{}^k \phi_k = +D_{\dot{B}}{}^k (\tfrac{1}{4} D^j{}_j \phi_k) = -i(\partial)_{\dot{B}}{}^A D_A{}^k \phi_k. \qquad (14.167)$$

Having arrived at a point where the application of D's yields space-time derivatives of fields we have encountered already we may list the field content of the hypermultiplet. The $\theta = 0$ components of the superfields

$$\phi_i, \, D_A{}^k \phi_k, \, D_{\dot{A}}{}^k \phi_k, \, i D_z \phi_i \qquad (14.168)$$

will be denoted by

$$A_i, \, \psi_A, \, \psi_{\dot{A}}, f_i \qquad (14.169)$$

respectively. The supersymmetry transformations of these component fields are given by

$$\delta \phi_i = (\varepsilon^A{}_k D_A{}^k + \varepsilon^{\dot{A}k} D_{\dot{A}k}) \phi_i$$

$$= \tfrac{1}{2} \varepsilon^A{}_i \psi_A - \tfrac{1}{2} \varepsilon^{\dot{A}}{}_i \psi_{\dot{A}}$$

$$\delta \psi_A = (\varepsilon^B{}_j D_B{}^j + \varepsilon^{\dot{B}j} D_{\dot{B}j}) D_A{}^k \phi_k$$

$$= \varepsilon_{Aj}(i D_z \phi^j) + \tfrac{4}{3} i(\partial)_{A\dot{B}} \phi^j \varepsilon^{\dot{B}}{}_j$$

$$= f^j \varepsilon_{Aj} + \tfrac{4}{3} i(\partial)_{A\dot{B}} A^j \varepsilon_j{}^{\dot{B}}$$

$$\delta f_j = (\varepsilon^A{}_k D_A{}^k + \varepsilon^{\dot{A}k} D_{\dot{A}k})(D_z) \phi_i$$

$$= \varepsilon^A{}_j i(\partial)_A{}^{\dot{B}} \psi_{\dot{B}} + i \varepsilon^{\dot{A}}{}_j (\partial)_{\dot{A}}{}^B \psi_B \qquad (14.170)$$

The $N = 2$ linear multiplet[59] is described by a real superfield L^{ij} which is a triplet of $SU(2)$ and has the defining condition

$$\sum_{(ijk)} D^i L^{jk} = 0 \qquad (14.171)$$

It is easily seen that it has no central charge dependence and its x-space component field content is $L^{ij}(x)$, $\lambda^i{}_\alpha$; S, P and V_μ where V_μ satisfies

$$\partial^\mu V_\mu = 0 \qquad (14.172)$$

In Chapter 15 we will find how the above superfields may be used to describe the x-space models of Chapter 13.

Chapter 15

Superspace Formulations of Rigid Supersymmetric Theories

15.1 $N = 1$ Superspace Theories. The Wess-Zumino Model

The component fields $(A, B; \chi_\alpha; F, G)$ which comprise the Wess-Zumino model are contained in a chiral superfield[46] $\varphi(x, \theta)$.

$$D_{\dot{A}}\varphi = 0 \qquad (15.1)$$

The identifications with the x-space fields are given by

$$\varphi(\theta = 0) = z \qquad D_A\varphi(\theta = 0) = \chi_A$$

$$-\frac{1}{2}D^2\varphi(\theta = 0) \equiv f \qquad (15.2)$$

The supersymmetric action is given by

$$A^{\text{w-z}} = \int d^4x\, d^4\theta\, \bar\varphi\varphi + \left\{ \int d^4x\, d^2\theta \left(\mu\varphi + \frac{m\varphi^2}{2} + \frac{\lambda\varphi^3}{3!} \right) + \text{h.c.} \right\} \quad (15.3)$$

The superfield $\bar\varphi\varphi$ is not chiral and to construct an invariant requires a $d^4\theta$; however, φ, φ^2, φ^3 are chiral superfields and the corresponding integration is only over $\int d^2\theta$. The integral of a chiral superfield over $\int d^4x\, d^4\theta$ vanishes.

The first term is the kinetic action which we may evaluate in component fields as follows. It is obtained by taking $\theta = 0$ in the expression

$$
\int d^4x \left(-\frac{1}{4}D^2 \right)\left(-\frac{1}{4}\bar D^2 \right)\bar\varphi\varphi = \int d^4x \left(-\frac{1}{4}D^2 \right)\left(-\frac{1}{4}\bar D^2 \bar\varphi \right)\varphi
$$

$$
= \frac{1}{16}\int d^4x \left\{ (D^2\bar D^2\bar\varphi)\varphi + 2(D^B\bar D^2\bar\varphi)D_B\varphi \right.
$$

$$
\left. + (\bar D^2\bar\varphi)D^2\varphi \right\}
$$

$$
= \int d^4x \left\{ (+\partial^2\bar\varphi)\varphi - \frac{i}{2}(\partial\!\!\!/)^{B\dot B}(D_{\dot B}\bar\varphi)D_B\varphi \right.
$$

$$
\left. + \frac{1}{16}(\bar D^2\bar\varphi)(D^2\varphi) \right\} \qquad (15.4)
$$

Using the above identifications we find the result

$$\int d^4x \left\{ -|\partial_\mu z|^2 - \frac{i}{2} \bar{\chi}_{\dot{B}} (\partial)^{B\dot{B}} \chi_B + \frac{1}{4} |f|^2 \right\} \tag{15.5}$$

The interaction terms are the most general possible, consistent with the model being renormalizable. The $\mu\varphi$ term can be removed by a shift of φ ($\varphi \to \varphi + $ constant). In terms of component fields they become

$$\int d^4x \left(\frac{\mu}{2} f + m \left(-\frac{1}{4} \chi^A \chi_A + \frac{1}{2} fz \right) + \lambda \left(+\frac{1}{4} fz^2 - \frac{1}{4} z\chi^A \chi_A \right) + \text{h.c.} \right) \tag{15.6}$$

Although, the superspace formalism is fairly obvious for the Wess-Zumino model, it is not so transparent for more complicated theories. However, there is a very simple method called "gauge completion" that enables one to pass with ease from the x-space formulation to the superspace formulation.[48] We will now illustrate this for the case of the Wess-Zumino model. The field content (z, χ_A, f) of this model has the following transformation laws in two component notation:

$$\delta z = \varepsilon^A \chi_A \qquad \delta \chi_A = f\varepsilon_A - 2i(\sigma^\mu)_{A\dot{B}} \varepsilon^{\dot{B}} \partial_\mu z$$

$$\delta f = -2i\varepsilon^{\dot{B}} (\sigma^\mu)_{A\dot{B}} \partial_\mu \chi^A \tag{15.7}$$

Let us regard z as the first component of a complex superfield $\varphi(x, \theta_A, \theta_{\dot{A}})$; in other words

$$\phi(x, \theta_A, \theta_{\dot{A}})|_{\theta=0} = z(x). \tag{15.8}$$

The supervariation of z is given by

$$\delta z = (\varepsilon^A Q_A + \varepsilon^{\dot{A}} Q_{\dot{A}}) \phi|_{\theta=0} = (\varepsilon^A D_A + \varepsilon^{\dot{A}} D_{\dot{A}}) \phi|_{\theta=0} \tag{15.9}$$

This follows, as taking $\theta = 0$, sets the $(\partial)_{A\dot{B}} \theta^{\dot{B}}$ term contained in Q_A and D_A to zero. Hence at $\theta = 0$ the action of D and Q are equivalent. This clearly holds for the lowest component of any superfield. Consequently

$$\delta z = (\varepsilon^A D_A + \varepsilon^{\dot{A}} D_{\dot{A}}) \phi|_{\theta=0} = \varepsilon^A \chi_A \tag{15.10}$$

Hence as δz does not involve $\varepsilon^{\dot{A}}$ we must conclude that

$$D_{\dot{A}} \phi|_{\theta=0} = 0 \tag{15.11}$$

But any superfield whose first component vanishes is zero, and hence we arrive at the well-known result that the Wess-Zumino multiplet is contained in a complex superfield which satisfies

$$D_{\dot{A}} \phi = 0 \tag{15.12}$$

It could, of course, involve other constraints, and to answer this question one must examine the field content of φ when subject to $D_{\dot{A}}\varphi = 0$. It will contain x-space fields which are the $\theta = 0$ components of the superfields

$$\phi, D_A\phi, -\frac{1}{2}D^2\phi \qquad (15.13)$$

where $D^2 = D^A D_A$. Clearly, using $D_{\dot{A}}$ only yields quantities which are space-time derivatives of the above fields.

The field content in Eq. (15.6) is none other than that of the Wess-Zumino multiplet and so we may conclude that no other superspace constraints are possible. One may check that the fields of Eq. (15.6) do indeed have the correct transformation properties. For example, for the auxiliary field f we find that

$$\delta f = -\frac{1}{2}(\varepsilon^A D_A + \varepsilon^{\dot{A}} D_{\dot{A}})(D^2\phi)|_{\theta=0}$$

$$= -\frac{1}{2}2\varepsilon^{\dot{A}}\{D_{\dot{A}}, D^B\}D_B\phi|_{\theta=0} = -2i\varepsilon^{\dot{A}}(\sigma^\mu)_{B\dot{A}}\partial_\mu\chi^B \qquad (15.14)$$

15.2 $N = 1$ Yang-Mills Theory[10]

The x-space component fields $(A_\mu; \lambda_\alpha; D)$ of $N = 1$ supersymmetric Yang-Mills theory are the highest dimension fields of a general superfield[53] $V(x^\mu, \theta^\alpha)$. The presence of other x-space component fields is to be expected as the gauge transformation of $A_\mu(\delta A_\mu = \partial_\mu\Lambda + \cdots)$ which ensures the absence in physical processes of the longitudinal part of A_μ must belong to a supermultiplet. This multiplet is a chiral supermultiplet Λ $(D_{\dot{A}}\Lambda = 0)$ which can be used to gauge away all the lower dimensional components of V leaving only the fields $(A_\mu; \lambda; D)$ and the gauge transformation of A_μ.

For simplicity we will first consider the Abelian case; the gauge transformation of V is

$$\delta V = (\bar{\Lambda} - \Lambda) \qquad (15.15)$$

The superfield V is real $V^+ = V$, but $\Lambda^+ \equiv -\bar{\Lambda}$. To construct an action we require a gauge covariant object which is given by

$$W_A = \bar{D}^2 D_A V \quad \text{and} \quad W_{\dot{A}} = D^2 D_{\dot{A}} V \qquad (15.16)$$

with $(W_{\dot{A}})^+ = W_A$. Under a gauge variation

$$\delta W_A = -\bar{D}^2 D_A\Lambda = -2i\bar{D}^{\dot{B}}(\partial)_{A\dot{B}}\Lambda = 0 \qquad (15.17)$$

Clearly W_A is chiral, i.e.,

$$D_{\dot{B}}W_A = 0 \qquad (15.18)$$

The reality of V implies that

$$D^A W_A - D^{\dot{A}} W_{\dot{A}} = (D^A \bar{D}^2 D_A - D^{\dot{A}} D^2 D_{\dot{A}}) V = 0 \qquad (15.19)$$

The chirality of W_A means that a supersymmetric gauge invariant action is given by

$$\frac{1}{64g^2} \int d^4x \, d^2\theta W^A W_A \qquad (15.20)$$

It is usually necessary to add the hermitian conjugate to the chiral integrals in order to ensure reality; however, the constraint of Eq. (15.19) implies that the above action is already real and so there is no need to add its hermitian conjugate. To see this note that the action equals

$$-\frac{1}{16g^2} \int d^4x \, d^4\theta D^A V \bar{D}^2 D_A V$$

while, taking its complex conjugate and using (15.19) gives

$$\int d^4x \, d^4\theta (D^A V \bar{D}^2 D_A V)^+ = \int d^4x \, d^4\theta D^2 D_{\dot{A}} V(-D^{\dot{A}} V)$$

$$= \int d^4x \, d^4\theta V(-D^{\dot{A}} D^2 D_{\dot{A}} V)$$

$$= \int d^4x \, d^4\theta V(-D^A \bar{D}^2 D_A V)$$

$$= \int d^4x \, d^4\theta (D^A V \bar{D}^2 D_A V) .$$

The reader may easily verify that this action, which has the correct dimensions (V must be dimensionless as $\dim[W_A] = \dim[\lambda_A] = 3/2$) and it does give the action of $N = 1$ supersymmetric Abelian gauge theory (see Chapter 6). The absence of the lower dimensional fields contained in V, from this action, is guaranteed by its gauge invariance.

The Yang-Mills generalization of these results starts with the gauge transformation

$$e^{V'} = e^{\bar{\Lambda}} e^V e^{-\Lambda} \qquad (15.21)$$

where $V \equiv V^i T_i$, $\Lambda = \Lambda^i T_i$ and T_i are the antihermitean generators of the gauge group. Since T_i are antihermitean, V^i is imaginary while for the chiral parameter the condition $\bar{\Lambda} \equiv \bar{\Lambda}^i T_i \equiv -\Lambda^+$ implies that $\Lambda^{i+} = \bar{\Lambda}^i$. Note in the Abelian case $T_1 = i$ and $V = V^1 i$ which reduces, as it must, for an Abelian gauge group to Eq. (15.15). Although, this transformation is more complicated, we can still use the chiral parameter to gauge away the $(C^i; \zeta_a^i; H^i, K^i)$ components of V. The gauge covariant field strength is given by

$$W_A = \bar{D}^2(e^{-V}D_A e^V) \tag{15.22}$$

It transforms under a gauge transformation as

$$W_A' = e^\Lambda W_A e^{-\Lambda} \tag{15.23}$$

and is chiral, that is,

$$D_{\dot{B}} W_A = 0 \tag{15.24}$$

The reality condition is somewhat more complicated, it is given by

$$\{\mathscr{D}'^A, W_A\} = \{\mathscr{D}'^{\dot{A}}, W_{\dot{A}}\} \tag{15.25}$$

where we define for this occasion only the primed derivatives by

$$\mathscr{D}'^A = e^{-V}D^A e^V = D^A + (e^{-V}D^A e^V)$$

$$\mathscr{D}'^{\dot{A}} = e^V D^{\dot{A}} e^{-V} = D^{\dot{A}} + (e^{+V}D^{\dot{A}} e^{-V}).$$

The action is given by

$$A^{\text{Y.M.}} = \frac{1}{64g^2 C_2(G)} \int d^4x \, d^2\theta \, \text{Tr} \, W^A W_A \tag{15.26}$$

which can be rewritten as

$$A^{\text{Y.M.}} = -\frac{1}{16g^2 C_2(G)} \int d^4x \, d^4\theta \, \text{Tr}\{(e^{-V}D^A e^V)\bar{D}^2(e^{-V}D_A e^V)\} \tag{15.27}$$

The generators T_i are normalized so that $\text{Tr} \, T_i T_j = C_2(G)\delta_{ij}$.

The above results could have easily been found by the method of gauge completion. Briefly, for the Abelian case, we identify λ_A with $\theta = 0$ component of W_A. Since the supervariation of λ_A involves ε_B but not $\varepsilon_{\dot{B}}$, we may conclude that

$$D_{\dot{B}} W_A = 0 \tag{15.28}$$

The component field D is the $\theta = 0$ component of $-2D^A W_A$ and as it is real we find that

$$D^A W_A - D^{\dot{A}} W_{\dot{A}} = 0 \tag{15.29}$$

Further analysis shows that the only other independent x-space component field contained in V is

$$F_{AB} = D_{(A} W_{B)}|_{\theta=0} \tag{15.30}$$

which with its complex conjugate $F_{\dot{A}\dot{B}}$ can be identified with F_{mn}. The above constraints of Eqs. (15.28) and (15.29) imply that F_{mn} satisfies the Bianchi identity

$$D_{[r}F_{mn]} = 0 \tag{15.31}$$

A chiral multiplet φ^a transforms in the representation R of gauge group according to

$$\varphi^a \rightarrow \varphi'^a = (e^\Lambda)^a_{\ b}\varphi^b \tag{15.32}$$

where $\Lambda^a_b = \Lambda^i(T_i)^a_{\ b}$ and $(T_i)^a_{\ b}$ represent the gauge group generators in the representation R. The chiral constraint on φ^a namely $D_{\dot{A}}\varphi^a = 0$ is still covariant as a result of $D_{\dot{A}}\Lambda = 0$. The gauge invariant supersymmetric action which generalies the Wess-Zumino action is

$$\int d^4x\, d^4\theta\, \bar\varphi_a (e^V)^a_{\ b}\varphi^b \tag{15.33}$$

where $(V)^a_{\ b} = V^i(T_i)^a_{\ b}$. The interaction term

$$\int d^4x\, d^2\theta \left(\frac{m_{ab}}{2}\varphi^a\varphi^b + \frac{d_{abc}}{3!}\varphi^a\varphi^b\varphi^c + \text{h.c.} \right) \tag{15.34}$$

is invariant provided

$$0 = d_{ebc}(T^i)^e_{\ a} + d_{aec}(T^i)^e_{\ b} + d_{abe}(T^i)^e_{\ c} \tag{15.35}$$

and similarly for m_{ab}. See Chapter (11) for an evaluation of these actions in terms of component fields.

Finally, let us derive in closed form the variation of V under a gauge transformation[54]. This result is necessary in order to find the Faddeev-Popov determinant. Now, the equation $[e^V, V] = 0$ implies that

$$\delta V e^V + V e^{\bar\Lambda} e^V e^{-\Lambda} - e^V \delta V - e^{\bar\Lambda} e^V e^{-\Lambda} V = 0 \tag{15.36}$$

using Eq. (15.21). Multiplying this result on the left and right by $e^{-V/2}$ we find that

$$e^{V/2}\delta V e^{-V/2} - e^{-V/2}\delta V e^{V/2} - e^{-V/2}[V,\bar\Lambda]e^{V/2} + e^{V/2}[V,\Lambda]e^{-V/2} = 0 \tag{15.37}$$

to lowest order in Λ, $\bar\Lambda$.

We introduce the notation

$$V \wedge X = [V, X]$$

$$V^2 \wedge X = [V,[V,X]] \tag{15.38}$$

with the definition $1 \wedge X = X$. Hence,

$$e^V X e^{-V} = e^V \wedge X \tag{15.39}$$

while

$$f(V) \wedge (g(V) \wedge X) = (f(V)g(V) \wedge X) \tag{15.40}$$

Equation (15.37), written in this notation, becomes

$$2\left(\sinh\frac{V}{2}\right) \wedge \delta V = \left(\cosh\frac{V}{2}\right) \wedge V \wedge (\bar{A} - A)$$

$$- \left(\sinh\frac{V}{2}\right) \wedge V \wedge (A + \bar{A}) \tag{15.41}$$

and also

$$\delta V = \frac{V}{2} \wedge \left\{ -(A + \bar{A}) + \coth\frac{V}{2} \wedge (\bar{A} - A) \right\}$$

$$= \bar{A} - A - \frac{1}{2}[V, A + \bar{A}] + O(V^2) \tag{15.42}$$

This concludes the deviation which will be used later.

15.3 A Geometrical Approach to $N = 1$ Supersymmetric Yang-Mills Theory

An interesting alternative approach to supersymmetric Yang-Mills is found by mimicking the usual steps for Yang-Mills theory.[55] We introduce super-connections $A_\pi{}^i$ which transform as

$$(\partial_\pi + A_\pi)' = e^{+K}(\partial_\pi + A_\pi)e^{-K} \tag{15.43}$$

where $K = K^i T_i$ and $A_\pi = T_j A^j{}_\pi$ and the matrices T_i represent the generators of the group G in any representation. In the above equation, and those that follow, the derivative ∂_π acts all the way to the right, that is, past the e^{-K} factor. We may therefore write the right-hand side as

$$(e^{+K}(\partial_\pi + A_\pi)e^{-K}) + \partial_\pi$$

where in the first term the ∂_π does not acts past the bracket. We note that ∂_π, A_π and $K = K^i T_i$ are antihermitean and so $A_\pi{}^i$ and K^i are real. Covariant derivatives are defined by

$$\mathbf{D}_\pi = \partial_\pi + A_\pi{}^i (T_i)^a{}_b \tag{15.44}$$

where $(T_i)^a{}_b$ are now the matrices appropriate to the field being acted upon. In particular, given a superfield ψ^a which transforms as

$$\psi'^a = (e^K)^a{}_b \psi^b \tag{15.45}$$

then $\mathbf{D}_\pi \psi$ is a gauge covariant superfield. We define super Yang-Mills field strengths by

$$[\mathbf{D}_N, \mathbf{D}_M\} = T_{NM}{}^R \mathbf{D}_R + F_{NM}{}^i T_i \tag{15.46}$$

where $\mathbf{D}_N = E_N{}^\pi \mathbf{D}_\pi$ and $T_{NM}{}^R$ are the flat torsion tensors, i.e., $T_{NM}{}^R$ are zero except for $T_{A\dot{B}}{}^n = -2i(\sigma^n)_{A\dot{B}}$.

We find that

$$F_{NM} = F_{NM}{}^i T_i$$

$$= +D_N A_M - (-1)^{NM} D_M A_N + [A_N, A_M\} - T_{NM}{}^R A_R \qquad (15.47)$$

As in the case of Yang-Mills we have the Bianchi identities, which follow from the identity

$$[\mathbf{D}_M, [\mathbf{D}_N, \mathbf{D}_R\}\} - [[\mathbf{D}_M, \mathbf{D}_N\}, \mathbf{D}_R\} + (-1)^{RN}[[\mathbf{D}_M, \mathbf{D}_R\}, \mathbf{D}_N\} = 0 \qquad (15.48)$$

We find that

$$I_{NMR} = [-(-1)^{(M+N)R} \mathbf{D}_R F_{MN} + T_{MN}{}^S F_{SR}]$$

$$-(-1)^{NR}(R \to N, N \to R, M \to M \text{ in first term})$$

$$-(-1)^{(N+R)M}(M \to N, N \to R, R \to M \text{ in first term}) \qquad (15.49)$$

We now examine the consequences of having chiral superfields in the presence of supersymmetric Yang-Mills. The original constant $D_A \varphi = 0$ must be modified to become gauge covariant. The unique generalization is $\mathbf{D}_A \varphi = 0$ However, this implies that

$$\{\mathbf{D}_A, \mathbf{D}_B\} \varphi = F_{AB}{}^i T_i \varphi = 0 \qquad (15.50)$$

and hence we are forced to conclude, since φ at $\theta = 0$ is arbitrary that

$$F_{AB} = 0 \quad \text{and} \quad F_{\dot{A}\dot{B}} = 0 \qquad (15.51)$$

With these constraints, all the other representations of supersymmetry also survive in the presence of super Yang-Mills. Such constraints are often called "representation preserving" constraints.[56,57]

We also adopt what is called a "conventional constraint"[56,57]. Consider the field strength

$$F_{A\dot{B}} = +D_A A_{\dot{B}} + D_{\dot{B}} A_A + \{A_A, A_{\dot{B}}\} + 2i(\sigma^n)_{A\dot{B}} A_n \qquad (15.52)$$

we find that A_n is redundant as a connection as it transforms exactly like the negative of the first three terms in $F_{A\dot{B}}$. Setting $F_{A\dot{B}} = 0$ expresses A_n in terms of A_A and $A_{\dot{B}}$

$$A_n = \frac{i}{4}(+D_A A_{\dot{B}} + D_{\dot{B}} A_A - \{A_A, A_{\dot{B}}\})(\sigma_m)^{A\dot{B}} \qquad (15.53)$$

This is reminiscent of the change from first to second order formalism in general relativity. We recall that in the first order formalism of general relativity we have a vierbein and a spin connection. To pass to second order formalism we eliminate the spin connection in terms of the vierbein.

To summarize we adopt the constraints

$$F_{AB} = F_{\dot{A}\dot{B}} = F_{A\dot{B}} = 0 \qquad (15.54)$$

In this approach there are now two ways to proceed. We could simply solve the above constraints or we could "solve the Bianchi identities." Let us first carry out the later program. Although the Bianchi identities are just identities, once we impose on them the above constraints they become non-trivial and can be used to investigate the consequences of the constraints. The best way to tackle the Bianchi identities is in order of increasing dimension. From their definition we find the dimensions of the Yang-Mills field strengths to be

$$[F_{\alpha\beta}] = 1 \qquad [F_{n\alpha}] = \tfrac{3}{2} \qquad [F_{nm}] = 2 \qquad (15.55)$$

The only nontrivial Bianchi identity with only fermionic indices is

$$I_{A\dot{B}C} = 0 = T_{A\dot{B}}{}^{n} F_{nC} + T_{C\dot{B}}{}^{n} F_{nA} \qquad (15.56)$$

Hence, we find that

$$F_{A\dot{B}C} + F_{C\dot{B}A} = 0 \qquad (15.57)$$

where $F_{A\dot{B}C} = (\sigma^{n})_{A\dot{B}} F_{nC}$. As such

$$F_{A\dot{B}C} = -\frac{1}{2}\varepsilon_{AC} F^{D}{}_{\dot{B}D} \equiv -\frac{i}{4}\varepsilon_{AC} W_{\dot{B}} \qquad (15.58)$$

By complex conjugation we find that

$$F_{B\dot{C}\dot{D}} = -\frac{1}{2}\varepsilon_{\dot{C}\dot{D}} F_{B}{}^{\dot{A}}{}_{\dot{A}} \equiv +\frac{i}{4}\varepsilon_{\dot{C}\dot{D}} W_{B}$$

where

$$F_{B\dot{C}\dot{D}} = (\sigma^{n})_{B\dot{C}} F_{n\dot{D}} \qquad (15.59)$$

We note that

$$(F_{An})^{*} = -F_{\dot{A}n} \qquad (W_{A})^{*} = +W_{\dot{A}} \qquad (15.60)$$

since

$$[(\sigma^{m})_{A\dot{B}}]^{*} = (\bar{\sigma}^{m})_{\dot{A}B} = (\sigma^{m})_{B\dot{A}}$$

Note $W_{A}(W_{\dot{A}})$ is the lowest dimension non-vanishing field strength.
Next we examine the Bianchi identity.

$$I_{ABn} = D_{A} F_{Bn} + D_{B} F_{An} = 0 \qquad (15.61)$$

which implies that

$$D_{A}\varepsilon_{BC} W_{\dot{C}} + D_{B}\varepsilon_{AC} W_{\dot{C}} = 0 \qquad (15.62)$$

Multiplying by ε^{BC} we find that

$$D_{A} W_{\dot{C}} = 0 \qquad (15.63)$$

and by complex conjugation

$$\mathbf{D}_{\dot{C}} W_B = 0 \qquad (15.64)$$

The $W_A(W_{\dot{A}})$ introduced in this section in Eqs. (15.58) satisfies gauge co-variant chiral constraints while the $W_A(W_{\dot{A}})$ introduced in Sec. 15.2 in Eqs. (15.16) satisfies ordinary chiral derivatives [see Eq. (15.18)]. The resolution of this paradox is that they are in fact not the same object although we have unfortunately used the same symbol in order to be in agreement with the literature. The $W_A(W_{\dot{A}})$ of Sec. 15.2 is in fact the $W_{A^0}(W_{\dot{A}^0})$ which we will introduce shortly in Eq. (15.94).

The identity

$$I_{A\dot{B}n} = + T_{\dot{B}A}{}^m F_{mn} + D_A F_{\dot{B}n} + T_{\dot{B}A}{}^m F_{mn} + D_{\dot{B}} F_{An} = 0 \qquad (15.65)$$

yields the result

$$-\varepsilon_{\dot{C}\dot{B}} \mathbf{D}_A W_C + \varepsilon_{CA} \mathbf{D}_{\dot{B}} W_{\dot{C}} + 16 F_{A\dot{B}C\dot{C}} = 0 \qquad (15.66)$$

where $F_{A\dot{B}C\dot{C}} = (\sigma^n)_{A\dot{B}}(\sigma^m)_{C\dot{C}} F_{nm}$. Tracing on $\dot{C}\dot{B}$ and CA yields the result

$$-\mathbf{D}^A W_A + \mathbf{D}^{\dot{A}} W_{\dot{A}} = 0 \qquad (15.67)$$

Tracing on just $\dot{C}\dot{B}$ however, yields the results $\mathbf{D}_{(A} W_{C)} = -8(\sigma^{nm})_{AC} F_{nm}$. The corresponding complex conjugated result is

$$\mathbf{D}_{(\dot{A}} W_{\dot{C})} = 8(\bar{\sigma}^{nm})_{\dot{A}\dot{C}} F_{nm} \qquad (15.68)$$

In fact, these three equations are the full content of this Bianchi identity as symmetrizing $F_{A\dot{B}C\dot{C}}$ on AC and $\dot{B}\dot{C}$ automatically yields zero.

The Bianchi identity I_{Amn} produces the results

$$\mathbf{D}_A F_{nm} = \frac{i}{8}(\sigma^n)_A{}^{\dot{C}} \mathbf{D}_m W_{\dot{C}} - \frac{i}{8}(\sigma^m)_A{}^{\dot{C}} \mathbf{D}_n W_{\dot{C}} \qquad (15.69)$$

which simply expresses the spinorial derivative of F_{nm} back in terms of the space-time derivative of $W_{\dot{A}}$. Complex conjugation yields an analogous result. In fact, Eq. (15.69) is a consequence of Eqs. (15.67), (15.63) and (15.64) and in this sense it was redundant to solve the I_{Amn} Bianchi identity.

The remaining Bianchi identity I_{nmr} is the usual type of Bianchi identity and it expresses the fact that F_{nm} is the curl of a field A_n:

$$F_{nm} = + \partial_n A_m - \partial_m A_n + [A_n, A_m] \qquad (15.70)$$

The above exposition can be summarized by noting that the non-zero field strengths F_{An}, F_{nm} and their complex conjugates can all be expressed in terms of the one superfield W_A and its complex conjugate $W_{\dot{A}}$ and their spinorial derivatives. The superfield W_A is subject to the constraints

$$\mathbf{D}_B W_{\dot{A}} = 0 = \mathbf{D}_{\dot{B}} W_A \qquad (15.71)$$

$$\mathbf{D}^A W_A - \mathbf{D}^{\dot{B}} W_{\dot{B}} = 0 \tag{15.72}$$

As we shall see shortly, the other results of the Bianchi analysis, i.e., Eqs. (15.69) and (15.70) are consequences of these constraints.

From this result we can extract the component-field content of supersymmetric Yang-Mills theory. The independent x-space components of a chiral superfield $W_A(W_{\dot{A}})$ are the $\theta = 0$ components of

$$W_A, \; \mathbf{D}_{(B} W_{A)}, \; -\tfrac{1}{2}\mathbf{D}^B W_B, \; \mathbf{D}^2 W_A. \tag{15.73}$$

which we call respectively

$$\lambda_A, \; F_{BA}, \; D \text{ and } \zeta_A. \tag{15.74}$$

However, the constraint of Eq. (15.72) tells us that D is real and that

$$\mathbf{D}^C(\mathbf{D}^B W_B) = -\tfrac{1}{2}\varepsilon^{CB}\mathbf{D}^2 W_B = -\tfrac{1}{2}\mathbf{D}^2 W^C$$

$$= -\mathbf{D}^C \mathbf{D}^{\dot{B}} W_{\dot{B}} = -2i\mathbf{D}^{C\dot{B}} W_{\dot{B}} \tag{15.75}$$

Hence at $\theta = 0$

$$\zeta^C = -4i(\mathbf{D})^{C\dot{B}} \lambda_{\dot{B}} \tag{15.76}$$

Further manipulations show that F_{mn} satisfies its usual Bianchi identity and so can be expressed as in Eq. (15.70) at $\theta = 0$ where F_{mn} is related to F_{AB} and $F_{\dot{A}\dot{B}}$ by

$$F_{AB} = \tfrac{1}{4}(\sigma^{mn})_{AB} F_{mn}$$

$$F_{\dot{A}\dot{B}} = \tfrac{1}{4}(\sigma^{mn})_{\dot{A}\dot{B}} F_{mn} \tag{15.77}$$

Hence, $N = 1$ supersymmetric Yang-Mills theory consists of the fields λ_α, A_μ, and D. Their super-transformations can be read off from above by the usual arguments; for example,

$$\delta\lambda_B = \varepsilon^A \mathbf{D}_A W_B|_{\theta=0} = D\varepsilon_A + \frac{1}{2}(\sigma^{mn})_{BA} F_{mn}\varepsilon^A$$

$$= D\varepsilon_B + F_{AB}\varepsilon^A \tag{15.78}$$

We now turn to the alternative approach, namely, to simply solve the constraints. That is, to solve

$$F_{AB} = F_{\dot{A}\dot{B}} = 0 \tag{15.79}$$

The most general solution is given by

$$\mathbf{D}_A = e^{-\Omega} D_A e^{\Omega} \qquad \mathbf{D}_{\dot{A}} = e^{-\bar{\Omega}} D_{\dot{A}} e^{+\bar{\Omega}} \tag{15.80}$$

where $\Omega = \Omega^i T_i$ and $\bar{\Omega} = \bar{\Omega}^i T_i$. In these equations the derivatives act to the right and so

$$\{\mathbf{D}_A, \mathbf{D}_B\} = \{e^{-\Omega} D_A e^{\Omega}, e^{-\Omega} D_B e^{\Omega}\}$$

$$= e^{-\Omega}\{D_A, D_B\}e^{\Omega} = 0$$

and similarly for $F_{\dot{A}\dot{B}}$. This is not a pure gauge as $(\Omega^i)^* \neq \Omega^i$. Hence we have expressed A_B and $A_{\dot{B}}$ in terms of Ω and $\bar{\Omega}$. To reproduce the transformation properties of A_A we require

$$e^{\Omega'} = e^{\Omega}e^{-K}(\text{or } e^{\bar{\Omega}'} = e^{\bar{\Omega}}e^{-K}) \tag{15.81}$$

We can also find a transformation of Ω under which the covariant derivatives are inert, namely,

$$e^{\Omega'} = e^{\bar{\Lambda}}e^{\Omega}(\text{or } e^{\bar{\Omega}'} = e^{+\Lambda}e^{\bar{\Omega}}) \tag{15.82}$$

where $D_A\bar{\Lambda} = 0$. These additional gauge invariances are to be expected when solving any type of covariant constraint.

Under an infinitesimal transformation, Ω is changed by

$$\delta\Omega = -K + \bar{\Lambda} \tag{15.83}$$

We can use K^i which is real to gauge away the real part of Ω^i. The easiest way to achieve this is to define the K-invariant real object V by

$$e^V = e^{\Omega}e^{-\bar{\Omega}} \tag{15.84}$$

If we choose K^i to gauge Ω^i to be real then $V = 2\Omega(\Omega = -\bar{\Omega})$. The corresponding Λ transformation of V is

$$e^{V'} = e^{\bar{\Lambda}}e^{V}e^{-\Lambda} \tag{15.85}$$

or, infinitesimally,

$$\delta V = (\bar{\Lambda} - \Lambda) + O(V, \Lambda, \bar{\Lambda}) \tag{15.86}$$

We could have also found this result by gauging away the real part of Ω and making a corresponding compensating K transformation with every $\Lambda(\bar{\Lambda})$ transformation in order to remain in the chosen gauge.

If we denote the component fields of V by $(C; \zeta; H, K, A_\mu; \lambda; D)$ then as discussed in Chapter 11 we can use the chiral field Λ to gauge away the $C; \zeta, H$ and K components leaving the gauge transformation of A_μ. Having made this Wess-Zumino gauge choice one must then make the appropriate compensating transformation to remain in the gauge. Hence again we are left with the x-space component field A_μ, λ and D.

It often turns out to be useful to redefine the chiral fields and covariant derivatives. Given a chiral field φ,

$$\mathbf{D}_{\dot{A}}\varphi = 0 \tag{15.87}$$

we can define a new field $\varphi_0 = e^{-\bar{\Omega}}\varphi$ which now satisfies the constraint

$$\mathbf{D}_{0\dot{A}}\varphi_0 = 0 \tag{15.88}$$

where $\mathbf{D}_{0\dot{A}} = D_{\dot{A}}$ and transforms like $\varphi'_0 = e^{i\Lambda}\varphi_0$. For antichiral fields, we make the transformations $\psi = e^{-\bar{\Omega}}\psi_0$. This satisfies the constraint

$$\mathbf{D}_{0A}\psi_0 = 0 \tag{15.89}$$

where $\mathbf{D}_{0A} = e^{-V}D_A e^V$ and transforms as $\psi'_0 = e^{\Lambda}\psi_0$. It is important to note, that, like the change to the chiral basis, these transformations of the spinorial derivatives are non-unitary and so do not have simple reality properties. However, often when constructing or evaluating actions these new objects which are often termed to be in the chiral representations are useful. The new derivatives are consistent with the definition

$$\mathbf{D}_{0N} = e^{+\bar{\Omega}}\mathbf{D}_N e^{-\bar{\Omega}} \tag{15.90}$$

We can define chiral field strengths by

$$[\mathbf{D}_{0N}, \mathbf{D}_{0M}\} = T_{NM}{}^R \mathbf{D}_{0R} + F_{0NM} \tag{15.91}$$

Clearly

$$F_{0NM} = e^{+\bar{\Omega}}F_{NM}e^{-\bar{\Omega}} \tag{15.92}$$

and so our constraints become

$$F_{0AB} = F_{0\dot{A}\dot{B}} = F_{0A\dot{B}} = 0 \tag{15.93}$$

while

$$W_{0A} = e^{+\bar{\Omega}}W_A e^{-\bar{\Omega}} \tag{15.94}$$

To calculate W_{0A} in terms of V is rather simple:

$$\{\mathbf{D}_{0A}, \mathbf{D}_{0\dot{A}}\} = \{D_A + A_{0A}, D_{\dot{A}}\}$$
$$= T_{A\dot{A}}{}^n \mathbf{D}_{0n} + \{+D_{\dot{A}}A_{0A} + 2i(\sigma^n)_{A\dot{A}}A_n\} \tag{15.95}$$

and as $F_{0A\dot{A}} = 0$ we find that

$$A_n = +\frac{i}{4}(D_{\dot{A}}A_{0A})(\sigma_n)^{A\dot{A}} \tag{15.96}$$

However,

$$W_{0A} = -2iF_0{}^{\dot{B}}{}_{A\dot{B}} = -2i[\mathbf{D}_0{}^{\dot{B}}, \mathbf{D}_{0n}](\sigma^n)_{A\dot{B}}$$
$$= D^{\dot{B}}D_{\dot{B}}(e^{-V}D_A e^V). \tag{15.97}$$

Since

$$\mathrm{Tr}\, W^A W_A = \mathrm{Tr}\, W_0{}^A W_{0A} \tag{15.98}$$

we recover the Yang-Mills action of Eq. (15.26).

15.4 $N = 2$ Superspace Theories

We now wish to repeat the previous discussion of $N = 1$ superspace theories for $N = 2$ superspace theories. We begin by deducing the superspace formulation of $N = 2$ Yang-Mills theory.[58] From its x-space formulation given in Chapter 12, we use the method of gauge completion. The x-space content is

$$(z, \lambda_{Ai}, F_{\mu\nu}, C^{ij}) \tag{15.99}$$

where $z = A + iB$. The supersymmetry transformation of z is of the form

$$\delta z = \varepsilon^A{}_i \lambda_A{}^i \tag{15.100}$$

As this does not involve $\varepsilon^{\dot A i}$ we may conclude that if z is the $\theta = 0$ component of a complex superfield $W(x^\mu, \theta^A{}_i, \theta^{\dot B j}, z)$, then W must be a chiral superfield

$$D_{\dot A i} W = 0 \tag{15.101}$$

This constraint implies that

$$\{ D_{\dot A i}, D_{\dot B j} \} W = 2\varepsilon_{\dot A \dot B} \varepsilon_{ij} D_z W = 0 \tag{15.102}$$

or that

$$D_z W = \frac{\partial}{\partial z} W = 0 \tag{15.103}$$

and consequently W has no dependence on the bosonic central charge coordinate z.

We must now determine if W must have any other superspace constraints in order to describe the $N = 2$ Yang-Mills theory. The x-space content of a complex superfield W which is chiral is found by applying $D_A{}^i$'s and evaluating all possible independent superfields at $\theta = 0$. We then arrive at the fields z, $\lambda_A{}^i$, t^{ij}_{AB}, t^{ijk}_{ABC}, t^{ijkl}_{ABCD}, which are, respectively, the $\theta = 0$ components of the following superfields:

$$W, \ D_A^i W, \ D_A^i D_B^j W, \ D_A^i D_B^j D_C^k W, \ D_A^i D_B^j D_C^k D_D^l W \tag{15.104}$$

As W has no central charge we have the symmetries

$$t^{ij}_{AB} = -t^{ji}_{BA}, \quad \text{etc.} \tag{15.105}$$

Utilizing this symmetry we may express t^{ij}_{AB} as

$$t^{ij}_{AB} = -\tfrac{1}{2}\varepsilon^{ij} F_{(AB)} - \tfrac{1}{2} C^{(ij)} \varepsilon_{AB} \tag{15.106}$$

Carrying out this analysis for all the fields we find that the x-space field content of the chiral W is

$$z, \ \lambda_A{}^i, \ C^{ij}, \ F_{AB}, \ \zeta_A{}^i, \ d \tag{15.107}$$

As such, another superspace constraint is required to reduce W so that it contains only the content of the $N = 2$ Yang-Mills theory.

The lowest dimensional component that is not of the correct form is C^{ij}, which is complex, while in $N = 2$ Yang-Mills it is real. We therefore impose C^{ij} real, which corresponds to the superspace constraint

$$D^{ij}W = \bar{D}^{ij}\overline{W} \qquad (15.108)$$

where $D^{ij} = D^{Ai}D_A{}^j$. The reader may check that this constraint also implies that

$$\zeta_A{}^i = -i(\not{\partial})_A{}^{\dot{B}}\lambda_{\dot{B}}{}^i \qquad (15.109)$$

$$d = 16\cdot 6(-\partial^2 z)$$

and that

$$\partial_{[\rho}F_{\mu\nu]} = 0 \qquad (15.110)$$

To summarize our results,[22] the fields of the $N = 2$ Yang-Mills theory are contained in a complex superfield W which is subject to

$$D_{\dot{A}i}W = 0 \qquad D^{ij}W = \bar{D}^{ij}\overline{W} \qquad (15.111)$$

The above calculation has been performed for the linearized $N = 2$ Yang-Mills theory. The results for the full non-Abelian theory must involve the same constraints in the limit that the gauge coupling goes to zero, and so they can only be

$$\mathbf{D}_{\dot{A}i}W = 0 \qquad \mathbf{D}^{ij}W = \bar{\mathbf{D}}^{ij}\overline{W} \qquad (15.112)$$

where $\mathbf{D}_N$ is the gauge covariant derivative and is given in terms of the gauge potential A_N by

$$\mathbf{D}_N = (D_N + gA_N) = E_N^\pi(\partial_\pi + gA_\pi) \qquad (15.113)$$

For a more geometric discussion of the superspace description of $N = 2$ Yang-Mills theory see the discussion toward the end of this section.

Let us now turn our attention to the $N = 2$ matter sector. In the formulation of the Sohnius hypermultiplet,[49] the x-space field content is

$$(A_i, \psi_A, \psi_{\dot{A}}^*, f_i) \qquad (15.114)$$

The lowest component A^i has the transformation law

$$\delta A^i = \tfrac{1}{2}\varepsilon^{Ai}\psi_A - \tfrac{1}{2}\varepsilon^{\dot{A}i}\psi_{\dot{A}} \qquad (15.115)$$

and so if we consider A^i to be the first component of the superfield ϕ^i then ϕ^i must possess the constraint

$$D_A{}^i \phi_j = \tfrac{1}{2} \delta^i_j D_A{}^k \phi_k$$

$$D_{\dot{A}}{}^i \phi_j = \tfrac{1}{2} \delta^i_j D_{\dot{A}}{}^k \phi_k \tag{15.116}$$

By raising the j index with ε^{ij} these constraints may be rewritten in the form

$$D_A{}^{(i} \phi^{j)} = 0 = D_{\dot{A}}{}^{(i} \phi^{j)} \tag{15.117}$$

The independent x-component fields in ϕ^i can be shown after a little thought (see Chapter 14) to be the lowest components of the superfields

$$\phi_i, D_A{}^k \phi_k, D^j{}_j \phi_i \tag{15.118}$$

and their complex conjugates. This has the same content as the Sohnius hypermultiplet and hence we may conclude that ϕ_i has only the constraints of Eq. (15.117).

Let us now consider the alternative formulation of the hypermultiplet that is described by the fields $(L^{ij}, \lambda^i{}_A, S, P, V_\mu)$, where $\partial^\mu V_\mu = 0$. The supertransformation of L^{ij} which is real must be of the form

$$\delta L^{ij} = \varepsilon^{Ai} \lambda_A{}^j + \varepsilon^{\dot{A}i} \lambda_{\dot{A}}{}^j + (i \leftrightarrow j) \tag{15.119}$$

and as a result, if L^{ij} is the first component of the real superfield also denoted L^{ij}, it must obey[59]

$$D_A{}^{(i} L^{jk)} = 0$$

$$D_{\dot{A}}{}^{(i} L^{jk)} = 0 \tag{15.120}$$

The reader may verify that no other superspace constraints are needed.

We can however relax Eq. (5.30), and so the constraint $\partial_\mu V^\mu = 0$, by introducing[50] the superfield $L^{ijkl} = L^{(ijkl)}$

$$D_A{}^{(i} L^{jk)} = D_{Al} L^{ijkl} \tag{15.121}$$

This multiplet does not have a conserved vector or any other such constraint and can be used to describe $N = 2$ matter. Of course it involves many more fields; however, the extra fields do not lead to further on-shell states provided one introduces an extra superfield G whose fields act as Lagrange multipliers. The superfield G satisfies the constraints

$$D_A{}^i D_{Bi} G = 0 = [D_A{}^i, D_{\dot{B}i}] G \tag{15.122}$$

For a description of this multiplet when matter belongs to a complex representation see Ref. 60. For further superspace descriptions of $N = 2$ matter see Ref. 16.

We now wish to construct actions for the above superspace theories. For a general $N = 2$ superfield ϕ which has $\partial_z \phi = 0$, an invariant quantity is given by

$$\int d^4x \, d^8\theta \phi \tag{15.123}$$

However, if this is to be an action ϕ must have dimension zero.

As such the actions for the above theories must be integrals over only a subspace of superspace or be constructed from superfields with subcanonical dimensions.

In the case of $N = 2$ Yang-Mills theories, the action can only involve $D_A{}^i$ and not $D_A{}^i$ acting on W, and the only candidate of the correct dimension is[58]

$$A^{\text{YM}} = \int d^4x D^{ij} D_{ij} W^2 + \text{h.c.} \tag{15.124}$$

Clearly this is invariant as

$$\delta A^{\text{YM}} = \int d^4x (\varepsilon^A{}_k D_A{}^k + \varepsilon^{Ai} D_{Ai}) D^{ij} D_{ij} W^2 = 0 \tag{15.125}$$

For the Sohnius hypermultiplet, the appropriate superspace action[49] is

$$\int d^4x D^{ij} (\bar{\phi}_i D^k{}_k \phi_j) \tag{15.126}$$

while for the version with the conserved vector V_μ the action[62] is

$$\int d^4x D^{(ij} D^{kl)} (L_{(ij} L_{kl)}) \tag{15.127}$$

The invariance of this action is not obvious, but may be verified by using the anticommutations for the D's.

The action for the relaxed formulation of the hypermultiplet is rather complicated.

For an interesting alternative formulation of $N = 2$ theories based on a different type of superspace to that considered here, see Ref. 63.

In quantizing a theory it is usual to work with unconstrained fields. In the case of superspace this means unconstrained superfields. As such it is necessary to solve the superspace constraints given above. The important exception to this rule is of course the chiral superfield ϕ, but here one can also solve the constraint with

$$\phi = \bar{D}^2 U \tag{15.128}$$

and quantize with the unconstrained prepotential U.

We recall that the constraints of $N = 1$ Yang-Mills

$$\mathcal{D}_A W_{\dot{B}} = 0 \qquad \mathcal{D}^A W_A = \mathcal{D}^{\dot{B}} W_{\dot{B}} \tag{15.129}$$

were solved by

$$W_{\dot{B}} = D^2(e^{-gV} D_{\dot{B}} e^{gV}) \tag{15.130}$$

The solution of the superspace constraints of extended superfields is rather complicated and analytic solutions are not known in the nonlinear case. For $N = 2$ Yang-Mills at the linearized level the solution of the constraints on W are given[64] in terms of an unconstrained dimension -2 superfield V^{ij} as

$$W = D^{ij} D_{ij} \bar{D}^{kl} V_{kl} \tag{15.131}$$

The solution of the constraints of the relaxed hypermultiplet formulation involves the prepotential ρ_{Ai} and X^{ijkl} of dimension $-3/2$ and -1 respectively.[66] It has been shown that the linearized constraints can be systematically iterated to solve the constraints in the non-Abelian theory.[65,66]

We will now consider the $N = 2$ Yang-Mills theory from a geometric viewpoint.[55,58] As in ordinary Yang-Mills theory we introduce potentials A_N which covariantize the superspace derivatives:

$$\mathbf{D}_N = D_N + gA_N \cdot Y \tag{15.132}$$

where Y is the Yang-Mills generator and $A_N = (A_A{}^i, A_{\dot{B}i}, A_c)$. We can then define super Yang-Mills field strengths by the equation

$$[\mathbf{D}_N, \mathbf{D}_M\} = T_{NM}{}^R \mathbf{D}_R + F_{NM} \cdot Y \tag{15.133}$$

where $T_{NM}{}^R$ is the torsion of rigid superspace and is zero except for the component

$$T_{A\dot{B}} = -2i(\sigma^m)_{A\dot{B}} \tag{15.134}$$

The F_{MN} must then obey the Bianchi identities

$$\sum_{(MNR)} (\mathbf{D}_M F_{NR} + T_{MN}{}^S F_{SR}) = 0 \tag{15.135}$$

the appropriate symmetrization being understood by the symbol $\sum_{(MNR)}$. Consider now $N = 2$ matter say in formulation (a). If it transforms under the gauge group the defining condition must be modified, in order to be gauge invariant, to

$$\mathbf{D}_A{}^{(i} \phi^{j)} = 0 = \mathbf{D}_A{}^{(i} \phi^{j)} = 0 \tag{15.136}$$

Consequently, we find that

$$\sum_{(ijk)} \{\mathbf{D}_A{}^i, \mathbf{D}_{\dot{B}}{}^j\} \phi^k = 0 = \sum_{(ijk)} F_A{}^{(i}{}_{\dot{B}}{}^{j)} \phi^k \tag{15.137}$$

and similarly for the other possible expressions. We must therefore conclude that

$$F_A{}^{(i}{}_B{}^{j)} = F_{\dot{A}}{}^{(i}{}_B{}^{j)} = F_{\dot{A}}{}^{(i}{}_{\dot{B}}{}^{j)} = 0 \qquad (15.138)$$

Hence, we have found that in order to have $N = 2$ matter in the presence of gauge fields we must constrain the super Yang-Mills field strengths.[57] An analysis of formulation (b) of $N = 2$ matter requires

$$D_A{}^{(i} L^{jk)} = D_{\dot{B}}{}^{(i} L^{jk)} = 0 \qquad (15.139)$$

and one finds the same constraints on F_{MN} as given above.

One can also eliminate,[57] by a covariant constraint, the potential A_m in terms of A_{Bi} and $A_{\dot{C}j}$. The appropriate constraint is

$$F_A{}^i{}_{\dot{B}i} = 0 \qquad (15.140)$$

This is similar to the change between the first and second order formalism in general relativity. The lowest dimension field strength remaining is of dimension one and is

$$W \equiv F^{Ai}{}_{Ai} \qquad (15.141)$$

The Bianchi identities, then imply, from the constraints of Eqs. (15.138) and (15.140), that

$$\mathscr{D}_{\dot{A}i} W = 0 \quad \text{and} \quad \mathscr{D}^{ij} W = \bar{\mathscr{D}}^{ij} \overline{W} \qquad (15.142)$$

which was our previous result.

Chapter 16

Superspace Formulation of $N = 1$ Supergravity

16.1 Geometry

The geometrical framework[69] of superspace supergravity has many of the constructions of general relativity, but also requires additional input. A useful guide in the construction of local superspace is that it should admit rigid superspace as a limit.

We begin with an eight-dimensional manifold $z^\pi = (x^u, \theta^{\underline{\alpha}})$ (x^u is a commuting coordinate, while $\theta^{\underline{\alpha}}$ is an anticommuting coordinate) which has a super-general coordinate reparametrization

$$z^\pi \to z'^\pi = z^\pi + \xi^\pi \tag{16.1}$$

where $\xi^\pi = (\xi^u, \xi^{\underline{\alpha}})$ are arbitrary functions of z^π.

Just as in general relativity we can consider scalar superfields, that is, fields for which

$$\phi'(z') = \phi(z) \tag{16.2}$$

and superfields with superspace world indices φ_A; for example

$$\varphi_A = \frac{\partial \phi}{\partial z^A} \tag{16.3}$$

The latter transform as

$$\varphi'_A(z') = \frac{\partial z^\pi}{\partial z'^A} \varphi_\pi(z) \tag{16.4}$$

The transformation properties of higher order tensors is obvious.

We must now specify the geometrical structure of the manifold. For reasons that will become apparent, the superspace formulation is essentially a vierbien formulation. We introduce supervierbiens $E_\pi{}^N$ which transform under the supergeneral coordinate transformations as

$$\delta E_\pi{}^N = \xi^A \partial_A E_\pi{}^N + \partial_\pi \xi^A E_A{}^N \tag{16.5}$$

The N-index transforms under the tangent space group which is taken to be just the Lorentz group; and so $\delta E_\pi{}^N = E_\pi{}^M \Lambda_M{}^N$ where

$$\Lambda_M{}^N = \begin{pmatrix} \Lambda_m{}^n & 0 & 0 \\ 0 & -\tfrac{1}{4}(\sigma_{mn})_A{}^B \Lambda^{mn} & 0 \\ 0 & 0 & +\tfrac{1}{4}(\sigma_{mn})_{\dot A}{}^{\dot B}\Lambda^{mn} \end{pmatrix} \qquad (16.6)$$

The matrix $\Lambda_m{}^n$ is an arbitrary function on superspace and it governs not only the rotation of the vector index, but also the rotation of the spinorial indices. Since we are dealing with an eight-dimensional manifold one could choose a much larger tangent space group. For example, $\Lambda_M{}^N$ could be an arbitrary matrix that preserves the metric

$$g_{NM} = \text{diag}(a_1 \eta_{mn}, a_2 \varepsilon_{AB}, a_3 \varepsilon_{\dot A \dot B}) \qquad (16.7)$$

where a_1, a_2 and a_3 are non-zero arbitrary constants. Demanding reality of the metric implies $a_2^* = a_3$ and we may scale away one factor. This corresponds to taking the tangent space group to be $Osp(4, 1)$. In such a formulation one could introduce a metric $g_{\pi A} = E_\pi{}^N g_{NM} E_A{}^M$ and one would have a formulation which mimicked Einstein's general relativity at every step.[70]

Such a formulation, however, would not lead to the x-space component $N = 1$ supergravity given earlier. One way to see this is to observe that the above tangent space group does not coincide with that of rigid superspace (super Poincaré/Lorentz), which has the Lorentz group, as given in Eq. (16.6) with $\Lambda_m{}^n$ a constant matrix, as its tangent space group. As linearized superspace supergravity must admit a rigid superspace formulation any formulation based on an $Osp(4, 1)$ tangent group will not coincide with linearized supergravity. In fact the $Osp(4, 1)$ formulation has a higher derivative action.

An important consequence of this restricted tangent space group is that tangent supervectors $V^N = V^\pi E_\pi{}^N$ belong to a reducible representation of the Lorentz group. This allows one to write down many more invariants. The objects $V^m V_m$, $V^A V^B \varepsilon_{AB}$, $V^{\dot A} V^{\dot B} \varepsilon_{\dot B \dot A}$ are all separately invariant.

In other words, in the choice of metric in Eq. (16.7) the constants a_1, a_2, a_3 can have any value including zero.

We define a Lorentz valued spin connection

$$\Omega_{\Lambda M}{}^N = \begin{pmatrix} \Omega_{\Lambda m}{}^n & 0 & 0 \\ 0 & -\tfrac{1}{4}\Omega_{\Lambda}{}^{mn}(\sigma_{mn})_A{}^B & 0 \\ 0 & 0 & \tfrac{1}{4}\Omega_{\Lambda}{}^{mn}(\bar\sigma_{mn})_{\dot A}{}^{\dot B} \end{pmatrix} \qquad (16.8)$$

This object transforms under super general coordinate transformations as

$$\delta\Omega_{\Lambda M}{}^N = \xi^\pi \partial_\pi \Omega_{\Lambda M}{}^N + \partial_\Lambda \xi^\pi \Omega_{\pi M}{}^N \qquad (16.9)$$

and under tangent space rotations as

$$\delta\Omega_{\Lambda M}{}^N = -\partial_\Lambda \Omega_M{}^N + \Omega_{\Lambda M}{}^s \Lambda_s{}^N + \Omega_{\Lambda R}{}^N \Lambda_M{}^R (-1)^{(M+R)(N+R)} \qquad (16.10)$$

The covariant derivatives are then defined by

$$D_A = \partial_A + \tfrac{1}{2}\Omega_A{}^{mn}J_{mn} \tag{16.11}$$

where J_{mn} are the appropriate Lorentz generators (see Appendix A). The covariant derivative with tangent indices is

$$D_N = E_N{}^A D_A \tag{16.12}$$

where $E_N{}^A$ is the inverse vierbien defined by

$$E_N{}^A E_A{}^M = \delta_N{}^M \tag{16.13}$$

or

$$E_A{}^M E_M{}^\pi = \delta_A^\pi \tag{16.14}$$

Equipped with super-vierbien and spin-connection we define the torsion and curvature tensors as usual

$$[D_N, D_M\} = T_{NM}{}^R D_R + \tfrac{1}{2}R_{NM}{}^{mn}J_{mn} \tag{16.15}$$

Using Eqs. (16.11) and (16.12) we find that

$$T_{NM}{}^R = E_M{}^A \partial_A E_N{}^\pi E_\pi{}^R + \Omega_{MN}{}^R - (-1)^{MN}(M \leftrightarrow N) \tag{16.16}$$

$$R_{MN}{}^{rs} = E_M{}^A E_N{}^\pi (-1)^{A(N+\pi)}\{\partial_A \Omega_\pi{}^{rs} + \Omega_A{}^{rk}\Omega_{\pi k}{}^s - (-1)^{A\pi}(A \leftrightarrow \pi)\} \tag{16.17}$$

The super-general coordinate transformations can be rewritten using these tensors

$$\delta E_A{}^M = -E_A{}^R \xi^N T_{NR}{}^M + D_A \xi^M \tag{16.18}$$

$$\delta \Omega_{AM}{}^N = E_A{}^R \xi^N R_{NRM}{}^N \tag{16.19}$$

where $\xi^N = \xi^A E_A{}^N$ and we have discarded a Lorentz transformation.

The torsion and curvature tensors satisfy Bianchi identities which follow from the identity

$$[\mathbf{D}_M, [\mathbf{D}_N, \mathbf{D}_R\}\} - [[\mathbf{D}_M, \mathbf{D}_N\}, \mathbf{D}_R\} + (-1)^{RN}[[\mathbf{D}_M, \mathbf{D}_R\}, \mathbf{D}_N\} = 0 \tag{16.20}$$

They read

$$0 = I^{(1)}{}_{RMN}{}^F = [-(-1)^{(M+N)R}\mathbf{D}_R T_{MN}{}^F + T_{MN}{}^S T_{SR}{}^F + R_{MNR}{}^F]$$

$$+ [+(-1)^{MN}\mathbf{D}_N T_{MR}{}^F - (-1)^{NR}T_{MR}{}^S T_{SN}{}^F - (-1)^{NR}R_{MRN}{}^F]$$

$$+ [-\mathbf{D}_M T_{NR}{}^F + (-1)^{(N+R)M}T_{NR}{}^S T_{SM}{}^F + (-1)^{(N+R)M}R_{NRM}{}^F] \tag{16.21}$$

and

$$I^{(2)}{}_{RMN}{}^{mn} = [-(-1)^{(M+N)R}\mathbf{D}_R R_{MN}{}^{mn} + T_{MN}{}^S R_{SR}{}^{mn}]$$

$$- (-1)^{NR}(R \to N, N \to R, M \to M \text{ in the first bracket})$$

$$- (-1)^{(N+R)M}(M \to N, N \to R, R \to M \text{ in the first bracket}) = 0$$

$$(16.22)$$

It can be shown that if $I^{(1)}{}_{MNR}{}^F$ holds then $I^{(2)}{}_{RMN}{}^{mn}$ is automatically satisfied. This result holds in the presence of constraints on $T_{MN}{}^R$ and $R_{MN}{}^{mn}$ and is a consequence of the restricted tangent space choice. We refer to this as Dragon's theorem.[71] For all fermionic indices we find that

$$I_{ABC}{}^N = -\mathbf{D}_A T_{BC}{}^N + T_{AB}{}^S T_{SC}{}^N + R_{ABC}{}^N - \mathbf{D}_C T_{AB}{}^N + T_{CA}{}^S T_{SB}{}^N$$

$$+ R_{CAB}{}^N - \mathbf{D}_B T_{CA}{}^N + T_{BC}{}^S T_{SA}{}^N + R_{BCA}{}^N = 0 \qquad (16.23)$$

and for fermionic indices with one bosonic index

$$I_{ABr}{}^N = -\mathbf{D}_A T_{Br}{}^N + T_{AB}{}^S T_{Sr}{}^N + R_{ABr}{}^N - \mathbf{D}_r T_{AB}{}^N + T_{rA}{}^S T_{SB}{}^N$$

$$+ R_{rAB}{}^N + \mathbf{D}_B T_{rA}{}^N - T_{Br}{}^S T_{SA}{}^N - R_{BrA}{}^N = 0 \qquad (16.24)$$

while

$$I_{Anr}{}^N = -\mathbf{D}_A T_{nr}{}^N + T_{An}{}^s T_{sr}{}^N + R_{Anr}{}^N - \mathbf{D}_r T_{An}{}^N + T_{rA}{}^s T_{sn}{}^N$$

$$+ R_{rAn}{}^N - \mathbf{D}_n T_{rA}{}^N + T_{nr}{}^s T_{sA}{}^N + R_{nrA}{}^N = 0 \qquad (16.25)$$

Clearly one can replace any undotted index by a dotted index and the signs remain the same. We recall that for rigid superspace all the torsions and curvatures vanish except for $T_{AB}{}^n = -2i(\sigma^n)_{A\dot{B}}$. Clearly this is inconsistent with an $Osp(4,1)$ tangent space group.

The dimensions of the torsions and curvature can be deduced from the dimensions of D_N. If F and B denote fermionic and bosonic indices respectively then

$$[D_F] = \tfrac{1}{2} \qquad [D_B] = 1 \qquad (16.26)$$

and

$$[T_{FF}{}^B] = 0 \qquad [T_{FF}{}^F] = [T_{FB}{}^B] = \tfrac{1}{2}$$

$$[T_{FB}{}^F] = [T_{BB}{}^B] = 1 \qquad [T_{BB}{}^F] = \tfrac{3}{2} \qquad (16.27)$$

while

$$[R_{FF}{}^{mn}] = 1 \qquad [R_{FB}{}^{mn}] = \tfrac{3}{2}$$

$$[R_{BB}{}^{mn}] = 2 \qquad (16.28)$$

It is useful to consider the notion of the geometric dimension of fields. This is the dimension of the field as it appears in the torsions and curvature. Such expressions never involve κ and as they are nonlinear in certain bosonic fields, such as the vierbien $e_\mu{}^n$, these fields must have zero geometric dimensions. The dimensions of the other field is determined in relation to $e_\mu{}^n$, say by the supersymmetry transformations. Hence

$$[e_\mu{}^n] = 0 \qquad [\psi_\mu{}^\alpha] = \tfrac{1}{2} \qquad [M] = [N] = [b_\mu] = 1 \qquad (16.29)$$

Note that these dimensions differ from the canonical assignment of dimension by one unit.

16.2 The Superspace Constraints

The above discussion is almost identical to the vierbein formulation of general relativity. The one difference of principle is the choice of a restricted tangent space group under which supertangent vectors belong to a reducible representation. To construct supergravity further input is required. The objects $E_A{}^N$ and $\Omega_A{}^{mn}$ contain many degrees of freedom. Although some of these are removed by the tangent space and supergeneral coordinate transformations, there still remain many degrees of freedom. A simple count shows that there remain

$$8(8 \times 8 + 8 \times 6) - 8(8 + 6) = 8 \times 7 \times 2 \qquad (16.30)$$

This number is many more degrees of freedom than the number of supergravity fields $e_\mu{}^n$, $\psi_\mu{}^\alpha$, M, N and b_μ minus their gauge invariances $(12 + 12)$. Furthermore, some of the fields have spin 3, i.e.,

$$E_\mu{}^m = \cdots + \theta^A \theta^{\dot{B}} h_\mu{}^m{}_{A\dot{B}} + \cdots \qquad (16.31)$$

These features are symptoms of an underlying disease. Given a chiral superfield $\varphi(x, \theta)$ of rigid superspace, we can insist that this representation survive in the presence of supergravity. The defining constraint of rigid supersymmetry $(D_A \varphi = 0)$ must be generalized to be covariant under supergeneral coordinate transformations. The simplest generalization is

$$\mathbf{D}_A \varphi = E_A{}^M \partial_M \varphi = 0 \qquad (16.32)$$

Equation (16.32) implies that

$$\{\mathbf{D}_A, \mathbf{D}_B\} \varphi = 0 = T_{AB}{}^N \mathbf{D}_N \varphi \qquad (16.33)$$

As φ at $\theta = 0$ is an arbitrary function of x, we must conclude that [56,57]

$$T_{AB}{}^{\dot{C}} = T_{AB}{}^n = 0 \qquad (16.34)$$

Complex conjugation implies that

$$T_{\dot{A}\dot{B}}{}^C = T_{\dot{A}\dot{B}}{}^n = 0 \tag{16.35}$$

The analysis we are carrying out has much in common with that for the rigid Yang-Mills theories and before proceeding further it is simpler (although not necessary) to impose the conventional constraints.[56,57] The spin connection. $\Omega_r{}^{mn}$ can be expressed in terms of the supervierbein $E_M{}^A$. For example

$$0 = T_{mn}{}^r = C_{mn}{}^r + \Omega_{mn}{}^r - (m \leftrightarrow n)$$

where

$$C_{MN}{}^R = E_M{}^A(\partial_A E_N{}^\pi)E_\pi{}^R - (-1)^{MN}(M \leftrightarrow N) \tag{16.36}$$

can be solved in the same way as one does when going from first to second order formalism in general relativity. Similarly, the constraint

$$0 = T_{AB}{}^C = C_{AB}{}^C + \Omega_{AB}{}^C + (A \leftrightarrow B) \tag{16.37}$$

can be used to solve for $\Omega_{AB}{}^C$. The solution is

$$\Omega_{ABC} = -\tfrac{1}{2}(C_{ABC} + C_{CBA} - C_{BCA}) \tag{16.38}$$

The constraint which expresses $\Omega_{A\dot{B}}{}^{\dot{C}}$ in terms of $E_M{}^A$ is

$$T_{A(\dot{B}}{}^{\dot{C})} = 0 \tag{16.39}$$

We recall that the spin connection is Lorentz valued and so $\Omega_{NA}{}^{\dot{B}} = 0$, etc. The spin connections $\Omega_{\dot{A}\dot{B}}{}^{\dot{C}}$ and $\Omega_{\dot{A}B}{}^C$ are solved for by the complex conjugates of the above constraints ($T_{\dot{A}\dot{B}}{}^{\dot{C}} = 0 = T_{\dot{A}(B}{}^{C)}$). It is important to realize that these constraints do not involve the solution of differential equations, but algebraically solve for $\Omega_N{}^{mn}$ in terms of $E_A{}^N$.

Two less obvious constraints solve algebraically for $E_m{}^A$ in terms of $E_A{}^A$ and $E_{\dot{A}}{}^A$. They are

$$T_{AB}{}^{\dot{C}} - \tfrac{1}{4}T_{Am}{}^n(\sigma_m{}^n)_B{}^{\dot{C}} = 0 \tag{16.40}$$

and

$$T_{AB}{}^n = -2i(\sigma^n)_{A\dot{B}} \tag{16.41}$$

The second term in Eq. (16.40) is to remove the spin connection from the equation. Note that the object $E_m{}^A$ has 4×8 degrees of freedom while each of the above two equations contain $4 \times 2 \times 2$ objects. For a detailed discussion of how to solve these constraints see Ref. 6 and 56.

There is some arbitrariness in the choice of conventional constraint; for example, rather than solve $\Omega_{A\dot{B}}{}^{\dot{C}}$ by $T_{A(\dot{B}}{}^{\dot{C})} = 0$, we could have used

$$T_{Am}{}^n(\sigma^m{}_n)_{\dot{B}}{}^{\dot{C}} = 0 \tag{16.42}$$

These different choices are related to each other by field redefinitions and so any choice is as good as any other. We have imposed the conventional constraints given in the table below:

Constraint	Superfield solved for
$T_{mn}{}^r = 0$	$\Omega_{mn}{}^r$
$T_{A(\dot{B}}{}^{\dot{C})} = 0$	$\Omega_{AB}{}^{\dot{C}}$
$T_{AB}{}^C = 0$	$\Omega_{AB}{}^C$
$\left.\begin{array}{l} T_{A\dot{B}}{}^n = -2i(\sigma^n)_{A\dot{B}} \\ T_{A\dot{B}}{}^{\dot{C}} - \frac{1}{4}T_{Am}{}^n(\bar{\sigma}^m{}_n)_{\dot{B}}{}^{\dot{C}} = 0 \end{array}\right\}$	$E_n{}^\pi$

We now consider what other representations of rigid supersymmetry generalize to supergravity. Of course, we must specify which rigid superspace we mean. The rigid anti-de Sitter superspace (Super Anti-de Sitter/$SO(3,1)$) which in the limit as the radius R goes to infinity becomes Poincaré superspace. In fact, not all representations of Poincaré superspace generalize to Anti-de Sitter (AdS) superspace, one example being $D_A\varphi_B = 0$ which now becomes

$$D_A{}^S\phi_B = 0 \qquad (16.43)$$

where $D_A{}^S$ is the anti de Sitter covariant derivative. The reason is that in anti de Sitter space derivatives obey a different algebra; in particular

$$\{D_A{}^S, D_B{}^S\} = \tfrac{1}{2}R^S{}_{AB}{}^{mn}J_{mn} \qquad (16.44)$$

where

$$R^S{}_{ABCD} = \tfrac{1}{6}(\varepsilon_{AC}\varepsilon_{BD} + \varepsilon_{BC}\varepsilon_{AD})m$$

$$T^S{}_{AB}{}^N = 0 \qquad (16.45)$$

while $R^S{}_{AB\dot{C}\dot{D}} = 0$. The other components of $R^S{}_{AB}{}^{mn}$ are found by complex conjugation and the parameter m is related to R by $m = \text{constant}/R$.

The representation of Eq. (16.43) does not generalize[72,73] to AdS superspace since

$$0 = \{D_A{}^S, D_B{}^S\}\phi_C = R^S{}_{ABC}{}^D\phi_D = \frac{m}{6}(\varepsilon_{AC}\phi_B + \varepsilon_{BC}\phi_A) \neq 0$$

The representation $D_{(A}\varphi_{B)} = 0$, however does generalize as

$$\sum_{(ABC)} \{D_A{}^S, D_B{}^S\}\phi_C = \sum_{(ABC)} R^S{}_{AB}{}^D{}_C\phi_D = 0 \qquad (16.46)$$

It makes sense only to generalize those representations that exist in AdS superspace in order that this superspace be included as a limit of the general supergravity superspace.

The most general irreducible representations of AdS superspace are chiral and linear corresponding to the constraints[72]

$$D^S_{(A} \phi_{BC\cdots)(\dot{A}\dot{B}\cdots)} = 0 \qquad (16.47)$$

$$D^{SA} \phi_{(AB\cdots)(\dot{A}\dot{B}\cdots)} = 0 \qquad (16.48)$$

respectively. For the case of no undotted spinor indices the first equation becomes $D_A \varphi_{(\dot{A}\dot{B}\cdots)} = 0$ and the second $D^2 \varphi_{(\dot{A}\dot{B}\cdots)} = 0$. The projectors corresponding to these representations are respectively[72]

$$\Pi_C(p,q) = +\frac{(D_A{}^S D^{SA} + mp/6)}{(1+p)m/3} \qquad (16.49)$$

$$\Pi_L(p,q) = -\frac{(D_A{}^S D^{SA} - (p/2+1)m/3)}{(1+p)m/3} \qquad (16.50)$$

where p and q are the number of undotted and dotted indices on the superfield on which the projector is acting. For example, consider a superfield with no spinor indices. The chiral and linear representations are given by

$$D_A{}^S D^{SA} \varphi = 0 \qquad (16.51)$$

$$\left(D_A{}^S D^{SA} - \frac{m}{3}\right)\varphi = 0 \qquad (16.52)$$

Applying a further D^S_A and using Eq. (16.47) we find that the first of these equations reduces to the usual result

$$D_A{}^S \varphi = 0 \qquad (16.53)$$

while the second is zero identically. The projectors obey the relations

$$\Pi_C(p,q)\Pi_C(p,q) = \Pi_C(p,q) \qquad \Pi_C(p,q)\Pi_L(p,q) = 0$$

and

$$\Pi_L(p,q)\Pi_L(p,q) = \Pi_L(p,q) \qquad \Pi_L(p,q)\Pi_C(p,q) = 0$$

and

$$\Pi_C(p,q) + \Pi_L(p,q) = 1 \qquad (16.54)$$

We now demand that all the irreducible representations of anti-de Sitter superspace Eqs. (16.47) and (16.48) generalize to local superspace. As the chiral representation $D_A \varphi = 0$ is one of these, we still learn that $T_{AB}{}^n = T_{AB}{}^{\dot{C}} = 0$. From the linear representation one can not derive any new constraints as its defining condition has a derivative with a contracted index. The other chiral representations in fact lead to no new constraints.[57] This results from the fact that the conventional constraints plus $T_{AB}{}^n = T_{AB}{}^{\dot{C}} = 0$ imply, using the Bianchi identities (see next section), that

$$R_{ABCD} = \tfrac{1}{6}(\varepsilon_{AC}\varepsilon_{BD} + \varepsilon_{BC}\varepsilon_{AD})R$$

$$R_{AB\dot{C}\dot{D}} = 0 \qquad \mathbf{D}_A R = 0. \tag{16.55}$$

The projectors of Eq. (16.54) then generalize to local superspace as follows:

$$\pi_L(p,q) = -\frac{(\mathbf{D}_A\mathbf{D}^A - \tfrac{1}{3}(1 + p/2)R)}{(1 + p)R/3} \tag{16.56}$$

$$\pi_C(p,q) = +\frac{(\mathbf{D}_A\mathbf{D}^A + (p/6)R)}{(1 + p)R/3} \tag{16.57}$$

The conditions of Eq. (16.55) are sufficient to ensure that the above projector also obeys Eq. (16.54). These projectors imply that the local representation obey the same defining conditions as in Eqs. (16.49) and (16.50) except with the replacement $D_A{}^S \to \mathbf{D}_A$.

As an illustration of the use of the above projectors let us consider a superfield with no spinor indices φ. The chiral and linear parts of this superfield are given respectively by

$$\mathbf{D}^2\varphi = 0 \tag{16.58}$$

and

$$(\mathbf{D}^2 - R/3)\varphi = 0 \tag{16.59}$$

where $\mathbf{D}^2 = \mathbf{D}^A\mathbf{D}_A$.

These conditions may be simplified as for the anti-de Sitter case. Consider

$$\mathbf{D}_A\mathbf{D}^2\varphi = \{\mathbf{D}_A, \mathbf{D}^B\}\mathbf{D}_B\varphi - \mathbf{D}^B\mathbf{D}_A\mathbf{D}_B\varphi$$

$$= R_A{}^B{}_B{}^C\mathbf{D}_C\varphi - \mathbf{D}^B(-\tfrac{1}{2}\varepsilon_{AB}\mathbf{D}^2\varphi) \tag{16.60}$$

Here use has been made of the constraint $T_{AB}{}^N = 0$. Using Eq. (16.55) we find that

$$\mathbf{D}_A\mathbf{D}^2\varphi = +\tfrac{1}{2}R\mathbf{D}_A\varphi - \tfrac{1}{2}\mathbf{D}_A\mathbf{D}^2\varphi \tag{16.61}$$

and so

$$\mathbf{D}_A\mathbf{D}^2\varphi = +\tfrac{1}{3}R\mathbf{D}_A\varphi \tag{16.62}$$

Applying $\mathbf{D}_A$ to Eq. (16.58) and dividing by R yields the result

$$\mathbf{D}_A\varphi = 0 \tag{16.63}$$

Applying $\mathbf{D}_A$ to Eq. (16.58) however yields no new result as

$$\mathbf{D}_A(\mathbf{D}^2 + R/3)\varphi = 0$$

vanishes identically.

The reader may be concerned that the generalization of the rigid representations to the local case is not unique. However, it can be shown that other defining conditions lead to exactly the same constraints.

Unlike the Yang-Mills case we must also impose an additional type of constraint.[57] Under a superconformal transformation, we get.

$$E_A{}^A \rightarrow e^L E_A{}^A \qquad E_{\dot{A}}{}^A = e^{L^*} E_{\dot{A}}{}^A$$

where L is a general superfield. The transformation properties of $E_m{}^A$ and $\Omega_A{}^{mn}$ are derived from those of $E_A{}^A (E_{\dot{A}}{}^A)$ using the conventional constraints which express the former in terms of the latter. Hence, by definition the conventional constraints are invariant under the superconformal transformations.

The representation-preserving constraints $T_{AB}{}^n = T_{AB}{}^{\dot{C}} = 0$ are also invariant under super-conformal transformations as can be checked for example by examining their explicit form in terms of $E_A{}^A (E_{\dot{A}}{}^A)$. The reader may consult Ref. 57 for a detailed calculation of the transformation properties of the curvature and torsion tensors.

In order to see the necessity for superconformal constraints let us consider now the action for $N = 1$ supergravity. It must be of the form,

$$A = \frac{1}{2\kappa^2} \int d^4x \, d^4\theta E f(E_A{}^A, E_{\dot{A}}{}^A)$$

where κ is the gravitational constant and

$$E = \det E_A{}^m$$

The $1/2\kappa^2$ is necessary in order to reproduce the $1/2\kappa^2$ in front of Einstein's action.

On dimensional grounds f must be a dimension-zero covariant tensor. However, there are no such non-zero covariant tensors and so the action can only be

$$A = \frac{1}{2\kappa^2} \int d^4x \, d^4\theta E$$

Now, if there is no superconformal constraint we can parametrize

$$E_A{}^m = \psi E_A'{}^m$$

where ψ is an arbitrary superfield. Varying the action with respect to ψ yields

$$\frac{\delta E_A{}^A}{\delta \psi} \frac{\delta A}{\delta E_A{}^A} = 0$$

which in turn implies that

$$E_A{}^\pi \frac{\delta E}{\delta E_A{}^\pi} = 2E = 0 \tag{16.64}$$

This is not an acceptable field equation as it means that $E_N{}^A$ is no longer invertible on-shell. Imposing superconformal constrains implies that ψ is not the most general superfield and that the equation of motion then involves a projection operator as it should.

The reader familiar with the new minimal formulation of supergravity will notice a flaw in the above argument; however, the conclusion, namely the need for superconformal constraints is always true. Clearly one should not break the superconformal group entirely, as this can also be achieved by a super-conformal gauge choice and in this case we would be left with the field content of conformal supergravity $(e_\mu{}^n, \psi_\mu{}^\alpha, b_\mu)$.

Given that one must choose constraints that break the superconformal invariance there are only two ways to do this. We can reduce $L(L^*)$ to be either a chiral or linear superfield. This requires constraints of dimension $\frac{1}{2}$ and 1 respectively. Examining the transformation properties, we find two candidates,[57] namely.

$$\delta T_{Am}{}^m = LT_{Am}{}^m - 2\mathbf{D}_A(2L + L^*) \tag{16.65}$$

and

$$\delta R = -6(\mathbf{D}_A\mathbf{D}^A - \tfrac{1}{3}R)L \tag{16.66}$$

In fact $T_{Am}{}^m$ is the only remaining non-zero dimension-$\frac{1}{2}$ tensor.

It can be shown as a consequence of the conventional and representation-preserving constraints that

$$\mathbf{D}_{(A}T_{B)} = 0 \tag{16.67}$$

The solution of (16.67), recalling our discussion on projectors, is

$$T_B = \mathbf{D}_B T \tag{16.68}$$

Hence, we may pick a superconformal gauge in which

$$T_B = 0 \tag{16.69}$$

The remaining superconformal invariance is given by a chiral parameter $\Sigma^* = (-2)(2L + L^*)$

$$\mathbf{D}_A\Sigma^* = 0 \tag{16.70}$$

Solving for L we have

$$L = -(2\Sigma^* - \Sigma) \tag{16.71}$$

and so the transformation of the supervierbein is

$$\delta E_A{}^A = -(2\Sigma^* - \Sigma)E_A{}^A \tag{16.72}$$

The constraint $T_B = 0$ leads (as we will see) to the old minimal formulation of $N = 1$ supergravity which is the one considered in this review (fields $e_\mu{}^n$, $\psi_\mu{}^\alpha$, M, N and b_μ). The remaining superconformal invariance is known as the Howe-Tucker group.

We could use the linear part of L to set $R = 0$. The remaining superconformal invariance is then given by the linear superfield $S = L$

$$\mathbf{D}^4\mathbf{D}_A\mathbf{S} = 0 \tag{16.73}$$

and the transformation of the vierbien is

$$\delta E_A{}^\pi = SE_A{}^\pi \tag{16.74}$$

The constraint $R = 0$ leads to the $20 + 20$ set of non-minimal Breitenlohner fields[75] with the above superconformal group.[76]

Actually, one can adopt a more complicated superconformal constraint.

$$Q = R + \zeta(\mathbf{D}^A T_A + \zeta T^A T_A) = 0 \tag{16.75}$$

where ζ is a parameter. Particular values of ζ lead to the old minimal, new minimal and Breitenlohner sets of auxiliary fields.

To summarize, the old minimal formulation of supergravity has the following constraints

Conventional

$$0 = T_{mn}{}^r = T_{AB}{}^C = T_{AB}{}^{\dot{C}}$$

$$0 = T_{AB}{}^n - 2i(\sigma^n)_{A\dot{B}} = T_{Am}{}^n(\bar{\sigma}_m{}^n)^{\dot{B}}{}_{\dot{C}}$$

Representation preserving

$$T_{AB}{}^{\dot{C}} = T_{AB}{}^n = 0$$

Super-Conformal choice

$$T_{Am}{}^m = 0$$

and their complex conjugates.

If we are dealing with conformal supergravity then we may use the superconformal invariance to choose $T_A = R = 0$.

16.3   Analysis of the Superspace Constraints

Given the constraints of supergravity there are two possible ways to proceed. We could simply solve the constraints[77,78] or alternatively we can investigate the consequences of the constraints using the Bianchi identities.[79] This is a very similar situation to that for supersymmetric Yang-Mills theories (see Chapter 15); however in the case of supergravity life is very much more complicated.

We begin by sketching how to utilize the Bianchi identities to find the consequence of the constraints. The constraints[80] of the previous section (conventional, representation preserving and superconformal) were

$$T_{AB}{}^{\dot{n}} - 2i(\sigma^n)_{A\dot{B}} = T_{AB}{}^m = 0$$

$$T_{Am}{}^m = T_{AB}{}^C = T_{AB}{}^{\dot{C}} = T_{Am}{}^n(\bar{\sigma}^m{}_n)_{\dot{B}}{}^{\dot{C}} = 0$$

$$T_{mn}{}^r = 0 \tag{16.76}$$

and their complex conjugates. The best way to analyze the Bianchi identities is to consider them in the order of increasing dimension. This strategy is particularly effective because it enables one to deal with linear relations until one encounters the dimension-2 Bianchi identities (see later).

The dimension-1/2 Bianchi identities are all identically satisfied except for $I_{AB\dot{C}}{}^m$. This identity has the form

$$0 = -\mathbf{D}_A T_{B\dot{C}}{}^m + T_{AB}{}^F T_{F\dot{C}}{}^m + R_{AB\dot{C}}{}^m - \mathbf{D}_{\dot{C}} T_{AB}{}^m + T_{\dot{C}A}{}^F T_{FB}{}^m$$

$$+ R_{\dot{C}AB}{}^m - \mathbf{D}_B T_{\dot{C}A}{}^m + T_{B\dot{C}}{}^F T_{FA}{}^m + R_{B\dot{C}A}{}^m \tag{16.77}$$

The curvature is Lorentz valued and so all the above terms involving the curvature vanish. The derivative of $T_{B\dot{C}}{}^n$ also vanishes as it is a numerically-invariant tensor. Using the constraint $T_{AB}{}^m = 0$ we are left with

$$T_{AB\dot{C}D\dot{D}} + T_{BA\dot{C}D\dot{D}} = 0 \tag{16.78}$$

where

$$T_{AB\dot{C}D\dot{D}} = T_{Am}{}^n(\sigma^m)_{B\dot{C}}(\sigma_n)_{D\dot{D}}$$

Hence we conclude that

$$T_{AB\dot{C}D\dot{D}} = -\tfrac{1}{2}\varepsilon_{AB}T^E{}_{E\dot{C}D\dot{D}} \tag{16.79}$$

The constraint $T_{Am}{}^n(\bar{\sigma}^m{}_n)_{\dot{B}\dot{C}} = 0$ becomes in this notation

$$T_{AE\dot{B}}{}^E{}_{\dot{C}} + T_{AE\dot{C}}{}^E{}_{\dot{B}} = 0 \tag{16.80}$$

and as a result

$$T_{AE\dot{B}}{}^E{}_{\dot{C}} = -\tfrac{1}{2}\varepsilon_{\dot{B}\dot{C}}T_{AE}{}^{\dot{D}E}{}_{\dot{D}} \tag{16.81}$$

Equation (16.79) then implies that

$$T^C{}_{C\dot{B}A\dot{C}} = -\tfrac{1}{2}\varepsilon_{\dot{B}\dot{C}}T^C{}_C{}^{\dot{D}}{}_{A\dot{D}}$$

substituting this result into Eq. (16.79) we find

$$T_{AB\dot{C}\dot{D}} = +\tfrac{1}{4}\varepsilon_{AB}\varepsilon_{\dot{B}\dot{C}}T^C{}_C{}^{\dot{D}}{}_{D\dot{D}} \tag{16.82}$$

Tracing on BC and $\dot{C}\dot{D}$ gives

$$T_A{}^{B\dot{C}}{}_{B\dot{C}} = -\tfrac{1}{2}T^C{}_C{}^{\dot{D}}{}_{A\dot{D}} \tag{16.83}$$

However, $T_{Am}{}^m$ is equivalent to

$$T_A{}^{B\dot{B}}{}_{B\dot{B}} = 0 \tag{16.84}$$

and hence $T^C{}_C{}^{\dot{D}}{}_{A\dot{D}} = 0$ and $T_{Am}{}^n = 0$. As such all dimension-1/2 or zero torsions vanish except $T_{A\dot{B}}{}^n = -2i(\sigma^n)_{A\dot{B}}$. Consequently, the lowest dimension at which $T_{RS}{}^F T_{FT}{}^N$ is bilinear in non-trivial torsions is dimension 2. This justifies our earlier statement concerning nonlinearities in the analysis of the Bianchi identities.

For the dimension-1 Bianchi identities the DT terms vanish and the T^2 term vanishes unless it is of the form $T\, 2i(\sigma^n)_{A\dot{B}}$. For example $I_{\dot{A}\dot{B}m}{}^n$ implies that

$$0 = R_{\dot{A}\dot{B}m}{}^n + T_{m\dot{A}}{}^C(-2i\sigma^n)_{\dot{B}C} - T_{\dot{B}m}{}^C(-2i\sigma^m)_{C\dot{A}}$$

Converting to spinor indices by multiplying with $(\sigma^m)_{\dot{C}C}$ and $(\sigma^n)_{D\dot{D}}$ we find that

$$R_{\dot{A}\dot{B}C\dot{C}D\dot{D}} + 4iT_{C\dot{C}\dot{A}D}\varepsilon_{\dot{D}\dot{B}} + 4iT_{C\dot{C}\dot{B}D}\varepsilon_{\dot{D}\dot{A}} = 0 \tag{16.86}$$

The Lorentz valued nature of $R_{PQm}{}^n$ allows us to express it in terms of $R_{PQA}{}^B$ and $R_{PQ\dot{A}}{}^{\dot{B}}$. Consider

$$R_{PQC\dot{C}}{}^{D\dot{D}} \equiv R_{PQm}{}^n(\sigma_m)_{C\dot{C}}(\sigma_n)_{D\dot{D}} \tag{16.87}$$

then we note that symmetrizing on C and D automatically antisymmetrizes on $\dot{C}$ and $\dot{D}$. However

$$(\sigma_m)_{(C|\dot{C}}(\sigma_n)_{|D)\dot{D}} - (m \leftrightarrow n) = -\varepsilon_{\dot{C}\dot{D}}(\sigma_{mn})_{CD}$$

while

$$(\sigma_m)_{C(\dot{C}}(\sigma_n)_{D|\dot{D})} - (m \leftrightarrow n) = \varepsilon_{CD}(\bar{\sigma}_{mn})_{\dot{C}\dot{D}} \tag{16.88}$$

and consequently

$$R_{PQC\dot{C}D\dot{D}} = -\tfrac{1}{2}\varepsilon_{\dot{C}\dot{D}}R_{PQm}{}^n(\sigma_{mn})_{CD} + \tfrac{1}{2}\varepsilon_{CD}R_{PQm}{}^n(\bar{\sigma}_{mn})_{\dot{C}\dot{D}}$$

$$= 2\varepsilon_{\dot{C}\dot{D}}R_{PQCD} + 2\varepsilon_{CD}R_{PQ\dot{C}\dot{D}} \tag{16.89}$$

We now decompose $T_{n\dot{A}}{}^D$ into its irreducible parts under the Lorentz group

$$T_{C\dot{C}\dot{A}D} = \varepsilon_{CD}\varepsilon_{\dot{C}\dot{A}}i\frac{R^*}{12} + \varepsilon_{CD}T_{(\dot{C}\dot{A})} + \varepsilon_{\dot{C}\dot{A}}T_{(CD)} + T_{(CD)(\dot{C}\dot{A})} \tag{16.90}$$

Substituting Eqs. (16.89) and (16.90) into Eq. (16.85) and multiplying by $\varepsilon^{\dot{C}\dot{D}}\varepsilon^{CD}$ implies that

$$T_{(\dot{C}\dot{A})} = 0 \tag{16.91}$$

Symmetrizing in DC, but multiplying by $\varepsilon^{\dot{C}\dot{A}}$ yields

$$R_{\dot{D}\dot{B}CD} + 6i\varepsilon_{\dot{D}\dot{B}}T_{(CD)} + 2iT_{(CD)(\dot{D}\dot{B})} = 0 \tag{16.92}$$

which implies that $T_{(CD)} = 0$.

Symmetrizing on CD and antisymmetrizing on $\dot{C}\dot{D}$ yields

$$R_{\dot{D}\dot{B}CD} - 2iT_{(CD)(\dot{D}\dot{B})} = 0 \tag{16.93}$$

Comparing Eqs. (16.92) and (16.93) we find

$$R_{\dot{A}\dot{B}CD} = 0 \tag{16.94}$$

and

$$T_{(CD)(\dot{A}\dot{B})} = 0 \tag{16.95}$$

The Bianchi identity then reduces to

$$R_{\dot{A}\dot{B}\dot{C}\dot{D}} = \tfrac{1}{6}(\varepsilon_{\dot{D}\dot{B}}\varepsilon_{\dot{C}\dot{A}} + \varepsilon_{\dot{C}\dot{B}}\varepsilon_{\dot{D}\dot{A}})R^* \tag{16.96}$$

and so

$$T_{C\dot{C}\dot{A}D} = \tfrac{1}{12}i\varepsilon_{CD}\varepsilon_{\dot{C}\dot{A}}R^* \tag{16.97}$$

The analysis proceeds along similar lines; it is lengthy and somewhat tedious. The net result[79] is that all torsions and curvatures can be expressed in terms of three superfields and their spinorial derivatives. These superfields are the R given above as well as $G_{A\dot{B}}$ and $W_{(ABC)}$. The superfields $G_{A\dot{B}}$ and W_{ABC} are given in terms of supertorsions by

$$T_{C\dot{C}DE} = +\frac{i}{4}(\varepsilon_{CE}G_{D\dot{C}} + 3\varepsilon_{CD}G_{E\dot{C}} - 3\varepsilon_{DE}G_{C\dot{C}})$$

$$T_{A\dot{A}B\dot{B}\dot{C}} = \varepsilon_{AB}(W_{\dot{A}\dot{B}\dot{C}} - \tfrac{1}{2}\varepsilon_{\dot{A}\dot{C}}\mathbf{D}^E G_{E\dot{B}} - \tfrac{1}{2}\varepsilon_{\dot{B}\dot{C}}\mathbf{D}^E G_{E\dot{A}}) + \varepsilon_{\dot{A}\dot{B}}\mathbf{D}_{(B}G_{C)\dot{A}} \tag{16.98}$$

where

$$T_{C\dot{C}D}{}^E = (\sigma^n)_{C\dot{C}}T_{nD}{}^E.$$

and

$$T_{C\dot{C}D\dot{D}}{}^E = (\sigma^n)_{C\dot{C}}(\sigma^m)_{D\dot{D}}T_{nm}{}^E \tag{16.99}$$

These superfields satisfy the constraints

$$\mathbf{D}_{\dot{A}}R^* = 0 \qquad \mathbf{D}_{\dot{A}}W_{(ABC)} = 0$$

$$\mathbf{D}^{\dot{B}}G_{A\dot{B}} = -\tfrac{1}{24}\mathbf{D}_A R^*$$

$$\mathbf{D}^A W_{(ABC)} = -i(\mathbf{D}_{B\dot{D}}G_C{}^{\dot{D}} + \mathbf{D}_{C\dot{D}}G_B{}^{\dot{D}}) \tag{16.100}$$

and G_n is real: $G_{A\dot{B}} = (\sigma^n)_{A\dot{B}}G_n = (G_{\dot{B}A})^*$.

As discussed earlier, on dimensional grounds, the action of simple supergravity is

$$\frac{1}{2\kappa^2}\int d^8z E \tag{16.101}$$

where $E = \det E_n{}^N$. As the supervierbiens are constrained, the equations of motion are not obvious. However, it can be shown[81] that this action and the superspace constraints of Eq. (16.76) imply that

$$R = G_{A\dot{B}} = 0$$

It will be shown in the next section that these are indeed the correct field equations, in the sense that they lead to the x-space equations of $N = 1$ supergravity.

The coupling of supergravity to a chiral matter field φ

$$\mathbf{D}_{\dot{A}}\varphi = 0 \tag{16.102}$$

is given by

$$\int d^8z E K(\varphi, \bar{\varphi}) \tag{16.103}$$

where K is an arbitrary function that begins with $K(\varphi, \bar{\varphi}) = \varphi\bar{\varphi} + \cdots$. The interaction terms are constructed using the chiral density[81]

$$\mathcal{E} = \frac{1}{4}\bar{\mathbf{D}}^2\left(\frac{E}{R^*}\right) \tag{16.104}$$

and are given by

$$\int d^4x\, d^2\theta \mathcal{E}f(\varphi) \tag{16.105}$$

where f is an analytic function of φ. The invariance of this action is established as follows. The function f being chiral can be written in the form

$$f(\varphi) = (\bar{\mathbf{D}}^2 - \tfrac{1}{3}R^*)U \tag{16.106}$$

where U is an arbitrary superfield. The expression

$$-\frac{1}{3}\int d^8zEU \tag{16.107}$$

is invariant and can be rewritten as

$$-\int d^8z\frac{E}{R*}\bar{\mathbf{D}}^2U + \int d^8z\frac{E}{R*}f(\varphi) = \int d^4x\,d^2\theta\mathcal{E}f(\varphi) \tag{16.108}$$

The first term vanishes as it is of the form of a divergence, i.e.,

$$\int d^8z E\mathbf{D}_{\dot{A}}V^{\dot{A}} = 0 \tag{16.109}$$

This follows after an integration by parts and using the torsion constraints (see the end of this section).

The introduction of Yang-Mills fields proceeds along the same lines as the rigid case. We introduce a superfield V which has the gauge transformation

$$e^{V'} = e^{\bar{A}}e^Ve^{-A} \tag{16.110}$$

where A is now locally chiral; $\mathbf{D}_{\dot{A}}\bar{A} = 0$. The covariant field strength is

$$W_A = (\bar{\mathbf{D}}^2 - \tfrac{1}{3}R*)(e^{-V}\mathbf{D}_Ae^V) \tag{16.111}$$

The corresponding action,[81] as W_A is chiral $(\mathbf{D}_{\dot{B}}W_A = 0)$, is of the form

$$A = \frac{1}{64C_2(G)g^2}\int d^4x\,d^2\theta\varepsilon\,\mathrm{Tr}\,W^AW_A \tag{16.112}$$

The coupling to matter is given by

$$\int d^4x\,d^4\theta\bar{\varphi}e^V\varphi \tag{16.113}$$

The alternative strategy was to solve the constraints.[77] This is a relatively complicated procedure. After using appropriate gauge transformations to remove certain superfields we find that all the vierbiens and spin connections can be expressed in terms of one superfield, $\mathbf{H}_\mu$. If we label the components of $\mathbf{H}_\mu$ by $(C_\mu; \zeta_\mu; M_\mu, N_\mu, e_\mu{}^n; \psi_\mu{}^\alpha; b_\mu)$ then the gauge invariance that $\mathbf{H}_\mu$ possesses can be used to remove C_μ and ζ_μ. The fields M_μ and N_μ transform as

$$\delta M_\mu = \partial^2 e_\mu - \partial_\mu\partial^\nu e_\nu, \qquad \delta N_\mu = \partial^2 t_\mu - \partial_\mu\partial^\nu t_\nu \tag{16.114}$$

which guarantees that they appear in the action as $\partial_\mu M^\mu = M'$ and $\partial_\nu N^\nu = N'$. The remaining gauge transformations are the usual ones on $e_\mu{}^n$ and $\psi_\mu{}^\alpha$. For a detailed discussion of the solution of the constraints see Refs. 6 and 77. In this way we discover that the constraints lead to the old minimal formulation of supergravity. Knowing the solution of the constraints in terms of $\mathbf{H}_\mu$ we may express the action and the non-zero torsions and curvatures in terms of H_μ. Consequently, we find the individual components of the torsions and curvatures in terms of $e_\mu{}^n$, $\psi_\mu{}^A$, M', N' and b_μ.

To find the x-space component fields within the context of "solving the Bianchi identities", one first identifies the independent x-space component fields contained in R, $G_{A\dot{B}}$ and $W_{ABC} = W_{(ABC)}$, taking account of Eq. (16.100). All these independent fields can be found in $T_{mn}{}^A$ and $R_{mn}{}^{rs}$ at $\theta = 0$ with the exception of R and $G_{A\dot{B}}$ at $\theta = 0$. We denote the components of R and $G_{A\dot{B}}$ at $\theta = 0$ by $M + iN$ and $b_{A\dot{B}} = (\sigma^n)_{A\dot{B}} b_n$. Then, one discovers that $T_{mn}{}^A$ and $R_{mn}{}^{rs}$ at $\theta = 0$ are given in terms of $E_\mu{}^n (\theta = 0) \equiv e_\mu{}^n(x)$ and $E_\mu{}^A (\theta = 0) \equiv \frac{1}{2}\psi_\mu{}^A$ and M, N and b_μ. Hence, the x-space fields in R, $G_{A\dot{B}}$ and W_{ABC} and so in all torsions and curvatures are $e_\mu{}^n$, $\psi_\mu{}^A$, M, N, and b_μ. How this is achieved in detail can be found by reading the next section backwards.

We now prove one of the results used above.

Lemma:[89] $\int d^8z E D_N V^N(-1)^N = 0.$

Proof: $\int d^8z E E_N{}^A \partial_A V^N(-1)^N$

$$= \int d^8z V^N \left(-E_N{}^A \partial_A E - (-1)^{A(A+N)} \partial_A E_N{}^A \right)$$

$$= \int d^8z V^N \left(E_N{}^A (\partial_A E_M{}^\pi) E_\pi{}^M (-1)^M - (-1)^{NM} E_M{}^A \partial_A E_N{}^\pi E_\pi{}^M (-1)^M \right)$$

$$= \int d^8z V^N T_{NM}{}^M (-1)^M = 0.$$

The last expression vanishes due to the constraints

$$T_{nr}{}^s = 0 \qquad \text{and} \qquad T_{nA}{}^A + T_{n\dot{A}}{}^{\dot{A}} = 0.$$

16.4 Superspace Supergravity from x-space Supergravity

In this section we will derive the superspace constraints and other results from the known x space formulation of supergravity.

A. On-Shell

Any supergravity theory which is on-shell, that is, one which satisfies its equations of motion, can be deduced purely from a knowledge of the number of on-shell states of a given spin. This can be achieved by writing in x-space, the unique linearized ghost free field equations corresponding to each on-shell state and Noether-coupling up the system of equations. (See Chapter 9.) This is a lengthy procedure. The result can be much more easily found in super-space, by using dimensional analysis and, of course, the number of on-shell states of a given spin. We now illustrate this procedure for $N = 1$ supergravity.

The on-shell states are represented by $h_{\mu\nu}$ ($h_{\mu\nu} = h_{\nu\mu}$) and $\psi_\mu{}^\alpha$ which have the gauge transformations

$$\delta h_{\mu\nu} = \partial_\mu \xi_\nu + \partial_\nu \xi_\mu$$

$$\delta \psi_{\mu\alpha} = \partial_\mu \eta_\alpha \tag{16.115}$$

We have omitted the nonlinear terms, as only the general form is important. The geometric dimension of $h_{\mu\nu}$ is zero while that of $\psi_\mu{}^\alpha$ is one half. The lowest dimension gauge covariant objects are of the form

$$\partial\psi \qquad \text{and} \qquad \partial\partial h \tag{16.116}$$

which have dimensions 3/2 and 2 respectively.

Consider now the super torsion and curvature; these objects at $\theta = 0$ must correspond to covariant x-space objects. If there is no such object then the corresponding tensor must vanish at $\theta = 0$ and hence to all orders in θ. The only dimension-0 tensors are $T_{AB}{}^n$ and $T_{A\dot{B}}{}^n$. There are no dimension-0 covariant objects except the numerically invariant tensor $(\sigma^n)_{A\dot{B}}$. Hence, we must conclude that

$$T_{AB}{}^n = 0 \qquad T_{A\dot{B}}{}^n = c(\sigma^n)_{A\dot{B}} \tag{16.117}$$

where c is a constant. We choose $c \neq 0$ in order to agree with rigid superspace. The reality properties of $T_{A\dot{B}}{}^n$ imply that c is imaginary and we can normalize it to take the value $c = -2i$.

There are no dimension-$\frac{1}{2}$ covariant tensors in x-space and so

$$T_{A\dot{B}}{}^{\dot{C}} = T_{AB}{}^C = T_{Am}{}^n = 0 \tag{16.118}$$

There are no dimension-1 covariant objects in x-space. This would not be the case if one had an independent spin connection, $w_\mu{}^{rs}$, for $\partial e + w + \cdots$ would be a covariant quantity. When $w_\mu{}^{rs}$ is not an independent quantity it must be given in terms of $e_\mu{}^n$ and $\psi_\mu{}^\alpha$ in such a way as to render the above dimension-1 covariant quantity zero. Hence, for a dependent spin connection, i.e., in second-order formalism, we have

$$T_{nA}{}^{\dot{B}} = T_{nA}{}^B = R_{AB}{}^{mn} = 0 = R_{A\dot{B}}{}^{mn} \tag{16.119}$$

In other words, every dimension-0, $\frac{1}{2}$, 1 torsion and curvature vanishes with the exception of $T_{A\dot{B}}{}^n = -2i(\sigma^n)_{A\dot{B}}$.

Comparing this set of constrains with the constraints of off-shell supergravity given in Section 16.2 or below we observe that the extra constraints are $T_{nA}{}^B = T_{nA}{}^{\dot{B}} = 0 = R_{AB}{}^{AB}$. In terms of the superfields R, $W_{(ABG)}$ and $G_{A\dot{B}}$ this is equivalent to

$$R = G_{A\dot{B}} = 0$$

The dimension-3/2 tensors can involve at $\theta = 0$ the spin-3/2 object $\partial\psi$ and so these will not all be zero. The only remaining non-zero tensors are $T_{mn}{}^A$, $R_{Ar}{}^{mn}$ and $R_{st}{}^{mn}$ and of course $T_{A\dot{B}}{}^n = -2i(\sigma^n)_{A\dot{B}}$. However, the previous constraints of Eqs. (16.117)–(16.119) are sufficient to specify the entire theory, as we will now demonstrate. The first nontrivial Bianchi identity has dimension 3/2 and is

$$I_{nB\dot{D}}{}^{\dot{C}} = -\mathbf{D}_n T_{B\dot{D}}{}^{\dot{C}} + T_{nB}{}^F T_{F\dot{D}}{}^{\dot{C}} + R_{nB\dot{D}}{}^{\dot{C}} + \mathbf{D}_{\dot{D}} T_{nB}{}^{\dot{C}} - T_{\dot{D}n}{}^F T_{FB}{}^{\dot{C}}$$
$$- R_{\dot{D}nB}{}^{\dot{C}} - \mathbf{D}_B T_{\dot{D}n}{}^{\dot{C}} + T_{B\dot{D}}{}^F T_{Fn}{}^{\dot{C}} + R_{B\dot{D}n}{}^{\dot{C}} = 0 \qquad (16.120)$$

Using the above constraints this reduces

$$-2i(\sigma^m)_{B\dot{D}} T_{mn}{}^{\dot{C}} - R_{nB\dot{D}}{}^{\dot{C}} = 0 \qquad (16.121)$$

Tracing on $\dot{D}$ and $\dot{C}$ then yields

$$(\sigma^m)_{B\dot{D}} T_{mn}{}^{\dot{D}} = 0 \qquad (16.122)$$

This is the Rarita-Schwinger equation as we will demonstrate shortly.

The spin-2 equation must have dimension two and is contained in the $I_{Bmn}{}^A$ Bianchi identity.

$$I_{Bmn}{}^A = -\mathbf{D}_B T_{mn}{}^A + T_{Bm}{}^F T_{Fn}{}^A + R_{Bmn}{}^A - \mathbf{D}_n T_{Bm}{}^A + T_{nB}{}^F T_{Fm}{}^A$$
$$+ R_{nBm}{}^A - \mathbf{D}_m T_{nB}{}^A + T_{mn}{}^F T_{FB}{}^A + R_{mnB}{}^A = 0 \qquad (16.123)$$

Application of the constraints gives

$$-\mathbf{D}_B T_{mn}{}^A + R_{mnB}{}^A = 0 \qquad (16.124)$$

On contracting with $(\sigma^m)_{\dot{B}A}$ we find

$$(\sigma^m)_{\dot{B}A}\mathbf{D}_B T_{mn}{}^A = 0 = R_{mnB}{}^A(\sigma^m)_{\dot{B}A} \qquad (16.125)$$

Using the fact that $R_{mnB}{}^A = -\frac{1}{4}R_{mn}{}^{pq}(\sigma_{pq})_B{}^A$ yields the result

$$R_{mn} - \frac{1}{2}\eta_{mn}R = 0$$

or

$$R_{mn} = 0 \quad \text{where } R_{mn} = R_{msn}{}^s \qquad (16.126)$$

We now wish to demonstrate that these are the spin-3/2 and spin-2 equations. The $\theta = 0$ components of $E_\mu{}^n$ and $E_\mu{}^A$ are denoted as follows:

$$E_\mu{}^n(\theta = 0) = e_\mu{}^n, \qquad E_\mu{}^A(\theta = 0) = \tfrac{1}{2}\psi_\mu{}^A \qquad (16.127)$$

At this stage the above equation is simply a definiton of the fields $e_\mu{}^n$ and $\psi_\mu{}^A$. The $\theta = 0$ components of $E_A{}^n$ may be gauged away by an appropriate super general coordinate transformation. As

$$\delta E_A{}^n(\theta = 0) = \xi^\pi \partial_\pi E_A{}^n|_{\theta=0} + \partial_A \xi^\pi E_\pi{}^n|_{\theta=0} = \cdots + \partial_A \xi^\mu e_\mu{}^n + \cdots \qquad (16.128)$$

we may clearly choose $\partial_A \xi^\mu$ so that $E_A{}^n = 0$. Similarly we may choose

$$E_{\dot{A}}{}^{\dot{B}} = \delta_{\dot{A}}{}^{\dot{B}}, \qquad E_A{}^B = \delta_A{}^B, \qquad E_A{}^{\dot{B}} = 0 \qquad (16.129)$$

To summarize

$$E_\pi{}^m(\theta = 0) = \begin{pmatrix} e_\mu{}^n & \tfrac{1}{2}\psi_\mu{}^A & \tfrac{1}{2}\psi_\mu{}^{\dot{A}} \\ 0 & \delta_B{}^A & 0 \\ 0 & 0 & \delta_{\dot{B}}{}^{\dot{A}} \end{pmatrix} \qquad (16.130)$$

For the spin connection $\Omega_\pi{}^m$ we define

$$\Omega_\mu{}^{mn}(\theta = 0) = w_\mu{}^{mn} \qquad (16.131)$$

and we use a Lorentz transformation to gauge

$$\Omega_\alpha{}^{mn}(\theta = 0) = 0 \qquad (16.132)$$

At $\theta = 0$ we then find

$$T_{\mu\nu}{}^{\dot{A}} = -\tfrac{1}{2}\partial_\mu \psi_\nu{}^{\dot{A}} + \Omega_{\mu\nu}{}^{\dot{A}} - (\mu \leftrightarrow \nu) = -\tfrac{1}{2}\psi_{\mu\nu}{}^{\dot{A}} \qquad (16.133)$$

where

$$\psi_{\mu\nu}{}^{\dot{A}} \equiv D_\mu \psi_\nu{}^{\dot{A}} - (\mu \leftrightarrow \nu)$$

and

$$D_\mu \psi_\nu{}^{\dot{A}} = \partial_\mu \psi_\nu{}^{\dot{A}} - \psi_\nu{}^{\dot{B}} w_{\mu\dot{B}}{}^{\dot{A}} \qquad (16.134)$$

Here we have used the results

$$\Omega_{\mu\nu}{}^{\dot{A}} = E_\nu{}^N w_{\mu N}{}^{\dot{A}} = \tfrac{1}{2}\psi_\nu{}^{\dot{B}} w_{\mu\dot{B}}{}^{\dot{A}} \qquad (16.135)$$

The torsion with all tangent indices is given in terms of $T_{\mu\nu}{}^{\dot{A}}$ by the relation

$$T_{\mu\nu}{}^{\dot{A}}(\theta = 0) = E_\mu{}^N(\theta = 0)E_\nu{}^M(\theta = 0)T_{NM}{}^{\dot{A}}(\theta = 0)(-1)^{NM}$$

$$= e_\mu{}^n e_\nu{}^m T_{nm}{}^{\dot{A}}(\theta = 0) \qquad (16.136)$$

where we have used the constraints $T_{Bn}{}^{\dot{A}} = T_{\dot{B}C}{}^{\dot{A}} = 0$. Consequently

$$0 = (\sigma^m)_{A\dot{B}}T_{mn}{}^{\dot{B}}(\theta = 0) = -\tfrac{1}{2}(\sigma^m)_{A\dot{B}}e_m{}^\mu e_n{}^\nu \psi_{\mu\nu}{}^{\dot{B}} \qquad (16.137)$$

and we recognize the Rarita-Schwinger equation on the right-hand side.

Actually to be strictly rigorous we must also show that $w_\mu{}^{mn}$ is the spin connection given in terms of $e_\mu{}^n$ and $\psi_\mu{}^A$. In fact, this follows from the constraint $T_{nm}{}^r = 0$. We note that

$$T_{\mu\nu}{}^r(\theta = 0) = -\partial_\mu e_\nu{}^r + w_{\mu\nu}{}^r - (\mu \leftrightarrow \nu) \qquad (16.138)$$

However,

$$T_{\mu\nu}{}^r(\theta = 0) = E_\mu{}^N(\theta = 0)E_\nu{}^M(\theta = 0)T_{NM}{}^r(\theta = 0)(-1)^{MN}$$

$$- \tfrac{1}{4}\psi_\mu{}^A\psi_\nu{}^B T_{AB}{}^r(\theta = 0) - \psi_\nu{}^B\psi_\nu{}^A T_{\dot{B}A}{}^r(\theta = 0)$$

$$= +\tfrac{1}{2}i\psi_\nu{}^{\dot{B}}(\sigma^r)_{A\dot{B}}\psi_\mu{}^A - (\mu \leftrightarrow \nu) \qquad (16.139)$$

Consequently we find that

$$w_{\mu n}{}^m e_\nu{}^n - \partial_\nu e_\mu{}^m - (\mu \leftrightarrow \nu) = +\frac{i}{2}\psi_\nu{}^{\dot{B}}(\sigma^m)_{A\dot{B}}\psi_\mu{}^A - (\mu \leftrightarrow \nu) \qquad (16.140)$$

which can be solved in the usual way to yield the correct expression for $w_{\mu n}{}^m$. The spin-2 equation is handled in the same way

$$R_{\mu\nu}{}^{mn}(\theta = 0) = \partial_\mu w_\nu{}^{mn} + w_\mu{}^{mr}w_{\nu r}{}^n - (\mu \leftrightarrow \nu) \qquad (16.141)$$

However

$$R_{\mu\nu}{}^{nm}(\theta = 0) = E_\mu{}^N(\theta = 0)E_\nu{}^M(\theta = 0)R_{NM}{}^{mn}(\theta = 0)(-1)^{mN}$$

$$= e_\mu{}^p e_\nu{}^q R_{pq}{}^{nm}(\theta = 0) + \tfrac{1}{2}(\psi_\mu{}^{\dot{A}}e_\nu{}^P R_{\dot{A}p}{}^{nm}$$

$$+ \psi_\mu{}^A e_\nu{}^P R_{Ap}{}^{nm}(\theta = 0) - (\mu \leftrightarrow \nu)) \qquad (16.142)$$

The object $R_{Ap}{}^{nm}$ can be found from the Bianchi identity $I_{Anr}{}^s$

$$0 = I_{Anr}{}^s = -D_A T_{nr}{}^s + T_{An}{}^F T_{Fr}{}^s + R_{Anr}{}^s - D_r T_{An}{}^s + T_{rA}{}^F T_{Fn}{}^s$$

$$+ R_{rAn}{}^s - D_n T_{rA}{}^s + T_{nr}{}^F T_{FA}{}^s + R_{nrA}{}^s \qquad (16.143)$$

Using the constraints we find that

$$R_{Anr}{}^s + R_{rAn}{}^s = +2iT_{nr}{}^{\dot{B}}(\sigma^s)_{A\dot{B}} \qquad (16.144)$$

From Eq. (16.137) we find that

$$R_{Anr}{}^s + R_{rAn}{}^s = -i(\sigma^s)_{A\dot{B}}e_n{}^\mu e_r{}^\nu \psi_{\mu\nu}{}^{\dot{B}} \qquad (16.145)$$

Contracting Eq. (16.142) with $e^\nu{}_m$ we find

$$e_m{}^\nu R_{\mu\nu}{}^{nm}(\theta = 0) \equiv R_\mu{}^n = e_\mu{}^p R_{pm}{}^{nm} + \tfrac{1}{2}(\psi_\mu{}^A R_{Am}{}^{nm} + \psi_\mu{}^{\dot{A}} R_{\dot{A}m}{}^{nm}) \qquad (16.146)$$

Equation (16.145) then gives

$$e_\mu{}^p R_{pm}{}^{nm} = R_\mu{}^n - \left(\frac{i}{2}\psi_\mu{}^A(\sigma^m)_{A\dot{B}}e_m{}^\lambda e^{n\tau}\psi_{\lambda\tau}{}^{\dot{B}} + \text{h.c.}\right) \qquad (16.147)$$

Equation (16.126) ($R_{mn} = 0$) then yields the spin-2 equation of $N = 1$ super-gravity, which is the left-hand side of the above equation.

At first sight it appears that the task is not finished; one should also analyze all the remaining Bianchi identities and show that they do not lead to any inconsistencies. However, it can be shown that the other Bianchi identities are now automatically satisfied.[71,82]

This technique can be used to find the on-shell theory of *any* supergravity theory. The only new feature is the appearance for $N \geqslant 4$ of scalar fields. Fortunately, these fields always appear in the coset space of a group G/H. That is they are contained in the objects $P_\mu{}^n$ and $Q_\mu{}^i$ where

$$e^{-\xi \cdot T} \partial_\mu e^{\xi \cdot T} = P_\mu{}^n K_n + Q_\mu{}^i H_i \qquad (16.148)$$

and H_i are the generators of H and K_i are the remaining generators. As $P_\mu{}^n$ and $Q_\mu{}^i$ have dimension one, one can still use dimensional arguments to find sufficient constraints.

The most recent application of this technique was to find the $N = 2, D = 10$ chiral supergravity theory.[82] Other literature on this technique may be found Ref. 83.

B. Off-Shell

The off-shell superspace theory of supergravity can be deduced from the supergravity theory in x-space.[48] Of course the superspace theory one obtains depends on the x-space theory one starts with. We will illustrate this for the old minimal theory of Chapter 9 which has the field content $e_\mu{}^n$, $\psi_\mu{}^\alpha$, M, N and b_μ. Their supersymmetry transformations are listed in Chapter 9.

The $\theta = 0$ components of $E_\mu{}^n$ and $\psi_\mu{}^\alpha$ may be unambiguously identified with the x-space fields $e_\mu{}^n$ and $\psi_\mu{}^\alpha$.

$$E_\mu{}^n(\theta = 0) = e_\mu{}^n \qquad (16.149)$$

$$E_\mu{}^\alpha(\theta = 0) = \tfrac{1}{2}\psi_\mu{}^\alpha \qquad (16.150)$$

For $\Omega_\mu{}^{rs}$ at $\theta = 0$ we identity the spin connection

$$\Omega_\mu{}^{rs}(\theta = 0) = w_\mu{}^{rs}(x)$$

These are the only fields of the correct dimension and with the correct transformation properties, as we shall see. The $\theta = 0$ components of the superparameter ξ^π can be identified with the general coordinate $\zeta^\mu(x)$ and supersymmetry parameters $\varepsilon^\alpha(x)$.

$$\xi^\mu(\theta = 0) = \zeta^\mu(x) \qquad \xi^\alpha(\theta = 0) = \varepsilon^\alpha(x) \qquad (16.151)$$

We now compare the transformations of Eq. (16.149). The left-hand result is evaluated by taking $\theta = 0$ in the expression

$$\delta E_\pi{}^M = \mathbf{D}_\pi \varepsilon^m - E_\pi{}^s \zeta^N T_{NR}{}^M \tag{16.152}$$

where

$$\mathbf{D}_\pi \varepsilon^m = \partial_\pi \varepsilon^M + \Omega_\pi{}^M{}_N \varepsilon^N$$

For the vierbien we find that

$$\delta E_\mu{}^n(\theta = 0) = \delta e_\mu{}^n = \mathbf{D}_\mu \zeta^n - e_\mu{}^r \zeta^m T_{mr}{}^n - e_\mu{}^r \varepsilon^\alpha T_{\alpha r}{}^n$$
$$- \tfrac{1}{2} \psi_\mu{}^\alpha \zeta^r T_{r\alpha}{}^n - \tfrac{1}{2} \psi_\mu{}^\alpha \varepsilon^\beta T_{\beta\alpha}{}^n \tag{16.153}$$

We compare this result with the known transformation of $e_\mu{}^u$ under supersymmetry:

$$\delta e_\mu{}^n = \kappa \bar{\varepsilon} \gamma^n \psi_\mu \tag{16.154}$$

We then conclude that

$$T_{\alpha r}{}^n = 0 \qquad T_{\alpha\beta}{}^n = 2(C\gamma^n)_{\alpha\beta} \tag{16.155}$$

Using these two constraints and the x-space expression for $w_\mu{}^{mn}(x)$ in terms of $e_\mu{}^n$ and $\psi_\mu{}^\alpha$ we find that $T_{mn}{}^r = 0$. This is achieved by examining Eqs. (16.138)–(16.140) above.

For the Rarita-Schwinger field

$$\tfrac{1}{2} \delta \psi_\mu{}^\alpha = \delta E_\mu{}^\alpha(\theta = 0)$$
$$= \mathbf{D}_\mu \varepsilon^\alpha - e_\mu{}^n \zeta^m T_{mn}{}^\alpha - e_\mu{}^n \varepsilon^\beta T_{\beta n}{}^\alpha - \tfrac{1}{2} \psi_\mu{}^\gamma \zeta^n T_{n\gamma}{}^\alpha - \tfrac{1}{2} \psi_\mu{}^\gamma \varepsilon^\beta T_{\beta\gamma}{}^\alpha \tag{16.156}$$

Comparing this with the known x-space result for the supersymmetry variation of $\psi_\mu{}^\alpha$:

$$\delta \psi_\mu = \frac{2}{\kappa} \mathbf{D}_\mu \varepsilon + (i\gamma_5 b_\mu + \gamma_\mu \eta)\varepsilon \tag{16.157}$$

where

$$\eta = -\tfrac{1}{3}(M + i\gamma_5 N + \not{b} i\gamma_5)$$

we find

$$T_{\alpha\beta}{}^\gamma = 0$$

$$T_{n\beta}{}^\alpha(\theta = 0) = \frac{\kappa}{2}[(i\gamma_5)b_n + \gamma_\mu \eta]^\alpha{}_\beta \tag{16.158}$$

Hence we have found the constraints

$$T_{mn}{}^r = T_{\alpha\beta}{}^\gamma = 0 = T_{\alpha n}{}^s = T_{\alpha\beta}{}^n - 2(c\gamma^n)_{\alpha\beta} \tag{16.159}$$

The reader will recognize these as the constraints of $N = 1$ supergravity found before.

Given the x-space transformation of $w_\mu{}^{mn}$ we can apply exactly the same procedure to find the constraints on $R_{MN}{}^{mn}$. Use is made of Eq. (16.132) and we leave this as an exercise for the reader. In fact, the components of $R_{MN}{}^{mn}$ are from Eq. (16.159) already determined the Bianchi identity.

Given the constraints we could reverse the above procedure and work from superspace to find the component field transformation laws in x-space.

Chapter 17

$N = 1$ Super-Feynman Rules

Perhaps the major use of the superfield description of supersymmetric theories is in the calculation of quantum effects. Superfields allow one to calculate the quantum behavior of supersymmetric theories, while keeping the supersymmetry manifest. The advantage this presents is analogous to calculating with usual theories in a Lorentz covariant formulation. In fact, super-Feynman rule calculations can, in some cases, be enormously easy to carry out; even easier than the corresponding x-space theory with only one of the x-space component fields.

Super-Feynman rules were first constructed for the Wess-Zumino model by Salam and Strathdee[84] and for $N = 1$ Yang-Mills in Ref. 85. In fact, it was using component fields that it was first shown that the Wess-Zumino model required only one wave function renormalization[86-88] which was in common to all the component fields and that in this theory the vacuum graphs vanished[212]. These results were later shown[89] using the original super-Feynman rules of Salam and Strathdee. These older super-Feynman rules were also used to show the vanishing of the quantum effective potential if supersymmetry was not broken at the tree level.[90] All these results[86-90,212] constitute what has became known as the non-renormalization theorems. However, the super-Feynman rules were reformulated.[91] Although the improved super-Feynman rules make the contact with the x-space component fields less transparent, the new rules are quite a bit faster in the evaluation of supergraphs. All the old results have been rederived by shorter proofs[91] and some new results have been found. As such, in this chapter we will discuss only the improved super-Feynman rules.

Before deriving these rules, however, we will recall from Chapter 14 same properties of integration and functional differentiation in superspace. Superspace integration,[52] as presently defined, is completely equivalent to differentiation. The full superspace integral is

$$\int d^8z = \int d^4x\, d^4\theta = \int d^4x(-\tfrac{1}{4}D^2)(-\tfrac{1}{4}\bar{D}^2) \tag{17.1}$$

where $D^2 = D^A D_A$ and $\bar{D}^2 = D^{\dot{A}}D_{\dot{A}}$. The chiral subintegral is given by

$$\int d^4x \, d^2\theta = \int d^4x (-\tfrac{1}{4}D^2) \tag{17.2}$$

The general superspace δ-function has the property

$$\int d^8z' f(z') \delta^8(z - z') = f(z) \tag{17.3}$$

and can be written as

$$\delta^8(z - z') = \delta^4(x - x') \delta^4(\theta - \theta') \tag{17.4}$$

The $\delta^4(\theta)$ is of the form $\sim \theta^2 \bar{\theta}^2$. For a general superfield V we define

$$\frac{\delta V(x, \theta)}{\delta V(x', \theta')} \equiv \delta^8(z - z') = \delta^4(x - x') \delta^4(\theta - \theta') \tag{17.5}$$

while for a chiral superfield we have

$$\frac{\delta \phi(x, \theta)}{\delta \phi(x', \theta')} = -\frac{\bar{D}^2}{4} \delta^4(x - x') \delta^4(\theta - \theta') = -\frac{\bar{D}^2}{4} \delta^8(z - z') \tag{17.6}$$

This preserves the chiral conditon $\bar{D}_{\dot{A}} \phi = 0$. We note the following formulae for covariant derivatives:

$$D^2 \bar{D}^2 D^2 = 16 \partial^2 D^2, \qquad \bar{D}^2 D^2 \bar{D}^2 = 16 \partial^2 \bar{D}^2, \tag{17.7}$$

and

$$\bar{D}^2 D^2 \phi = 16 \partial^2 \phi \tag{17.8}$$

for a chiral superfield ϕ. We can write a chiral integral as a full superspace integral as follows:

$$\int d^2\theta \, d^4x \, \phi \cdot j = \int d^2\theta \, d^4x \, \phi \left(\frac{\bar{D}^2 D^2 j}{16 \partial^2} \right)$$

$$= \int d^2\theta \left(-\frac{1}{4} \bar{D}^2 \right) d^4x \, \phi \left(\frac{-D^2 j}{4 \partial^2} \right)$$

$$= -\int d^4\theta \, d^4x \, \phi \frac{\bar{D}^2 j}{4 \partial^2}, \tag{17.9}$$

where ϕ and j are chiral superfields.

The D's and δ functions obey $\delta_{12} D_1^2 \bar{D}_1^2 \delta_{12} = 16 \delta_{12}$ under a $\int d^4x$, but a term with fewer than four D's in between the δ's will vanish.

17.1   General Formalism

The construction of super-Feynman rules follows very closely the derivation of the usual Feynman rules from a path integral. The general superfield action is of the form

$$S = \int d^4x \, d^4\theta \, \mathcal{L}(V, \phi, \bar{\phi}) + \left\{ \int d^4x \, d^2\theta \, W(\phi) + \text{h.c.} \right\} \qquad (17.10)$$

where V is a general superfield and ϕ is a chiral multiplet. For example V could carry the x-space component fields of $N = 1$ Yang-Mills while ϕ could carry those of the Wess-Zumino model.

We define the vacuum-to-vacuum amplitude in the presence of the external sources J for V and j for ϕ (j is a chiral superfield with the same chirality as ϕ):

$$Z[J,j] = N \int [DV D\phi D\bar{\phi}] \exp\left\{\frac{i}{\hbar}(S + V \cdot J + \phi \cdot j)\right\} \qquad (17.11)$$

where

$$V \cdot J = \int d^4x \, d^4\theta \, VJ$$

$$\phi \cdot j = \int d^4x \, d^2\theta \, \phi j + \text{h.c.} \qquad (17.12)$$

$$N = [Z(J = 0, \quad j = 0)]^{-1}$$

We then have the usual result

$$\frac{\delta^m Z[J,j]}{\delta J(1) \ldots \delta J(n) \ldots \delta j(m)}\bigg|_{\substack{J=0 \\ j=0}} = \left(\frac{i}{\hbar}\right)^m G'_m(1, \ldots, n, n+1, \ldots m) \qquad (17.13)$$

where

$$G'_m(1, \ldots n; n+1, \ldots m) = \langle 0|TV(1) \ldots V(n)\phi(n+1) \ldots \phi(m)|0\rangle \qquad (17.14)$$

are the Green's functions.

We then define

$$\frac{i}{\hbar} W[J,j] = \ln Z[J,j] \qquad (17.15)$$

and $W[J,j]$ is the generating functional for the connected Green's functions $G_m(1, \ldots n; n+1, \ldots m)$,

$$\frac{i}{\hbar} \frac{\delta^m W[J,j]}{\delta J(1) \ldots \delta J(n) \delta j(n+1) \ldots \delta j(m)} = \left(\frac{i}{\hbar}\right)^m G(1, \ldots, n; n+1, \ldots, m) \qquad (17.16)$$

The fields Φ and v are defined by

$$\Phi = \frac{\delta W}{\delta j} \qquad \bar{\Phi} = \frac{\delta W}{\delta \bar{j}} \qquad v = \frac{\delta W}{\delta J} \qquad (17.17)$$

and these relations can be inverted to give j and J in terms of Φ, $\bar{\Phi}$ and σ. The effective action $\Gamma(\Phi, \bar{\Phi}, v)$, the generator of one-particle irreducible graphs, $\Gamma^m(1, \ldots, m)$, is given by

$$\Gamma(\Phi, \bar{\Phi}, v) = W[j[\Phi, v], J[\phi, v]] - (j \cdot \phi + J \cdot v) \tag{17.18}$$

In the above expressions the $W[J, j]$ are understood to depend on J, $\bar{j}$ and j, although the latter dependence of $\bar{j}$ is often suppressed. As is common in the literature we will often denote v by V when discussing the effective action.

We divide the action up into a free part and an interacting part

$$S = S^0 + S^{\text{int}} \tag{17.19}$$

Then we may write, setting $\hbar = 1$ for convenience,

$$Z[J, j] = \exp\left\{iS^{\text{int}}\left(\frac{1}{i}\frac{\delta}{\delta j}, \frac{1}{i}\frac{\delta}{\delta J}\right) Z_0[j, J]\right\} \tag{17.20}$$

where

$$Z_0[j, J] = N_0 \int D\phi D\bar{\phi} DV \exp\left\{i(S_0 + J \cdot V + j \cdot \phi)\right\} \tag{17.21}$$

By explicitly performing the integration in $Z_0[j, J]$ we obtain the propagators of the theory. These integrals are of the form

$$I[y] = \int \exp i(\tfrac{1}{2}x^T A x + x^T y) \, dx \, d\bar{x} \tag{17.22}$$

These are Gaussian and may be evaluated by a shift of variable to give

$$I[y] = \text{constant} \exp\left(-\frac{i}{2}y^T A^{-1} y\right) \tag{17.23}$$

In terms of the two-point free connected Green's function we find

$$Z_0[j, J] = \exp\left(-\frac{i}{2}\int J(x_1)G(x_1, x_2)J(x_2)\,dx_1\,dx_2\right.$$

$$\left. + \text{ terms involving } j \text{ and } \bar{j}\right) \tag{17.24}$$

17.2 The Wess-Zumino Multiplet

We will now derive the super-Feynman rules for the Wess-Zumino multiplet beginning with the superpropagator. This proceeds much like the case of a complex scalar field in the usual Minkowski space. The major complication is induced by the chiral nature of ϕ which means it is not a general function on superspace, but in effect depends on a subspace of superspace. To obtain the super-Feynman rules in their most useful form we swap the chiral sub-superspace integrals which occur in the potential for full superspace integrals as was done in Eq. (17.9).

For the contribution of the Wess-Zumino multiplet ϕ to Z_0 we have

$$Z_0[j] = \int [D\phi D\bar{\phi}] \exp i \left\{ \int d^8z\, \bar{\phi}\phi + \int d^4x\, d^2\theta(j\phi + m\phi^2 + \text{h.c.}) \right\} \quad (17.25)$$

Using the fact that

$$\bar{D}^2 D^2 \phi = +16\partial^2 \phi \quad (17.26)$$

we can rewrite the integral as a full integral over superspace,

$$Z_0[j] = \int D\phi D\bar{\phi} \exp i \int d^8z \left(\bar{\phi}\phi - \frac{1}{2}\frac{m}{4}\phi\frac{D^2}{\partial^2}\phi - \frac{1}{2}\frac{m}{4}\bar{\phi}\frac{\bar{D}^2}{\partial^2}\bar{\phi} \right.$$

$$\left. - \frac{\phi}{4}\frac{D^2}{\partial^2}j - \frac{\bar{\phi}}{4}\frac{\bar{D}^2}{\partial^2}\bar{j} \right)$$

$$= \int D\phi D\bar{\phi} \exp i \int d^8z \left\{ \frac{1}{2}(\phi, \bar{\phi}) \begin{bmatrix} -\dfrac{m}{4}\dfrac{D^2}{\partial^2} & 1 \\ 1 & -\dfrac{m}{4}\dfrac{\bar{D}^2}{\partial^2} \end{bmatrix} \begin{bmatrix} \phi \\ \bar{\phi} \end{bmatrix} \right.$$

$$\left. - (\phi, \bar{\phi}) \begin{bmatrix} \dfrac{D^2}{4\partial^2}j \\ \dfrac{\bar{D}^2}{4\partial^2}\bar{j} \end{bmatrix} \right\} \quad (17.27)$$

This is an integral of the form of Eq. (17.22) and so equals

$$\exp \left\{ -i \int d^8z \frac{1}{2}\left(\frac{D^2}{4\partial^2}j, \frac{\bar{D}^2}{4\partial^2}\bar{j} \right) A^{-1} \begin{bmatrix} \dfrac{D^2}{4\partial^2}\bar{j} \\ \dfrac{\bar{D}^2}{4\partial^2}\bar{j} \end{bmatrix} \right\} \quad (17.28)$$

where A^{-1} is the inverse of the matrix

$$A = \begin{bmatrix} -\dfrac{m}{4}\dfrac{D^2}{\partial^2} & 1 \\ 1 & -\dfrac{m}{4}\dfrac{\bar{D}^2}{\partial^2} \end{bmatrix} \quad (17.29)$$

Using the relations

$$D^2\bar{D}^2 D^2 = 16\partial^2 D^2 \quad (17.30)$$

and

$$\bar{D}^2 D^2 \bar{D}^2 = 16\partial^2 \bar{D}^2 \quad (17.31)$$

one may check that

$$
A^{-1} = \begin{bmatrix} \dfrac{m\bar{D}^2}{4(\partial^2 - m^2)} & 1 + \dfrac{m^2\bar{D}^2 D^2}{16\partial^2(\partial^2 - m^2)} \\ 1 + \dfrac{m^2 D^2 \bar{D}^2}{16\partial^2(\partial^2 - m^2)} & \dfrac{mD^2}{4(\partial^2 - m^2)} \end{bmatrix} \tag{17.32}
$$

The vacuum-to-vacuum amplitude for the free theory then takes the form

$$
Z_0[j] = \exp\left\{ -i \int d^8 z \left[j \frac{1}{\partial^2 - m^2} \bar{j} \right. \right.
$$

$$
\left. \left. + \frac{1}{2}\left(\bar{j} \frac{m\bar{D}^2}{4\partial^2(\partial^2 - m^2)} \bar{j} + j \frac{mD^2}{4\partial^2(\partial^2 - m^2)} j \right) \right] \right\} \tag{17.33}
$$

For this free theory the connected Green's functions are generated by

$$
W_0[j] = -\int d^8 z \left[j \frac{1}{\partial^2 - m^2} \bar{j} + \frac{1}{2}\bar{j} \frac{m\bar{D}^2}{4\partial^2(\partial^2 - m^2)}\bar{j} + \frac{1}{2}j \frac{mD^2}{4\partial^2(\partial^2 - m^2)}j \right] \tag{17.34}
$$

Therefore the only non-zero connected Green functions are the two-point functions, for which we find, upon using Eq. (17.16)

$$
\langle \phi(1)\bar{\phi}(2) \rangle = \langle 0|T\phi(1)\bar{\phi}(2)|0 \rangle
$$

$$
= +\frac{i}{16} \frac{\bar{D}_1^2 D_1^2 \delta_{12}}{(\partial^2 - m^2)} \tag{17.35}
$$

$$
\langle \phi(1)\phi(2) \rangle = \langle 0|T\phi(1)\phi(2)|0 \rangle
$$

$$
= +\frac{i}{4} \frac{m\bar{D}_1^2 \delta_{12}}{(\partial^2 - m^2)} \tag{17.36}
$$

Taking the $\theta_1 = \theta_2 = 0$ component of equation (17.35) we find the usual x-space result

$$
\langle 0|Tz(1)\bar{z}(2)|0 \rangle = \frac{i}{\partial^2 - m^2} \delta^4(x_1 - x_2) \tag{17.37}
$$

The higher component results can be extracted by applying covariant derivatives. Applying D_{A1}, $D_{\dot{B}2}$ and taking $\theta_1 = \theta_2 = 0$ yields

$$
\langle 0|T\chi_A(1)\chi_{\dot{B}}(2)|0 \rangle = +\frac{i}{16} D_{A1} \frac{\bar{D}_1^2 D_{\dot{B}2} D_2^2 \delta_{12}}{(\partial^2 - m^2)}
$$

$$
= \frac{2(\partial_1)_{A\dot{B}} \delta^4(x_1 - x_2)}{\partial^2 - m^2} \tag{17.38}
$$

The last propagator is that for the auxiliary fields; we find that

$$\langle 0|Tf(1)\bar{f}(2)|0\rangle = +\frac{1}{4}\cdot\frac{i}{16}\cdot\frac{D_1^2\bar{D}_1^2\bar{D}_2^2 D_2^2\delta_{12}}{(\partial^2 - m^2)}$$

$$= \frac{4i\partial^2}{(\partial^2 - m^2)}\delta^4(x_1 - x_2) \tag{17.39}$$

We recall that $f = -\frac{1}{2}(D^2\varphi)(\theta = 0)$.

The interacting Wess-Zumino theory is given by Eq. (17.10) when

$$W(\phi) = +\frac{\lambda\phi^3}{3!} \tag{17.40}$$

We must therefore evaluate terms like

$$S_{\text{int}}\left(\frac{1}{i}\frac{\delta}{\delta j}\right)[ij(1)\cdot ij(2)\cdot ij(3)]$$

$$= +\frac{i\lambda}{3!}\int d^4x_4\, d^2\theta_4\,\frac{\delta^3}{i^3\delta^3 j(4)}[ij(1)\cdot ij(2)\cdot ij(3)]$$

$$= +i\lambda\int d^4x_4\, d^2\theta_4\left\{-\frac{\bar{D}_4^2}{4}\delta_{14}\right\}\left\{-\frac{\bar{D}_2^2}{4}\delta_{24}\right\}\left\{-\frac{\bar{D}_3^2}{4}\delta_{34}\right\}$$

$$= +i\lambda\int d^8z_4\,\delta_{14}\left\{-\frac{\bar{D}_2^2}{4}\delta_{24}\right\}\left\{-\frac{\bar{D}_3^2}{4}\delta_{34}\right\} \tag{17.41}$$

Examining Eq. (17.41) we find that one has an option as to how one may formulate the super-Feynman rules. We could associate an integral $d^2\theta$ with each vertex as in the second to the last line and consider the propagators to be given by Eqs. (17.35) and (17.36). The factors of D^2 come from the functional differentiation. However, it is preferable to use the last line of Eq. (17.41) and to associate a $d^4\theta$ with each vertex and hence not to include a factor D^2 with one of the three propagators leaving the vertex. We will adopt this latter approach.

We move to momentum space by taking the Fourier transform

$$\phi(x) = \int\frac{d^4k}{(2\pi)^4}e^{-ik\cdot x}\phi(k) \tag{17.42}$$

and adopting the convention that $\phi(+k)$ corresponds to a momentum $+k$ leaving a vertex. As a result, the action of $i\partial_\mu$ on $\phi(x)$ becomes $k_\mu\phi(k)$, and so the action of the covariant derivative $D_A\phi(x)$ becomes in momentum space

$$D_A\phi(k) = \left\{\frac{\partial}{\partial\theta^A} - (\sigma^\mu)_{A\dot{A}}\bar{\theta}^{\dot{A}}k_\mu\right\}\phi(k) \tag{17.43}$$

while $D_{\dot{A}}\phi(x)$ becomes

$$D_{\dot{A}}\phi(k) = \left\{\frac{\partial}{\partial\theta^{\dot{A}}} - (\sigma^{\mu})_{A\dot{A}}\theta^{A}k_{\mu}\right\}\phi(k) \tag{17.44}$$

The anticommutator of two covariant derivatives then becomes

$$\{D_{A}, D_{\dot{A}}\}\phi(k) = -2(\sigma^{\mu})_{A\dot{A}}k_{\mu}\phi(k) \tag{17.45}$$

Using the fact that

$$\delta^{8}(z_{1} - z_{2}) = \int\frac{d^{4}k}{(2\pi)^{4}}e^{-ik\cdot(x_{1}-x_{2})}\delta^{4}(\theta_{1} - \theta_{2}) \tag{17.46}$$

the action of D_{A} on $\delta^{8}(z_{1} - z_{2})$ becomes

$$D_{1A}\delta^{4}(\theta_{1} - \theta_{2}) = \left(\frac{\partial}{\partial\theta_{1}^{A}} - (\sigma^{\mu})_{A\dot{A}}\theta_{1}^{\dot{A}}k_{\mu}\right)\delta^{4}(\theta_{1} - \theta_{2})$$

$$= -D_{2A}\delta^{4}(\theta_{1} - \theta_{2}) \tag{17.47}$$

the $\delta^{4}(x_{1} - x_{2})$ factor being understood. In the last line, we have used the fact that in momentum space

$$D_{2A} = \frac{\partial}{\partial\theta_{2}^{A}} - (\sigma^{\mu})_{A\dot{A}}(-k_{\mu})\theta_{2}^{\dot{A}} \tag{17.48}$$

When acting on δ_{12} this result agrees with our convention that the momentum k flows out of the point labeled 1, but into the point labeled 2.

The super-Feynman rules in momentum space are as follows:

(i) The propagators are found by putting the previous equations in momentum space. They are

$$\langle\phi(1)\bar{\phi}(2)\rangle = \frac{-i}{p^{2} + m^{2}}\delta_{12} \tag{17.49}$$

$$\langle\phi(1)\phi(2)\rangle = \frac{imD^{2}}{4p^{2}(p^{2} + m^{2})}\delta_{12} \tag{17.50}$$

(ii) The vertices have a factor $i\lambda$ (or an equivalent factor read off from S) as well as a factor of $-\frac{1}{4}\bar{D}^{2}$ $(-\frac{1}{4}D^{2})$ for two $(n-1)$ of the three (n) chiral (antichiral) lines. If one, or more of the legs of the vertex is an external line then, we associate one, or more less factors of $-\frac{1}{4}\bar{D}^{2}$ $(-\frac{1}{4}D^{2})$ for each external line.

(iii) We associate a factor

(a) $\int\frac{d^{4}k}{(2\pi)^{4}}$ with each independent loop

(b) $\prod_{P_{ext}}\int\frac{d^{4}p_{ext}}{(2\pi)^{4}}(2\pi)^{4}\delta\left(\sum_{ext}p_{ext}\right)$ for the external momenta

and

(c) $\int d^4\theta_{\text{vert}}$ with each vertex.

(iv) Finally, one has the usual combinatoric factors.

The origin of these rules is apparent from the above discussion. The propagators come from the free propagators with the absence of the $\bar{D}^2(D^2)$ factors. The $\bar{D}^2(D^2)$ factors in rule (ii) result from the chiral functional derivative. The removal of one of them is required to convert the vertex θ-space integral to a full superspace integral; see Eq. (6.15). External lines are not differentiated and so we remove the extra $\bar{D}^2(D^2)$ factor.

The above rules yield $Z[j]$ from which the connected Green's functions may be calculated.

We can convert the above rules to Euclidean space. This is equivalent to doing the Wick rotation which is specified by the replacements $m^2 \rightarrow m^2 - i\varepsilon$ in the propagators. We define $x_E = (x_4, \mathbf{x})$ where $x_4 = ix_0$. This implies that

$$d^4x = -id^4x_E \qquad x_E^2 = x_0^2 + \mathbf{x}^2 \qquad (17.51)$$

As usual, we then define $k_E = (k_4, \mathbf{k})$ where $k_4 = +ik_0$ so

$$d^4k = id^4k_E \qquad k_E^2 = k_0^2 + \mathbf{k}^2 \qquad (17.52)$$

Often the super-Feynman rules are given after the change to Euclidean space. Making the above changes we find the Euclidean rules to be given by:

(i) The propagators are

$$\langle \phi(1)\bar{\phi}(2) \rangle = \frac{\delta_{12}}{p^2 + m^2}$$

$$\langle \phi(1)\phi(2) \rangle = \frac{mD^2}{4p^2(p^2 + m^2)}\delta_{12} \qquad (17.53)$$

(ii) The vertices have a factor λ with a factor $-\frac{1}{4}\bar{D}^2$ $(-\frac{1}{4}D^2)$ for all but one of the chiral (antichiral) lines unless if they are external legs. Then there is no factor of $-\frac{1}{4}\bar{D}^2(-\frac{1}{4}D^2)$ on any external line.

(iii) A factor

(a) $\int \frac{d^4k}{(2\pi)^4}$ for each loop

(b) $\prod_{P_{\text{ext}}} \int \frac{d^4 P_{\text{ext}}}{(2\pi)^4} (2\pi)^4 \delta \left(\sum_{\text{ext}} P_{\text{ext}} \right)$ for the external momenta

(c) $\int d^4\theta_{\text{vert}}$ for each vertex.

(iv) The usual combinatoric factors.

In the above, use has been made of the relation $L = I - V + 1$ to eliminate factors of i from the propagators, vertices and loop integrations and final δ-function. The above rules have omitted a factor of $-i$ from each propagator, $+i$ from each vertex, $+i$ from each loop integration and $-i$ from the δ-function. This gives an overall missing factor of $(i)^{(-I+V+L-1)}$ which equals 1.

The effective action can be found after applying the above rules to one-particle irreducible graphs, external momenta with appropriate external fields, and introducing appropriate symmetry factors.

17.3 Super Yang-Mills Theory

Let us now treat the super Yang-Mills multiplet contained in the superfield V. The action is given by

$$A^{YM} = \frac{1}{64g^2} \text{Tr} \int d^4x \, d^2\theta \, W^A W_A \tag{17.54}$$

where

$$W_A = \bar{D}^2(e^{-gV} D_A e^{+gV}) \tag{17.55}$$

The trace is understood to contain the factor $1/C_2(G)$. Although it is usual to add the hermitian conjugate to chiral integrals, this term is real by itself as a result of the constraints on W_A. It can be rewritten as

$$A^{YM} = -\frac{\text{Tr}}{16g^2} \int d^8z \, W^A(e^{-gV} D_A e^{gV}) \tag{17.56}$$

$$A^{YM} = \frac{\text{Tr}}{16} \int d^8z \left\{ V D^A \bar{D}^2 D_A V + \frac{1}{16} gV\{D^A V, \bar{D}^2 D_A V\} \right.$$
$$\left. - \frac{g^2}{16}\left(\frac{1}{4}[V, D^A V]\bar{D}^2[V, D_A V] + \frac{1}{3}(D^A V)\bar{D}^2[V,[V, D_A V]] \right) + O(V^5) \right\} \tag{17.57}$$

To perturbatively quantize the theory we must add a gauge-fixing term which is taken to be (we will discuss the ghosts shortly)

$$-\frac{1}{\alpha}\text{Tr} \int d^8z \frac{1}{2} V \Pi_0 \partial^2 V = -\frac{1}{16\alpha}\text{Tr} \int d^8z \, D^2 V \bar{D}^2 V \tag{17.58}$$

where Π_0 is one of the projectors $\Pi_i = (\Pi_{0+}, \Pi_{1/2}, \Pi_{0-})$, $i = 1, 2, 3$, given as

$$\Pi_{0+} \equiv \frac{\bar{D}^2 D^2}{16\partial^2} \qquad \Pi_{0-} \equiv \frac{D^2 \bar{D}^2}{16\partial^2} \tag{17.59}$$

$$\Pi_0 = \Pi_{0+} + \Pi_{0-} \tag{17.60}$$

and

$$\Pi_{1/2} = -\frac{D^A \bar{D}^2 D_A}{8\partial^2} = -\frac{D^{\dot{A}} D^2 D_{\dot{A}}}{8\partial^2} \qquad (17.61)$$

As their name implies they satisfy

$$\sum_{i=1}^{3} \Pi_i = 1 \quad \text{and} \quad \Pi_i \Pi_j = \delta_{ij} \Pi_j \qquad (17.62)$$

The gauge-fixing term when expressed in component fields contains the term $(\partial_\mu A^\mu)^2$ as well as its superpartners.

With this gauge fixing term the part of the action quadratic in V becomes

$$A_{(2)}^{YM} = \text{Tr} \int d^8 z \left[\frac{1}{2} V \left\{ -\Pi_{1/2} - \frac{\Pi_0}{\alpha} \right\} \partial^2 V \right] \qquad (17.63)$$

In the super-Fermi-Feynman gauge, $\alpha = +1$, this takes the particularly simple form

$$A_{(2)}^{YM} = -\text{Tr} \int d^8 z \frac{1}{2} V \partial^2 V \qquad (17.64)$$

The vacuum-to-vacuum amplitude in the presence of the source then has the form

$$Z^0[J] = N_0 \int [dV] \exp i\{A_{(2)}^{YM} + J \cdot V\} \qquad (17.65)$$

Performing the V integration we find that

$$Z^0[J] = \exp\left\{ +\frac{i}{2} \int d^8 z J [\Pi_{1/2} + \alpha \Pi_0] \frac{1}{\partial^2} J \right\} \qquad (17.66)$$

We find that the two-point connected Green's function is

$$G(1, 2) = -(\Pi_{1/2} + \alpha \Pi_0) \frac{i}{\partial^2} \delta^8(z_1 - z_2) \qquad (17.67)$$

which in momentum space becomes

$$G(p) = (+\Pi_{1/2} + \alpha \Pi_0) \frac{i}{p^2} \delta_{12} \qquad (17.68)$$

Performing the same transition, as before, to Euclidean space, the propagator is given by

$$G(p) = -(\Pi_{1/2} + \alpha \Pi_0) \frac{1}{p^2} \delta_{12} \qquad (17.69)$$

which in the super Fermi-Feynman gauge has the simple form

$$G(p) = -\frac{1}{p^2} \delta_{12} \qquad (17.70)$$

The vertices can easily be read off from the action.

Finally, we must restore unitarity by adding ghosts. This is a straightforward application of the usual methods.[92] Corresponding to the gauge transformations which have a chiral parameter we choose the gauge-fixing terms

$$D^2 V - \bar{f} = 0 \qquad \bar{D}^2 V - f = 0 \qquad (17.71)$$

The Faddeev-Poppov determinant, Δ is defined by

$$\Delta \int [\mathcal{D}\Lambda \mathcal{D}\bar{\Lambda}] \delta(D^2 V^\lambda - \bar{f}) \delta(\bar{D}^2 V^\lambda - f) = 1 \qquad (17.72)$$

where V^λ is the transformed V with parameters $(\Lambda, \bar{\Lambda})$, i.e.,

$$V^\lambda = V + \frac{V}{2} \wedge \{-(\Lambda + \bar{\Lambda}) + (\coth V/2) \wedge (\bar{\Lambda} - \Lambda) + \cdots\}$$

$$= V + H(V)\Lambda + \bar{H}(V)\bar{\Lambda} + \cdots \qquad (17.73)$$

The vacuum-to-vacuum amplitude is given by

$$Z[j] = \int \mathcal{D}V \exp(iA^{\text{YM}} + j \cdot V) \cdot \Delta \delta(D^2 V - \bar{f}) \delta(\bar{D}^2 V - f) \qquad (17.74)$$

As Δ multiplies the δ functions we only require its value Δ' when $D^2 V = \bar{f}$, $\bar{D}^2 V = f$. The only contribution to Δ' arising from the integral over Λ, $\bar{\Lambda}$ in Eq. (17.74) is when Λ, $\bar{\Lambda}$ are very close to zero. Consequently, we may use the above equation for V^λ and neglect the order λ^2 terms, to yield the result

$$\Delta' \int [\mathcal{D}\Lambda \mathcal{D}\bar{\Lambda}] \delta(D^2(H(V)\Lambda + \bar{H}(V)\bar{\Lambda})) \cdot \delta(\bar{D}^2(H(V)\Lambda + \bar{H}(V)\bar{\Lambda})) = 1$$

$$(17.75)$$

Using the properties of the δ function we find that Δ' may be represented by

$$\Delta' = \int [\mathcal{D}c \mathcal{D}c' \mathcal{D}\bar{c} \mathcal{D}\bar{c}'] \exp i \{d^4 x \, d^2 \theta c' \bar{D}^2 (H(V)c + \bar{H}(V)\bar{c})$$

$$+ \int d^4 x \, d^2 \bar{\theta} \bar{c}' D^2 (H(V)c + \bar{H}(V)\bar{c}) + \text{h.c.}\}$$

$$\equiv \int [\mathcal{D}c \mathcal{D}c' \mathcal{D}\bar{c} \mathcal{D}\bar{c}'] \exp(iA_{\text{FP}})$$

$$= \int [\mathcal{D}c \mathcal{D}c' \mathcal{D}\bar{c} \mathcal{D}\bar{c}'] \exp i \int d^8 z \left\{ (c' + \bar{c}') \cdot \left\{ \left(-\frac{V}{2} \right) \wedge (c + \bar{c}) \right.\right.$$

$$\left.\left. + \left(\coth \frac{V}{2} \right) \wedge (c - \bar{c}) \right\} \right\} \qquad (17.76)$$

The ghosts c and c' are chiral ($D_{\dot{A}} c = 0 = D_{\dot{A}} c'$) while $\bar{c}$ and $\bar{c}'$ are antichiral.

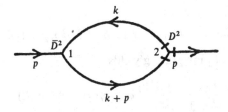

Fig. 17.1.

Finally we may average over the gauge-fixing condition by integrating over

$$\int [\mathscr{D}f \mathscr{D}\bar{f}] \exp\left(-\frac{i}{16}\left(\int \bar{f}f d^8 z\right)\right) \tag{17.77}$$

As this is field independent no Nielson-Kallash ghost is required.

The final answer is

$$Z[j] = \int [\mathscr{D}V \mathscr{D}c \mathscr{D}\bar{c} \mathscr{D}c' \mathscr{D}\bar{c}'] \exp\left\{iA^{\text{YM}} + iA_{\text{FP}} + ij \cdot V \right.$$

$$\left. - \frac{i}{2.16} \int d^8 z V (D^2 \bar{D}^2 + \bar{D}^2 D^2) V\right\} \tag{17.78}$$

The Yang-Mills vertices may easily be read off from this action. The Yang-Mills coupling to matter involves the action

$$\int d^8 z \bar{\varphi}_a (e^{gV})^a{}_b \varphi^b \tag{17.79}$$

and it is straightforward to read off the appropriate vertices.

17.4 Applications of $N = 1$ Super-Feynman Rules

As our first example of the application of $N = 1$ super-Feynman rules, we calculate the one-loop correction to the propagator in the massless Wess-Zumino model whose action is of the form

$$A = \int d^4 x \, d^4\theta \, \phi\bar{\phi} + \frac{\lambda}{3!}\left\{\int d^4 x \, d^2\theta \, \phi^3 + \text{h.c.}\right\} \tag{17.80}$$

The relevant diagram is given in Fig. 17.1. The bar corresponds to the $\bar{\varphi}$ end of the $\langle\varphi\bar{\varphi}\rangle$ propagator. Using the super-Feynman rules in Euclidean space given earlier we find that this graph gives

$$\frac{1}{2}(\lambda)^2 \int \frac{d^4 p}{(2\pi)^4} \frac{d^4 k}{(2\pi)^4} d^4\theta_1 d^4\theta_2 \phi(-p, \theta_1)\left(-\frac{\bar{D}_1^2}{4}\right)\left(-\frac{D_2^2}{4}\right)\frac{\delta_{12}}{(p+k)^2} \cdot \frac{\delta_{12}}{k^2} \bar{\phi}(p, \theta_2)$$

$$\tag{17.81}$$

Using the results

$$D_2^2 \delta_{12} = D_1^2 \delta_{12} \quad \text{and} \quad \delta_{12} \bar{D}_1^2 D_1^2 \delta_{12} = 16 \delta_{12} \tag{17.82}$$

and performing the θ_2 integrating yields

$$\frac{1}{2} \lambda^2 \int \frac{d^4 p \, d^4 \theta_1}{(2\pi)^4} \phi(-p, \theta_1) A(p) \bar{\phi}(p, \theta_1) \tag{17.83}$$

where

$$A(p) = \int \frac{d^4 k}{(2\pi)^4} \frac{1}{k^2} \frac{1}{(p+k)^2} \tag{17.84}$$

The factor $\frac{1}{2}$ is the combinatoric factor of the graph $(\frac{1}{2} = (3 \times 3 \times 2)/(3! \cdot 3!))$.

Since this graph is logarithmically divergent we must regulate the theory. We use the method of dimensional reduction.[93-95] This involves doing all the γ-matrix algebra, and so D algebra, in four dimensions while all the momentum integrals are done in $\nu = 4 - 2\varepsilon$ dimensions. That is, if $i = 1$ to $4 - 2\varepsilon$ and $\sigma = 4 - 2\varepsilon$ to 4 we set k_σ or ∂_σ equal to zero. This is in contrast to dimensional regularization,[96] where both γ-matrix algebra and momentum integrals are done in ν dimensions. Since the numbers of fermions and bosons change by different amounts as one changes dimension the method of dimensional regularization clearly does not manifestly preserve supersymmetry. In fact, dimensional reduction does not manifestly preserve supersymmetry either as to make it consistent one must drop the Fierz identity (see the latter article in Ref. 94). The lack of supersymmetry does not, however, show up at low loop orders. For a discussion of anomalies in the supercurrent see Ref. 97.

Regulating by the method of dimensional reduction means evaluating all D algebra in four dimensions, but doing the momentum integrals in ν dimensions. Hence, in the above example $A(p)$ becomes

$$A(p) = (\mu)^{2\varepsilon} \int \frac{d^\nu k}{(2\pi)^\nu} \frac{1}{k^2} \frac{1}{(p+k)^2}$$

$$= \frac{1}{(4\pi)^{\nu/2}} \frac{\Gamma\left(2 - \frac{\nu}{2}\right) \left(\Gamma\left(\frac{\nu}{2} - 1\right)\right)^2 (p^2)^{\nu/2 - 2}}{\Gamma(\nu - 2)}$$

$$= \frac{1}{(4\pi)^2} \frac{1}{\varepsilon} + O(\varepsilon^0) \tag{17.85}$$

When calculating in component fields one can either work with A_μ, $\mu = 1$ to 4 or decompose A_μ into $\{A_i, S_\sigma\}$ where S_σ are the so-called ε scalars. Performing this split in the component action one finds that in general the S_σ acquire gauge-covariantized kinetic terms $-1/2 \, (D_i S^\sigma)^2$ and interaction terms

of the form

$$-S_\sigma^2(A^2 + B^2) \quad \text{and} \quad \bar\chi\gamma_\sigma S^\sigma\chi \tag{17.86}$$

There is, however, no term of the form $S^\sigma\partial_\sigma A$ as $\partial_\sigma A = 0$. The contribution of the S_σ "scalar" fields gives the difference between the dimensional reduction and dimensional regularization schemes.

The most general massless, renormalizable, $N = 1$ supersymmetric theory is of the form

$$A = \frac{1}{64g^2} \int d^4x\, d^2\theta\, \text{Tr}\, W^A W_A + \int d^4x\, d^4\theta\, \bar\phi_a(e^{gV})^a{}_b \phi^b$$

$$+ \left\{ \int d^4x\, d^2\theta\, d_{abc}\, \phi^a\phi^b\phi^c + \text{h.c.} \right\} + \text{gauge fixing} + \text{ghosts} \tag{17.87}$$

where $V^a{}_b = V^s(T_s)^a{}_b$ and the $(T_s)^a{}_b$ are the generators of the gauge group G in the representation R to which the chiral matter belongs. In the Wess-Zumino gauge this theory is a standard renomalizable field theory. Its renormalization in superspace is complicated by the non-polynomial nature of V.[98]

We will now evaluate the $N = 1$ Yang-Mills contribution to the $\phi\bar\phi$ propagator and assume for simplicity that R is an irreducible representation. The relevant graph is given in the figure below.

Fig. 17.2.

This graph gives

$$g^2(T^s)_a{}^b(T^s)_b{}^c \int \frac{d^4p}{(2\pi)^4} d^4\theta_1\, d^4\theta_2 \frac{d^4k}{(2\pi)^4} \phi^a(-p, \theta_1)$$

$$\cdot \left(-\frac{\delta_{12}}{(p+k)^2} \right) \left(-\frac{D_1^2}{4} \right) \left(-\frac{\bar D_2^2}{4} \right) \left(\frac{\delta_{12}}{k^2} \right) \bar\phi_c(p, \theta_2) \tag{17.88}$$

Using the same relations as before we find the result

$$-g^2 C(R) \int \frac{d^4p}{(2\pi)^4} d^4\theta_1\, \phi^a(-p, \theta_1)\bar\phi_a(p, \theta_1) A(p) \tag{17.89}$$

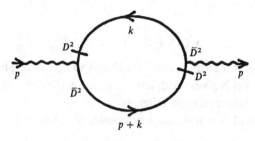

Fig. 17.3.

where

$$(T^s)_a{}^b (T^s)_b{}^c = \delta_a{}^c C(R) \tag{17.90}$$

The minus sign can be traced to the fact that the $\langle VV \rangle$ propagator is equal to minus the $\phi\bar\phi$ propagator.

Let us now do a slightly more complicated example; we will find the one-loop correction to the VV propagator that is due to internal ϕ lines which are represented by the graph in Fig. 17.3.

The resulting contribution to the effective action, which requires a factor of $1/2$ due to its containing V^2, is

$$\frac{g^2}{2} T(R) \int \frac{d^4p}{(2\pi)^4} \frac{d^4k}{(2\pi)^4} d^4\theta_1 \, d^4\theta_2 \, V^s(-p, \theta_1)$$

$$\cdot \frac{\bar{D}_1^2 D_1^2}{16} \frac{\delta_{12}}{(p+k)^2} \frac{D_1^2 \bar{D}_1^2}{16} \frac{\delta_{12}}{k^2} V^s(p, \theta_2) \tag{17.91}$$

where $(T^s)^a{}_b (T^t)^b{}_a = \delta^{st} T(R)$. We now integrate by parts the $\bar{D}_1^2$ off the first δ_{12} and onto $V^s(-p, \theta_1)$ and the other δ_{12}. This manoeuvre yields the expression

$$+\frac{g^2}{2} T(R) \int \frac{d^4p}{(2\pi)^4} \frac{d^4k}{(2\pi)^4} d^4\theta_1 \, d^4\theta_2 \frac{1}{k^2(p+k)^2} \frac{1}{(16)^2}$$

$$\cdot \{ \bar{D}_1^2 V^s(-p, \theta_1) D_1^2 \delta_{12} D_1^2 \bar{D}_1^2 \delta_{12} V^s(p, \theta_2)$$

$$+ 2D_1{}^{\dot{B}} V^s(-p, \theta_1) D_1^2 \delta_{12} D_{1\dot{B}} D_1^2 \bar{D}_1^2 \delta_{12} V^s(p, \theta_2)$$

$$+ V^s(1) D_1^2 \delta_{12} \bar{D}_1^2 D_1^2 \bar{D}_1^2 \delta_{12} V^s(p, \theta_2) \} \tag{17.92}$$

Next we integrate by parts the D_1^2 factor off the first δ_{12} factor in the first of the terms above. When doing this we remember that a loop requires a $D^2 \bar{D}^2$ factor to be non-zero. This follows from the equations

$$\delta_{12} D^2 \bar{D}^2 \delta_{12} = 16 \delta_{12} \tag{17.93}$$

but

$$\delta_{12} D_A \bar{D}^2 \delta_{12} = \delta_{12} \bar{D}^2 \delta_{12} = \delta_{12} D_{\dot{B}} \delta_{12}$$
$$= \delta_{12} D_{\dot{B}} D^2 \delta_{12} = \delta_{12} D^2 \delta_{12} = \delta_{12} D_A \delta_{12}$$
$$= \delta_{12} D_{\dot{B}} D_A \delta_{12} = 0 \tag{17.94}$$

The contribution from this graph then equals

$$+ \frac{g^2}{2} \frac{T(R)}{(16^2)} \int \frac{d^4 p}{(2\pi)^4} \frac{d^4 k}{(2\pi)^4} \frac{d^4 \theta_1 \, d^4 \theta_2}{k^2 (p+k)^2}$$
$$\cdot [D_1^2 \bar{D}_1^2 V^s(-p, \theta_1) \delta_{12} D_1^2 \bar{D}_1^2 \delta_{12} V^s(2)$$
$$+ 8 D_1^{\dot{B}} V^s(1) D_1^2 \delta_{12} (+k)^A{}_{\dot{B}} D_A \bar{D}_1^2 \delta_{12} V^s(2)$$
$$- V^s(1) D_1^2 \delta_{12} 16 k^2 \bar{D}_1^2 \delta_{12} V^s(2)] \tag{17.95}$$

In the above equation, we evaluated the products of more than two D's or $\bar{D}$'s using the D anticommutation algebra. Similar manipulations allow us to leave a naked δ_{12} and we may then evaluate the θ_2 integral to yield

$$\frac{1}{2} T(R) g^2 \int \frac{d^4 p}{(2\pi)^4} \frac{d^4 k}{(2\pi)^4} \frac{d^4 \theta_1}{16 p^2 (p+k)^2}$$
$$\cdot [V^s(-p, \theta_1)(\bar{D}^2 D^2 + \bar{D}_{\dot{B}} D_A (k)^{A\dot{B}} - 16 k^2) V^s(p, \theta_1)] \tag{17.96}$$

Using the fact that

$$\int d^\nu k \frac{1}{p^2} \frac{k^2}{(p+k)^2} = 0 = \int d^\nu k \frac{(p_\mu + 2 k_\mu)}{p^2 (p+k)^2} \tag{17.97}$$

we obtain

$$\frac{1}{2} T(R) g^2 \int d^4 p \, d^4 \theta \, A(p) V(1) \Pi_{1/2} p^2 V(1) \tag{17.98}$$

where

$$\Pi_{1/2} = \frac{D^A \bar{D}^2 D_A}{8 p^2} \tag{17.99}$$

The other contributions to the VV propagator from V and ghost loops are given in the graphs below.

Fig. 17.4.

These can be evaluated to contribute to the effective action the term

$$= \frac{1}{2} C_2(G) \int \frac{d^4 p}{(2\pi)^4} d^4\theta \, V^s(-p, \theta) \left[\left(-\frac{5}{2} \Pi_{1/2} + \frac{1}{2} \Pi_0 \right) \right.$$

$$\left. + \left(-\frac{1}{2} \Pi_{1/2} - \frac{1}{2} \Pi_0 \right) \right] V^s(p, \theta) A(p)$$

$$= -\frac{3}{2} C_2(G) \int \frac{d^4 p}{(2\pi)^4} d^4\theta \, V^s(-p, \theta) p^2 \Pi_{1/2} V^s(p, \theta) A(p) \qquad (17.100)$$

where

$$f_{rst} f_{r'st} = \delta_{rr'} C_2(G)$$

17.5   Divergence in Super-Feynman Graphs

Using the above super-Feynman rules we can calculate the degree of divergence of any super-Feynman graph. Consider a super-Feynman graph with P propagators of any kind, V vertices of any kind and E external ϕ or $\bar{\phi}$ lines and any number of external V lines. We note that no matter what the vertex is it will have four D's. If the vertex is V^n it has four D's, while a $\phi\phi V^n$ or ϕ^3 has four D's from the differentiation of the chiral sources. The only exception to this rule is when the vertex has a chiral external line. No D's are present on these external chiral lines as the external fields are not functionally differentiated. A massive $\bar{\phi}\bar{\phi}$ or $\phi\phi$ propagator has an additional $D^2/q^2 \sim 1/q$ factor; hence the overall degree of divergence is

$$D = 4L - 2P + 2V - C - E - 2L \qquad (17.101)$$

where C is the number of $\bar{\phi}\bar{\phi}$ or $\phi\phi$ propagators. The final factor comes from the fact that for each loop we require four D's in order to gain a non-zero answer when carrying out the final θ integration connected with the loop. Using the topological formula $L - P + V = 1$, we recover the well-known result[85]

$$D = 2 - C - E \qquad (17.102)$$

The above degree of divergence takes no account of the fact that if we calculate in a gauge invariant way then graphs with only external V lines ($E = 0$) must have four D factors on the external lines to be gauge invariant. Hence if there are only external V lines the degree of divergence is in effect

$$D = -C$$

which is at most logarithmically divergent and convergent if $C \neq 0$. Thus the VV propagator can possess a logarithmic divergence.

For graphs with two external lines of opposite chirality, that is the $\bar{\phi}\phi$ propagator and with $C = 0$, we can have a logarithmic divergence. These two results agree with our previous explicit one-loop calculations.

However, if we have only lines of a given chirality then the θ integral in the effective action will give zero unless there are at least a further two factors of D on the external lines; that is, the final result is of the form

$$\int d^4x \, d^4\theta (\phi)^n D^2 \phi \tag{17.103}$$

and not of the form $\int d^4x \, d^4\theta \, \phi^{n+1}$, which vanishes. The degree of divergence of this graph is

$$D = 1 - C - (n + 1) \tag{17.104}$$

and so for $n > 1$ these graphs are finite. This means that for graphs where all external lines are of the same chirality, there are no divergences. Put another way, we require no infinite $m\phi^2$ or $\lambda\phi^3$ counterterms!

We now wish to prove the result in another way by proving the following theorem.

Non-Renormalization Theorem[86-90,212,91]: Any perturbative quantum contribution to the effective action must be expressible as one integral over the whole of $N = 1$ superspace. That is, the contribution must be of the form

$$\Gamma = \sum_n \int d^4x_1 \ldots d^4x_n \, d^4\theta \, G(x_1, \ldots, x_n) f(\phi(x_1, \theta), V(x_1, \theta),$$

$$\ldots D^A \phi(x_1, \theta), D^A V(x_1, \theta) \ldots) \tag{17.105}$$

We note that this quantity involves only one θ integration over quantities that are local in θ.

Proof: This theorem follows in a straightforward manner from the super-Feynman rules given earlier. Consider the θ structure of any super-Feynman graph; from the propagators we get δ_{ij} factors, and from vertices we have $\int d^4\theta_i$ and D's and/or D's acting on the δ_{ij} of the propagators. Let us focus our attention on one particular propagator connecting vertices i_1 and i_2. We may integrate any D's or $\bar{D}$'s that may be on the $\delta_{i_1 i_2}$ factor onto other internal propagators or onto external fields (or current sources if we calculate $W[J, j]$).

After these integrations are done we are left with a naked $\delta_{i_1 i_2}$ and so may evaluate the θ integral at the i_1 vertex. This involves replacing i_1 by i_2 at all other points in the supergraph and so is equivalent, in θ space, to simply contracting the propagator to a point.

We may carry out this procedure for all but one of the propagators that belong to a given loop and so shrink the θ-space expression for the particular loop to be of the form

$$\int d^4\theta_{j_2} \int d^4\theta_{j_1} \delta_{j_1 j_2} \bar{D}_{j_1} \ldots D_{j_1} \ldots \delta_{j_1 j_2} \cdots \qquad (17.106)$$

That is, the loop has shrunk to contain only two vertices. If there are fewer than two D's and two $\bar{D}$'s in this expression, or in any similar expression which occurred earlier, then the graph as a whole will vanish. If there are more than two D's or two $\bar{D}$'s we may use the commutation relations of the D's and $\bar{D}$'s to convert the expression to momentum factors and two D's and two $\bar{D}$'s. For the remaining two D's and two $\bar{D}$'s we may use the result

$$\delta_{j_1 j_2} \bar{D}_{j_1}^2 D_{j_1}^2 \delta_{j_1 j_2} = 16\delta_{j_1 j_2} \qquad (17.107)$$

to evaluate the expression and so carry out the $d^4\theta_{j_1}$ integration leaving only the $d\theta_{j_2}$ integration.

Performing this procedure for every loop we are left with an expression involving momenta, external fields, D's, $\bar{D}$'s acting on an external field and one final vertex integration; in other words an expression of the form of Eq. (17.105). QED.

Applying this theorem to a general $N = 1$ supersymmetric theory we find that the mass and interaction counter-terms which are of the form $\int d^2x\, d^2\theta(m\phi^2 + \lambda\phi^3) +$ h.c. cannot arise as they are superspace sub-integrals. Hence, there are no independent mass and coupling infinities and we recover the results of Refs. 86, 87 and 88.

Another theorem that has important consequences for the construction of realistic models and which is a simple consequence of the non-renormalization theorem is:

Theorem[90]: The effective potential vanishes for those supersymmetric configurations that preserve supersymmetry classically.

Proof: A field configuration that preserves supersymmetry at the classical level has zero-vacuum energy. Since the classical potential is a sum of auxiliary fields it must vanish.

The quantum contributions to the effective potential, which is the x-independent part of the effective action, must be for the form

$$\int d^4\theta\, d^4x\, f(\langle\varphi\rangle, \langle V\rangle, D\langle V\rangle, \ldots).$$

However, for classical configurations that preserve supersymmetry $\langle\phi\rangle$ and

$\langle V \rangle$ are independent of θ since their spinors have zero-vacuum expectation values, and by assumption the expectation values of the auxiliary fields vanish. Hence the above integral and so the effective potential vanish. QED.

This result means that the quantum effective potential can attain a value that is the same as the absolute minimum of the classical supersymmetric theory for those theories that do not break supersymmetry at the classical level. In particular, any degeneracies at the minimum of the classical potential, i.e., vacuum configurations that have the same absolute minimum of the potential, are not removed by quantum corrections. These degeneracies are in fact a common feature of supersymmetric theories.

This result also implies that if supersymmetry is not broken at the tree level it will not be broken perturbatively by quantum correction. This is a simple consequence of the above statement, namely if the quantum effective potential is to break supersymmetry it must have a minimum that is below zero. Since the value zero is attained by the quantum effective potential, such a minimum would require a distortion of the classical effective potential that is outside the validity of perturbation theory.[99] For a discussion of the non-perturbative breaking of supersymmetry see Ref. 100.

17.6   One-loop Infinities in a General *N* = 1 Supersymmetric Theory

The most general $N = 1$ supersymmetric renormalizable theory invariant under a gauge group G contains chiral superfields ϕ^a in a representation R of G and $N = 1$ Yang-Mills fields contained in the general superfield V. Its action is

$$A = \int d^4x \, d^4\theta \, [\bar{\phi}_a(e^{gV})^a{}_b \phi^b] + \frac{\mathrm{Tr}}{64g^2} \int d^4x \, d^2\theta \, W^A W_A$$

$$+ \left[\int d^4x \, d^2\theta \left(\frac{m_{ab}}{2!} \phi^a \phi^b + \frac{1}{3} d_{abc} \, \phi^a \phi^b \phi^c \right) + \mathrm{h.c.} \right]$$

$$+ \text{gauge fixing} + \text{ghosts} \qquad\qquad (17.108)$$

The group indices a may be written as $\{i, A\}$ where A labels the irreducible representation and i its members.

The possible renormalization constants, assuming supersymmetry is preserved, are

$$\phi_0^a = Z^{\frac{1}{2}a}{}_b \phi^b \qquad V_0 = Z_V^{1/2} V$$

$$g_0 = Z_g g, \qquad d_{abc}^0 = Z_{abc}{}^{a'b'c'} d_{a'b'c'}$$

$$m_{ab}^0 = Z_{ab}{}^{a'b'} m_{a'b'} \qquad\qquad (17.109)$$

The $N = 1$ non-renormalization theorem ensures that there are no mass or ϕ^3 interaction-term infinities and so

$$Z_{abc}{}^{a'b'c'} Z_{a'}^{1/2a''} Z_{b'}^{1/2b''} Z_{c'}^{1/2c''} = \delta_{(a}{}^{a''} \delta_b{}^{b''} \delta_{c)}{}^{c''} \tag{17.110}$$

and

$$Z_{ab}{}^{a'b'} Z_{a'}^{1/2a''} Z_{b'}^{1/2b''} = \delta_{(a}{}^{a''} \delta_{b)}{}^{b''} \tag{17.111}$$

In the background field method (see below) one has

$$Z_g Z_V{}^{1/2} = 1 \tag{17.112}$$

If one is not working with the background-field method then the above relation may not hold, but it will differ by gauge-dependent infinities.

As a result the only surviving possible infinities are $Z'_a{}^b$ and Z_V, that is one infinity with each superfield. We will now evaluate these renormalization constants beginning with $Z'_a{}^b$. The one-loop corrections to the $\phi^b \bar{\phi}_a$ propagator are shown in Fig. 17.5 below.

Fig. 17.5.

These graphs lead to the expression

$$g^2 \int \frac{d^4p}{(2\pi)^4} d^4\theta \, A(p)\phi^b(-p,\theta)(S_b{}^a - C_A\delta^a{}_b)\bar{\phi}_a(p,\theta) \tag{17.113}$$

where

$$d^{bce} \bar{d}_{ace} = 2S^b{}_a g^2 \tag{17.114}$$

This is finite if and only if $C_A\delta_b^a = S^a{}_b$.

The VV propagator at the one loop level is given by the graphs of Fig. 17.6 which yield the expression

Fig. 17.6.

$$\frac{1}{2}g^2 \int \frac{d^4p}{(2\pi)^4} V^s \Pi_{1/2} p^2 V^s \left[-3C_2(G) + \sum_A T_A \right] A(p) \qquad (17.115)$$

where

$$\sum_A T_A \delta^{rs} = (T^r)_a{}^b (T^s)_b{}^a \qquad (17.116)$$

This is finite if and only if

$$3C_2(G) = \sum_A T_A \qquad (17.117)$$

Hence the necessary and sufficient conditions for finiteness [101,102] at the one-loop level are given by Eqs. (17.114) and (17.117). We notice that if the propagators are finite at the one-loop level then their one-loop finite parts also vanish.

From the above expressions we may read off the one-loop β-functions

$$\beta(g) = \mu \frac{\partial}{\partial \mu} g = -\frac{1}{16\pi^2} \left[3C_2(G) - \sum_A T_A(R) \right] \qquad (17.118)$$

$$\beta_{abc} = \mu \frac{\partial}{\partial \mu} d_{abc} = 3d_{c(ef}\gamma_{g)}{}^c \qquad (17.119)$$

where $\gamma_a{}^b$ is the anomolous dimension of ϕ_a and is given by

$$\gamma_a{}^b = Z^{-1/2}{}_a{}^c \mu \frac{\partial}{\partial \mu} Z^{+1/2}{}_c{}^b$$

$$= \frac{1}{16\pi^2} (S_a{}^b - C_A \delta_a{}^b) \qquad (17.120)$$

Equation (17.119) is a simple consequence of Eq. (17.110).

The occurrence of one-loop finite $N = 1$ supersymmetric theories leads one to ask if this finiteness persists to higher orders. It is a remarkable fact that

Theorem: One-loop finiteness of an $N = 1$ supersymmetric theory ensures two-loop finiteness.

This result has been found by explicit calculation[101] and also by a variant of the type of anomalies argument which will be given shortly.[102,103] The situation for three loops is unknown.

The two-loop β-functions for a general $N = 1$ supersymmetric theory are given in Refs. 103 and 104.

It has been shown that two-loop finiteness ensures that the three-loop gauge β function vanishes.[105] And it has been shown that the Yukawa β functions also vanish.[213] More recently, it has been found that one can adjust the

Yukawa couplings so that these $N = 1$ theories are finite to all orders of perturbation theory.[214]

17.7   The Background-Field Method[106]

Let us first illustrate the background-field method within the context of Yang-Mills theory. We begin with the classical action $A[A_\mu]$ and write the vector potential A_μ as a classical (background) part $A_\mu{}^c$ and a quantum part $\hat{A}_\mu$,

$$A_\mu = A_\mu{}^c + \hat{A}_\mu \qquad (17.121)$$

The action $A = A[A_\mu{}^c + \hat{A}_\mu]$ now contains vertices which involve both classical and quantum fields. As we shall see, we wish only to consider graphs with only internal $\hat{A}_\mu$ lines and only external $A_\mu{}^c$ lines. The gauge transformation of A_μ which leaves the action invariant is as usual

$$\delta A_\mu = \frac{1}{g}\partial_\mu \Lambda + [A_\mu, \Lambda] \qquad (17.122)$$

We now wish to express this transformation in terms of transformations of $A_\mu{}^c$ and $\hat{A}_\mu$. Since we have in effect only one equation for δA_μ for two unknowns $\delta A_\mu{}^c$ and $\delta \hat{A}_\mu$, this is not unique. However, one set of transformations is given by

$$\delta A_\mu{}^c = \frac{1}{g}\partial_\mu \Lambda + [A_\mu{}^c, \Lambda] \equiv D_\mu{}^c \Lambda$$

$$\delta \hat{A}_\mu = [\hat{A}_\mu, \Lambda] \qquad (17.123)$$

We refer to this possibility as the background gauge transformation.

We now wish to calculate quantum corrections using the action $A[A_\mu{}^c + \hat{A}_\mu]$, but quantizing only the $\hat{A}_\mu$ field. To do this we gauge fix the transformations that reproduce those of Eq. (17.122), with the exception that the gauge-fixing term preserves the background gauge transformation of Eq. (17.123). An example of such a gauge-fixing term is given by

$$F = D_\mu{}^c \hat{A}_\mu \equiv \partial_\mu \hat{A}^\mu + g[A_\mu{}^c, \hat{A}^\mu] \qquad (17.124)$$

We must then add the corresponding Fadeev-Popov ghosts. We note that the ghost terms S_{FP} will involve $A_\mu{}^c$ as the gauge fixing term involves $A_\mu{}^c$. All this amounts to calculating the vacuum-to-vacuum amplitude in the presence of the background field $A_\mu{}^c$ and the source term j_μ for $\hat{A}_\mu$

$$Z[A_\mu{}^c, j_\mu] = \int [d\hat{A}_\mu] \exp i\left(A + \frac{1}{2\alpha}\int F^2\, d^4x + S_{FP} + j^\mu \cdot \hat{A}_\mu \right)$$

where

$$j^\mu \cdot \hat{A}_\mu = \int d^4x\, j^\mu(x)\hat{A}_\mu(x) \qquad (17.125)$$

Although the above procedure is a little more complicated than a straight-forward computation it has one enormous advantage. Namely, any graph, which now has only external $A_\mu{}^c$ fields, will be manifestly background field gauge-invariant. This is in contrast to the usual situation where the gauge-fixing term breaks gauge invariance and one then has to work with B.R.S. invariance which is not manifestly realized. In this case the individual graphs are not in general gauge-invariant. Of course in the background-field method the quantum gauge invariances are broken, but all the effects of this are hidden inside the graphs.

From Z we can define the generator Γ of one-particle irreducible graphs,

$$\Gamma[A_\mu{}^c, \tilde{A}_\mu] = W - j^\mu \cdot \tilde{A}_\mu \tag{17.126}$$

where

$$Z = \exp iW \tag{17.127}$$

and

$$\tilde{A}_\mu = \frac{\delta W}{\delta j_\mu} \tag{17.128}$$

The one-particle irreducible graphs with only background external lines are then given by $\Gamma[A_\mu{}^c]$ and must be manifestly gauge-invariant. The effective action given above is related to the usual effective action in that they give the same predictions for gauge-invariant quantities and are equal if a special choice of gauge is made in the usual formulations.[107]

Let us now apply the background field method to calculate the infinities in Yang-Mills theories. They must be local and can only be of the form

$$\int -\frac{1}{4} Z \operatorname{Tr}\{F_{\mu\nu}^c\}^2 d^4x \tag{17.129}$$

where

$$F_{\mu\nu}^c = \partial_\mu A_\nu^c - \partial_\nu A_\mu^c - g[A_\mu^c, A_\nu^c] \tag{17.130}$$

This is because $F_{\mu\nu}^c$ is the only possible background field gauge covariant object. Consequently, if we have the renormalizations

$$A_{\mu0} = Z_\mu^{1/2} A_\mu, \qquad g_0 = Z_g g \tag{17.131}$$

then we must impose the relation

$$Z_{A_\mu}^{1/2} Z_g = 1 \tag{17.132}$$

in order to agree with the counterterm of Eq. (17.129). When calculating not using the background-field method one finds that Z_{A_μ} and Z_g are not, in general, related by Eq. (17.132); however the behavior of the divergences is controlled by the Ward identity.

17.8   The Superspace Background-Field Method[108]

Unlike the Yang-Mills case it is not desirable to make a linear split in the Yang-Mills prepotential. This is because the gauge transformation of V is very non-linear in V itself and such a split would not lead to transformation properties for the quantum and classical parts that were simple. A better split is found by mimicking the solution of the constraints. We split the general superfield V_T into the background superfields Ω and $\bar{\Omega}$ and the quantum field V as follows

$$e^{V_T} = e^{\Omega} e^V e^{-\bar{\Omega}} \tag{17.133}$$

Under quantum symmetries these fields transform as

$$e^{V'} = e^{\bar{\Lambda}} e^V e^{-\Lambda} \qquad e^{\Omega'} = e^{\Omega} \qquad e^{\bar{\Omega}'} = e^{\bar{\Omega}} \tag{17.134}$$

and under background symmetries as

$$V' = e^K V e^{-K} \qquad e^{\Omega'} = e^{\bar{\Lambda}_0} e^{\Omega} e^{-K}$$

$$e^{-\bar{\Omega}'} = e^K e^{-\bar{\Omega}} e^{-\Lambda_0} \tag{17.135}$$

In these transformations Λ_0 is a chiral superfield

$$D_{\dot{A}} \Lambda_0 = 0 \tag{17.136}$$

The parameter Λ is background covariantly chiral

$$\mathscr{D}_{\dot{A}} \Lambda = 0 \tag{13.137}$$

where the background covariant derivations are defined by

$$\mathscr{D}_A = e^{-\Omega} D_A e^{\Omega} \qquad \mathscr{D}_{\dot{A}} = e^{-\bar{\Omega}} D_{\dot{A}} e^{\bar{\Omega}} \tag{17.138}$$

We observe that both quantum and classical transformations reproduce the correct transformation on V_T. In the quantum case

$$e^{V_T'} = \exp(e^{\Omega} \bar{\Lambda} e^{-\Omega}) \cdot \exp(V_T) \cdot \exp(-e^{-\bar{\Omega}} \Lambda e^{\bar{\Omega}})$$

$$= \exp \bar{\Lambda}_0 \cdot \exp V_T \cdot \exp(-\Lambda_0) \tag{17.139}$$

where $\Lambda_0 = e^{-\bar{\Omega}} \Lambda e^{\bar{\Omega}}$ in the background case

$$e^{V_T'} = (e^{\bar{\Lambda}_0} e^{\Omega} e^{-K})(e^K e^V e^{-K})(e^K e^{-\bar{\Omega}} e^{-\Lambda_0})$$

$$= e^{\bar{\Lambda}_0} e^{V_T} e^{-\Lambda_0} \tag{17.140}$$

It is most useful to consider the covariant derivatives in their chiral representation

$$V_A = e^{-V_T} D_A e^{V_T}, \qquad V_{\dot{A}} = D_{\dot{A}} \tag{17.141}$$

Substituting for V_T we find

$$V_A = e^{\bar{\Omega}} e^{-V} \mathcal{D}_A e^{V} e^{-\bar{\Omega}}; \qquad V_{\dot{A}} = D_{\dot{A}} \tag{17.142}$$

It is useful to work with derivatives that are symmetric with respect to the background field, and so we make the non-unitary transformation

$$V_N \rightarrow e^{-\bar{\Omega}} V_N e^{\bar{\Omega}} \tag{17.143}$$

and the covariant derivatives become

$$V_N = (e^{-V} \mathcal{D}_A e^{V}, \mathcal{D}_{\dot{A}}, V^n) \tag{17.144}$$

where

$$\{V_A, V_{\dot{A}}\} = -2i(\sigma^n)_{A\dot{A}} V_n \tag{17.145}$$

These derivatives have particularly simple transformation properties:

quantum: $\qquad\qquad V_N \rightarrow e^{\Lambda} V_N e^{-\Lambda}$

background $\qquad\quad V_N \rightarrow e^{K} V_N e^{-K} \tag{17.146}$

The lowest dimensional non-zero field strength is given as usual by

$$W_A = -2i[V^{\dot{B}}, \{V_A, V_{\dot{B}}\}] \tag{17.147}$$

For chiral superfields φ_T we first multiply by $e^{-\bar{\Omega}}$ appropriately in order that they satisfy a background chiral constraint

$$\mathcal{D}_{\dot{A}} \varphi_T = 0 \tag{17.148}$$

As φ_T transforms linearly we then make a linear split.

$$\varphi_T = \varphi_B + \varphi_Q \tag{17.149}$$

Both these fields transform under quantum transformations as

$$\varphi_Q \rightarrow e^{\Lambda} \varphi_Q \tag{17.150}$$

where $\Lambda^i T_i$ is in an appropriate representation and is background covariantly chiral. Under background transformations they transform as

$$\varphi_Q \rightarrow e^{K} \varphi_Q \tag{17.151}$$

The most general renormalizable action is given by

$$A = \frac{1}{64g^2} \int d^4x \, d^2\theta \, \mathrm{Tr} \, W^A W_A + \int d^4x \, d^4\theta \, \bar{\varphi}_{Ta} (e^{gV})^a{}_b \varphi_T{}^b$$

$$+ \left\{ \int d^4x \, d^2\theta \, \frac{1}{3!} d_{abc} \, \varphi_T{}^a \varphi_T{}^b \varphi_T{}^c + \frac{m_{ab}}{2} \varphi_T{}^a \varphi_T{}^b + \text{h.c.} \right\} \tag{17.152}$$

To quantize the theory we choose a background covariant gauge-fixing term, for example

$$\mathcal{D}^2 V - \bar{f} = 0 \qquad \bar{\mathcal{D}}^2 V - f = 0 \qquad (17.153)$$

The corresponding Faddeev-Poppov determinant $\varDelta$ is given by

$$\varDelta \int d\varLambda\, d\bar{\varLambda}\, \delta(\mathcal{D}^2 V^\varLambda - \bar{f})\delta(\bar{\mathcal{D}}^2 V^\varLambda - f) = 1 \qquad (17.154)$$

where $V^\varLambda$ is a gauge transformed V according to Eq. (17.134). The vacuum-to-vacuum amplitude is

$$Z[j] = \int [\mathcal{D}V\mathcal{D}\varphi_Q\mathcal{D}\bar{\varphi}_Q] \exp\left\{ iA + i \int d^8 z VJ \right.$$

$$\left. + \left(\int d^4 x\, d^2\theta j\varphi_Q + \text{h.c.} \right) \right\} \left\{ \varDelta\delta(\mathcal{D}^2 V^\varLambda - \bar{f})\delta(\bar{\mathcal{D}}^2 V^\varLambda - f) \right\} \quad (17.155)$$

The factor $\varDelta$ may be replaced by the representation

$$\int [\mathcal{D}c\mathcal{D}c'\mathcal{D}\bar{c}\mathcal{D}\bar{c}'] \exp i \int d^8 z$$

$$\times \left\{ (\bar{c}' + c') \wedge \left(-\frac{V}{2} \right) \wedge \left[(c + \bar{c}) + \left(\coth \frac{V}{2} \right) \wedge (c - \bar{c}) \right] \right\} \quad (17.156)$$

Although this looks much as before we note that now, the ghosts are background covariantly chiral.

We may average over the gauge-fixing condition by integrating over

$$\int [\mathcal{D}f\mathcal{D}\bar{f}\mathcal{D}b\mathcal{D}\bar{b}] \exp\left(-\frac{i}{16}(f\bar{f} + b\bar{b}) \right) \qquad (17.157)$$

The background covariantly chiral fields b and $\bar{b}$ are the Nielson-Kallosh ghosts which anticommute and contribute only at the one loop level.

The Yang-Mills part of the action is given by

$$\frac{1}{16C_2(G)} \int d^8 z\, \text{Tr}\{ -e^{-V}\mathcal{D}^\varLambda e^V \bar{\mathcal{D}}^2 e^{-V}\mathcal{D}_\varLambda e^V$$

$$-\frac{1}{2} V(\mathcal{D}^2\bar{\mathcal{D}}^2 + \bar{\mathcal{D}}^2\mathcal{D}^2)V\} \qquad (17.158)$$

and so the part bilinear in V which yields all one-loop graphs is given by

$$\frac{1}{16C_2(G)} \int d^8 z\, \text{Tr}\left\{ V\left(\mathcal{D}^\varLambda\bar{\mathcal{D}}^2\mathcal{D}_\varLambda - \frac{1}{2}(\mathcal{D}^2\bar{\mathcal{D}}^2 + \bar{\mathcal{D}}^2\mathcal{D}^2) \right)V \right\} \quad (17.159)$$

One may verify that this is equal to

$$-\frac{1}{16C_2(G)} \int d^8z \operatorname{Tr}\left\{ V\left(\mathscr{D}^n\mathscr{D}_n - \frac{1}{16} W^A \mathscr{D}_A - \frac{1}{16\cdot 2}(\mathscr{D}^A W_A) \right) V \right\} \quad (17.160)$$

The reader will observe that there is only one spinorial covariant derivative in this expression. As a loop requires four internal spinorial derivatives in order not to be zero it follows that the first non-zero one-loop graph with internal Yang-Mills lines has four external legs and is of the form $W^2 \overline{W}^2$.

In general the effective action Γ must be a gauge-invariant function of the background potentials A_N. The effective action also obeys the $N = 1$ non-renormalization theorem and so is expressed as an integral over the whole of $N = 1$ superspace.

Consequently, the counterterms must be of the form

$$Z \int d^4x \, d^4\theta \, f(A_B^c, \mathscr{D}_A^c A_B^c, \ldots) \quad (17.161)$$

The only possible candidate on dimensional grounds is

$$Z \int d^4x \, d^4\theta \, A_B^c \overline{\mathscr{D}}^{c^2} A^{cB} + \cdots \quad (17.162)$$

This term can be rewritten in the form

$$Z \int d^4x \, d^2\theta \, W^{cB} W_B^c \quad (17.163)$$

where $W_B^c = \overline{\mathscr{D}}^{c^2} A_B^c$. This shows that $N = 1$ Yang-Mills can possess an infinite wavefunction renormalization. For the same reasons as in the case of ordinary Yang-Mills theory, the coupling and wave-function infinites will obey the relation $Z_g Z_V^{1/2} = 1$. This of course agrees with our explicit calculation where such an infinity occurred.

Chapter 18

Ultra-violet Properties of the Extended Rigid
Supersymmetric Theories

In Chapter 17 we noted how the interacting Wess-Zumino model had even fewer infinite renormalization factors than one might naively expect by demanding that the renormalization procedure preserve supersymmetry. In fact, it only has one infinite renormalization factor as does the $N = 1$ Yang-Mills theory (in the background-field method or in a preferred gauge). As one studies theories with more supersymmetries one might expect an even more remarkable ultraviolet behaviour.

The most spectacular renormalization properties of supersymmetric theories are the finiteness of a large class of extended rigid supersymmetric theories. At first, attention was focused entirely on the maximally extended $N = 4$ supersymmetric Yang-Mills theory. The β-function was shown to vanish for this theory for one,[109] two[110] and three loops.[111] Soon after the three-loop calculation, an argument for finiteness to all orders was made.[112,113] This argument was based on the anomaly structure of supersymmetric theories. More recently two more arguments for the finiteness of $N = 4$ Yang-Mills theory have been found. One argument relies on a generalization of the $N = 1$ non-renormalization theorem to extended supersymmetry,[114,115] while the other relies on putting $N = 4$ Yang-Mills theories in a light-cone gauge.[116,117]

It was noticed[118] using the results of Ref. 119 that the β-function of $N = 2$ Yang-Mills theory vanished for two loops. It was then argued using the non-renormalization argument that $N = 2$ Yang-Mills theory was finite above one loop.[114] While re-examining the anomalies argument it was realized that any $N = 2$ rigid supersymmetric theory was finite above one loop.[120] In fact, it is possible to arrange the representation content of $N = 2$ rigid theories so that there are finite $N = 2$ theories.[120] A modern account of the anomalies argument in a form which applies to $N = 2$ theories can be found in Ref. 121, while the application of the non-renormalization argument[114,115] to $N = 2$ theories is given also in Ref. 121. In the following discussion we will only consider the anomaly and non-renormalization arguments.

These results have been confirmed by explicit calculation using the $N = 1$ superfield formalism of $N = 2$ theories. It has been found that the two-loop β-function of any rigid $N = 2$ theory vanishes.[122]

18.1 The Anomalies Argument[112,113]

The strategy that this argument employs rests on the fact that in any supersymmetric theory, the energy-momentum tensor $\theta_{\mu\nu}$, some of the internal currents $j_\mu{}^i{}_j$, and the supercurrent $j_{\mu\alpha i}$, lie in a supermultiplet. Consequently, any superconformal anomalies which these currents possess must also lie in a supermultiplet. Typically this supermultiplet of anomalies will include $\theta_\mu{}^\mu$, $(\gamma^\mu j_{\mu i})_\alpha$ and $\partial^\mu j_\mu{}^i{}_j$ for some i,j, corresponding to the breaking of dilation, special supersymmetry and some of the internal currents respectively. Clearly, if some of the relevant internal symmetries are preserved (i.e., $\partial_\mu j^{\mu i}{}_j = 0$ for some i, j) then the anomaly multiplet, if it is irreducible, will vanish, and consequently $\theta_\mu{}^\mu = 0$. However, $\theta_\mu{}^\mu$ is proportional to an operator $(F_{\mu\nu}F^{\mu\nu} + \cdots)$ times the β-function and so the β-function must vanish. From this result one can argue in specific formalisms such as the background-field method for the finiteness of the theory being considered (see below).

To illustrate how the argument goes, we will first argue for the finiteness of $N = 4$ Yang-Mills theory using a simplified version of the anomalies argument. To do this we must make the following assumption.

The quantum corrections of $N = 4$ Yang-Mills preserve one supersymmetry and the $SU(4)$ internal symmetry.

Let us first establish that all the chiral currents of the theory are preserved. The maximal symmetry is $U(4) = SU(4) \times U(1)$; however, this $U(1)$ factor, whose generator is denoted by B, has for general N the following commutation relation with $Q^j{}_A$:

$$[Q_A{}^j, B] = \frac{i(N-4)}{4} Q^j{}_A \qquad (18.1)$$

In the case of $N = 4$, B and $Q_A{}^j$ commute; and consequently B, which is a chiral rotation, has the same action on all the states of any $N = 4$ multiplet. However, $N = 4$ Yang-Mills is a CPT self-conjugate multiplet; so B must have the same action on the $+1/2$ and $-1/2$ helicity states, which is possible only if the action of B is zero. Hence, the model only has $SU(4)$ symmetry, which decomposes under $O(4)$ as follows:

$$\mathbf{15} \text{ of } SU(4) = \mathbf{6} + \mathbf{9} \text{ of } O(4) \qquad (18.2)$$

where the **9** of $O(4)$ are chiral currents, while the **6** of $O(4)$ are currents that do not involve γ_5. Hence if $SU(4)$ is preserved the **9** chiral currents are preserved and so all chiral currents are preserved.

In the $O(4)$ formulation, which has **4** Majorana spinors $\chi_{\alpha i}$, the **9** chiral currents contain the term $\bar{\chi}^{(i}\gamma^\mu\gamma^5\chi^{j)}$ while the **6** currents contain the term $\bar{\chi}^{[i}\gamma^\mu\chi^{j]}$.

Consider now the $N = 4$ Yang-Mills theory when decomposed into $N = 1$ representations. The R current $j_\mu^{(5)}$ will be one of the 9 and will be preserved, i.e., $\partial_\mu j^{\mu(5)} = 0$.

We must now consider the form of the anomaly equation. It has a right-hand side constructed from the $N = 1$ Yang-Mills field strengths W and the chiral matter fields ϕ^i. On dimensional grounds it must be of the form (see Chapter 20).

$$D^{\dot{A}} J_{A\dot{A}} = -\frac{1}{3}\frac{\beta(g)}{g} D_A W^2 \quad \text{plus terms involving } \phi^i \qquad (18.3)$$

However, as previously discussed, the first term is a chiral anomaly and so contains $\theta_\mu{}^\mu$ and $\partial^\mu j_\mu^{(5)}$. Since $\partial^\mu j^{\mu(5)} = 0$ we must conclude that $\beta(g) = 0$.

A similar but somewhat stronger argument[112] can be employed starting from the assumption that the quantum $N = 4$ Yang-Mills theory preserves $N = 1$ supersymmetry and $O(4)$ symmetry. In this case one must consider $N = 2$ anomaly multiplets.

We now wish to apply[121] the anomalies argument to $N = 2$ supersymmetric rigid theories and prove the following theorem.

Theorem[120]: The β-function in $N = 2$ rigid supersymmetric theories vanishes above one loop.

Proof: We shall first establish that the quantum theory preserves $N = 2$ supersymmetry and $U(2)$ internal symmetry. At the classical level the two supersymmetries are manifestly preserved by the $N = 2$ matter formulations given earlier. However, in order to quantize the theory in a straightforward manner let us consider those formulations that do not involve constraints in x-space. That is, we consider the relaxed hypermultiplet formulation consisting of the superfields L, L^{ij}, L^{ijkl}; or we could also use the new relaxed version of the Sohnius hypermultiplet discussed in Chapter 15.

We now require a method of regularization that preserves these symmetries. The safest method is that of higher derivatives.[123] That is, we consider the action which in component fields contains

$$-\frac{1}{4}F_{\mu\nu}^2 - \frac{1}{4}\frac{1}{\Lambda^{2r}} F^{\mu\nu}(\partial^2)^r F_{\mu\nu} + \text{supersymmetric extension} \qquad (18.4)$$

In the above, r is an integer which is usually equal to 1 or 2. This method clearly preserves $N = 2$ supersymmetry and $U(2)$ and regulates all graphs except primitive one-loop diagrams. In general, introducing higher derivatives into a theory alters the infinity structure at one loop, i.e., 'the β-function' of the theory before the higher derivative is removed. However, for the $N = 2$ theories expressed in the formalism given above the additional massive states introduced by the higher derivatives must belong to the $N = 2$ supermultiplets with the content one spin 1, four spin 1/2 and five spin 0. This follows from the fact that this is the only available massive multiplet which does not possess a central charge and involves spins less than or equal to one. Consequently, if 'the β-function' at one loop for the theory with higher derivatives is given by[124,125]

$$\beta(g) = \frac{g^3}{96\pi^2} \sum_\lambda (-1)^{2\lambda} C_\lambda (1 - 12\lambda^2) \qquad (18.5)$$

where the sum is over all helicity states, then the β-function is the same as in the theory without higher derivative. It is argued in Ref. 126 that this formula for 'the β-function' is indeed the correct one.

Hence if the theory is finite at the one-loop level, the theory when higher derivative regulated will be finite at one loop and so be rendered finite to all orders by higher derivatives.

As such an $N = 2$ one-loop finite theory can preserve $N = 2$ supersymmetry and $U(2)$ at the quantum level.

It therefore only remains to establish that the multiplet of currents of $N = 2$ supersymmetric theories has the required form. The $N = 2$ supercurrent is an object J which has dimension 2; and its anomaly equation has the form

$$D^{ij} J = \text{anomaly} \qquad (18.6)$$

where $D^{ij} \equiv D^{A(i} D_A^{j)}$. The right-hand side of the above equation must be of dimension 3 and must be constructed out of the gauge invariant superfields W of $N = 2$ Yang-Mills and the $N = 2$ matter superfields and covariant derivatives of these fields. Let us focus on the term involving the $N = 2$ Yang-Mills fields W. The only candidate is

$$D^{ij} J = -\frac{1}{3} \frac{\beta(g)}{g} \bar{D}^{ij} \overline{W}^2 + \text{terms involving matter fields} \qquad (18.7)$$

where $\beta(g)$ is the β-function for the gauge coupling constant, g. Now the first term is a set of anomalies that contains a contribution to $\theta_\mu{}^\mu$ as well as to a divergence of one of the chiral $U(2)$ currents. However, these chiral currents are preserved; and so $\beta(g) = 0$.

If, on the other hand, the theory has infinities at the one-loop level, then one must regulate these infinities separately. One then expects that the $U(2)$ currents will be preserved above one loop and going through the same argument as above one finds that the β-function vanishes above one loop.

There exists a variant of the anomalies argument for finiteness which involves the Adler-Bardeen theorem for the chiral current rather than appealing directly to chiral current conservation.[127] This argument[128] has also been applied to show the finiteness above one loop of $N = 2$ theories.

18.2   The Non-Renormalization Argument

We can now present the non-renormalization argument for finiteness. We first give the extended non-renormalization theorem.

Theorem[114]: The effective action Γ for any extended supersymmetric theory that possesses an unconstrained superfield formalism can be written as one integral over the whole of extended superspace:

$$\Gamma = \int d^4x_1 \ldots d^4x_n \, d^{4N}\theta f(\phi(x_1,\theta_1),\ldots,\phi(x_n,\theta_n), D\phi(x_1,\theta_1),\ldots)$$

$$\cdot g(x_1,\ldots,x_n) \qquad\qquad\qquad\qquad (18.8)$$

where ϕ is a generic superfield.

Proof: When a supersymmetric theory admits an unconstrained formalism we can take the propagators to be δ_{12} and the vertices must have a factor $\int d^{4N}\theta$. One will also have various D factors on the lines leaving the vertices. We can now follow exactly the same argument as was used to prove the $N = 1$ non-renormalization theorem: we integrate the D's off the δ_{12}'s and shrink the θ-space loops until only one integral remains.

Counterterms are known to be local, and their contribution to Γ is therefore of the form

$$\Gamma = \int d^4x \, d^{4N}\theta f(\phi, D\phi,\ldots) \qquad\qquad\qquad\qquad (18.9)$$

Of course the effective action must, in the absence of any anomalies, also obey the Ward identities corresponding to the symmetries of the theory. The simplest method to implement these symmetries is to use the background-field method. The background-field formalism for the extended theories is very similar to that for $N = 1$ Yang-Mills, except for the gauge-fixing procedure and corresponding ghosts. This latter point, which is discussed later, leads to the one-loop exception clause in the following theorem.

Theorem[114]: Consider any supersymmetric theory that possesses an unconstrained superfield formulation; then using the background-field formalism, we can express the quantum contributions, *above one loop*, to the effective action Γ as one integral over the whole of superspace of a gauge-invariant function of the background potentials and matter fields.

Consequently, the counterterms must be of the form

$$\Gamma = \int d^4x \, d^{4N}\theta f(A_N^c, X^c, D_M^c A_N^c, D^c X^c) \qquad\qquad (18.10)$$

where f is a gauge-invariant function of the background gauge potential A_N^c and the matter fields X^c. We will return to the one-loop exception later.

The strategy[114] of the non-renormalization argument is simply to see if dimensional analysis allows any counterterms of the form of Eq. (18.33). In the absence of dimensional coupling constants this is unlikely as the measure has a dimension of $-4 + 2N$. It was argued in Ref. 114 that $N = 2$ Yang-Mills theory was finite above one loop and that $N = 4$ Yang-Mills theory would be finite if an $N = 4$ superfield formalism existed. In Ref. 115 it was argued that $N = 4$ Yang-Mills theories were finite as a result of the $N = 2$ extended superfield formalisms discussed in Chapter 5. The finiteness of $N = 2$ rigid supersymmetric theories above one loop and the criterion for finiteness at one loop were established in Ref. 120.

We will now apply the non-renormalization argument to $N = 2$ rigid supersymmetric theories. As discussed in Chapter 15, the $N = 2$ Yang-Mills theory is represented by A_N, which has the same dimension as D_N, namely, one if $N = m$ and one-half otherwise. The $N = 2$ matter is represented by the fields L, L^{ij}, L^{ijkl} which all have dimension 1. Using the above theorem we find that, above one loop in the background-field formalism, the local counterterms are of the form

$$\Gamma = \int d^8\theta \, d^4x f(A_N, D_{Bj}A_N, L, L^{ij}, L^{ijkl}, \ldots)$$

$$= \int d^8\theta \, d^4x (A_{Bi}D_{Aj}\ldots D_{Ck}A_{Dl} + LD_{Ai}\ldots D_{Ck}L + \cdots) \qquad (18.11)$$

However on dimensional grounds none of the above terms is allowed, and so we must conclude that the $N = 2$ rigid supersymmetric theories are finite above one loop and have correspondingly a vanishing β-function above one loop.

The above statement must be interpreted carefully when the theories have one-loop infinities, i.e., a non-zero one-loop β-function. In that case one will find the inevitable higher loop $1/\varepsilon^n$ poles, where $n > 1$, which are a consequence of the $1/\varepsilon$ pole at one loop. Also the subtraction procedure must be minimal in order that the β-function does not receive higher loop contributions due to finite counter terms inserted in the divergent one-loop graphs.

In order to make the above discussion more concrete, let us do the superspace power counting for $N = 2$ Yang-Mills theory and verify that it is indeed finite above one loop.[129] The prepotential of the $N = 2$ Yang-Mills theory is V^{ij} and is of dimension -2. As such, the action must have the generic form

$$A = \int d^4x \, d^8\theta (VD^8V + V^2D^{12}V + \cdots) \qquad (18.12)$$

Consequently, the propagators are of the form $D^{-8}\delta_{12}$ and the vertices involve a D^{12} factor as well as a full superspace integration. The external lines in the background formalism above one loop are in terms of $A_{Bi} = D_{Bi}D^4 V$. Let there be E external lines, P propagators and V cubic vertices.

The degree of divergence for a graph with only *cubic* vertices is

$$D = 4L - 4P - E\left(2 + \frac{1}{2}\right) + 6V - 4L \tag{18.13}$$

The last factor results from requiring eight D's for every internal loop. Using the relation $3V = 2P + E$ we find that

$$D = -\frac{E}{2} \tag{18.14}$$

and we conclude that above one loop all graphs involving only cubic vertices are finite. The reader may easily generalize this result to include graphs involving any type of vertices.

Let us now return to the one-loop exception mentioned above. It results from the fact that in the extended theories one must introduce ghosts as usual; however, in these theories the ghost action has a residual invariance. Fixing this invariance and introducing ghosts for ghosts one finds that these new ghosts still possess a gauge invariance. This goes on indefinitely and one finds an infinite number of ghosts for ghosts. Fortunately, for super Yang-Mills fields the second-generation ghosts and all further ghosts only couple to the background fields; and for matter, even the first generation matter ghosts do not couple to the quantum fields. Hence it is a problem that only affects one-loop graphs. To define the theory one must truncate the infinite sequence of ghosts for ghosts. This is achieved at the expense of introducing the background gauge prepotential, and so one finds that at one loop one can have an explicit occurrence of the background prepotential.

In $N = 2$ Yang-Mills theories the field strength is given in terms of the prepotential by

$$W = \bar{\mathcal{D}}^{kl}\bar{\mathcal{D}}_{kl}\mathcal{D}^{ij}V_{ij} \tag{18.15}$$

Gauge invariance

$$\delta V^{ij} = \mathcal{D}_{Ak}\chi^{ijkA} + \mathcal{D}_{\dot{A}k}\chi^{ijk\dot{A}} \tag{18.16}$$

leaves W invariant as

$$D_{(i}{}^{A}D_{j}{}^{B}D_{k)}{}^{C} = 0 \tag{18.17}$$

Any gauge-fixing term involving V^{jk} must, however, also be invariant under the gauge transformation

$$\delta\chi^{(ijk)A} = \mathcal{D}_{Bl}\chi^{(ijkl)(AB)} \tag{18.18}$$

and so on. The origin of the one-loop exception was given in Ref. 114, and a detailed discussion can be found in Ref. 115.

18.3 Finite $N = 2$ Supersymmetric Rigid Theories[120]

We have seen in the previous section that $N = 2$ supersymmetric rigid theories are finite if they have vanishing one-loop β-functions. For an $N = 1$ theory consisting of $N = 1$ Yang-Mills theory and Wess-Zumino multiplets ϕ_σ in the representation R_σ of the gauge group, the one-loop β-function is[130]

$$\beta(g) = \frac{1}{(4\pi)^2}g^3\left(\sum_\sigma T(R_\sigma) - 3C_2(G)\right) \tag{18.19}$$

An $N = 2$ rigid theory consists of $N = 2$ Yang-Mills and $N = 2$ matter. The $N = 2$ Yang-Mills theory consists of $N = 1$ Yang-Mills and one Wess-Zumino multiplet ϕ in the adjoint representations. The $N = 2$ matter, on the other hand, consists of Wess-Zumino multiplets X_σ and Y_σ in the representations R_σ and $\bar{R}_\sigma$ respectively. Adjusting the one-loop β-function Eq. (18.19), we find that the one-loop β-function for $N = 2$ rigid theories is

$$\beta(g) = \frac{2g^3}{(4\pi)^2}\left(\sum_\sigma T(R_\sigma) - C_2(G)\right) \tag{18.20}$$

There are many cases for which the $\beta(g)$ vanishes. For example, in the case of $SU(N)$, $C_2(N) = N$, while for the fundamental representation $T(R) = 1/2$. Hence one can have $2N$ fundamental representations and the theory will be finite.

It is interesting[131] that most of the groups that have been proposed for grand unification belong to a single sequence of groups. Indeed, of the five sequences of groups into which all Lie groups were classified by Cartan, this is the only finite sequence. This sequence is $E_8, E_7, E_6, E_5 = SO(10), E_4 = SU(5)$, and $E_3 = SU(3) \times SU(2)$. The fact that $E_3 \times U(1)$ is the group of low energy physics is intriguing, as the $U(1)$ factor is a typical remnant of some higher symmetry breaking. We will now investigate[132] which $N = 2$ theories that have the above gauge groups are finite. For $E_4 = SU(5)$ we must have $p + 3q + 7r = 10$ where p, q and r are the number of hypermultiplets in the $5 + \bar{5}$, $10 + \overline{10}$ and $15 + \overline{15}$ representations, respectively. For $SO(10)$ one can have p and q hypermultiplets in the $10 + \overline{10}$ and $16 + \overline{16}$ representations, respectively, provided $p + 2q = 8$. The group E_6 will result in a finite theory if we use four hypermultiplets in the $27 + \overline{27}$ representation, while the group E_7 requires three hypermultiplets in the $56 + \mathbf{56}$ representation. For E_8 the lowest dimensional representation is the adjoint, and so one is only allowed three chiral fields in the adjoint representations. We observe that for the groups $SO(10)$, E_6, E_7 and E_8 the theories that are finite contain the observed fermions. In fact, they contain three or more generations plus their mirrors. Of course it is not clear how many generations would continue to remain massless when the gauge groups are spontaneously broken. Recently, it has been pointed out that there are other possible finite theories when the $N = 2$ matter is a CPT self-conjugate multiplet.[151]

In Ref. 133 it has been shown that an $N = 2$ mass term introduces no infinities.

It is appropriate at this point to discuss the strengths and weaknesses of the above arguments. The anomalies argument in its strongest form relies on the use of the higher derivative method of regularization. However, the implementation of this method to $N = 2$ superspace theories has not been worked out in every detail. There is a variant of the anomalies argument that uses the Adler-Bardeen theorem,[128] and in this version it has been used to establish the finiteness of a class of $N = 1$ supersymmetric theories up to two loops,[101,102] and in this sense the anomalies argument has a wider applicability.

The non-renormalization argument, and in its detailed use the anomalies argument, relies on the $N = 2$ super-Feynman rules. A detailed account of these rules has not been given; but it is clear that they will lead to severe *off-shell* infrared divergences, that is, divergences of the form $\int d^4k/k^4$. In order to make sense out of this formalism, these divergences must be firstly regulated and secondly removed. In doing this one may lose the manifest nature of some of the symmetries of the theory. For example, adding mass terms breaks gauge invariance, while putting the theories in a periodic box[134] breaks Lorentz invariance. It is not clear that infrared divergences will not lead to an evasion of the non-renormalization theorem.[135]

A further problem concerns the fact that when $N = 2$ matter belongs to a complex representation, the $N = 2$ matter must be doubled in order to admit a relaxed hypermultiplet superspace formulation.[120] However, it has been verified[122] by explicit computation that even with odd numbers of complex representations, the theories are finite at two loops. Finally, there are no detailed studies on how regulating one-loop infinities by non-supersymmetric schemes may pollute higher order diagrams containing these one-loop divergences. These latter objections of course apply to both the anomalies and non-renormalization arguments.

There has been little discussion of the shortcomings of the light-cone approach. However, it is far from clear how one recovers Lorentz invariance. Presumably, one must verify that the non-linear Ward identities for Lorentz invariance are valid. This may, however, involve adding finite local counter terms to the theory. This argument has yet to be extended to $N = 2$ theories.

Here we have stressed these weaknesses not because of a mistrust in the arguments for finiteness, but to show that they are not proofs in a mathematical sense and that there is still room for further work.

It is possible that the non-renormalization and anomalies argument will apply to two-dimensional extended σ-models, which also have remarkable finiteness properties.[136] However, since supergravity theories involve dimensional coupling constants, and both the anomalies and non-renormalization arguments rely on dimensional analysis, it is easily seen that these arguments can be evaded. For a review of the status of infinities in supergravity theories see Refs. 121 and 137. An interesting recent calculation in this context is given in Ref. 138.

18.4   Explicit Breaking and Finiteness

Given the large class of finite theories discussed in the previous section, we now wish to investigate whether adding soft terms can preserve their finiteness. By soft terms we mean terms of dimension 3 or less that are gauge invariant and parity conserving. Inserting a soft term into any graph lowers its degree of divergence; however, these terms also explicitly break the supersymmetry and internal symmetry that are responsible for the finiteness of these very special theories. We will find that not all soft terms preserve finiteness, but only certain combinations of soft terms.

The first soft terms which were found to maintain finiteness were the addition of $N = 1$ supersymmetric mass terms to $N = 4$ Yang-Mills theory.[139] A general analysis giving the necessary and sufficient conditions for a soft term to maintain finiteness in $N = 4$ Yang-Mills theory was given using the spurion technique in Ref. 140. However, some authors[141] have independently found finiteness-preserving soft terms for $N = 4$ Yang-Mills by using the light-cone formalism, while an analysis at the level of component fields which found some of the soft terms that preserve finiteness was given in Ref. 142.

One particular soft term of the form $(A^2 - B^2)$ which was found in Refs. 140 and 141 was also later found in Ref. 143 by using an extension of the light-cone formalism of Ref. 141.

The analysis of the necessary and sufficient conditions for a soft term to preserve finiteness in the finite class of $N = 2$ theories was given in Ref. 144 by using the spurion technique. Some of these results were also found by calculating at the level of component fields.[145]

We will follow Refs. 140 and 144 and use the spurion technique[146] to investigate the divergences induced by soft terms. Let us consider, to begin with, a general $N = 1$ super-symmetric theory consisting of $N = 1$ Yang-Mills,

V and Wess-Zumino multiplets ϕ. We will denote the x-space component fields of ϕ by (A, B, χ_A, F, G) and those of $N = 1$ Yang-Mills by (A_μ, λ, D).

As an example, consider adding the term $\mu^2(A^2 - B^2)$ to the supersymmetric action. In order to still work with the superfield formalism, and in particular with the super-Feynman rules, we will rewrite this addition in superspace by introducing the spurion superfield,

$$S = \mu^2 \theta^2 \qquad (18.21)$$

where $\theta^2 = \theta^A \theta_A$ and μ^2 is a constant. The superfield S is a chiral superfield in the sense that $D_{\dot{A}} S = 0$, but it is not a scalar superfield in that it does not transform correctly under supersymmetry. The $\mu^2(A^2 - B^2)$ term is now added by including in the action the term

$$\int d^4x \, d^2\theta \, S\phi^2 + \text{h.c.} = \int d^4x \, \mu^2(A^2 - B^2) \qquad (18.22)$$

We can now calculate quantum processes using the super-Feynman rules used previously. We have the same propagators and vertices as before except for the extra vertex given in Eq. (18.22). The S superfield is only an external field, and the additional vertex has only one $-1/4 \, \bar{D}^2$ factor associated with one of the two chiral lines as shown below.

Fig. 18.1.

This assignment of $\bar{D}^2$ factors is so as to have an integral over the whole of $N = 1$ superspace at the vertex. Consequently, the non-renormalization theorem is still valid. Counting dimensions we find that the dimension of S is one. The spurions necessary to introduce all other possible soft terms are listed in Table 18.1.

In Table 18.1, μ_1, m, n, ξ and e are constants and the group indices are not displayed, but are understood to be present in the appropriate places. The factors of V are necessary to maintain gauge invariance and are important for obtaining the correct answer when calculating in superspace. Their component expressions, however, have been evaluated for simplicity in the Wess-Zumino gauge.

Table 18.1.

Spurion	Addition	Dimension of spurion
$U_1 = \mu_1^2 \theta^2 \bar{\theta}^2$	$\int d^4x\, d^4\theta\, U_1 \bar{\phi}(e^{gV})\phi = \int d^4x\, \mu_1^2(A^2 + B^2)$	0
$S_1 = \theta^2 \xi$	$\int d^4x\, d^4\theta\, S_1 \phi^3 + \text{h.c.} = \int d^4x\, \xi(A^3 - 3AB^2)$	0
$M = \theta^2 \bar{\theta}^2 m$	$\int d^4x\, d^4\theta\, M \mathscr{D}^A \phi \mathscr{D}_A \phi + \text{h.c.} = \int d^4x\, m\chi^A \chi_A + \text{h.c.}$	-1
$N = \theta^2 \eta$	$\int d^4x\, d^2\theta\, N W^A W_A + \text{h.c.} = \int d^4x\, \eta\lambda^A \lambda_A + \text{h.c.}$	0
$E = \theta^2 \bar{\theta}^2 e$	$\int d^4x\, d^4\theta\, E(e^{gV}\phi e^{-gV}\bar{\phi}^2) + \text{h.c.} = \int d^4x\, eA(A^2 + B^2)$	-1

All of the above insertions produce new vertices which can be used to construct super-Feynman graphs. The $\chi^A \chi_A$ insertion, for example, gives the additional vertices shown in the diagram below.

Fig. 18.2.

Any new infinities induced by the addition of the above insertions will obey the non-renormalization theorem and so be of the generic form

$$\int d^4x\, d^4\theta\, S(D^2 S)^r U(D^2 \bar{D}^2 U)^P f(\phi, V, D\phi, \dots) \qquad (18.23)$$

where S is a chiral spurion superfield, $S = s\theta^2$, U is a general spurion superfield, $U = u\theta^2 \bar{\theta}^2$, and f is a gauge-invariant function of ϕ and V.

Let us return to the addition of only the $A^2 - B^2$ term for the purposes of illustration. The induced infinity will be of the form

$$\int d^4x\, d^4\theta\, S(D^2 S)^r f(\phi, V, D\phi, \dots) \qquad (18.24)$$

As the dimension of S is one, the only possible infinity is of the form

$$\int d^4x\, d^4\theta\, S\bar{\phi} \tag{18.25}$$

However, this term is only gauge-invariant if ϕ is a gauge singlet. Consequently in the absence of gauge singlets any $A^2 - B^2$ insertion induces no new types of infinity. Using an inductive argument of Weinberg,[152] let us assume that there are no induced infinities at n loops; then the induced infinities at $n + 1$ loops arise either as an overall superficial divergence or as subdivergences. The latter infinity is absent by assumption while the former type of infinity is absent since there are no S-dependent infinities. Consequently we arrive at the following result.

Theorem: In any $N = 1$ supersymmetric theory an $A^2 - B^2$ term induces no additional infinities if there are no gauge singlets in the theory.

Let us now consider an $A^2 + B^2$ addition. The resulting induced infinity can only be of the form

$$\int d^4x\, d^4\theta\, U\bar{\phi}\phi + \left(\int d^4x\, d^4\theta\, u\phi^2 + \text{h.c.} \right) \tag{18.26}$$

These infinities are of the form $A^2 + B^2$ and $A^2 - B^2$; however, the latter is often forbidden by symmetry arguments.

The $\bar{\chi}\chi$ insertion can induce in a general theory the infinities

$$\int d^4x\, d^4\theta [U(\phi\bar{\phi}^2) + U\phi D^2\phi + U(\phi^3) + U(D^2\bar{D}^2 U)\bar{\phi}\phi$$

$$+ U(D^2\bar{D}^2 U)\bar{\phi}^2 + \text{h.c.}] \tag{18.27}$$

plus possible linear terms.

The reader is referred to Ref. 140 for the infinities produced by the other additions.

We will now return to the task at hand and consider which soft terms preserve finiteness in those $N = 2$ rigid supersymmetric theories that are finite. In terms of an $N = 1$ superfield description, $N = 2$ Yang-Mills is composed of $N = 1$ Yang-Mills V and one Wess-Zumino multiplet ϕ in the adjoint representation, while $N = 2$ matter consists of σ chiral multiplets X^a_σ in the representation R_σ and σ chiral multiplets $Y_{a\sigma}$ in the representation $\bar{R}_\sigma$. The index a labels the elements of the representation R_σ. The action of $N = 2$ rigid supersymmetric theories written in terms of these $N = 1$ superfields is

$$A = \text{Tr} \int d^4x \, d^2\theta \frac{W^A W_A}{64g^2}$$

$$+ \int d^4x \, d^4\theta [\bar{\phi}^s (e^{gV})_s{}^t \phi_t + \bar{X}_{a\sigma} (e^{gV^\sigma})^a{}_b X^b{}_\sigma + Y_{a\sigma} (e^{-gV^\sigma})^a{}_b \bar{Y}^b{}_\sigma]$$

$$+ g \int d^4x \, d^2\theta \, \phi_s (R_s^\sigma)^a{}_b X^b{}_\sigma Y_{a\sigma} + \text{h.c.}$$

$$+ \text{ gauge fixing } + \text{ ghosts} \tag{18.28}$$

In the above $(V^\sigma)^a{}_b = V^s (R_s^\sigma)^a{}_b$ and $(R_s^\sigma)^a{}_b$ are the generators of the group G in the representation R^σ. For the adjoint representation we have, for example,

$$(T_s)_{lk} = -f_{slk} \tag{18.29}$$

where f_{slk} are the structure constants of G. We recall that these theories are finite if and only if $C_2(G) = \sum_\sigma T(R^\sigma)$.

Clearly, from our previous discussion the addition of any term of the form $A^2 - B^2$ maintains finiteness. Let us consider adding an $A^2 + B^2$ term, which is achieved by adding the terms

$$\int d^4x \, d^4\theta [U_1 \bar{\phi}^s (e^{gV})_s{}^k \phi_k + U_{2\sigma} \bar{X}_{a\sigma} (e^{gV^\sigma})^a{}_b X^b{}_\sigma + U_{3\sigma} Y_{a\sigma} (e^{-gV^\sigma})^a{}_b \bar{Y}^b{}_\sigma] \tag{18.30}$$

where the spurion superfields are of the form

$$U_1 = \mu_1 \theta^2 \bar{\theta}^2 \qquad U_{i\sigma} = \mu_{i\sigma} \theta^2 \bar{\theta}^2 \qquad i = 2, 3 \tag{18.31}$$

The induced infinities can only be of the generic form

$$\int d^4x \, d^4\theta \, U(\bar{\phi}\phi + \bar{X}X + Y\bar{Y}) \tag{18.32}$$

as a ϕ^2, X^2 or Y^2 term is ruled out by the symmetry

$$\phi \to e^{2i\alpha}\phi, \qquad X \to e^{-i\alpha}X, \qquad Y \to e^{-i\alpha}Y \tag{18.33}$$

which is a symmetry both of the $N = 2$ action and of the spurion insertion of Eq. (18.30).

We now consider these induced infinities at one loop. The relevant diagrams are given in Fig. 18.3. These graphs can be evaluated most easily using the following argument. Consider any propagator that contains a U insertion and is part of a larger graph (see Fig. 18.4). Using the super-Feynman rules we may evaluate this part of the graph and see that it contributes the factor

$$\int d^4\theta_2 \left(-\frac{1}{4}\bar{D}_1^2\right)\left(-\frac{1}{4}D_2^2\right)\frac{\delta_{12}}{k^2} U(2) \left(-\frac{1}{4}\bar{D}_2^2\right)\frac{\delta_{23}}{k^2} \tag{18.34}$$

Fig. 18.3.

$$\begin{array}{c} 2 \\ \underset{1}{\overline{}} \overset{}{\underset{I\,U}{\vert}} \underset{3}{\overline{}} \end{array}$$

Fig. 18.4.

to the total expression for the graph. Integrating the D's by parts and using the fact that any infinity cannot contain a D acting on U, we find the infinite part of the graph contains the factor

$$\int d^4\theta_2 \frac{\delta_{12}}{k^2} U(2) \frac{\bar{D}_2^2 D_2 \bar{D}_2^2}{16\cdot(-4)} \frac{\delta_{23}}{k^2} = -\int d^4\theta_2\, \delta_{12} U(2)\left(-\frac{1}{4}\bar{D}_2^2\right)\frac{\delta_{23}}{k^2}$$

$$= -U(1)\left(-\frac{1}{4}\bar{D}_1^2\right)\frac{\delta_{13}}{k^2} \tag{18.35}$$

That is, the graph with the U insertion is equal to $(-U)$ times the graph with no U insertion.

Similarly we find that a U insertion on a $\bar{\phi}(gV)^n\phi$ vertex yields a graph which is equal to $(+U)$ times the graph with no U insertion.

Taking the case of external $X\bar{X}$ lines; the one-loop graphs are shown in Fig. 18.5.

Fig. 18.5.

Using the above discussion and the one-loop finiteness condition which is diagramatically given below,

Fig. 18.6.

we find the induced infinity is of the form

$$C(R) \int d^4x \, d^4\theta \, X_{a\sigma} \bar{X}^{a\sigma} [(-U_1 - U_{3\sigma}) - (-U_{2\sigma}) - (U_{2\sigma}) - (U_{2\sigma}) + (0)]$$

$$= -C(R) \int d^4x \, d^4\theta \, X_{a\sigma} \bar{X}^{a\sigma} \{U_1 + U_{2\sigma} + U_{3\sigma}\} \tag{18.36}$$

We note that the last graph vanishes as there are not four D's contained in the vector loop. Hence, there are no $\bar{X}X$ infinities at one loop provided

$$U_1 + U_{2\sigma} + U_{3\sigma} = 0 \qquad \forall \sigma \tag{18.37}$$

The $\bar{Y}Y$ infinities also vanish in this case due to the $X \leftrightarrow Y$, $U_{2\sigma} \leftrightarrow U_{3\sigma}$ interchange symmetry. In fact, the $\bar{\phi}\phi$ one-loop infinities also vanish when Eq. (18.37) holds and one uses the finiteness condition for $N = 2$ theories.

We can summarize the above discussion by the statement that $A^2 + B^2$ additions preserve finiteness at one loop if and only if Eq. (12.17) holds. By a straightforward generalization of the above discussion it can be shown that an $A^2 + B^2$ term preserves finiteness to all orders provided Eq. (18.37) holds. The reader is referred to Ref. 144 for an account of this argument.

For the case of $N = 4$ Yang-Mills theories, we have only one species of $N = 2$ matter fields which are in the adjoint representation, and then the finiteness condition becomes $U_1 + U_2 + U_3 = 0$. This is equivalent to the statement that

$$S \operatorname{Tr} m^2 = \sum_j m_j^2 (-1)^{2j+1} = 0$$

In $N = 2$ theories, however, $S \operatorname{Tr} m^2$ is in general non-zero.

The addition of all possible soft terms and the resulting induced one-loop infinities, including infinities due to mixed insertions, are given in Ref. 108. The following is a schematic description of this analysis. Denoting the physical field component content of any of the chiral fields $X^a{}_\sigma$, $Y_{a\sigma}$, ϕ by A, B, χ, and

Table 18.2.

Insertion	Infinity Produced					
	$A^2 - B^2$	$A^2 + B^2$	$\bar{\chi}\chi$	$\bar{\lambda}\lambda$	$A(A^2 + B^2)$	$A^3 - 3AB^2$
$A^2 - B^2$						
$A^2 + B^2$		✓				
$\bar{\chi}\chi$	✓	✓			✓	
$\bar{\lambda}\lambda$	✓	✓				✓
$A(A^2 + B^2)$	✓	✓			✓	
$A^3 - 3AB^2$	✓	✓				✓

the spinor in the Yang-Mills multiplet by λ, the one-loop infinities are given in Table 18.2. A tick indicates the appearance of an infinity. Consider adding a $\chi\chi$ insertion; it gives rise to an infinity of the form $A(A^2 + B^2)$. The only way this infinity can be cancelled is by adding an appropriate $A(A^2 + B^2)$ soft term and arranging its coefficient such that the $A(A^2 + B^2)$ infinities cancel. Once this has been carried out, it is found that the $A^2 - B^2$ infinities cancel automatically. The remaining $A^2 + B^2$ infinities do not cancel, but they can be cancelled by adding an appropriate $A^2 + B^2$ soft term. The resulting soft insertions that produce no infinities are of the form

$$m^2(A^2 + B^2) + m\bar{\chi}\chi + mA(A^2 + B^2)$$

Examination of the coefficients reveals that this term is none other than an $N = 1$ supersymmetric mass term, and so can be rewritten in the form

$$m \int d^2\theta \, d^4x \, \text{Tr} \, \phi^2 + \text{h.c.} \tag{18.38}$$

for the case of ϕ, and similarly for X_σ and Y^σ.

An alternative set of soft insertions that also induces no infinities is found by adding a mass term for the gaugino $\bar{\lambda}\lambda$. The resulting $A^3 - 3AB^2$ infinity can only be cancelled by adding a term of the same form, i.e., $A^3 - 3AB^2$ with an exactly chosen coefficient. Again the $A^2 - B^2$ infinity cancels automatically and the remaining $A^2 + B^2$ infinity can be cancelled by adding an appropriate $A^2 + B^2$ term. The resulting combination of terms is of the generic form

$$m\bar{\lambda}\lambda + m^2(A^2 + B^2) + m(A^3 - 3AB^2) \tag{18.39}$$

Although this is not an $N = 1$ mass term, it is like an $N = 1$ mass term in the sense that it is related by $O(2)$ invariance to the mass term of Eq. (12.19); as such one would expect this term to preserve finiteness.

The $N = 1$ supersymmetric mass terms preserve finiteness to all orders. This results from the fact that a term like $\int d^4x \, d^2\theta \, m\phi^2$ as well as similar terms for X and Y cannot be generated as a result of the non-renormalization theorem. We can also be confident that the "$N = 1$ like mass terms" preserve finiteness to all orders as they are related to $N = 1$ mass terms by an $O(2)$ rotation.

Consequently we may summarize this section by listing the necessary and sufficient conditions that preserve finiteness in $N = 2$ theories: The soft terms must be expressible as a linear combination of

(1) $N = 1$ supersymmetric masses,
(2) $N = 1$ like mass terms, [i.e., $(m\bar{\lambda}\lambda + \cdots)$],
(3) any $A^2 - B^2$ mass,
(4) $A^2 + B^2$ masses provided they satisfy

$$U_1 + U_{2\sigma} + U_{3\sigma} = 0 \qquad \forall \sigma \qquad (18.40)$$

One can also consider whether soft terms preserve the finiteness of the $N = 1$ theories that have been shown to be finite for two-loops.[101,102] Using arguments similar to those presented above for the $N = 2$ case, we find that of the terms listed above terms of type (1), (3) and (4) preserve finiteness to two loops,[101] while more recently it has been shown that terms of type (3) preserve one-loop finiteness.[163] Softly broken realistic models of $N = 2$ supersymmetry have been considered in Ref. 164.

Chapter 19

Spontaneous Breaking of Supersymmetry and Realistic Models

In this chapter, we wish to explain some of the general features of the present realistic models of sypersymmetry. In particular, we would like to explain why supersymmetry is broken at some intermediate scale (i.e. 10^{10} GeV) and why it was so desirable to include supergravity. We do this by discussing possible ways of breaking supersymmetry; it will emerge that there are tight theoretical constraints on how supersymmetry can be broken and on the resulting mass spectra. The need to find a model, in which the mechanism of supersymmetry breaking results in sufficiently large masses for the unobserved scalar super-partners of the quarks and leptons, leads to the form of the present realistic models.

Supersymmetry is broken if and only if the supercharge Q_α does not annihilate the vacuum

$$Q_\alpha |0\rangle \neq 0 \tag{19.1}$$

Given a field φ,

$$\langle 0|\delta\varphi|0\rangle = \langle 0|[\varphi, Q_\alpha\}|0\rangle \tag{19.2}$$

and so supersymmetry is broken if and only if

$$\langle 0|\delta\varphi|0\rangle \neq 0 \qquad \text{for some } \delta\varphi \tag{19.3}$$

Consider the fields of the Wess-Zumino model $(A, B, \chi_\alpha, F, G)$; the Lorentz invariance of the vacuum implies that $\langle \chi_\alpha \rangle = \langle \partial_\mu A \rangle = \langle \partial_\mu B \rangle = 0$ and so the only field variation with a possible non-zero vacuum expectation value is

$$\langle 0|\delta\chi_\alpha|0\rangle = ((\langle F \rangle + i\gamma_5\langle G \rangle)\varepsilon)_\alpha \tag{19.4}$$

Similarly, for the fields of $N = 1$ supersymmetric Yang-Mills theory, the only possible non-zero variation is

$$\langle 0|\delta\lambda_\alpha|0\rangle = i(\gamma_5)_\alpha{}^\beta \varepsilon_\beta \langle 0|D|0\rangle \tag{19.5}$$

Consequently, in any rigid theory supersymmetry is broken if and only if one or more of the auxiliary fields (F, G or D) acquires a vacuum expectation value.

As with the spontaneous breaking of any rigid symmetry one can establish the presence of a massless (Goldstone) mode, in this case a Goldstone spinor.

The Goldstone spinor can be identified once the vacuum expectation values of the auxiliary fields are known. One redefines the spinors such that only one spinor has a non-zero vacuum expectation value when varied; this spinor, associated with the non-zero vacuum expectation value, is the Goldstone spinor. These considerations hold if supersymmetry is broken at the quantum or classical level. We now consider the latter case in more detail.

19.1   Tree-level Breaking of Supersymmetry

The most general renormalizable rigid model of $N = 1$ supersymmetry is given in terms of component fields in Eq. (11.44). The tree-level potential has the form

$$V = |f^a|^2 + \frac{1}{2}(D^s)^2. \tag{19.6}$$

where

$$\begin{aligned} f_a^* &= m_{ab}z^b + d_{abc}z^bz^c + \mu_a \\ D^s &= -g(T^s)^a{}_b z_a^* z^b + \xi^s \end{aligned} \tag{19.7}$$

The addition of field independent terms in f^a and D^s requires the existence of singlet chiral fields or $U(1)$ factors in the gauge group respectively. Clearly V is positive semi-definite and, recalling the considerations of the previous section, supersymmetry is broken if and only if V is greater than zero. This result can also be shown directly using the supersymmetry algebra. The energy P_0 can be expressed as

$$\sum_A \{(Q^{Ai})^* Q^{Ai} + Q^{Ai}(Q^{Ai})^*\} = 2P_0 \qquad \forall i \tag{19.8}$$

Hence the vacuum energy is given by

$$V = \langle 0|P_0|0 \rangle = \sum_A \{\|Q^{Ai}|0\rangle\|^2 + \|(Q^{Ai*}|0\rangle\|^2\} \tag{19.9}$$

Hence supersymmetry is broken if and only if $V > 0$.

Examining Eq. (19.9) we observe that if one supersymmetry is preserved i.e., $Q^{A1}|0\rangle = 0$ then the vacuum energy vanishes (i.e., $\langle 0|P_0|0 \rangle = 0$. However, in this case

$$Q^{Ai}|0\rangle = 0 \qquad \forall i \tag{19.10}$$

and supersymmetry is preserved by all the supercharges. This result is summarized in the following theorem

Theorem[99]: In any rigid supersymmetric theory, either all the supersymmetries are broken or none at all.

Examining Eq. (19.7) we observe that if there are no terms linear in the auxiliary fields in the action (i.e., $\mu^\alpha = \xi^S = 0$) then an absolute minimum of the potential is given by taking all the spin-zero fields to have zero vacuum expectation value (i.e., $\langle z^\alpha \rangle = 0$). As such, supersymmetry is not broken in these models. It is often the case that the absolute minimum is degenerate. In other words, there exists non-zero vacuum expectation value field configurations that also have $V = 0$. In this latter case, although supersymmetry is not broken, gauge invariance may be broken.

Clearly, adding a term linear in any auxiliary field to the action does not in general lead to the breaking of supersymmetry. The matter fields can often take vacuum expectation values that still ensure the vanishing of the auxiliary fields.

However, by a judicious choice of matter representations and the addition of suitable linear terms one can break supersymmetry. There are two well known ways of doing this, corresponding to the possibility of adding terms linear in the matter or the gauge sector. We now give these two mechanisms.

The Fayet-O'Raifeartaigh Mechanism[147]

Consider a Wess-Zumino model that contains only three chiral multiplets $\varphi_i(i = 1, 2, 3)$ and has the superpotential

$$W(\varphi_i) = \lambda \varphi_1 \varphi_2 + g \varphi_3 (\varphi_2^2 - m^2) \qquad (19.11)$$

Clearly, supersymmetry must be broken as

$$f_1 = \frac{\partial W}{\partial z_1} = \lambda z_2 \quad f_2 = \frac{\partial W}{\partial z_2} = \lambda z_1 + g z_2 z_3 \quad f_3 = \frac{\partial W}{\partial z_3} = g(z_2^2 - m^2) \quad (19.12)$$

and there is no choice of z^i that can simultaneously make f_i vanish for all i. The potential energy is given by

$$V = \lambda^2 |z_2|^2 + g^2 |z_2^2 - m^2|^2 + |\lambda z_1 + 2g z_2 z_3|^2 \qquad (19.13)$$

The minimum of this potential occurs when

$$\lambda^2 z_2 + 2g^2 z_2^*(z_2^2 - m^2) = 0$$

$$(\lambda z_1 + 2g z_2 z_3) = 0 \qquad (19.14)$$

Clearly the position of the minimum is degenerate and is only determined up to

$$\lambda z_1 + 2g z_2 z_3 = 0 \qquad (19.15)$$

with the value of z_2 being given by

$$\lambda^2 z_2 + 2g^2 z_2^*(z_2^2 - m^2) = 0 \qquad (19.16)$$

The occurrence of a degeneracy at the minimum of a potential in a supersymmetric theory is not an uncommon feature. Although this degeneracy is not removed, by quantum corrections, when supersymmetry is not broken at the tree level,[90] in this case however, we expect it to be removed, since supersymmetry is broken at the tree level. It has been suggested[99,148,149] that the vacuum expectation values of the degenerate fields may be very large and this might be used to solve the hierarchy problem.

It is easy to extend the above mechanism to theories with gauge invariance and have the gauge invariance spontaneously broken.

The Fayet-Iliopoulos Mechanism[150]

Consider the action of super QED as given in Eq. (11.43). To this we add a mass term and the Fayet-Iliopoulos term

$$\left[-\frac{m}{2}(S_1 \cdot S_1) + (1 \leftrightarrow 2) \right]_F + [\xi V]_D$$

$$= \int d^4x \left\{ \left[m \left(F_1 A_1 + G_1 B_1 - \frac{1}{2}\bar{\chi}_1 \chi_1 \right) + (1 \leftrightarrow 2) \right] + \xi D \right\} \quad (19.17)$$

The latter term is invariant as D is inert under gauge transformations. The expressions for the auxiliary fields are

$$F_i = -mA_i \qquad G_i = -mB_i \qquad i = 1, 2$$
$$D = -(g(A_1 B_2 - B_2 A_1) + \xi) \qquad (19.18)$$

Again there is no value of the spin-zero fields that leads to vanishing auxiliary fields and so supersymmetry must be broken. We refer the reader to the original references for the mass spectrum in these two models.

We now make some general observations concerning the masses produced by these supersymmetry breaking mechanisms. An important quantity is

$$S \operatorname{Tr} M^2 = \sum_j (-1)^{2j}(2j + 1)m_j^2. \qquad (19.19)$$

which is the sum of the mass splittings in each supermultiplet. In a theory in which the only auxiliary fields that have non-zero vaccum expectation values are from chiral multiplets, we have the relation.

$$S \operatorname{Tr} M^2 = 0 \qquad (19.20)$$

This result[151] is a consequence of the fact that in these theories the mass splitting originates from terms of the form

$$\int d^4x \, d^2\theta \langle \varphi \rangle \varphi^2 + \text{h.c.} = \int \langle F \rangle (A^2 - B^2) d^4x \qquad (19.21)$$

and hence the mass splitting in each supermultiplet vanishes. These supermultiplets have the problem that one of the scalar masses is lower than the fermion mass while the other is higher.

In a general theory we find that[151]

$$S\,\mathrm{Tr}\,M^2 = -2\xi_a \langle D^a \rangle \,\mathrm{Tr}\,Y_a \tag{19.22}$$

In the above equation Y_a are the generators of the additional $U(1)$ in the gauge group, $\langle D^a \rangle$ are the vacuum-expectation values of the corresponding gauge auxiliary fields and $\mathrm{Tr}\,Y_a$ is the sum of the $U(1)$ charges that the chiral multiplets possess. These mass splittings originate from the term

$$\sum_a \int d^8 z\,\bar\varphi \langle V^a \rangle \varphi Y_a = \int \sum_a Y_a \langle D^a \rangle (A^2 + B^2)\,d^4x \tag{19.23}$$

A problem that occurs with the Fayet-Iliopoulos mechanism is the occurrence of anomalies associated with the $U(1)$ factors needed to break the supersymmetry. There will be no gauge anomalies if

$$\mathrm{Tr}\,Y_a{}^3 = 0 \tag{19.24}$$

However, there are also mixed anomalies arising from a triangle graph with one gauge and two gravity vertices. These anomalies vanish if and only if

$$\mathrm{Tr}\,Y_a = 0 \tag{19.25}$$

The $U(1)$ of the standard model has $\mathrm{Tr}\,Y = 0$ to avoid mixed $U(1)$ and $SU(2)$ gauge anomalies as well as to avoid mixed gravitational anomalies and $U(1)$ anomalies. Hence this $U(1)$, if it is anomaly free, can not give a contribution to $S\,\mathrm{Tr}\,M^2$ and the above discussion demonstrates that for anomaly free theories

$$S\,\mathrm{Tr}\,M^2 = 0 \tag{19.26}$$

at the tree level. We note that it is natural to have $\mathrm{Tr}\,Y = 0$ as $\mathrm{Tr}\,Y = 0$ ensures that no D term is generated by quantum corrections.[152]

19.2 Quantum Breaking of Supersymmetry

In view of the restricted nature of the mass pattern achieved by the tree-level breaking of supersymmetry it might be hoped that supersymmetry could be broken by quantum corrections. However, as discussed in Chapter 17 it can be shown that [90,99] "If supersymmetry is not broken at the classical level then it will not be broken by any perturbative corrections." It follows from the proof of this theorem that the quantum effective potential must be of the form

$$V \sim |\langle f \rangle|^2 g(\langle z \rangle) + \langle D \rangle^2 h(\langle z \rangle) \tag{19.27}$$

where g and h are arbitrary functions.

Consequently it is necessary either to break supersymmetry at the tree level or to find some non-perturbative mechanism for breaking supersymmetry. As will be explained later, $S\operatorname{Tr} M^2 = 0$ is not phenomenologically acceptable. Since the tree-level breaking of supersymmetry does lead, at tree level to $S\operatorname{Tr} M^2 = 0$, we must hope that $S\operatorname{Tr} M^2$ is modified by quantum corrections. Fortunately, this is the case, when supersymmetry is broken at the tree level, $S\operatorname{Tr} M^2$ acquires quantum corrections. Clearly, these corrections must be proportional to the supersymmetry-breaking parameter and as the contribution to the effective action is an integral over the whole of superspace (see chapter 17) they are of the form

$$S\operatorname{Tr} M^2 \sim \left(\frac{\langle F \rangle^2 + \langle D \rangle^2}{M^2}\right)\lambda^n \qquad (19.28)$$

where M is the dominant mass in the process and λ are some appropriate coupling constants.

19.3 The Gauge Hierarchy Problem[153,154]

Let us first consider a theory with a momentum cutoff Λ. All quantities in the theory including the parameters (masses, couplings, etc.) are functions which can be calculated in terms of Λ. If the theory is renormalizable the only terms which diverge as Λ goes to infinity can be absorbed into the parameters of the theory. The coupling constants and wave-function renormalizations, on dimensional grounds, can only diverge as $\ln \Lambda$. The same is true of fermion masses; as a result of chiral transformations the fermion mass corrections must be proportional to the fermion mass. There is in a general theory no symmetry to protect the scalar mass and they will, in general, diverge as Λ^2. Examples of one loop graphs that display this type of divergence are given below

Fig. 19.1.

where $---$ are scalar propagators, $\sim\!\sim$ are vector propagators and $-$ are fermion propagtors. These diagrams lead to

$$\mu^2(\Lambda) = \mu^2_0 + \Lambda^2(c_1 \lambda + c_2 g^2 + \cdots) \qquad (19.29)$$

Consider, now a theory with two cutoffs Λ_1 and Λ_2 where $\Lambda_1 \gg \Lambda_2$ and $\Lambda_2 \sim 0$. The masses calculated with these different cutoffs are related by

$$\mu^2(\Lambda_1) \sim \mu^2(\Lambda_2 \sim 0) + \Lambda_1^2(c_1\lambda + c_2 g^2 + \cdots) \qquad (19.30)$$

This equation means that the scalar masses calculated with and without the high energy effects of the theory differ by a term $\sim \Lambda^2$ (in GUT theories it is natural to have $\Lambda^2 \sim 10^{26}\ (\text{GeV})^2$). Now $\mu^2(\Lambda_2 \sim 0)$ is required to be of order the weak scale if it is the Higgs which is responsible for the symmetry breaking $SU(2) \times U(1) \to U(1)$. Consequently, we must tune $\mu^2(\Lambda_1)$ very finely to cancel almost exactly the Λ^2 term. This is considered unnatural in the sense that one must tune very finely the microscopic parameters of the theory in order to have a theory that reproduces the macroscopic world as we see it. This is the technical gauge hierarchy problem. A deeper question is why there should be such disparate scales in nature. As we shall see supersymmetry can solve the technical gauge hierarchy problem.

The above discussion can be rephrased without reference to the possibly unphysical cutoffs, Λ. Consider a change Δm^2 in the high energy behavior of the masses of the theory. Although this effect will be suppressed by Λ^{-2} the change in $\Delta\mu^2$ is

$$\Delta\mu^2 \sim \Lambda^2 \frac{\Delta m^2}{\Lambda^2} \sim \Delta m^2 \qquad (19.31)$$

Hence the change in $\Delta\mu^2$ is not protected in any way from the high energy behavior of the theory. Put another way, the problem is that the theory does not possess a systematic perturbation theory. That is, if $\mu^2(0)$ is small at the classical level, its one-loop correction is large.

Of course, we usually take the cutoff Λ to infinity and remove all reference to it, in physical quantities, by renormalizing the theory. The above discussion still survives as we now show. To be concrete, let us consider $SU(5)$ and let Σ be the **24** and H the **5**. At the tree level $\langle\Sigma\rangle = M_x/g$. The one-loop effective action contains a term $\lambda\,\bar{H}\Sigma^2 H$ from the graph below.

Fig. 19.2.

Taking the momenta in the external H line to be μ^2 and M_x^2 respectively we find that

$$\lambda(\mu^2) \sim \lambda(M_x^2) + \frac{g^4}{16\pi^2} \ln \frac{M_x^2}{\mu^2} \tag{19.32}$$

The induced H mass is $\lambda\langle\Sigma\rangle^2$. Even if $\lambda(\mu^2)$ is very small at tree level it is of order g^4 at one loop and unless we tune $\lambda(M_x^2)$ very finely the induced Higgs mass will be unacceptably large.

In supersymmetric theories the scalars are in the same supermultiplet as the spinors. Namely one may suppose that as spinors have only logarithmic divergences the scalars must also have only logarithmic divergences or less. Indeed, this is the case, the graphs which lead to the quadratic divergences are partially canceled by additional graphs. In the example above, the two graphs which cancel are

Fig. 19.3.

In this case, we find that

$$\lambda(\mu^2) \sim \lambda(M_x^2) \ln\left(\frac{\mu^2}{M_x^2}\right)$$

The theoretical reason for this is the well-known renormalization properties of supersymmetric theories. In particular, if supersymmetry is preserved then the effective potential vanishes.[90] As such, the graphs considered above, which are proportional to the gauge coupling g^4, contribute to the effective potential and so must vanish. Consequently, the renormalization of masses and chiral couplings must be balanced by the corresponding chiral wave function renormalizations according to Eqs. (17.109), (17.110) and (17.111). Thus, couplings in the chiral potential are renormalized in such a way that the bare coupling is proportional to renormalized coupling times the chiral renormalizations which, being dimensionless, have at most a logarithmic dependence on the mass scales of the theory. Thus they scale like those of the above equation and do not contain the analogue of the second term in Eq. (19.32). Similarly, bare and renormalized masses are related through the chiral wave function renormalization and as a result, the second terms with a quadratic on Λ, in Eq. (19.29), are absent.

Of course, in general we do not expect the above properties to hold much below the scale of any supersymmetry breaking since the above graphs will no longer cancel and will result in contributions proportional to the super-symmetry-breaking scale. We wish however, still to maintain the solution to the technical gauge-hierarchy problem while still having supersymmetry broken. This places restrictions on the scales and possible ways in which supersymmetry is broken.

Roughly speaking the Higgs mass is of order 10^2 GeV and we expect the contribution from radiative corrections to be $g^2 M_s$ where M_s is the scale of supersymmetry breaking. To preserve the value of the Higgs mass we then find that $M_s \sim 10^3$ GeV for $g^2 \sim 10^{-1}$. However, this simplistic analysis may not be valid if the sector of the theory that breaks supersymmetry is isolated, by some special mechanism, from the standard model sector. We now give a more detailed discussion of how the supersymmetry-breaking scale is controlled by demanding that the technical gauge-hierarchy problem is solved.

If supersymmetry breaking leads to a mass splitting Δm in the superpartners (i.e. the gauge boson and gaugino) circulating round the above loop, then the induced Higgs mass is of the form

$$\frac{g^2}{(4\pi)^2} \Delta m \tag{19.33}$$

A similar result holds if there is a mass splitting in a chiral supermultiplet that Yukawa couples to the Higgs field. The relevant graphs are

Fig. 19.4.

As this quantity $\leqslant 1\,\mathrm{TeV}$ we find that

$$\Delta m \leqslant 1\,\mathrm{TeV}\frac{(4\pi)^2}{g^2} \tag{19.34}$$

Hence if Δm denotes the effective supersymmetry breaking scale we expect $\Delta m \sim m_\mathrm{W}$ in order not to have to make any unnatural fine tuning. This argument suggests that if the solution of the technical gauge hierarchy problem is due to supersymmetry then the superpartners of the observed particles must not have masses very much higher than m_W.

Supersymmetry, however, must be broken in some sector of the theory at the tree level. In any chiral supermultiplet with non-zero $\langle F \rangle$ we find contributions to the gauge gauging mass splitting which arise from the super graph below (Fig. 19.5).

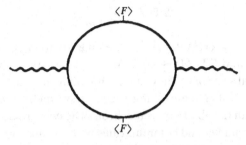

Fig. 19.5.

The contribution is of the form

$$g^2 \frac{\langle F \rangle^2}{M^2} = g^2 \frac{M_s^4}{M^2}. \tag{19.35}$$

By the above discussion this mass splitting can not be too large. There are two ways to avoid this: either, we make[148] M very heavy ($\sim M_X$) or we do not allow the chiral multiplet to couple to gauge fields.[149] This leads to the result that, if the scale of supersymmetry breaking is large, then the sector which breaks the supersymmetry should be isolated from the observed sector of the theory.[155]

19.4   Comments on the Construction of Realistic Models

We saw in the previous chapter that we must break supersymmetry at the tree level in order to break supersymmetry at all. At the tree level, in order to avoid anomalies we must have $S \operatorname{Tr} M^2 = 0$. However, examining the spectrum of observed particles, with the knowledge that there are no scalar fermionic partners below 20 GeV, we must conclude that in the observed sector

$$S \operatorname{Tr} M^2 > 20 \text{ GeV} \tag{19.36}$$

This consideration is somewhat sensitive to the value of the mass of the Higgs scalars.

Further, in the Fayet-O'Raifeartaigh mechanism the tree-level scalar masses are split such that one is above and the other below the mass of the spinor in the same multiplet. This is clearly phenomenologically unacceptable. In the Fayet-O'Raifearartaigh mechanism once one has ensured anomaly freedom it is very difficult to find a superpotential that breaks supersymmetry rather than electric change or color.

Fortunately, $S \operatorname{Tr} M^2$ receives quantum corrections when supersymmetry is broken at the tree level. We recall that these corrections are of the form

$$S \operatorname{Tr} M^2 \sim \lambda^n \frac{M_S^4}{M^2} \tag{19.37}$$

where $M_S^4 \sim \langle F \rangle^2 + \langle G^2 \rangle + \langle D \rangle^2$. If we take M^2 to be of order $M_X \sim 10^{15}$ GeV, $\lambda^n \sim 10^{-4}$ and $S \operatorname{Tr} M^2 \sim 20$ GeV then $M_S \sim 10^{+13}$ GeV. In fact, 10^{10} GeV is the typical scale of supersymmetry breaking in realistic models of rigid supersymmetry. Rather than taking the supersymmetry-breaking sector to communicate with the observed sector through heavy particles[148] it is possible that it is a gauge singlet and communicates with the observed world through small Yukawa couplings.[149] In the latter case, it is possible to postpone the coupling to an arbitrary order. Hence, rather than break supersymmetry around 10^3 GeV it is necessary, in order to achieve a realistic mass spectrum, to break supersymmetry at 10^{10} GeV. As discussed in the previous section this still solves the gauge-hierarchy problem since the supersymmetry-breaking sector is isolated from the standard model sector. The supersymmetry breaking feeds down to the observed sector giving an effective scale of supersymmetry breaking of order 10^3 GeV in the standard model sector.

It is instructive to consider the cosmological constant in these models. To cancel the cosmological constant we must add to the action the term

$$e m_{3/2} \{ M m_{pl} + \tfrac{1}{4} \bar{\psi}_\mu \sigma^{\mu\nu} \psi_\nu \} \tag{19.38}$$

where $\kappa^{-1} = m_{pl}$ and $m_{3/2}$ is an arbitrary parameter. Eliminating the auxiliary field M yields a cosmological constant $\sim m_{pl}^2 m_{3/2}^2$. To cancel the matter induced cosmological constant, which is of the form M_s^4, we require[156]

$$m_{3/2} \sim \frac{m_S^2}{m_{pl}} \tag{19.39}$$

Substituting $m_S \sim 10^{10}$ GeV, $m_{pl} \sim 10^{19}$ GeV we find $m_{3/2} \sim m_W$.

The occurrence of a non-negligible spin-3/2 mass makes one suspicious that one may have overlooked some other supergravity effects. Examining the F density formula in the tensor calculus we find the term κN and κM times the lowest component of the chiral multiplet whose action is being evaluated. The kinetic action when written as $[S \cdot T(S)]_F$ has as its lowest components $FA - GB$ and $GA + FB$.

Eliminating F and G then leads to the terms

$$\frac{\kappa^2}{2}(BM - AN)^2 + \frac{\kappa^2}{2}(BN + AM)^2 \tag{19.40}$$

replacing M by $m_{3/2} m_{pl}$ we find scalar masses of the form

$$m_{3/2}^2(A^2 + B^2) \tag{19.41}$$

As every field has a kinetic term, this is a universal feature for all spin-zero fields. The supergravity effects are not negligible and indicate that when $M_s \sim 10^{10}$ GeV we cannot neglect supergravity.

As these effects are what one wants, one might start with a supergravity plus matter theory to begin with. Although, this is a non-renormalizable theory it has certain advantages; when supergravity is present the auxiliary fields occur in the action in the form

$$V = -\tfrac{1}{3}M^2 - \tfrac{1}{3}N^2 + \tfrac{1}{2}F^2 + \tfrac{1}{2}G^2 + \tfrac{1}{2}D^2 \tag{19.42}$$

plus interaction terms. It is not positive definite! As a result it is rather easy to break supersymmetry. Consider, the simple Polonyi superpotential[157]

$$W = de + f \tag{19.43}$$

where d and f are constants. Using the tensor calculus this leads to the bosonic coupling

$$e\{d(F - (MA + NB)) - fM\} \tag{19.44}$$

Eliminating the auxiliary fields we find the supersymmetry-breaking potential

$$\tfrac{1}{2}d^2 - \tfrac{3}{4}|dA + f|^2 - \tfrac{3}{4}(dB)^2 \tag{19.45}$$

The criterion for the breaking of supersymmetry in supergravity theories are different from those in rigid theories; however, the reader may verify, by examining the mass spectrum, that the above potential does break supersymmetry and may be used to cancel the cosmological constant.

The formula for supertrace M^2 also looks more promising; it has been found to be[158]

$$S\operatorname{Tr} M^2 = 2(N - 1)m_{3/2}^2 \tag{19.46}$$

where N is the number of chiral matter multiplets and we have taken $\langle D^a \rangle = 0$. Recalling the above discussion it is easy to understand why this expression takes the form it does.

As a result the more recent realistic models involve supergravity from the outset. Supersymmetry is broken at the tree level in some hidden sector and this breaking feeds down by supergravitational effects to the observed sector. In a further interesting development it has been shown that the Higgs mass which is initially positive may change sign at some lower energy and so $SU(2) \times U(1)$ is broken by radiative effects.[159]

This is a heuristic account of the rather complicated subject of realistic models and we refer the reader to Ref. 160 for a much more complete account.

Chapter 20

Currents in Supersymmetric Theories

20.1 General Considerations

The story of currents in supersymmetric theories began with a paper of Ferrara and Zumino[165] who explicitly calculated the supercurrent, energy-momentum tensor and chiral current in the context of the Wess-Zumino model. They found that these currents belonged to one supermultiplet. In this section we wish to discuss what are the possible structures (i.e., supermultiplet) that the multiplet of currents in a supersymmetric theory can belong to. We will begin by discussing the currents in superconformal theories and then examine how the various currents can acquire superconformal anomalies.

The reason for this approach is that the structure of the superconformal currents is much simpler and, as we shall see, their multiplet structure is in fact unique. Although the structure of the currents of super Poincaré theories is not unique it can, in all known cases, be found by adding suitable multiplets of superconformal anomalies to the superconformal currents.

We can deduce the possible structures of the current multiplet without regard to a particular model by demanding that the currents and their transformations give rise to the correct algebra, namely the supersymmetry algebra. The following discussion of currents is similar to that of Refs. 166 and 167.

To every symmetry of an action there corresponds a conserved current $j_\mu{}^k$ which in turn generates a conserved charge

$$Q^k = \int d^3x \, j_0{}^k \tag{20.1}$$

If the symmetry transformations form a closed algebra then we have the relation

$$[Q^k, Q^l] = f^{klm} Q^m \tag{20.2}$$

The variation of the current under the symmetry can be written in the form

$$\delta j_\mu{}^l = [Q^k, j_\mu{}^l] \tag{20.3}$$

However, taking $\mu = 0$ and integrating we find that

$$\int d^3x \, \delta j_0{}^l = \int d^3x \, [Q^k, j_0{}^l] = [Q^k, Q^l]$$

$$= f^{klm} Q^m = \int d^3x \, f^{klm} j_0{}^m \qquad (20.4)$$

Clearly, this equation places restrictions on the variation of $\delta j_\mu{}^l$.

For the case of an internal symmetry group we have one current for each generator and as the transformations of the currents do not involve the space-time derivatives we may remove the integral and the solution of the above equation can only be

$$\delta j_\mu{}^l = f^{klm} j_\mu{}^m \qquad (20.5)$$

Hence, the currents belong to the adjoint representation of the internal symmetry group.

In order to illustrate the more complicated features that can occur for space-time groups, let us now consider the Poincaré group, which has generators P_μ and $J_{\rho\kappa}$. The current for P_μ is the energy-momentum tensor $\theta_{\mu\nu}$ which is symmetric. Under the translations $\theta_{\mu\nu}$ transforms as $\delta\theta_{\mu\nu} = \partial_\rho \theta_{\mu\nu}$. Using this relation we find that

$$[P_\mu, P_\nu] = \int d^3x \, [P_\mu, \theta_{0\nu}] = \int d^3x \, \partial_\mu \theta_{0\nu}$$

$$= \int d^3x \, \partial_0 \theta_{0\nu} = -\int d^3x \, \partial_i \theta^i{}_\nu = 0 \qquad (20.6)$$

Under Lorentz rotations $J_{\mu\nu}$, the energy-momentum is a second-rank tensor, that is, it transforms as

$$\delta_{J_{\rho\kappa}} \theta_{\mu\nu} = -(x_\rho \partial_\kappa - x_\kappa \partial_\rho)\theta_{\mu\nu} - [+(\delta_{\rho\nu}\theta_{\mu\kappa} - \delta_{\kappa\nu}\theta_{\mu\rho}) + (\mu \leftrightarrow \nu)] \quad (20.7)$$

From this variation we can deduce the relation

$$[P_\mu, J_{\rho\kappa}] = \eta_{\mu\rho} P_\kappa - \eta_{\mu\kappa} P_\rho \qquad (20.8)$$

Now let us consider the current for $J_{\rho\kappa}$. In fact it is a moment of $\theta_{\mu\nu}$, namely,

$$L_{\rho\mu\nu} = -x_\nu \theta_{\rho\mu} + x_\mu \theta_{\rho\nu}. \qquad (20.9)$$

It is conserved by virtue of the fact that $\theta_{\mu\nu}$ is symmetric. From the variation of $\theta_{\mu\nu}$ under Lorentz rotations we can deduce the final commutation relation of the Poincaré group:

$$[J_{\mu\nu}, J_{\rho\kappa}] = (\eta_{\mu\rho} J_{\nu\kappa} + \cdots) \qquad (20.10)$$

Let us now consider the case when $\theta_{\mu\nu}$ is not only symmetric but also traceless, $\theta_\mu{}^\mu = 0$. We now have new conserved currents $d_\mu = x^\nu \theta_{\nu\mu}$ and $K_{\mu\nu} = 2x_\nu x^\lambda \theta_{\lambda\mu} - x^2 \theta_{\mu\nu}$. These currents then give rise to new charges D and K_μ. Using the variations of $\theta_{\mu\nu}$ we can calculate the commutators of these new generators D and K_μ with P_ρ and $J_{\rho\kappa}$. We note $\theta_\mu{}^\mu = 0$ is the only algebraic constraint we may place on $\theta_\mu{}^\mu$.

For example, we find that

$$[D, P_\nu] = \int d^3x [x^\kappa \theta_{\kappa 0}, P_\nu]$$

$$= + \int d^3x \, x^\kappa \partial_\nu \theta_{\kappa 0} = - \int d^3x \, \theta_{\nu 0} = -P_\nu \qquad (20.11)$$

while

$$[D, J_{\rho\mu}] = 0$$

$$[K_\mu, J_{\rho\kappa}] = \eta_{\mu\rho} K_\kappa - \eta_{\mu\kappa} K_\rho$$

$$[K_\mu, P_\nu] = -2\eta_{\mu\nu} D + 2J_{\mu\nu} \qquad (20.12)$$

The final commutators that remain to be found are $[D, K_\nu]$ and $[K_\mu, K_\nu]$. On grounds of Lorentz invariance, i.e., the Jacobi identities found by taking the above commutators with $J_{\mu\nu}$, we must conclude that

$$[D, K_\nu] = aK_\nu + bP_\nu$$

$$[K_\mu, K_\mu] = cJ_{\mu\nu} \qquad (20.13)$$

where a, b and c are constants.

Using the (D, K, P) Jacobi identity we find that $a = +1$ and the (D, J, K) Jacobi identity implies $b = 0$ and $c = 0$ giving the results

$$[D, K_\nu] = K_\nu \qquad [K_\mu, K_\nu] = 0 \qquad (20.14)$$

Consequently, we arrive at a group with generators P_ν, $J_{\rho\kappa}$, D and K_μ, which is of course the conformal group. Hence, a theory which has $\theta_\mu{}^\mu = 0$ admits the conformal symmetry.

Indeed one may reverse the above argument. Starting from the conformal group with generators P_μ, $J_{\mu\nu}$, D and K_ρ and their commutation relations, one may ask what is the form of the currents that give rise to this algebra. One can deduce that the translation generator P_μ has a current $\theta_{\mu\nu}$ which is symmetric and traceless and that the other generators have currents which are formed in terms of $\theta_{\mu\nu}$ by the equations above. This result rests on the following facts. The currents that generate these charges must

(a) belong to representations of the conformal group,
(b) have transformation properties that generate the correct algebra in the sense described above.

The representations of the conformal group are known and can be constructed according to the theory of induced representations. In this case the coset space of interest is

$$\frac{\text{conformal group}}{(J_{\mu\nu}, D, K_\rho)}$$

This space is four dimensional and contains the well known transformations of the Poincaré group on Minkowski space. Given any representation of $J_{\mu\nu}$, D and K_ρ we can induce it to the whole group.

We will now give a simple method for deducing the action of the generators on the fields of these induced representations.[168] Their action at the origin is a representation of the group with generators $(J_{\mu\nu}, D, K_\rho)$. We take the action of these generators to be

$$J_{\mu\nu}\phi(0) = \Sigma_{\mu\nu}\phi(0)$$

$$D\phi(0) = \Delta\phi(0)$$

$$K_\mu\phi(0) = \kappa_\mu\phi(0) \tag{20.15}$$

where $\Sigma_{\mu\nu}$ are matrix representations of the generators $J_{\mu\nu}$ and correspond to the spin of the field ϕ being considered, the Δ is the dilatation weight. Since $[J_{\mu\nu}, D] = 0$ we find that

$$[\Delta, \Sigma] = 0 \tag{20.16}$$

if Σ is an irreducible representation and from Schur's Lemma we must conclude that Δ is a number. The κ_μ tensor is generally not found in actual applications, but we keep it for completeness. The field at a general point x is given by

$$\phi(x) = \exp(x^\mu P_\mu)\phi(0)$$

The action of any generator S is now given by

$$S\phi(x) = \exp(+x^\mu P_\mu)\exp(-x^\mu P_\mu)S\exp(x^\mu P_\mu)\phi(0)$$

$$= \exp(x^\mu P_\mu)\{S + [S, x^\mu P_\mu] + \tfrac{1}{2}[[S, x^\mu P_\mu], x^\nu P_\nu] + \cdots\}\phi(0) \tag{20.17}$$

This can be evaluated using only the group commutation relations and Eq. (20.15). For example, for the dilatation generator

$$D\phi(x) = \exp(x^\mu P_\mu)[D - x^\mu P_\mu]\phi(0)$$

$$= \exp(x^\mu P_\mu)[\Delta - x^\mu P_\mu]\phi(0)$$

$$= \Delta\phi(x) - x^\mu P_\mu\phi(x)$$

$$= -x^\mu \partial_\mu\phi(x) + \Delta\phi(x) \tag{20.18}$$

For the K_ρ generator one finds the result

$$K_\mu \phi(x) = (2x_\mu x^\nu \partial_\nu - x^2 \partial_\mu)\phi - x_\mu \Delta \phi - 2x^\nu \Sigma_{\mu\nu} \phi + \kappa_\mu \phi \qquad (20.19)$$

The Lorentz rotations take their standard form:

$$J_{\rho\kappa} \phi(0) = \exp(x^\mu P_\mu)(\Sigma_{\rho\kappa} + [J_{\rho\kappa}, x^\mu P_\mu])\phi(0)$$

$$= (\Sigma_{\rho\kappa} + (x_\kappa \partial_\rho - x_\rho \partial_\kappa))\phi(x)$$

Applying this to $\theta_{\mu\nu}$ we find after taking the appropriate Lorentz matrix $\Sigma_{\mu\nu}$ that

$$\delta\theta_{\mu\nu} = K_\rho \theta_{\mu\nu} = (2x_\rho x^\tau \partial_\tau \theta_{\mu\nu} - x^2 \partial_\rho \theta_{\mu\nu}) - 2\Delta x_\rho \theta_{\mu\nu}$$

$$- 2x^\kappa(\eta_{\rho\kappa}\theta_{\mu\nu} - \eta_{\kappa\mu}\theta_{\rho\nu} + \eta_{\rho\nu}\theta_{\mu\kappa} - \eta_{\kappa\nu}\theta_{\mu\rho}) \qquad (20.20)$$

where Δ is the dilatation weight of $\theta_{\mu\nu}$. We have omitted the κ_μ terms, referred to above by assumption. Enforcing the conservation condition $\partial^\mu \theta_{\mu\nu} = 0$ implies that

$$\tfrac{1}{2}(2\Delta - 8)(\theta_{\rho\nu} + \theta_{\nu\rho}) + \tfrac{1}{2}[\theta_{\rho\nu} - \theta_{\nu\rho}] - 2\eta_{\rho\nu}\theta_\mu{}^\mu = 0 \qquad (20.21)$$

Taking $\Delta = -4$ as the dilatation weight of $\theta_{\mu\nu}$, leads to

$$\theta_{\rho\nu} - \theta_{\nu\rho} = 0 = \theta_\mu{}^\mu \qquad (20.22)$$

It only remains to show that the currents for $J_{\mu\nu}$, D and K_ρ are the appropriate moments of $\theta_{\mu\nu}$. Consider the commutator of P_ν and K_ρ

$$[P_\nu, K_\rho] = \left[\int d^3x\, \theta_{0\nu}, K_\rho\right] = \int d^3x (2x_\rho x^\tau \partial_\tau \theta_{0\nu} - x^2 \partial_\rho \theta_{0\nu}$$

$$- 2\Delta x_\rho \theta_{0\nu} + 2x^\kappa \eta_{\rho 0}\theta_{\kappa\nu}$$

$$- 2x_0 \theta_{\rho\nu} + 2x^\kappa \eta_{\rho\nu}\theta_{0\kappa} - 2x_\nu \theta_{0\rho}) \qquad (20.23)$$

Integrating by parts and using the conservation of $\theta_{\mu\nu}$ we find

$$[P_\nu, K_\rho] = +2\eta_{\nu\rho}D - 2J_{\nu\rho} \qquad (20.24)$$

where $J_{\nu\rho}$ and D have currents which are the moments of $\theta_{\mu\nu}$ given in the equations above. The current for K_ρ can be deduced by considering the $[J_{\mu\nu}, K_\rho]$ commutator. Thus we have shown the desired result, the current $\theta_{\mu\nu}$ for P_μ is symmetric and conserved and all other generators D, K_ρ, $J_{\mu\nu}$ are the appropriate moments of $\theta_{\mu\nu}$.

Hence for space-time groups, the structure of the currents is more complicated than in the case of internal symmetry groups for which there is one independent current for each generator.

Bearing this discussion in mind let us now examine the $N = 1$ supersymmetry group which has the generators P_μ, Q_α, R and $J_{\mu\nu}$. The currents for P_μ, Q_α and R are $\theta_{\mu\nu}$ (symmetric), $j_{\mu\alpha}$, and $j_\mu^{(5)}$ respectively. As before $J_{\mu\nu}$ has a current which is a moment of $\theta_{\mu\nu}$ given by Eq. (20.9). We now consider the relation

$$\{Q_\alpha, Q_\beta\} = +2(\gamma^\nu C)_{\alpha\beta} P_\nu \tag{20.25}$$

This equation can be written as

$$\int d^3x \{Q_\alpha, j_{0\beta}\} = 2(\gamma^\nu C)_{\alpha\beta} \int d^3x\, \theta_{0\nu} \tag{20.26}$$

and so the variation of $j_{\mu\beta}$ must be of the form

$$\delta\varepsilon_\alpha j_{\mu\beta} = 2(\gamma^\nu C)_{\alpha\beta}\theta_{\mu\nu} + \partial^\rho R_{\mu\rho\alpha\beta} \tag{20.27}$$

where $R_{\mu\rho\alpha\beta} = -R_{\rho\mu\alpha\beta}$. The last term must vanish when $\mu = 0$ and integrated over $\int d^3x$ as it does not contribute to Eq. (20.12). It is not clear that it must be of the form given above, but this always seems to be the case for such quantities that lead to vanishing charges.

Similarly from the relation

$$[Q_\alpha, R] = i(\gamma_5)_\alpha{}^\beta Q_\beta \tag{20.28}$$

we can deduce that

$$\delta j_\mu{}^{(5)} = i(\gamma_5)_\alpha{}^\beta j_{\mu\beta} + \partial^\nu R_{\mu\nu\alpha} \tag{20.29}$$

where $R_{\mu\nu\alpha} = -R_{\nu\mu\alpha}$. Hence we have found that $j_\mu{}^{(5)}$, $j_{\mu\alpha}$ and $\theta_{\mu\nu}$ must belong to a supermultiplet.

As before we now consider the case when some of the currents satisfy further algebraic constraints. Consider the case when

$$\theta_\mu{}^\mu = 0, \qquad (\gamma^\mu j_\mu)_\alpha = 0 \tag{20.30}$$

These in fact are the only possible algebraic constraints. Then we have the new conserved currents d_μ and $K_{\nu\mu}$ as before but also

$$S_{\mu\alpha} = (\not x j_\mu)_\alpha \tag{20.31}$$

The corresponding conserved charges are D, K_ν and S_α. The currents which give rise to this new algebra must belong to a representation of the algebra and in particular the super Poincaré subalgebra.

Consider the currents we have discussed above

$$\theta_{\mu\nu}, j_{\mu\alpha}, j_\mu{}^{(5)} \tag{20.32}$$

subject to the constraints of Eq. (20.17) plus moments of these currents. Carrying out a degree of freedom count we find for these currents 5, 8 and

3 degrees of freedom, respectively. This matching of fermion and boson degrees of freedom leads us to suspect that this set of currents constitutes a supermultiplet without the addition of other quantities. Writing down the most general possible transformations consistent with linearity, matching dimensions and demanding that they give rise to the correct algebra, we have

$$\delta j_\mu^{(5)} = i \bar{\varepsilon} \gamma_5 j_\mu$$

$$\delta j_\mu = \left(+2\gamma^\nu \theta_{\mu\nu} - ia_1 \gamma_5 \partial j_\mu^{(5)} - \frac{ia_3}{2} \varepsilon_{\mu\nu\rho\kappa} \gamma^\nu \partial^\rho j^{\kappa(5)} \right) \varepsilon$$

$$\delta\theta_{\mu\nu} = -\frac{a_2}{2} (\bar{\varepsilon}\sigma^{\mu\kappa} \partial_\kappa j^\nu + \bar{\varepsilon}\sigma^{\nu\kappa} \partial_\kappa j^\mu) \tag{20.33}$$

where a_1, a_2 and a_3 are constants.

The above transformations do indeed close on all currents provided $a_1 = +1, a_2 = 1/2, a_3 = -1$. By construction these currents give rise to $\{Q, Q\} \sim P$, $\{Q, R\} \sim Q$ but one may also check the other relations of the super-Poincaré group; for example

$$[P_\mu, Q_\alpha] = \int d^3x \, \delta\theta_{0\mu} = 0 \tag{20.34}$$

Using the now, know transformations of the currents under the super-Poincaré group and the Jacobi identities we may deduce, in a similar way, for the Poincaré group, the (anti) commutation of all the generators. This is achieved as follows. We find that

$$[Q, K_\mu] = \int d^3x (2x_\mu x^\nu \delta_Q \theta_{\nu 0} - x^2 \delta_Q \theta_{\mu 0}) \tag{20.35}$$

Using the $\delta_Q \theta_{\mu\nu}$ given above, we can evaluate this and find that

$$[Q, K_\mu] = -\gamma_\mu S \tag{20.36}$$

where S_α is the moment of $j_{\mu\alpha}$ given above. Given, the super-Poincaré and the conformal algebra and Eq. (20.33) one can deduce the whole of the super conformal group using the Jacobi identities. Hence, one finds that the generators

$$P_\mu, \ Q^\alpha, \ R, \ J_{\mu\nu}, \ S^\alpha, \ D \text{ and } K_\mu \tag{20.37}$$

satisfy the $N = 1$ super conformal algebra given in Chapter 2. Consequently we have shown that a theory that admits the constraints of Eq. (20.30) on its energy-momentum tensor and super current is superconformally invariant.

Further, we have found that the currents[165]

$$(j_\mu^{(5)}, j_{\mu\alpha}, \theta_{\mu\nu}) \tag{20.38}$$

and their appropriate moments belong to a supermultiplet which, by examining Eq. (20.33), is seen to be irreducible.

As for the Poincaré group one can reverse the above procedure. Starting the $N = 1$ superconformal algebra it can be shown[167] under rather general assumptions that the above multiplet of superconformal currents is unique. The proof of this statement is based on the fact that the currents must belong to a representation of the full superconformal group as well as generating that algebra.

Let us now consider a theory that is $N = 1$ Poincaré supersymmetric, but not invariant under the full superconformal group. In particular, the theory may not be dilatation invariant ($\theta^\mu_{\ \mu} \neq 0$), S supersymmetric ($\gamma^\mu j_\mu \neq 0$) or R invariant ($\partial_\mu j^{\mu(5)} \neq 0$) or not invariant under any combination of these symmetries. Relaxing the constraints on the currents of the superconformal theory clearly destroys the Fermi-Bose balance of the multiplet and as such we must add further degrees of freedom. As a result we will find a supermultiplet of currents which consists of the multiplet of superconformal currents, denoted A, and a multiplet of anomalies, denoted B. The transformation of these multiplets under supersymmetry must be such that when $B = 0$ we recover the transformation laws of the superconformal currents alone; that is, it is of the form

$$\delta A = A\varepsilon + B\varepsilon$$

$$\delta B = B\varepsilon \tag{20.39}$$

Let us consider the case for which $\theta_\mu^{\ \mu} \neq 0$, $\gamma^\mu j_\mu \neq 0$ and $\partial^\mu j_\mu^{(5)} \neq 0$. We require 2 bosonic degrees of freedom to restore Fermi-Bose balance. The minimal possibility is that the multiplet of anomalies is a chiral supermultiplet P, Q, χ_α, F, G such that

$$F = 2\theta_\mu^{\ \mu} \qquad \chi_\alpha = (\gamma^\mu j_\mu)_\alpha \qquad G = -3\partial^\mu j_\mu^{(5)} \tag{20.40}$$

and P and Q are new quantities of dimension 3. The new multiplet of currents can be obtained by adding the anomalies into the transformations of the currents in the most general way consistent with dimension matching, linearity, and generating the $N = 1$ super-Poincaré algebra. For example,

$$\delta j_\mu^{(5)} = i\bar{\varepsilon}\gamma_5 j_\mu + ia\varepsilon\gamma_5 \gamma^\mu \gamma \cdot j, \quad \text{etc.} \tag{20.41}$$

The constants that arise can then be fixed by demanding closure for the enlarged multiplet of currents, or more simply by demanding that the multiplet of anomalies be a chiral multiplet. For example, we find that

$$\delta(\partial^\mu j_\mu^{(5)}) = \delta\left(\frac{G}{-3}\right) = ia\bar{\varepsilon}\gamma_5 \partial\!\!\!/\gamma \cdot j = -\frac{1}{3}i\bar{\varepsilon}\gamma_5 \partial\!\!\!/\chi \tag{20.42}$$

and consequently $a = -1/3$.

The resulting transformations[165] are

$$\delta j_\mu^{(5)} = i\bar\varepsilon\gamma_5 j_\mu - \frac{1}{3}i\bar\varepsilon\gamma_5\gamma_\mu\gamma\cdot j$$

$$\delta j_\mu = \left[+2\gamma^\nu\theta_{\mu\nu} - i\gamma_5\not\partial j_\mu^{(5)} + i\gamma_5\gamma_\mu\partial_\nu j^{\nu(5)} + \frac{i}{2}\varepsilon_{\mu\nu\rho\kappa}\gamma^\nu\partial^\rho j^{\kappa(5)} \right.$$

$$\left. + \frac{i}{3}\sigma_{\mu\nu}\partial^\nu(P + i\gamma_5 Q) \right]\varepsilon$$

$$\delta\theta_{\mu\nu} = -\frac{1}{4}(\bar\varepsilon\sigma^{\mu\kappa}\partial_\mu j^\nu + \bar\varepsilon\sigma^{\nu\kappa}\partial_\kappa j^\mu)$$

$$\delta P = \bar\varepsilon\gamma\cdot j \qquad \delta Q = i\bar\varepsilon\gamma_5\gamma\cdot j \tag{20.43}$$

Let us now consider the case when

$$\theta_\mu{}^\mu \neq 0 \qquad (\gamma^\mu j_\mu)_\alpha \neq 0 \tag{20.44}$$

but R-invariance is preserved (i.e., $\partial^\mu j_\mu^{(5)} = 0$). In this case we require 3 bosonic degrees of freedom to restore Fermi-Bose balance. The minimal possibility is to have a linear multiplet of anomalies $(\zeta, C, t_{\mu\nu})$ where $\partial^\nu t_{\mu\nu} = 0$ and we identify $C = \frac{1}{2}\theta_\mu{}^\mu$ and $\zeta_\alpha = (\gamma^\mu j_\mu)_\alpha$.

The reader will recall that a linear multiplet had the x-space field content $(c, a_{\mu\nu}, \chi_\alpha)$ where $a_{\mu\nu} = -a_{\nu\mu}$ has the gauge invariance $\delta a_{\mu\nu} = \partial_\mu\Lambda_\nu - \partial_\nu\Lambda_\mu$. The so-called linear anomaly has the field content $(\xi_\alpha; d; t_{\mu\nu})$ $t_{\mu\nu} = -t_{\nu\mu}$ which are the dual of $(c, a_{\mu\nu}, \chi_\alpha)$ in the sense that the following action is supersymmetric

$$\int d^4x(cd + \bar\chi\xi + a^{\mu\nu}t_{\mu\nu})$$

Note that corresponding to the gauge invariance of $a_{\mu\nu}$ we have $\partial^\nu t_{\mu\nu} = 0$.

The resulting multiplet of currents contains $j_\mu^{(5)}, j_{\mu\alpha}, \theta_{\mu\nu}$ and $t_{\mu\nu}$ and has the transformation properties[166,169]

$$\delta j_\mu^{(5)} = i\bar\varepsilon\gamma_5 j_\mu$$

$$\delta j_\mu = \left(+2\gamma^\nu\theta_{\mu\nu} - i\gamma_5\not\partial j_\mu^{(5)} + \frac{i}{2}\varepsilon_{\mu\nu\rho\kappa}\gamma^\nu\partial^\rho j^{\kappa(5)} + 2\gamma^\nu t_{\mu\nu} \right)\varepsilon$$

$$\delta\theta_{\mu\nu} = -\frac{1}{4}(\bar\varepsilon\sigma^{\mu\kappa}\partial_\mu j^\nu + \bar\varepsilon\sigma^{\nu\mu}\partial_\mu j^\mu)$$

$$\delta t_{\mu\nu} = +\frac{i}{4}\bar\varepsilon\gamma_5\partial_\kappa\gamma_\lambda\gamma\cdot j\varepsilon^{\mu\nu\kappa\lambda} \tag{20.45}$$

In fact, there are other ways to achieve the above Fermi-Bose balance using much larger anomaly multiplets.

We will now reformulate the above supercurrent multiplets in superspace. The superconformal currents $(j_\mu^{(5)}; j_{\mu\alpha}; \theta_{\mu\nu})$ have as their lowest component the chiral current $j_{A\dot{A}}^{(5)}$, which we identify with the $\theta = 0$ component of the real supercurrent $J_{A\dot{A}}$. The $(\gamma^\mu j_\mu)_\alpha = 0$ constraint then translates into the superspace constraint

$$D^{\dot{A}} J_{A\dot{A}} = 0 \qquad (20.46)$$

The reader may verify that no other superspace constraints are necessary and $J_{A\dot{A}}$ subject to the above constraint has as its content only the superconformal currents subject to their constraints.

The currents which have superconformal anomalies must therefore have

$$D^{\dot{A}} J_{A\dot{A}} \neq 0 \qquad (20.47)$$

For a chiral multiplet of anomalies we had $(\gamma^\mu j_\mu)_A = \chi_A$. Identifying χ_A as the spinor in the chiral supermultiplet S $(ieD_{\dot{A}}S = 0)$, we find that

$$D^{\dot{A}} J_{A\dot{A}} = D_A S \qquad (20.48)$$

The linear multiplet of anomalies must, by similar considerations, be given by

$$D^{\dot{A}} J_{A\dot{A}} = L_A \qquad (20.49)$$

where $D_{\dot{B}} L_A = 0$ and $D^A L_A = D^{\dot{A}} L_{\dot{A}}$. The reader will recognize L_A as having the same x-space content as the Maxwell superfield strength W_A which satisfies the same constraints. This constraint can be solved by $L_A = \bar{D}^2 D_A T$ where T is an unconstrained superfield. A useful exercise is to show that R-invariance is indeed preserved:

$$\{D^A, D^{\dot{A}}\} J_{A\dot{A}} = -2i(\partial)^{A\dot{A}} J_{A\dot{A}}$$
$$= D^A L_A - D^{\dot{A}} L_{\dot{A}} = (D^A \bar{D}^2 D_A - D^{\dot{A}} D^2 D_{\dot{A}}) T = 0 \qquad (20.50)$$

The discussion in this section could be carried out at a more pedagogical level by starting with the currents belonging to a general supermultiplet whose first component is the chiral current. If the next component is $\zeta_{\mu\alpha}$ we can demand that $j_{\mu\alpha} = \zeta_{\mu\alpha} - c(\gamma_\mu \gamma^\nu \zeta_\nu)_\alpha$, where c is a constant, be the supercurrent and so be conserved. Evaluating the consequences of this conservation condition yields the chiral and linear anomalies as well as other possible anomaly sets. The methods given here can be generalized to find the currents of extended supersymmetric theories. The conformal supercurrent[170] for $N = 2$ theories is a real superfield J of dimension 2 which is subject to

$$D^{ik} J = 0 \qquad (20.51)$$

The conformal supercurrents for $N = 3$ and 4 theories are given in Ref. 171.

Again it can be shown that for $N \leqslant 4$ these superconformal currents are unique.[167]

Having obtained the superconformal currents one can then feed in anomalies as for the $N = 1$ case.

20.2   Currents in the Wess-Zumino Model[165]

The massless but interacting Wess-Zumino model has the form

$$A = \int d^4x \left\{ -\frac{1}{2}(\partial_\mu A)^2 - \frac{1}{2}(\partial_\mu B)^2 - \frac{1}{2}\bar{\chi}\partial\!\!\!/\chi \right.$$

$$\left. + \frac{1}{2}F^2 + \frac{1}{2}G^2 + \lambda[F(A^2 - B^2) + 2ABG - \bar{\chi}(A - i\gamma_5 B)\chi] \right\}$$

$$\equiv \int d^4x L \tag{20.52}$$

This theory is invariant under the full $N = 1$ superconformal group which contains the usual super-Poincaré group $J_{\mu\nu}, P_\mu, Q_\alpha$, chiral transformations R, dilatations D, special translations K_μ and special supersymmetry transformations S_α.

The supersymmetry transformations are given in Chapter 5 while the translations and Lorentz rotations are of the usual form. For example, on A

$$\delta_{P_\mu} A = \partial_\mu A \qquad \delta_{J_{\mu\nu}} A = -(x_\mu \partial_\nu - x_\nu \partial_\mu)A \tag{20.53}$$

The superconformal transformations can be found in the same way as the conformal ones which were discussed above. In this case, however, the coset involved is (superconformal group)/$(J_{\mu\nu}, D, S_\alpha, K_\mu, R)$. From a more detailed discussion of the transformation properties of these fields under the full superconformal group see Ref. 172.

Using the well known Noether procedure, we can calculate the currents corresponding to these transformations. For example, one finds that

$$\theta_{\mu\nu} = \partial_\mu A \partial_\nu A + \partial_\mu B \partial_\nu B - \frac{1}{4}\bar{\chi}\gamma_\mu \partial_\nu \chi + \eta_{\mu\nu}L + a_1(\partial_\mu \partial_\nu - \eta_{\mu\nu}\partial^2)(A^2 + B^2)$$

$$j_{\mu\alpha} = -\partial\!\!\!/(A - i\gamma_5 B)\gamma^\mu \chi - \lambda(A^2 - B^2 + 2iAB\gamma_5)\gamma_\mu \chi + a_2\sigma_{\mu\nu}\partial^\nu(A + i\gamma_5 B)\chi$$

$$j_\mu^{(5)} = \frac{2}{3}i(B\partial_\mu A - A\partial_\mu B - \frac{i}{4}\bar{\chi}\gamma_5\gamma_\mu\chi) \tag{20.54}$$

To the energy-momentum tensor $\theta_{\mu\nu}$ and the supercurrent $j_{\mu\alpha}$ we have explicitly added certain improvement terms. These terms are identically conserved and so occur with arbitrary coefficients. Choosing $a_1 = -1/6$ and $a_2 = -2/3$ we find that

$$\theta_\mu{}^\mu = 0 \quad \text{and} \quad \gamma^\mu j_\mu = 0 \tag{20.55}$$

As discussed previously this allows us to define new conserved currents which are moments of $\theta_{\mu\nu}$. These in fact correspond to the Noether currents corresponding to the symmetries D, K_ρ, $J_{\mu\nu}$ and S_α.

Given the explicit expression in Eq. (20.54) for $\theta_{\mu\nu}$, $j_{\mu\alpha}$ and $j_\mu{}^{(5)}$ we may use the supersymmetry variations of the fields to find the supersymmetry variations of these currents. The result of this calculation is the same as the result found previously, namely

$$\delta j_\mu{}^{(5)} = i\bar\varepsilon \gamma_5 j_\mu$$

$$\delta j_{\mu\alpha} = \left(2\gamma^\nu \theta_{\mu\nu} - i\gamma_5 \partial\!\!\!/ j_\mu{}^{(5)} + \frac{i}{2}\varepsilon_{\mu\nu\kappa\lambda}\gamma^\nu \partial^\kappa j^{\lambda(5)} \right)\varepsilon$$

$$\delta\theta_{\mu\nu} = -\frac{1}{4}(\bar\varepsilon \sigma^{\mu\kappa}\partial_\kappa j^\nu + \bar\varepsilon \sigma^{\nu\kappa}\partial_\kappa j^\mu) \tag{20.56}$$

This supermultiplet of currents is an irreducible multiplet of not only the super-Poincaré group, but also of the larger superconformal group. The transformations for the dilatations and special conformal transformations on the fields of the theory are the usual ones, while those for S_α are similar to the ones for supersymmetry, but with ε_α replaced by $(\not x \eta)_\alpha$, i.e., $\delta_s A = -\bar\eta \not x \chi$. However, when the ordinary supersymmetry transformations involve derivatives of fields, the S supersymmetry transformations contain additional terms proportional to η alone, i.e., $\delta_s \chi = [F + i\gamma_5 G + \partial\!\!\!/(A + i\gamma_5 B)](\not x \eta) + 2(A + i\gamma_5 B)\eta$. The reader may verify that the corresponding currents are then just the moments of θ_μ and $j_{\mu\alpha}$ discussed above.

To the interacting Wess-Zumino theory of Eq. (20.36) we may add the Poincaré supersymmetric mass term

$$\int d^4 x\, m\left(AF + GB - \frac{1}{2}\chi\bar\chi \right) \tag{20.57}$$

The theory is now no longer invariant under dilatations D, the conformal chiral transformations R, the special translations K_μ, and the special supersymmetry transformations S_α. As such

$$\partial^\mu S_{\mu\alpha} = \partial^\mu(\not x j_\mu)_\alpha = (\gamma^\mu J_\mu) \neq 0$$

$$\partial^\mu(x^\nu \theta_{\mu\nu}) = \theta_\mu{}^\mu \neq 0 \tag{20.58}$$

and similarly $\partial^\mu j_\mu{}^{(5)} \neq 0$.

The Noether currents are given by

$$j_\mu = \delta\phi \frac{\delta L}{\partial(\partial_\mu \phi)} - k_\mu \tag{20.59}$$

where $\delta\phi$ is the variation of ϕ under the appropriate symmetry and the Lagrangian which varies as $\delta L = \partial^\mu k_\mu$ now receives contributions proportional to m which lead to a chiral multiplet of anomalies proportional to m

$$\theta_\mu{}^\mu = -m[(A^2 + B^2)] - \frac{m}{2}\bar\chi\chi \qquad \partial^\mu j_\mu{}^{(5)} = +i\frac{m}{3}(\bar\chi_1\gamma_5\chi_2)$$

$$\gamma\cdot j = 2m(A - i\gamma_5 B)\chi \qquad P = m(A^2 - B^2) \qquad Q = 2mAB \qquad (20.60)$$

The multiplet of currents transforms according to Eq. (20.27).

In terms of superspace the supercurrent multiplet is given by

$$J_{A\dot A} = -\tfrac{1}{3}D_{\dot A}\bar\phi D_A\phi - \tfrac{2}{3}\bar\phi i(\overleftrightarrow{\partial})_{A\dot A}\phi \qquad (20.61)$$

where $\overleftrightarrow{\partial} = \overrightarrow{\partial} - \overleftarrow{\partial}$. The reader may check that

$$D^{\dot A} J_{A\dot A} = -\frac{m}{6}D_A\phi^2 \qquad (20.62)$$

20.3   Currents in $N = 1$ Super Yang-Mills Theory

Since this theory only involves as its physical fields A_μ and $\lambda_A(\lambda_{\dot A})$ the R transformations can only act on the spinors and hence the chiral current can only be of the form

$$j^{(5)}{}_{A\dot A} = \lambda_A\lambda_{\dot A} \qquad (20.63)$$

The supercurrent, having as its lowest component $j^{(5)}{}_{A\dot A}$, must then be of the form

$$J_{A\dot A} = W_A W_{\dot A} \qquad (20.64)$$

The reader may verify that for the classical theory the currents are of the superconformal multiplet, i.e.,

$$D^{\dot A} J_{A\dot A} = -W_A D^{\dot A} W_{\dot A} + (D^{\dot A} W_A) W_{\dot A} = 0 \qquad (20.65)$$

This follows from the constraints

$$D_{\dot B} W_A = 0 \qquad (20.66)$$

and the classical equation of motion (i.e., the auxiliary field $D = 0$),

$$D^{\dot A} W_{\dot A} = 0 \qquad (20.66)$$

20.4   Quantum Generated Anomalies

As is well known, even though a theory may be conformally invariant at the classical level, its quantum corrections do not, in general, respect conformal invariance. This is a consequence of the renormalization of the theory; here one is forced to introduce a scale in order to regulate the infinities and so give any meaning to the theory. Hence in a general theory one finds that the quantum corrections provide $\theta_{\mu\nu}$ with a non-zero trace. For example in massless QCD it is well known that[173]

$$\theta_\mu{}^\mu = \frac{\beta(g)}{g}\left(+\frac{1}{2}F^2{}_{\mu\nu}\right) \tag{20.67}$$

where $\beta(g)$ is the Callan-Symanzik β-function. Another symmetry that is often violated by quantum corrections is the γ_5 transformation. In supersymmetric theories the divergence of the chiral current and the trace of the energy-momentum tensor often belong to the same anomaly multiplet and hence we can expect these quantum generated anomalies to be related by supersymmetry. As an example of this phenomenon let us consider the quantum corrections to $N = 1$ Yang-Mills theory. We recall from earlier in this chapter that the supercurrent is of the form

$$J_{A\dot{A}} = W_A W_{\dot{A}} \tag{20.68}$$

We expect in the quantum theory that $\theta_\mu{}^\mu \neq 0$ and/or $\partial_\mu j^{\mu(5)} \neq 0$, and so in superspace that

$$D^{\dot{A}} J_{A\dot{A}} = N_A \tag{20.69}$$

The term on the right-hand side of the above equation, N_A, will be composed out of $W_A(W_{\dot{A}})$. Now W_A has dimension 3/2 and $J_{A\dot{A}}$ and N_A must have dimensions 3 and 7/2 respectively. If it is a chiral multiplet of anomalies then $N_A = D_A S$, where $D_{\dot{A}} S = 0$ and S is of dimension 3. Clearly, the only candidate for S is $W_A W^A$. In fact, since $J_{A\dot{A}}$ is gauge invariant, it must be composed of W^2 or $W\overline{W}$; and as such the chiral multiplet of anomalies given above is the only possibility. For example, one cannot have a linear set of anomalies.

It only remains to find the coefficient in front of N_A. This involves evaluating graphs with the insertion $J_{A\dot{A}}$; for example, one has to evaluate the supergraphs given in the diagrams below.

Fig. 20.1.

Taking care to insert the current of the regularized theory into the graphs one finds the result[174,177]

$$D^{\dot{A}} J_{A\dot{A}} = -\frac{1}{3} \frac{\beta(g)}{g} D_A (W^B W_B) \tag{20.70}$$

In the case of dimensional regularization one must use the current in $4 - 2\varepsilon$ dimensions.

Applying successive spinorial covariant derivatives to this equation and evaluating the resulting expressions at $\theta = 0$ one finds the following x-space results:

$$(\gamma^\mu j_\mu)_\alpha = \frac{-2\beta(g)}{g} \left[\frac{-1}{2} (\sigma^{\mu\nu} F_{\mu\nu}) + i\gamma_5 D \right] \lambda$$

$$\theta_\mu{}^\mu = -2 \frac{\beta(g)}{g} \left(-\frac{1}{4} F^{\mu\nu} F_{\mu\nu} - \frac{1}{2} \bar{\lambda} \partial \!\!\!/ \lambda + \frac{1}{2} D^2 \right)$$

$$\partial_\mu j^{\mu(5)} = \frac{+2}{3} \frac{\beta(g)}{g} \left[-\frac{1}{4} F^{\mu\nu *} F_{\mu\nu} + \frac{1}{2} \partial^\mu (\bar{\lambda} \gamma_\mu \gamma_5 \lambda) \right]$$

$$P = -\beta(g) \bar{\lambda} \lambda, \qquad Q = -i\beta(g) \bar{\lambda} \gamma_5 \lambda \tag{20.71}$$

We note that $\theta_\mu{}^\mu$ agrees with the well-known results concerning this quantity.

The situation for the Wess-Zumino multiplet is rather more complicated and the reader is referred to Refs. 174 and 177 for a discussion.

20.5   Currents and Supergravity Formulations

There is a very useful and interesting connection[178] between the multiplet of currents and supergravity formulations. Einstein's equations of gravity coupled to matter are of the form

$$R_{\mu\nu} - \tfrac{1}{2} g_{\mu\nu} R = \kappa \theta_{\mu\nu} \tag{20.72}$$

where κ is the gravitational constant and $\theta_{\mu\nu}$ is the energy-momentum tensor of the matter. As such the linearized coupling between matter and gravity is of the form

$$2\kappa \theta_{\mu\nu} h^{\mu\nu} \tag{20.73}$$

where $2\kappa h_{\mu\nu} \equiv g_{\mu\nu} - \eta_{\mu\nu}$.

For a supersymmetric theory $\theta_{\mu\nu}$ belongs to a multiplet of currents that contains $j_{\mu\alpha}, j_\mu{}^{(5)}$ and other objects. Therefore one must have supergravity field equations for each of these objects and the linearized coupling will be of the form

$$\kappa (2\theta^{\mu\nu} h_{\mu\nu} - \bar{j}^{\mu\alpha} \psi_{\mu\alpha} + b^\mu j_\mu{}^{(5)} + \cdots) \tag{20.74}$$

Hence for every set of currents we may read off a corresponding set of supergravity fields.

Generally these lead to a formulation of supergravity; however, it may happen that one can not construct an action out of the resulting supergravity fields. An obvious case where this can happen is when one has only one auxiliary fermion. In this case it is necessary to add further anomaly multiplets or corresponding supergravity fields.

The constraints on the currents are now reflected by gauge transformations on supergravity fields. For example, translational invariance ($\partial^\mu \theta_{\mu\nu} = 0$) implies the gauge transformation

$$\delta h_{\mu\nu} = \partial_\mu \xi_\nu + \partial_\nu \xi_\mu \tag{20.75}$$

and conservation of supersymmetry ($\partial^\mu j_{\mu\alpha} = 0$) implies that

$$\delta \psi_{\mu\alpha} = \partial_\mu \eta_\alpha(x) \tag{20.76}$$

Let us now apply this strategy to the super multiplets of currents found above. The $N = 1$ super conformal currents $(\theta_{\mu\nu}, j_{\mu\alpha}, j_\mu^{(5)})$ lead to the corresponding fields

$$(h_{\mu\nu}, \psi_{\mu\alpha}, b^{(5)}) \tag{20.77}$$

The gauge transformations are those of Eqs. (20.75) and (20.76) as well as

$$\delta h_{\mu\nu} = \eta_{\mu\nu} d(x) \qquad \delta \psi_\mu = \gamma_\mu e(x) \quad \text{and} \quad \delta b^{(5)} = \partial_\mu \Lambda(x) \tag{20.78}$$

corresponding, respectively, to the constraints:

$$\theta_\mu^{\ \mu} = 0 = (\gamma^\mu)_\alpha^{\ \beta} j_{\mu\beta} = \partial^\mu J_\mu^{(5)} \tag{20.79}$$

These are the fields of $N = 1$ conformal supergravity[179] which is discussed in Chapter 3.

For the non-super conformal currents that possessed a chiral multiplet of anomalies

$$(\theta_{\mu\nu}, j_{\mu\alpha}, j_\mu^{(5)}, P, Q) \tag{20.80}$$

we find the corresponding supergravity fields[180]

$$(h_{\mu\nu}, \psi_{\mu\alpha}, b_\mu, M, N) \tag{20.81}$$

The gauge invariances are only those of Eqs. (20.75) and (20.76). These fields form the minimal auxiliary field formulation of $N = 1$ supergravity. The auxiliary fields being b_μ, M and N.

The Poincaré supercurrents with a linear multiplet of anomalies were

$$(\theta_{\mu\nu}, j_{\mu\alpha}, j_\mu^{(5)}, t_{\mu\nu}) \tag{20.82}$$

where $\partial^\mu j_\mu^{(5)} = \partial^\mu t_{\mu\nu} = 0.$ $t_{\mu\nu} = -t_{\nu\mu}$ and the resulting supergravity fields were

$$(h_{\mu\nu}, \psi_{\mu\alpha}, b_\mu, a_{\mu\nu})$$

These have the gauge invariances of Eqs. (20.75) and (20.76) as well as

$$\delta b_\mu = \partial_\mu \Lambda(x), \qquad \delta a_{\mu\nu} = \partial_\mu \Lambda_\nu - \partial_\nu \Lambda_\mu \qquad (20.83)$$

These fields are the new minimal formulation of supergravity.[181] The fields b_μ and $a_{\mu\nu}$ are auxiliary fields despite their gauge transformations. They occur in the action as

$$\int d^4x \varepsilon^{\mu\nu\rho\kappa} b_\mu \partial_\nu a_{\rho\kappa} \qquad (20.84)$$

This set of currents preserves R chiral invariance and the resulting formulation is locally R invariant, b_μ playing the role of the connection. In fact, only R invariant matter can be coupled to this formulation of supergravity.

Other anomaly multiplets lead to other formulations of supergravity. In fact there are three (20 + 20) formulations of supergravity.[13, 182, 183]

Chapter 21

Introduction to Two-Dimensional Supersymmetric Models and Superstring Actions

21.1 2-Dimensional Models of Rigid Supersymmetry

In this chapter we will discuss the possible 2-dimensional supersymmetric models. We do this partly because these models are much simpler than their higher dimensional analogues, and one can use them to illustrate general features and techniques without becoming involved with very complicated expressions. However, our main motivation is that these models underlie superstring theories, in the sense that string theories can be viewed as 2-dimensional models living on the world sheet of the string. A much more detailed treatment of the topics considered in this chapter will be given in later chapters.

Before discussing the possible models, we will review our conventions and some peculiarities of 2-dimensional Dirac algebra. The coordinates of our 2-dimensional space are labelled by ξ^a $(\alpha = 0, 1)$ and we take our spinor indices to be $A, B, \dots$. The reader should be aware that these conventions are virtually orthogonal to those we have used in higher dimensions; however, the reason for this departure will become apparent when we write down string theories. We take as our metric $\eta_{\alpha\beta} = (-, +)$ and choose $\gamma^\beta = (i\sigma^2, \sigma^1)$ where σ^1 and σ^2 are two of the three Pauli matrices and as a result γ_β satisfies $\gamma_\beta \gamma_\delta + \gamma_\delta \gamma_\beta = 2\eta_{\beta\delta}$.

We choose

$$\gamma_5 = +\gamma^0 \gamma^1 = \begin{pmatrix} 1 & 0 \\ 0 & -1 \end{pmatrix}. \tag{21.1}$$

The charge conjugation matrix C which satisfies

$$C\gamma^\beta C^{-1} = -\tilde{\gamma}^\beta \tag{21.2}$$

is chosen to be

$$C_{AB} = -\begin{pmatrix} 0 & 1 \\ -1 & 0 \end{pmatrix} = -\gamma^0. \tag{21.3}$$

Note that $(C\gamma^\beta)$ and $C\gamma_5$ are symmetric while C is antisymmetric.

We define the Dirac conjugate of a spinor χ_A to be $\bar\chi^D = -\chi^\dagger \gamma^0$. We can also define conjugate spinors by

$$\chi^c = C\bar\chi \quad \text{and} \quad \bar\chi^c = C^{-1}\chi = \tilde\chi C. \tag{21.4}$$

We observe that applying the above "conjugation" operation twice yields the identity operation, as it must. The following relations are often useful

$$\gamma_\beta \gamma^\delta \gamma^\beta = 0 \qquad \gamma^{\alpha\beta} \equiv \frac{1}{2}[\gamma^\alpha, \gamma^\beta] = \varepsilon^{\alpha\beta}\gamma_5 \tag{21.5}$$

where $\varepsilon^{01} = +1$. Finally we record the Fierz identity

$$\delta_A{}^B \delta_C{}^D = \frac{1}{2}\delta_A{}^D \delta_C{}^B + \frac{1}{2}(\gamma_\beta)_A{}^D (\gamma^\beta)_C{}^B + \frac{1}{2}(\gamma_5)_A{}^D (\gamma_5)_C{}^B. \tag{21.6}$$

The supersymmetry algebra which has one unconstrained complex supercharge Q_A is of the form

$$\{Q_A, \bar Q^B\} = -\frac{i}{2}(\gamma^\delta)_A{}^B P_\delta \quad \{Q_A, Q_B\} = \{\bar Q^A, \bar Q^B\} = 0 \quad [Q_A, P_\beta] = 0$$

$$[Q_A, J_{\alpha\beta}] = \frac{1}{2}(\gamma_{\alpha\beta})_A{}^B Q_B \tag{21.7}$$

plus the commutators of the Poincaré group.

We could derive all the irreducible representations of this algebra, as we did for the 4-dimensional super-Poincaré group. From these on-shell states, we could then find the corresponding supersymmetry transformation laws and invariant actions as before. We simply give the results of this analysis. The basic model contains the fields (A, χ_A, N), which are all complex. The off-shell supersymmetry transformation laws are

$$\delta A = i\bar\varepsilon\chi \qquad \delta\chi = \slashed\partial A\varepsilon + N\varepsilon^c \qquad \delta N = i\bar\varepsilon^c\slashed\partial\chi \tag{21.8}$$

and for their complex conjugates

$$\delta A^* = i\bar\chi\varepsilon, \quad \delta\bar\chi = -\bar\varepsilon\slashed\partial A^* + \bar\varepsilon^c N^*, \quad \delta N^* = -i\bar\chi\slashed\partial\varepsilon^c.$$

The closure on A is given by

$$[\delta_1, \delta_2]A = i\bar\varepsilon_2(\slashed\partial A\varepsilon_1 + N\varepsilon_1{}^c) - (1\leftrightarrow 2) = i\{\bar\varepsilon_2\slashed\partial\varepsilon_1 - (1\leftrightarrow 2)\}A. \tag{21.9}$$

The N dependent term vanishes as C_{AB} is antisymmetric and the ε's anticommute.

The invariant free action is given by

$$\mathscr{A} = \frac{1}{2}\int d^2\xi\{-|\partial_\alpha A|^2 - \bar\chi\slashed\partial\chi + |N|^2\}. \tag{21.10}$$

It is instructive to construct the currents of this theory corresponding to the generators P_α, Q_A and the chiral rotation generator R. One finds these currents are given respectively by

$$\theta_{\alpha\beta} = -\partial_{(\alpha}A\partial_{\beta)}A^* + \frac{1}{2}\eta_{\alpha\beta}\delta_\delta A\partial^\delta A^* - \frac{i}{8}\bar{\chi}(\gamma_\alpha\overleftrightarrow{\partial}_\beta + \gamma_\beta\overleftrightarrow{\partial}_\alpha)\chi$$

$$j_{\beta A} = (\partial\!\!\!/A^*\gamma_\beta\chi)_A$$

$$j_\beta{}^5 = i\bar{\chi}\gamma_\beta\chi. \tag{21.11}$$

These currents are conserved and also satisfy the additional properties

$$\theta_\alpha{}^\alpha = 0 = (\gamma^\beta j_\beta)_A. \tag{21.12}$$

These relations correspond to the existence of new symmetries, namely dilatations and S-supersymmetry, which have the current $\xi^\beta\theta_{\alpha\beta}$ and $(\xi j_\alpha)_A$ respectively.

By varying the currents given in Eq. (21.11) under the known supersymmetry transformations of Eq. (21.8) we find that

$$(\theta_{\alpha\beta}, j_{\alpha A}, j_\alpha{}^{(5)}) \tag{21.13}$$

forms a supermultiplet with the transformation properties

$$\delta\theta_{\alpha\beta} = \frac{i}{4}\bar{\varepsilon}\gamma^{(\alpha|\delta}\overleftrightarrow{\partial}_\delta j^{|\beta)} + \frac{i}{4}\bar{j}^{(\alpha|}\overleftrightarrow{\partial}_\delta\gamma^{\delta|\beta)}\varepsilon$$

(the notation means: symmetrize on indices, except those between vertical bars).

$$\delta j_{\beta A} = 2(\gamma^\delta\theta_{\beta\delta}\varepsilon)_A + \frac{1}{2}(\partial\!\!\!/j_\beta{}^{(5)}\varepsilon)_A$$

$$\delta j_\beta{}^{(5)} = i\bar{\varepsilon}j_\beta + i\bar{j}_\beta\varepsilon. \tag{21.14}$$

That the currents form a supermultiplet is an inevitable consequence of the fact that the transformation properties of the current must give rise to the supersymmetry algebra. For example, the equation

$$\delta j_{\alpha A} = \{j_{\alpha A}, \bar{Q}^B\} = \left(2\gamma^\delta\theta_{\alpha\delta} + \frac{1}{2}\partial\!\!\!/j_\alpha{}^{(5)}\right)^B_A \tag{21.15}$$

after taking $\alpha = 0$ integrating over ξ^1 and using a suitable identification of P_δ implies that

$$\{Q_A, \bar{Q}^B\} = -\frac{i}{2}(\gamma^\delta)^B_A P_\delta \tag{21.16}$$

as expected. The rest of the algebra can be deduced in a similar manner. It is clear from the above discussion that the chiral current, supersymmetry current, and energy-momentum tensor must lie in a supermultiplet in any supersymmetric theory. This phenomenon was first found in Ref. 109, and for a discussion of the relationship between the algebra and its currents see Chapter 20. The reader who examines the count of fermionic and bosonic degrees of freedom in the supercurrent multiplet will realize that outside two dimensions the supercurrent multiplet will in general have to contain other objects.

It is instructive to consider the various possible constraints on an unconstrained spinor χ_A which has components $\chi = \begin{pmatrix} a \\ b \end{pmatrix}$

(a) The Majorana constraint is given by $\chi^c = \chi$ and implies the equation

$$C\bar{\chi} = C(-\widetilde{\chi^\dagger \gamma^0}) = C\gamma^0 \tilde{\chi}^\dagger = \chi \tag{21.17}$$

whence

$$a = a^* \qquad b = +b^*. \tag{21.18}$$

We could impose this constraint on the previous supersymmetry algebra to give a Majorana supercharge and corresponding Majorana supersymmetry parameter. In order to impliment this change we decompose Q_A and $\bar{Q}^A$ into two Majorana spinors $Q_A{}^i, i = 1, 2$ as follows:

$$Q_A = \frac{1}{2}(Q^1{}_A - iQ^2{}_A), \qquad \bar{Q}^A = \frac{1}{2}(\bar{Q}^{1A} + i\bar{Q}^{2A}).$$

The inverse transformation is

$$Q^1{}_A = (Q_A + Q^c_A), \qquad Q^2{}_A = +i(Q_A - Q^c_A).$$

In terms of these Majorana supercharges, the algebra of Eq. (21.7) becomes

$$\{Q_A{}^i, \bar{Q}^{Bj}\} = -i\delta^{ij}(\gamma^\delta)_A{}^B P_\delta, \qquad [Q_A{}^i, P_\delta] = 0$$

$$[Q_A{}^i, J_{\alpha\beta}] = \frac{1}{2}(\gamma_{\alpha\beta})_A{}^B Q_B{}^i.$$

Setting $Q_A{}^2$ to be zero we find the algebra with one Majorana supercharge $Q_A = Q_A{}^1$:

$$\{Q_A, \bar{Q}^B\} = -i(\gamma^\delta)_A{}^B P_\delta, \qquad [Q_A, P_\delta] = 0$$

$$\{Q_A, J_{\alpha\beta}\} = \frac{1}{2}(\gamma_{\alpha\beta})_A{}^B Q_B$$

We will refer to such a superalgebra and the models which represent it as being of type $(1, 1)$. The multiplet of this algebra consists of (A, χ, N) where A and N are real and χ is a Majorana spinor. The off-shell transformation laws are easily deduced from Eq. (21.8) to be

$$\delta A = i\bar{\varepsilon}\chi \qquad \delta\chi = (\not\partial A + N)\varepsilon \qquad \delta N = i\bar{\varepsilon}\not\partial\chi \qquad (21.19)$$

The invariant action is given by

$$A = \frac{1}{2}\int d^2\xi\{-\partial_\alpha A\partial^\alpha A - i\bar{\chi}\not\partial\chi + N^2\}$$

(b) Alternatively as in any even dimensional space time we could impose a Weyl constraint

$$\chi = \gamma_5\chi. \qquad (21.20)$$

This implies that χ is of the form

$$\chi = \begin{pmatrix} a \\ 0 \end{pmatrix}. \qquad (21.21)$$

Imposing this constraint on the supercharge Q_A, it then only has its upper component, which we denote by Q_+. The corresponding supersymmetry parameter then only has the opposite chirality in order that $\bar{\varepsilon}Q$ does not vanish, and its lower component we denoted by ε^*. The Weyl supersymmetry algebra found by imposing these restrictions in Eq. (21.7) is

$$\{Q_+, Q_+^*\} = iP_z \qquad \{Q_+, Q_+\} = 0 \qquad [Q_+, P_z] = 0 = [Q_+, P_{\bar{z}}] \qquad (21.22)$$

We call such a superalgebra a $(2,0)$ superalgebra. It is useful to introduce the notation $R_z = \frac{1}{2}(R_0 + R_1)$ and $R_{\bar{z}} = \frac{1}{2}(R_0 - R_1)$ which results in the equations

$$R_\alpha P^\alpha = (-2)(R_z P_{\bar{z}} + R_{\bar{z}} P_z) \qquad (21.23)$$

while

$$\gamma_z\gamma_{\bar{z}} + \gamma_{\bar{z}}\gamma_z = -1 \qquad \gamma_z^2 = \gamma_{\bar{z}}^2 = 0. \qquad (21.24)$$

The supermultiplet (A, χ, N) transforms under Weyl supersymmetry as

$$\delta A = i\varepsilon\chi \qquad \delta\chi = 2\partial_z A\varepsilon, \qquad \delta N = 0 \qquad (21.25)$$

and hence a multiplet is given by (A, χ) since we can consistently set $N = 0$. In the above we have taken $\varepsilon = \begin{pmatrix} 0 \\ +\varepsilon^* \end{pmatrix}$ and

$$\partial_z = \frac{1}{2}\left(\frac{\partial}{\partial\zeta^0} + \frac{\partial}{\partial\zeta^1}\right), \qquad \partial_{\bar{z}} = \frac{1}{2}\left(\frac{\partial}{\partial\zeta^0} - \frac{\partial}{\partial\zeta^1}\right).$$

We observe that, unlike the case for four dimensions, the operator that appears on the right-hand side of $\{Q_+, Q_+^*\}$ is iP_z, and it is not in general a one-to-one operator. As such, we cannot demand that supermultiplets have equal numbers of fermionic and bosonic degrees of freedom. It is easy to find an example of this phenomenon: consider a field ϕ subject to $\partial_z \phi = 0$ which is the first component of the above supermultiplet. Under $(2, 0)$ supersymmetry the spinor component of this multiplet will be inert, and so can be set to zero. As such, ϕ is inert under supersymmetry and forms a supermultiplet by itself; we refer to such objects as singletons.

(c) The other possibility is to impose both a Majorana and Weyl constraint. In this case, if $\gamma_5 \chi = \chi$ then χ has only its upper component, which is real. A spinor which is both Majorana and Weyl can only exist when the dimension of space-time is $8n + 2$ for n an integer. This statement refers to spaces with only one time; for space with different signatures these spinors exist in different dimensions.

Let us suppose we are dealing with a Weyl spinor (i.e., $\chi = \gamma_5 \chi$), then we may rewrite the Weyl condition as $\gamma_{\bar{z}} \chi = 0$. If χ satisfies the Dirac equation, $\partial\!\!\!/ \chi = 0$, we find that

$$\gamma_z \partial_{\bar{z}} \chi = 0 \tag{21.26}$$

which implies

$$\partial_{\bar{z}} \chi = 0. \tag{21.27}$$

As such, χ is a function of $\xi^0 + \xi^1$ alone, and so is left-moving only. In two dimensions a particle must either go to the left or the right, and which way it travels is a Lorentz invariant statement. A spin 0 particle may also be left moving only if it satisfies $\partial_{\bar{z}} A = 0$. In terms of the more usual language this is equivalent to the equation

$$\partial_\alpha A = \varepsilon_\alpha{}^\beta \partial_\beta A.$$

The spinor action can be written in the form

$$-\frac{i}{2} \int d^2\xi \, \bar{\chi} \partial\!\!\!/ \chi = -i \int d^2\xi \chi^* \partial_{\bar{z}} \chi. \tag{21.28}$$

While the action for a scalar field in this language becomes

$$-\frac{1}{2} \int d^2\xi (\partial_\alpha A \partial^\alpha A^*) = +2 \int d^2\xi \, \partial_z A \partial_{\bar{z}} A^*.$$

However, the action for a scalar which is subject to the constraint of Eq. (21.28) vanishes identically. This is quite a general problem that applies to any odd rank field strength that is self-dual. Its general solution is unknown, but in two dimensions it has been proposed, see Ref. 184 to consider an action with a Lagrange multiplier $\lambda_{\bar{z}\bar{z}}$, namely

$$2 \int d^2\xi \left(\partial_z A \partial_{\bar{z}} A + \frac{1}{2} \lambda_{\bar{z}\bar{z}} \partial_z A \partial_z A \right).$$

A superalgebra with one Majorana Weyl supercharge $Q_A = \begin{pmatrix} Q_+ \\ 0 \end{pmatrix}$ is called a $(1,0)$ superalgebra and it is found from the above $(1,1)$ algebra to satisfy

$$\{Q_+, Q_+\} = 2iP_z, \qquad [Q_+, P_z] = 0. \tag{21.29}$$

Note that $(Q_+)^* = Q_+$ and that the corresponding supersymmetry parameter ε is also real. This algebra is represented on the real fields (A, χ) by the transformations

$$\delta A = i\varepsilon\chi_+, \qquad \delta\chi_+ = 2\partial_z A\varepsilon \tag{21.30}$$

The invariant action being

$$\int d^2\xi \left\{ \partial_z A \partial_{\bar z} A - \frac{i}{2}\chi_+ \partial_{\bar z}\chi_+ \right\}.$$

There exist another representation of $(1,0)$ supersymmetry which is of interest; it contains the fields (ζ_-, b) where ζ_- is the one component of a Majorana spinor ζ_A which is also Weyl in the sense $(\gamma_5)_A^B\zeta_B = -(\zeta)_A$ and b is a bosonic field. The corresponding transformation rules are

$$\delta\zeta_- = b\varepsilon, \qquad \delta b = +i2\partial_z\zeta_-\varepsilon.$$

The invariant action is given by

$$\int d^2\xi \left\{ i\zeta_- \partial_z\zeta_- + \frac{1}{2}b^2 \right\}.$$

Its corresponding on-shell states are one right-moving Majorana Weyl spinor; the field b being auxiliary. Thus the on-shell state is a singlet under $(1,0)$ supersymmetry.

21.2   Coupling of 2-Dimensional Matter to Supergravity

We now wish to couple the Majorana supersymmetric theory of Eq. (21.19) to supergravity. Our physical motivation for this construction is that this theory forms the basis for the superstring, which is required to be invariant under general coordinate and local sypersymmetry as these symmetries will correspond to symmetries of the world sheet of the string.

Before we do this, however, we will briefly discuss the superparticle which is described by the fields

$$(\phi^\mu(\tau), \chi^\mu(\tau)) \qquad (\mu = 0, \dots, D-1) \tag{21.31}$$

where χ^μ is a real Grassmann variable. The free action is given by

$$\frac{1}{2}\int d\tau(\dot\phi^\mu\dot\phi^\nu - \chi^\mu\dot\chi^\nu)\eta_{\mu\nu} \tag{21.32}$$

where $\dot{\phi}^\mu = \partial\phi^\mu/\partial\tau$. It is invariant under the supersymmetry transformations

$$\partial\phi^\mu = \varepsilon\chi^\mu \qquad \delta\chi^\mu = \dot{\phi}^\mu\varepsilon \tag{21.33}$$

as well as under rigid translations in τ.

We now couple this free action to the supergravity multiplet, which has the field content

$$(e, \psi). \tag{21.34}$$

These fields are the 1-dimensional analogues of the vierbein and gravitino. The final result may be obtained by a straightforward application of the Noether coupling method discussed previously, and is

$$\frac{1}{2}\int d\tau e^{-1}\{\dot{\phi}^\mu\dot{\phi}^\nu - \chi^\mu\dot{\chi}^\nu - \kappa\psi\chi^\mu\dot{\phi}^\nu\}\eta_{\mu\nu}. \tag{21.35}$$

This action is invariant under the local supersymmetry transformations

$$\delta\phi^\mu = \varepsilon\chi^\mu \qquad \delta\chi^\mu = \left(\partial_\tau\phi^\mu - \frac{\kappa}{2}\psi\chi^\mu\right)\varepsilon$$

$$\delta e = \varepsilon\psi \qquad \delta\psi = \frac{2}{\kappa}\partial_\tau\varepsilon \tag{21.36}$$

as well as under "general coordinate" transformations in τ.

We now turn to the 2-dimensional matter theory whose free action is given by

$$\mathcal{A}_0 = \int d^2\xi\left\{-\frac{1}{2}(\partial_\alpha A)^2 - \frac{i}{2}\bar{\chi}\partial\chi + \frac{1}{2}N^2\right\} \tag{21.37}$$

and which is invariant under the Majorana $(1, 1)$ supersymmetry, i.e.,

$$\delta A = i\bar{\varepsilon}\chi \qquad \delta\chi = \partial\!\!\!/A\varepsilon + N\varepsilon \qquad \delta N = i\bar{\varepsilon}\partial\!\!\!/\chi. \tag{21.38}$$

The fields A and N are real, while χ is a Majorana spinor. The supergravity multiplet is given by

$$h_{\alpha\beta}, \psi_{\alpha A}. \tag{21.39}$$

Their rigid supersymmetry transformations are

$$\delta h_{\alpha\beta} = \frac{i\kappa}{2}(\bar{\varepsilon}\gamma_\beta\psi_\alpha + \bar{\varepsilon}\gamma_\alpha\psi_\beta) \qquad \delta\psi_\alpha = -\gamma^{\beta\delta}\partial_\beta h_{\delta\alpha}\varepsilon \tag{21.40}$$

and they also possess Abelian local transformations

$$\delta\psi_\alpha = \partial_\alpha\eta \qquad \delta h_{\alpha\beta} = \partial_\alpha f_\beta + \partial_\beta f_\alpha.$$

The Noether procedure begins by letting the supersymmetry become local $\varepsilon \to \varepsilon(\xi)$. The free action is no longer invariant, but transforms as

$$\delta \mathcal{A}_0 = i \int d^2 \xi \{-(\partial_\alpha \bar{\varepsilon}) \partial\!\!\!/ A \gamma^\alpha \chi\}. \tag{21.41}$$

We cancel this term and gain an action invariant to order κ^0 by adding a term to the action:

$$\mathcal{A}_1 = \mathcal{A}_0 + i \int d^2 \xi \frac{\kappa}{2} \bar{\psi}_\alpha \partial\!\!\!/ A \gamma^\alpha \chi. \tag{21.42}$$

This is invariant provided we identify the Abelian local fermionic transformation with the (now local) supersymmetry transformation by

$$\eta = \frac{2}{\kappa} \varepsilon \tag{21.43}$$

The next step is to vary $\mathcal{A}_1$ under supersymmetry; we find the result

$$\delta \mathcal{A}_1 = -\frac{\kappa}{2} \int d^2 \xi \{ \bar{\psi}_\alpha \gamma^\beta \gamma^\alpha \chi (\partial_\beta \bar{\varepsilon}) \chi + \bar{\psi}_\alpha \gamma^\beta \gamma^\alpha \chi \bar{\varepsilon} (\partial_\beta \chi)$$
$$+ \bar{\psi}_\alpha \gamma^\beta \gamma^\alpha \gamma^\delta \varepsilon \partial_\beta A \partial_\delta A + \bar{\psi}_\alpha \partial\!\!\!/ A \gamma^\alpha \varepsilon N \}. \tag{21.44}$$

We will now deal with these four terms one by one. Fierzing the first term yields the result

$$+\frac{\kappa}{4} \int d^2 \xi (\bar{\psi}_\alpha \gamma^\beta \gamma^\alpha \partial_\beta \varepsilon) \bar{\chi} \chi. \tag{21.45}$$

In this last step we have used the fact that

$$\bar{\chi} \gamma_\alpha \chi = \bar{\chi} \gamma_5 \chi = 0 \tag{21.46}$$

if χ is a Majorana spinor. We cancel this term by adding a piece to the action, which now becomes

$$\mathcal{A}_2 = \mathcal{A}_1 - \frac{\kappa^2}{16} \int d^2 \xi \bar{\psi}_\alpha \gamma^\beta \gamma^\alpha \psi_\beta \bar{\chi} \chi. \tag{21.47}$$

The third term is cancelled by the variations of the vierbeins which must occur in the scalar kinetic term, namely

$$\int d^2 \xi \left\{ -\frac{1}{2} \delta(e e^\alpha{}_a e^{\beta a}) \partial_\alpha A \partial_\beta A \right\} \tag{21.48}$$

where $e_a{}^\alpha = \eta_a{}^\alpha + \kappa h_a{}^\alpha$. The last term is cancelled by adding a term to δN. The second term is cancelled by adding terms to $\delta \chi$ as well as varying the vierbeins in the spinor kinetic term.

Finally, the reader may vary $\mathscr{A}_2$ and verify that no further additions to the action are necessary, but complete invariance is gained by adding terms to the transformation laws of the fields. We summarize the final action (see Refs. 185, 186).

$$\mathscr{A} = \int d^2\xi e \left\{ -\frac{1}{2} e_a{}^\alpha e^{\beta a} \partial_\alpha A \partial_\beta A - \frac{i}{2} \bar{\chi} e_a{}^\alpha \gamma^a \partial_\alpha \chi + \frac{1}{2} N^2 \right.$$

$$\left. + \frac{i\kappa}{2} \bar{\psi}_\alpha \partial\!\!\!/ A \gamma^\alpha \chi - \frac{\kappa^2}{16} \bar{\psi}_\alpha \gamma^\beta \gamma^\alpha \psi_\beta \bar{\chi}\chi \right\}. \tag{21.49}$$

This action possesses general coordinate invariance, local Lorentz invariance, local supersymmetry, but also two other symmetries, namely Weyl invariance and S supersymmetry. Weyl invariance has a parameter $\Lambda(\xi)$ and the transformations are

$$\delta e_\alpha{}^a = \Lambda e_\alpha{}^a \qquad \delta\psi_\alpha = \frac{1}{2}\Lambda\psi_\alpha \quad \text{and}$$

$$\delta A = 0 \qquad \delta\chi = -\frac{1}{2}\Lambda\chi \qquad \delta N = -\Lambda N. \tag{21.50}$$

The S supersymmetry has a spinor parameter $\zeta_A(\xi)$, and all fields are inert except for

$$\psi_\alpha \to \psi_\alpha + \gamma_\alpha \zeta. \tag{21.51}$$

This is an invariance as a result of the 2-dimensional identity

$$\gamma^\beta \gamma_\alpha \gamma_\beta = 0. \tag{21.52}$$

In fact, the action is invariant under the local 2-dimensional superconformal group.

It is useful to count the degrees of freedom of the supergravity fields. The vierbein $e_\alpha{}^a$ has four degrees of freedom, but it has four gauge transformations. There is one local Lorentz, two general coordinate and one Weyl transformation. As a result we may gauge $e_\alpha{}^a$ to be of the form $e_\alpha{}^a = \delta_\alpha{}^a$. The gravitino $\psi_{\alpha A}$ has four degrees of freedom, but it also has four gauge transformations, consisting of two Q supersymmetries and two S supersymmetries. We may use these symmetries to gauge $\psi_\alpha{}^1 = 0$, see Ref. 186. Carrying out the closure of these symmetries on the supergravity fields is somewhat academic since one is bound to find a local symmetry; in this sense auxiliary fields are not required.

The equations of motion in these gauges become the free equations

$$\partial^2 A = 0 \qquad \partial\!\!\!/ \chi = 0. \tag{21.53}$$

However, we must not forget the equations of motion of $e_\alpha{}^a$ and $\psi_{\alpha A}$ which, in this gauge, reduce to

$$+\partial_\alpha A \partial_\beta A - \frac{i}{2}\bar{\chi}\gamma_{(\alpha}\partial_{\beta)}\chi - \frac{1}{2}\eta_{\alpha\beta}(\partial_\delta A)^2 = 0 \tag{21.54}$$

as well as

$$\gamma^\beta \gamma^\delta \partial_\beta A \chi = 0. \tag{21.55}$$

Thus the action is a free action subject to constraints.

We have yet to write down the local supersymmetry transformations of the fields which were deduced by the Noether procedure on the action. It is instructive, however, to deduce their form by also using the Noether procedure on the algebra. We begin with the rigid supersymmetry algebra and make the supersymmetry parameter local, i.e., $\varepsilon \to \varepsilon(\xi)$. The algebra no longer closes; for example on χ_A we find that

$$[\delta_1, \delta_2]\chi = i\{\gamma^\beta \varepsilon_2 \bar{\varepsilon}_1(\partial_\beta \chi) + \bar{\varepsilon}_1 \partial\!\!\!/\chi \varepsilon_2 + \gamma^\beta \varepsilon_2(\partial_\beta \bar{\varepsilon}_1)\chi\} - (1 \leftrightarrow 2). \tag{21.56}$$

The first two terms will Fierz as before, to give a translation. The last term contains a derivative of ε_1 and will yield, in the closure, terms not recognizable as a symmetry. We remove this term by replacing the $\partial_\alpha A$ in $\delta\chi$ by a super-covariant derivative

$$\partial_\alpha A \to \hat{D}_\alpha A = \partial_\alpha A - \frac{i\kappa}{2}\bar{\psi}_\alpha \chi \tag{21.57}$$

and so

$$\delta\chi = (\gamma^\beta \hat{D}_\beta A + N)\varepsilon. \tag{21.58}$$

For N we also find $\partial_\beta \varepsilon$ terms and the solution is similar. We take

$$\delta N = i\bar{\varepsilon}\gamma^\beta \hat{D}_\beta \chi \tag{21.59}$$

where

$$\hat{D}_\beta \chi = D_\beta \chi - \frac{\kappa}{2}(\partial\!\!\!/A + N)\psi_\beta \tag{21.60}$$

and

$$D_\beta \chi = \left(\partial_\beta + \frac{1}{4}\omega_{\beta ab}\gamma^{ab}\right)\chi.$$

Here we have included a suitable spin connection, which is required for closure at higher order in κ. With this covariantization of the derivatives, the algebra closes at order κ^0. For example, on A we find that

$$[\delta_1, \delta_2]A = i2\bar{\varepsilon}_2 \gamma^\beta \varepsilon_1 \hat{D}_\beta A$$

$$= i2\bar{\varepsilon}_2 \gamma^\beta \varepsilon_1 \partial_\beta A + \left(2\bar{\varepsilon}_2 \gamma^\beta \varepsilon_1 \frac{\kappa}{2} \bar{\psi}_\beta\right)\chi. \tag{21.61}$$

The first term is a general coordinate transformation and the second term is a field dependent supersymmetry transformation. In fact, the reader may verify that the closure works to all orders without further additions (which are sometimes necessary) to the variations of the fields. The final results are

$$\delta A = i\bar{\varepsilon}\chi \qquad \delta\chi = (\hat{\slashed{D}}A + N)\varepsilon \qquad \delta N = i\bar{\varepsilon}\hat{\slashed{D}}\chi \tag{21.62}$$

and

$$\delta e_\alpha{}^a = i\kappa\bar{\varepsilon}\gamma^a\psi_\alpha \qquad \delta\psi_\alpha = \frac{2}{\kappa}D_\alpha\varepsilon \tag{21.63}$$

A peculiarity of two dimensions is that

$$\frac{1}{4}\omega_\alpha{}^{bc}\gamma_{bc} = -\frac{1}{2}\omega_\alpha\gamma_5 \tag{21.64}$$

where $\omega_\alpha = -\frac{1}{2}\varepsilon^{ab}\omega_{\alpha ab}$. As usual, the spin connection contains gravitino terms which guarantee that its transformation does not contain $\partial\varepsilon$ terms; thus it is a sum of the usual vierbein terms, $\omega_\alpha{}^{ab}(e)$ and the gravitino terms:

$$\omega_\alpha = \omega_\alpha(e) - \frac{\kappa^2}{2}\bar{\psi}_\alpha\gamma_5\gamma^\delta\psi_\delta. \tag{21.65}$$

It is now straightforward to write down the actions that define the usual superstring theories. These live on the world sheet of the string, whose coordinates are $\zeta^\alpha = (\tau, \sigma)$. The superstring has fields $x^\mu(\zeta)\chi^\mu{}_A(\zeta)$ and the auxiliary field N^μ where $\mu = 0, \ldots, 9$. These μ indices are those of space-time, and appear like "internal labels" on the world sheet fields without complicating any of the previous analysis. The action is

$$\mathscr{A} = +\frac{1}{2\pi\alpha'}\int d^2\xi\, e\left\{-\frac{1}{2}e_a{}^\alpha\partial_\alpha x^\mu e^{\beta a}\partial_\beta x^\nu - \frac{i}{2}\bar{\chi}^\mu\gamma^a e_a{}^\alpha\partial_\alpha\chi^\nu + N^\mu N^\nu\right.$$

$$\left. +\frac{i\kappa}{2}\bar{\psi}_\alpha\slashed{\partial}x^\mu\gamma^\alpha\chi^\nu - \frac{\kappa^2}{16}\bar{\psi}_\alpha\gamma^\beta\gamma^\alpha\psi_\beta\bar{\chi}^\mu\chi^\nu\right\}\eta_{\mu\nu}. \tag{21.66}$$

It is invariant under 2-dimensional general coordinate transformations, Q supersymmetry, S supersymmetry, Weyl and local Lorentz transformations. These transformations are as before with an appropriate distribution of μ, ν indices (i.e., $\delta x^\mu = i\bar{\varepsilon}\chi^\mu, \ldots$). It is also invariant under the 10-dimensional (rigid) Poincaré group. Of course, it was in this model that supersymmetry was independently discovered, and it leads to the Ramond-Neveu-Schwarz formulation of the superstring, see Ref. 7.

As before we may choose a gauge $e_\alpha{}^a = \delta_\alpha{}^a$ and $\psi_{\alpha A} = 0$; the theory then becomes a free theory subject to the constraints given previously.

Let us now construct the local coupling of the (A, χ) multiplet of $(1,0)$ supersymmetry. We may obtain this result from the Majorana, i.e., $(1,1)$ superalgebra case given above. Examining Eqs. (21.62) and (21.63) the reduction to $(1,0)$ is obtained by taking χ_A to be Weyl in the sense $\gamma_5\chi = \chi$ and ε_A Weyl in the opposite sense $\gamma_5\varepsilon = -\varepsilon$. Due to the relation $\bar{\varepsilon}\gamma^\delta\gamma_5\chi = \bar{\varepsilon}\gamma^\delta\chi = -\bar{\varepsilon}\gamma_5\gamma^\delta\chi = -\bar{\varepsilon}\gamma^\delta\chi = 0$, the variation of N vanishes and we may consistently set N to zero. The gravitino $\psi_{\alpha A}$ has the same handedness as ε_A, i.e., $\gamma_5\psi_\alpha = -\psi_\alpha$. The transformations then read

$$\delta A = i\bar{\varepsilon}\chi, \qquad \delta\chi = \hat{\slashed{D}}A\varepsilon \tag{21.67}$$

with

$$\hat{D}_\alpha A = \partial_\alpha A - \frac{i}{2}\kappa\bar{\psi}_\alpha\chi.$$

While for the $(1,0)$ supergravity multiplet we have

$$\delta e_\alpha^a = i\kappa\bar{\varepsilon}\gamma^a\psi_\alpha; \qquad \delta\psi_\alpha = \frac{2}{\kappa}D_\alpha\varepsilon. \tag{21.68}$$

The coupling of this $(1,0)$ multiplet to supergravity is found by making the above changes in Eq. (21.49) yield the result

$$A = \int d^2\xi e\left\{-\frac{1}{2}e_a^\alpha e^{\beta a}\partial_\alpha A\partial_\beta A - \frac{i}{2}\bar{\chi}e_a^\alpha\gamma^a\partial_\alpha\chi + \frac{i\kappa}{2}\bar{\psi}_\alpha\slashed{\partial}A\gamma^\alpha\chi\right\}. \tag{21.69}$$

The final term does not contribute as $\bar{\chi}\chi = 0$ if χ is a Weyl spinor.

The above action is Weyl invariant, the transformations being those of Eq. (21.50). It also possess local S, supersymmetry under which all fields are inert except for

$$\delta\psi_\alpha = \gamma_\alpha\lambda \tag{21.70}$$

where

$$\gamma_5\lambda = \lambda$$

The other multiplet of $(1,0)$ supersymmetry contains the fields (ζ, b) where ζ satisfies $\gamma_5\zeta = -\zeta$ and is a real spinor, while b is a real bosonic field. This multiplet can also be obtained from a reduction of the $(1,1)$ multiplet of Eq. (21.38). Since $\gamma_5\varepsilon = -\varepsilon$ for $(1,0)$ supersymmetry the variation of A vanishes and we may set A equal to zero. Identifying $\chi = \zeta$ and $N = b$, we find the transformations of Eq. (21.62) are

$$\delta\zeta = b\varepsilon, \qquad \delta b = i\bar{\varepsilon}\hat{\cancel{D}}\zeta \tag{21.71}$$

where

$$\hat{D}_\beta\zeta = \left(\partial_\beta + \frac{1}{4}\omega_{\beta ab}\gamma^{ab}\right)\zeta - \frac{\kappa}{2}b\psi_\beta, \tag{21.72}$$

the transformations of $(1,0)$ supergravity of Eq. (21.68) being of course unchanged. The coupling to supergravity is deduced from Eq. (21.49) to be

$$A = \int d^2\xi e\left\{-\frac{i}{2}\bar{\zeta}e_a^\alpha\gamma^a\partial_\alpha\zeta + \frac{1}{2}b^2\right\}. \tag{21.73}$$

We observe that the action does not in fact contain the gravitino and as such, it possesses S-supersymmetry in a trivial way. It is of course Weyl invariant.

The heterotic string[187] is a superstring that has $(1,0)$ supersymmetry on its world-sheet. The string moves through a ten-dimensional super space-time with the co-ordinates $x^\mu(\zeta), \chi_A^\mu(\zeta) (\mu = 0, 1, \ldots, 9)$ and $\chi_A^I(\zeta); I = 1, \ldots, 32$ where $x^\mu(\zeta)$ is a real bosonic field but $\chi_A^\mu(\zeta)$ and χ_A^I are Majorana Weyl spinors with Weyl handedness

$$\gamma_5\chi^\mu = \chi^\mu, \qquad \gamma_5\chi^I = -\chi^I. \tag{21.74}$$

In terms of representations of $(1,0)$ supersymmetry we may identify fields to belong to the multiplets

$$(x^\mu, \chi_A^\mu) \tag{21.75}$$

and

$$(\chi_A^I, b) \tag{21.76}$$

where we have added the auxiliary field b^I.

The action for the heterotic string is the coupling of the above super-multiplets to supergravity:

$$A = A_1 + A_2$$

where

$$A_1 = \frac{1}{2\pi\alpha'}\int d^2\xi e\left\{-\frac{1}{2}e_a^\alpha e^{\beta a}\partial_\alpha x^\mu\partial_\alpha x^\nu\right.$$

$$\left. -\frac{i}{2}\bar{\chi}^\mu e_a^\alpha\gamma^a\partial_\alpha\chi^\nu + \frac{i\kappa}{2}\bar{\psi}_\alpha\cancel{\partial}x^\mu\gamma^\alpha\chi^\nu\right\}\eta_{\mu\nu} \tag{21.77}$$

$$A_2 = \frac{1}{2\pi\alpha'} \int d^2\xi e \left\{ -\frac{i}{2} \bar{\chi}^I \gamma^a e_a^\alpha \partial_\alpha \chi^I + \frac{1}{2} (b^I)^2 \right\}. \tag{21.78}$$

The transformations of the fields are obtained by scattering μ and I indices on Eqs. (21.67) and (21.71)

$$\delta x^\mu = i\bar{\varepsilon}\chi^\mu, \qquad \delta\chi^\mu = \hat{D}x^\mu\varepsilon \tag{21.79}$$

where

$$\hat{D}_\alpha x^\mu = \partial_\alpha x^\mu - \frac{i\kappa}{2} \bar{\psi}_\alpha \chi^\mu \tag{21.80}$$

and

$$\delta\chi^I = b^I\varepsilon, \qquad \delta b^I = i\bar{\varepsilon}\hat{D}\chi^I \tag{21.81}$$

where

$$\hat{D}_\alpha \chi^I = \left(\partial_\alpha + \frac{1}{4} \omega_\alpha{}^{ab} \gamma_{ab} \right) \chi^I - \frac{\kappa}{2} b^I \psi_\alpha \tag{21.82}$$

as well as the transfomrations of e_α^a and ψ_α of Eq. (21.68).

Varying the above fields we find the equations of motion, which upon choosing the gauge $e_a^\alpha = \delta_a^\alpha$, $\psi_\alpha = 0$, become

$$\partial^2 x^\mu = 0, \qquad \hat{\partial}\chi^\mu = 0 \tag{21.83}$$

$$\hat{\partial}\chi^I = 0, \qquad b^I = 0 \tag{21.84}$$

as well as the constraints

$$\hat{\partial}x^\mu\gamma^\alpha\chi^\nu\eta_{\mu\nu} = 0$$

$$\left\{ \partial_\alpha x^\mu \partial_\beta x^\nu - \frac{1}{2}\eta_{\alpha\beta}(\partial_\delta x \partial^\delta x^\nu) - \frac{i}{2}\bar{\chi}^\mu\gamma_{(\alpha}\partial_{\beta)}\chi^\nu \right\}\eta_{\mu\nu} - \frac{i}{2}\bar{\zeta}^I\gamma_{(\alpha}\partial_{\beta)}\zeta^I = 0. \tag{21.85}$$

Chapter 22

Two-Dimensional Supersymmetry Algebras

Before beginning the task in front of us it will be useful to construct a set of conventions and adopt a particularly convenient set of coordinates and basis for the spinors that displays in an obvious way their Lorentz properties. We will carry out this construction both in Minkowski and Euclidean space. The latter space occurs naturally in two-dimensional statistical systems and after a Wick rotation from Minkowski space. This discussion of conventions will be in considerable detail as it leads to a simplicity of the resulting formulae that allows for their easy manipulation.

22.1   Conventions in Two-Dimensional Minkowski and Euclidean Space

The rotation group in two dimensions is $SO(1, 1)$ in Minkowski space or $SO(2)$ in Euclidean space. In contrast to the Lorentz group in higher dimensions, in two dimensions this group is an Abelian group with one generator and its irreducible representations are one-dimensional. This leads to a considerable simplification and it is desirable to have a formalism which takes this into account. That is, we wish to decompose all representations of the Lorentz group, such as tensors into irreducible one-dimensional representations of the Lorentz group. The appropriate formalism in Minkowski space is the adaptation of light-cone coordinates which are given by

$$z \equiv x^0 + x^1, \qquad \bar{z} \equiv x^0 - x^1 \tag{22.1}$$

where z and $\bar{z}$ are independent coordinates. The inverse coordinate transformation is

$$x^0 = \frac{1}{2}(z + \bar{z}), \qquad x^1 = \frac{1}{2}(z - \bar{z}). \tag{22.2}$$

Under the Lorentz group with $\omega = \omega^{01}$ we have

$$\delta x^0 = \omega x^1, \qquad \delta x^1 = \omega x^0$$

and so

$$\delta z = \omega z, \qquad \delta \bar{z} = -\omega \bar{z}. \tag{22.3}$$

In these new coordinates the Minkowski line element becomes

$$ds^2 = -(dx^0)^2 + (dx^1)^2 = -dz\,d\bar{z} \tag{22.4}$$

and writing it as

$$ds^2 = g_{zz}\,dz\,dz + g_{z\bar{z}}\,dz\,d\bar{z} + g_{\bar{z}z}\,d\bar{z}\,dz + g_{\bar{z}\bar{z}}\,d\bar{z}\,d\bar{z} \tag{22.5}$$

we deduce that the metric in these new coordinates is

$$g_{zz} = g_{\bar{z}\bar{z}} = 0; \qquad g_{z\bar{z}} = g_{\bar{z}z} = -\tfrac{1}{2}. \tag{22.6}$$

The inverse metric is given by

$$g^{zz} = g^{\bar{z}\bar{z}} = 0; \qquad g^{z\bar{z}} = g^{\bar{z}z} = -2. \tag{22.7}$$

Corresponding to the coordinate transformation of Eq. (22.1), we must also transform tensors by the usual rules for a coordinate change. For a tensor with components S^μ in the coordinate system x^μ we find, using Eq. (22.1) that it has components

$$S^z = S^0 + S^1, \qquad S^{\bar{z}} = S^0 - S^1 \tag{22.8}$$

in $z, \bar{z}$ coordinates.

For a tensor L_μ with a lower index we find using Eq. (22.2) that

$$L_z = \frac{1}{2}(L_0 + L_1), \qquad L_{\bar{z}} = \frac{1}{2}(L_0 - L_1). \tag{22.9}$$

Examples of such tensors are provided by dz and $d\bar{z}$ corresponding to Eq. (22.8) and

$$\frac{\partial\phi}{\partial z} \quad \text{and} \quad \frac{\partial\phi}{\partial \bar{z}}$$

where ϕ is a scalar corresponding to those in Eq. (22.9). Indeed one easily verifies that

$$dz = dx^0 + dx^1, \qquad d\bar{z} = dx^0 - dx^1 \tag{22.10}$$

while

$$\partial_z \equiv \frac{\partial}{\partial z} = \frac{1}{2}\left(\frac{\partial}{\partial x^0} + \frac{\partial}{\partial x^1}\right), \qquad \partial_{\bar{z}} \equiv \frac{\partial}{\partial \bar{z}} = \frac{1}{2}\left(\frac{\partial}{\partial x^0} - \frac{\partial}{\partial x^1}\right) \tag{22.11}$$

in accordance with the above formulae.

We recognize that each *component* of a tensor in light-cone coordinates forms an irreducible tensor of the Lorentz group, for example

$$\delta S^z = \omega S^z, \qquad \delta S^{\bar{z}} = -\omega S^{\bar{z}}. \tag{22.12}$$

The components of light-cone tensors with more indices are straightforward to find and we give the example of a tensor of type $T_{\mu\nu}$

$$T_{zz} = \frac{1}{4}\{T_{00} + T_{01} + T_{10} + T_{11}\}$$

$$T_{z\bar{z}} = \frac{1}{4}\{T_{00} - T_{01} + T_{10} - T_{11}\}$$

$$T_{\bar{z}z} = \frac{1}{4}\{T_{00} + T_{01} - T_{10} - T_{11}\}$$

$$T_{\bar{z}\bar{z}} = \frac{1}{4}\{T_{00} - T_{01} - T_{10} + T_{11}\}. \tag{22.13}$$

The most general tensor in light-cone coordinates is of the form

$$T \overset{\overbrace{z\ldots z}^{n} \overbrace{\bar{z}\ldots\bar{z}}^{\bar{n}}}{\underset{\underbrace{z\ldots z}_{m} \underbrace{\bar{z}\ldots\bar{z}}_{\bar{m}}}{}} \equiv T_{m,\bar{m}}^{n,\bar{n}} \tag{22.14}$$

and it forms an irreducible representation of the Lorentz group which transforms as

$$\delta T_{m,\bar{m}}^{n,\bar{n}} = w[(n-m) - (\bar{n}-\bar{m})]. \tag{22.15}$$

The scalar product in light-cone language becomes

$$L \cdot S \equiv S^{\mu} L_{\mu} = S^{z} L_{z} + S^{\bar{z}} L_{\bar{z}}$$

$$= -2(S_{\bar{z}} L_{z} + S_{z} L_{\bar{z}}) = -\frac{1}{2}(S^{z} L^{\bar{z}} + S^{\bar{z}} L^{z}). \tag{22.16}$$

As such for any tensor with two indices

$$T^{z\bar{z}} + T^{\bar{z}z} = -2T^{\mu}{}_{\mu}$$

Using the metric we may always express any tensor in terms of only upper and lower z indices or equivalently upper $\bar{z}$ indices and lower $\bar{z}$ indices etc. For example the tensor of Eq. (22.14) has the same tensor character as the tensor

$$T_{m-\bar{n},0}^{n-\bar{m},0} \quad \text{and as} \quad T_{0,\bar{m}-n}^{0,\bar{n}-m}.$$

To illustrate the above let us rewrite the action for a scalar field in light-cone coordinates

$$-\frac{1}{2}\int d^{2}\xi \partial_{\alpha}A\partial^{\alpha}A = -\int dz\,d\bar{z}\,\partial_{z}A\partial_{\bar{z}}A.$$

In many applications of two-dimensional systems one works in Euclidean space and we now find the analogue of the light-cone coordinates above. The change to Euclidean space is achieved from Minkowski space by the usual Wick notation

$$x^0 \to -ix^0, \qquad x^1 \to x^1. \tag{22.17}$$

Let us then re-label $x^0 \to x^1$ and $x^1 \to x^2$ since x^1 and x^2 are not distinguished in any way in Euclidean space. The line element of Eq. (22.4) becomes

$$ds^2 = (dx^1)^2 + (dx^2)^2. \tag{22.18}$$

After making such a change it is advantageous to work with the coordinates

$$z \equiv x^1 + ix^2, \qquad \bar{z} \equiv x^1 - ix^2 \tag{22.19}$$

where $\bar{z}$ is the complex conjugate of z. The inverse coordinate transformation being

$$x^1 = \frac{1}{2}(z + \bar{z}), \qquad x^2 = \frac{i}{2}(z - \bar{z}). \tag{22.20}$$

The line element of Eq. (22.18) becomes

$$ds^2 = dz \, d\bar{z} \tag{22.21}$$

and so the metric is given by

$$g_{zz} = g_{\bar{z}\bar{z}} = 0, \qquad g_{z\bar{z}} = g_{\bar{z}z} = \frac{1}{2} \tag{22.22}$$

with inverse

$$g^{zz} = g^{\bar{z}\bar{z}} = 0, \qquad g^{z\bar{z}} = g^{\bar{z}z} = 2. \tag{22.23}$$

Under a 'Euclidean' rotation

$$x^1 \to x^1 \cos \omega + x^2 \sin \omega, \qquad x^2 = x^2 \cos \omega - x^1 \sin \omega \tag{22.24}$$

but we find that

$$z \to e^{-i\omega} z, \qquad \bar{z} \to e^{i\omega} \bar{z}. \tag{22.25}$$

Given a tensor with components L^μ and S^μ in the $z, \bar{z}$ coordinates they have components

$$L^z = L^1 + iL^2, \qquad L^{\bar{z}} = L^1 - iL^2 \tag{22.26}$$

and

$$S_z = \frac{1}{2}(S_1 - iS_2), \qquad S_{\bar{z}} = \frac{1}{2}(S_1 + iS_2). \tag{22.27}$$

One particular example of the latter are the derivatives

$$\partial_z \equiv \partial \equiv \frac{\partial}{\partial z} = \frac{1}{2}\left(\frac{\partial}{\partial x^1} - i\frac{\partial}{\partial x^2}\right) \tag{22.28}$$

$$\partial_{\bar{z}} = \bar{\partial} \equiv \frac{\partial}{\partial \bar{z}} = \frac{1}{2}\left(\frac{\partial}{\partial x^1} + i\frac{\partial}{\partial x^2}\right) \tag{22.29}$$

22.1a Minkowski Spinors

It is also desirable to introduce the analogue of light-cone coordinates for the spinors. The Lorentz group is generated by $\gamma_{\alpha\beta}/2 \equiv [\gamma_\alpha, \gamma_\beta]/4$ and its action on a spinor is given by

$$\delta\psi = \frac{1}{2}\omega^{\alpha\beta}\gamma_{\alpha\beta}\psi. \tag{22.30}$$

Consequently, to decompose spinors into irreducible representations of the Lorentz group is the same as decomposing them into eigenstates of γ_5. Let us just carry out this exercise in Minkowski space using the γ matrices, charge conjugation matrices and their properties given in Eqs. (21.1) to (21.6).

In this representation

$$\delta\psi = -\frac{1}{2}\omega\gamma_5\psi$$

where $\omega = \omega^{01}$.

Since γ_5 is diagonal, the upper and lower components of ψ are separately representation of the Lorentz group. Correspondingly we denote the components of ψ by

$$\psi = \begin{pmatrix} \psi_+ \\ \psi_- \end{pmatrix} \tag{22.31}$$

and similarly for any spinor.

A metric in spinor space is provided by the charge conjugation matrix. We can regard the Majorana conjugate spinor as one with indices raised.

$$\bar{\psi}^C \equiv (\psi^+, \psi^-) \equiv \bar{\psi}C = -(\psi_+, \psi_-)\begin{pmatrix} 0 & 1 \\ -1 & 0 \end{pmatrix} = (\psi_-, -\psi_+) \tag{22.32}$$

or

$$\psi^+ = \psi_-, \qquad \psi^- = -\psi_+$$

The corresponding metric is given by

$$\eta^{++} = \eta^{--} = 0, \qquad \eta^{+-} = +1, \qquad \eta^{-+} = -1 \tag{22.33}$$

with its inverse

$$\eta_{++} = \eta_{--} = 0, \qquad \eta_{-+} = +1, \qquad \eta_{+-} = -1. \tag{22.34}$$

Another method of raising the index is by using the Dirac conjugate

$$\bar{\psi}^D \equiv -\psi^+ \gamma^0 \equiv (\bar{\psi}^+, \bar{\psi}^-) = -((\psi_+)^*, (\psi_-)^*) \begin{pmatrix} 0 & 1 \\ -1 & 0 \end{pmatrix}$$

$$= ((\psi_-)^*, -(\psi_+)^*). \tag{22.35}$$

Written more explicitly we have

$$\bar{\psi}^+ = (\psi_-)^*, \qquad \bar{\psi}^- = -(\psi_+)^*. \tag{22.36}$$

Consider now the Dirac operator

$$\gamma^\alpha \partial_\alpha = \gamma^z \partial_z + \gamma^{\bar{z}} \partial_{\bar{z}} \tag{22.37}$$

where

$$\gamma^z = \gamma^0 + \gamma^1, \qquad \gamma^{\bar{z}} = \gamma^0 - \gamma^1.$$

Using the explicit form for the γ matrices we find

$$\gamma^\alpha \partial_\alpha = 2 \begin{pmatrix} 0 & \partial_z \\ -\partial_{\bar{z}} & 0 \end{pmatrix}. \tag{22.38}$$

The Dirac equation $\partial\!\!\!/\psi = 0$ therefore becomes

$$\bar{\partial}\psi_+ = 0, \qquad \partial\psi_- = 0 \tag{22.39}$$

where

$$\partial = \partial_z, \qquad \bar{\partial} = \partial_{\bar{z}}.$$

We will talk of ψ_+ and ψ_- as being left- and right-moving respectively.

The Dirac action takes the form

$$\int d^2x \frac{i}{2} \bar{\psi} \partial\!\!\!/\psi = -\frac{i}{2} \int dz\, d\bar{z}\, \{\bar{\psi}^+ \partial_z \psi_- - \bar{\psi}^- \partial_{\bar{z}} \psi_+\}. \tag{22.40}$$

We can now summarize the discussion on the various types of spinor given in Chapter 21. Given any spinor χ, it contains two irreducible representations of the Lorentz group χ_+ and χ_- and their complex conjugates $(\chi_+)^*$ and $(\chi_-)^*$. If the spinor is Majorana then Eqs. (22.32) and (22.36) imply that $(\chi_-)^* = \chi_-$ and $(\chi_+)^* = \chi_+$. Thus, we can consistently keep only χ_+ and demand it to be real or complex, while setting $\chi_- = 0$ and vice versa. Demanding $\chi_- = 0$ and χ_+ real, we have one Majorana-Weyl spinor.

As a final example let us consider the supersymmetry algebra for a complex supercharge Q_A which contains the anticommutators

$$\{Q_A, \bar{Q}^B\} = -\frac{i}{2}(\gamma^\delta)_A{}^B P_\delta$$

$$\{Q_A, Q_B\} = 0 = \{\bar{Q}^A, \bar{Q}^B\}. \tag{22.41}$$

Using Eqs. (22.31) and (22.36) we find that it becomes

$$\{Q_+, (Q_+)^*\} = iP_z$$

$$\{Q_-, (Q_-)^*\} = iP_{\bar{z}} \tag{22.42}$$

and all remaining anticommutators vanishing.

In this form it is particularly simple to check the adjoint properties of the anticommutators; for example

$$[\{Q_+, (\bar{Q}_+)\}]^* = (iP_z)^* \sim \{Q_+, \bar{Q}_+\} = iP_z \tag{22.43}$$

as P_z is anti-Hermitian.

It is straightforward to work directly with light-cone coordinates and chiral spinors. Lorentz invariance is easily checked when one realizes that for any two-spinors the combination $\psi_+\chi_+$ transforms like a tensor T_z, while the combination $\psi_-\chi_-$ transforms like a $T_{\bar{z}}$ and $\psi_-\psi_+$ is invariant. This has led in some quarters to the adoption of the notation

$$T_z = T_+, \qquad T_{\bar{z}} = T_- \tag{22.44}$$

and similarly for more complicated spinor tensors. In later chapters we will also use this notation.

22.1b Euclidean Spinors

We now repeat the above exercise for spinors in two-dimensional Euclidean space. We could transfer the above results from Minkowski space to Euclidean space however, let us begin directly in Euclidean space.

We choose as our Euclidean γ matrices the real matrices

$$(\gamma^1)_A{}^B \equiv \begin{pmatrix} 1 & 0 \\ 0 & -1 \end{pmatrix} \qquad (\gamma^2)_A{}^B \equiv \begin{pmatrix} 0 & 1 \\ 1 & 0 \end{pmatrix} \tag{22.45}$$

and

$$\gamma_5 \equiv \gamma^1\gamma^2 = \begin{pmatrix} 0 & 1 \\ -1 & 0 \end{pmatrix} \equiv \varepsilon_{AB}. \tag{22.46}$$

They satisfy the relations

$$\gamma^a\gamma^b = \delta^{ab} + \varepsilon^{ab}\gamma_5 \tag{22.47}$$

$$\varepsilon^{ab}\gamma_b = \gamma^a\gamma_5, \qquad \gamma_5^2 = -1. \tag{22.48}$$

The Lorentz generator is $\gamma_{ab}/2 \equiv [\gamma_a, \gamma_b]/4$ and the one generator corresponding to the parameter $\omega = \omega^{12}$ is given by $\gamma_{12}/2 = \gamma_5/2$. The transformation of any spinor ψ is given by

$$\delta\psi = \frac{\omega}{2}\gamma_5\psi.$$

The Dirac conjugate of a spinor ψ_A is defined by

$$\bar{\psi}^D = (\psi)^+ \quad \text{or} \quad \bar{\psi}^A = (\psi_A)^*. \tag{22.49}$$

Under a Lorentz transformation

$$\delta\bar{\psi}^D = -\bar{\psi}^D\gamma_5\frac{\omega}{2} \tag{22.50}$$

and one readily verifies that

$$\bar{\psi}^D\psi = \sum_{A=1}^{2} (\psi_A)^*\psi_A \tag{22.51}$$

is 'Euclidean' invariant.

The charge conjugation matrix C which is taken to satisfy

$$C\gamma^\mu C^{-1} = +\tilde{\gamma}^\mu$$

and it can be chosen to be

$$C^{AB} = \delta^{AB} \equiv \mathbb{1} \tag{22.52}$$

It has the property that its transpose satisfies $\tilde{C} = +C$. The corresponding Majorana conjugate of a spinor ψ is defined by

$$\bar{\psi}^C \equiv \psi C \quad \text{or} \quad (\bar{\psi}^A)^C = \psi_A \tag{22.53}$$

A Majorana spinor is one whose Majorana conjugate equals its Dirac conjugate viz $\bar{\psi}^C = \bar{\psi}^D$. For the components of $\psi = (\psi_1, \psi_2)$ this implies that

$$\psi_1^* = \psi_1, \qquad \psi_2^* = \psi_2 \tag{22.54}$$

Since $(\gamma_5)^2 = -1$ in Euclidean space, a Weyl spinor ψ must satisfy

$$\gamma_5\psi = \pm i\psi \tag{22.55}$$

which implies for the components that

$$\psi_2 = \pm i\psi_1. \tag{22.56}$$

In two-dimensional Euclidean space we cannot have a Majorana-Weyl spinor, since imposing Eq. (22.54) and either one of the two possibilities in Eq. (22.56) implies that $\psi_1 = \psi_2 = 0$. The situation in Minkowski space is different, as explained in the previous section, as there one can have Majorana-Weyl spinors. These results are in agreement with the general theorems which state that in Euclidean space Majorana-Weyl spinors exist only in 8, 16, 22, ... dimensions, while in Minkowski space they exist in 2, 10, 18, 24, ... dimensions.

We now illustrate the fact that we could have chosen a different charge conjugation matrix with different properties provided we also modify our definitions of Dirac and Majorana conjugate. Let us choose a C such that

$$C\gamma^\mu C^{-1} = -\tilde{\gamma}^\mu$$

which has a sign charge with respect to our previous definition.

One particular such choice is $C = \gamma_5$ which satisfies $\tilde{C} = -C$. As usual the Majorana condition is $\bar{\psi}^C = \psi C$ while the Dirac conjugate is

$$\bar{\psi}^D \equiv \psi^+ \gamma_5 \quad \text{or} \quad (\bar{\psi}^D)^B = (\psi_A)^*(\gamma_5)_A{}^B. \tag{22.57}$$

One easily verifies that Majorana spinors but not Majorana Weyl spinors exist for this choice and satisfy Eq. (22.54). We will not discuss this possibility for C further.

Given a spinor $\psi = (\psi_1, \psi_2)$, it transforms under a $\omega\gamma_5/2$ or Euclidean rotation as

$$\delta\psi_1 = \tfrac{1}{2}\omega\psi_2$$

while

$$\delta\psi_2 = -\tfrac{1}{2}\omega\psi_1.$$

Hence, the combinations

$$\psi_+ \equiv \frac{1}{2}(\psi_1 - i\psi_2) \quad \text{and} \quad \psi_- \equiv \frac{1}{2}(\psi_1 + i\psi_2) \tag{22.58}$$

are eigenstates of γ_5 and transform as

$$\delta\psi_+ = \frac{i\omega}{2}\psi_+, \qquad \delta\psi_- = -\frac{i\omega}{2}\psi_-.$$

For any two spinors ψ and χ a Euclidean invariant in spinor space is given by

$$\bar{\psi}^C\chi = \sum_A \psi^A \chi_A \tag{22.59}$$

Hence we may interpret $C^{AB} = \delta^{AB}$ as a metric in spinor space. This invariant can be expressed in terms of the basis of the irreducible spinor representations of the Lorentz group given in Eq. (22.58) by the equation

$$\bar{\psi}^C\chi = \psi_+\chi_+\eta^{++} + \psi_+\chi_-\eta^{+-} + \psi_-\chi_+\eta^{-+} + \psi_-\chi_-\eta^{--} = \psi_1\chi_1 + \psi_2\chi_2. \tag{22.60}$$

We read off the inverse metric to be

$$\eta^{++} = \eta^{--} = 0; \qquad \eta^{+-} = \eta^{-+} = 2 \tag{22.61}$$

and so the metric is given by

$$\eta_{++} = \eta_{--} = 0; \qquad \eta_{+-} = \frac{1}{2} = \eta_{-+}. \tag{22.62}$$

Using this metric we may lower indices as follows

$$\psi^+ \equiv \eta^{+-}\psi_- = 2\psi_-$$
$$\psi^- \equiv \eta^{-+}\psi_+ = 2\psi_+. \tag{22.63}$$

Another method of raising the index is by using Dirac conjugation which in the representation chosen is straightforward complex conjugation. We define the objects $\bar{\psi}^+$ and $\bar{\psi}^-$ by the equation

$$\bar{\psi}^D\chi = \psi_1{}^*\chi_1 + \psi_2{}^*\chi_2 \equiv \bar{\psi}^+\chi_+ + \bar{\psi}^-\chi_- \tag{22.64}$$

where χ is any spinor.

Using Eq. (22.58) we find that

$$\bar{\psi}^+ = \psi_1{}^* + i\psi_2{}^*, \qquad \bar{\psi}^- = \psi_1{}^* - i\psi_2{}^* \tag{22.65}$$

We define $\bar{\psi}_+$ and $\bar{\psi}_-$ by using the metric of Eq. (22.62):

$$\bar{\psi}_+ = \frac{1}{2}\bar{\psi}^-, \qquad \bar{\psi}_- = \frac{1}{2}\bar{\psi}^+. \tag{22.66}$$

It is often useful to use the relations

$$\bar{\psi}^+ = (\psi^-)^*, \qquad \bar{\psi}^- = (\psi^+)^* \tag{22.67}$$

Note that for a Majorana spinor $\bar{\psi}^+ = \psi^+$ and $\bar{\psi}^- = \psi^-$ and hence $(\psi^-)^* = \psi^+$ and $(\psi^+)^* = \psi^-$ or equivalently $(\psi_-)^* = \psi_+$.

In order to get a feeling for the simplicity that the above formalism brings let us rewrite some common expressions. Consider first the kinetic energy for a scalar field A which is given by

$$\int d^2x \frac{1}{2} \partial_\alpha A \partial_\beta A \delta^{\alpha\beta}. \tag{22.68}$$

We can put this in $z, \bar{z}$ coordinates directly; the Jacobian J is given by

$$dx^1 dx^2 \equiv J \, dz \, d\bar{z} = \frac{i}{2} dz \, d\bar{z} \tag{22.69}$$

while

$$4\partial_z A \partial_{\bar{z}} A = (\partial_1 A)^2 + (\partial_2 A)^2. \tag{22.70}$$

Putting all this together we have

$$i \int dz \, d\bar{z} \, \partial_z A \partial_{\bar{z}} A. \tag{22.71}$$

Another way of achieving the same result is to write the action as

$$\frac{1}{2} \int d^2x \sqrt{\det g_{\alpha\beta}} \, \partial_\alpha A \partial_\beta A g^{\alpha\beta} \tag{22.72}$$

which for $g_{\alpha\beta} = \delta_{\alpha\beta}$ reduces to Eq. (22.68). However, the above action is valid in any coordinate system and using the metric of Eq. (22.62) and noticing that $\sqrt{\det g_{\alpha\beta}} = +i/2$ we again recover Eq. (22.71).

As another example let us rewrite the Dirac equation in the new basis. To achieve this it is necessary to put the $(\gamma^\alpha)_A{}^B$ matrix into the new basis on its vector index α and its spinor indices A and B. On the vector index we define

$$\gamma^z \equiv \gamma^1 + i\gamma^2, \qquad \gamma^{\bar{z}} = \gamma^1 - i\gamma^2. \tag{22.73}$$

For the spinor indices in accord with Eqs. (22.58) and (22.63) and for any matrix we have

$$\Gamma_\pm{}^+ = \tfrac{1}{2}\{\Gamma_1{}^1 + i\Gamma_1{}^2 \mp i\Gamma_2{}^1 \pm \Gamma_2{}^2\} \tag{22.74}$$

$$\Gamma_\pm{}^- = \tfrac{1}{2}\{\Gamma_1{}^1 - i\Gamma_1{}^2 \mp i\Gamma_2{}^1 \mp \Gamma_2{}^2\}. \tag{22.75}$$

As a result we find that

$$\gamma^z = \begin{bmatrix} (\gamma^z)_+{}^+ & (\gamma^z)_+{}^- \\ (\gamma^z)_-{}^+ & (\gamma^z)_-{}^- \end{bmatrix} = \begin{bmatrix} 0 & 2 \\ 0 & 0 \end{bmatrix} \tag{22.76}$$

$$\gamma^{\bar{z}} = \begin{bmatrix} (\gamma^{\bar{z}})_+{}^+ & (\gamma^{\bar{z}})_+{}^- \\ (\gamma^{\bar{z}})_-{}^+ & (\gamma^{\bar{z}})_-{}^- \end{bmatrix} = \begin{bmatrix} 0 & 0 \\ 2 & 0 \end{bmatrix} \tag{22.77}$$

and consequently, we have in this basis

$$\partial\!\!\!/ = 2 \begin{pmatrix} 0 & \partial \\ \bar{\partial} & 0 \end{pmatrix} = \gamma^z \partial_z + \gamma^{\bar{z}} \partial_{\bar{z}} \tag{22.78}$$

where

$$\partial = \partial_z, \qquad \bar{\partial} = \partial_{\bar{z}}$$

Acting on a spinor the two components are given by

$$\psi_+ = 2\partial \chi_- \qquad \psi_- = 2\bar{\partial}\chi_+ \tag{22.79}$$

where $\psi = \partial\!\!\!/ \chi$.

Putting all this together the spinor action becomes

$$\frac{i}{2} \int d^2x \bar{\chi} \partial\!\!\!/ \chi = -\int dz \, d\bar{z} \{\bar{\chi}^+ \partial \chi_- + \bar{\chi}^- \bar{\partial}\chi_+\} \tag{22.80}$$

22.2 Superalgebras in Two-Dimensions

The existence of complex, Majorana, Weyl and Majorana-Weyl spinors in two-dimensional Minkowski space allow more types of superalgebras than one finds in say four-dimensional Minkowski space. Given a supercharge it can be complex, Majorana, Weyl or Majorana-Weyl. In the latter two cases we have two types of Weyl spinors corresponding to the two possible eigenvalues of γ_5. Conversely, given a complex supercharge Q, we may decompose it into Weyl supercharges Q_+ and Q_-. Further we may decompose Q_+ into its real and imaginary pieces which are each Majorana-Weyl spinors and similarly for Q_-. Given a Majorana spinor we may decompose it into two Majorana-Weyl spinors. Of course, a superalgebra may have more than one supercharge and to each we may apply the above discussion.

A general superalgebra can therefore be labelled by (N, M) where N are the numbers of left-handed Majorana-Weyl spinors which are of type Q_+ $(Q_+{}^* = Q_+)$ and M are the numbers of right-handed Majorana-Weyl spinors of type Q_- $(Q_-{}^* = Q_-)$. We will also refer to the generators which carry $+(-)$ and $z(\bar z)$ indices as the left (right) parts of the algebra. The $(1, 1)$ superalgebra in two-dimensions was found in the context of the superstring in the seminal paper of Gervais and Sakita given in Ref. 7. Indeed this algebra and the $N = 1$, $D = 4$ superalgebras of Ref. 1, were the first algebras to be discovered. The $(2, 2)$ superalgebras were given in Ref. 214, while algebras for $N \neq M$ were first given in Refs. 216–218 but the general case was given in Ref. 219.

Consider first a $(1, 0)$ superalgebra, i.e. a left algebra. It contains a Majorana Weyl spinor Q_+ $(Q_+{}^* = Q_+)$, while the Poincaré group has the anti-Hermitian generators P_z, $P_{\bar z}$ and $J = J_{01}$. We will, in outline, apply the discussion of Chapter 2 to find the most general such algebra.

The Poincaré group generators satisfy their usual relations which in our notation take the form.

$$[P_z, J] = -P_z \qquad [P_{\bar z}, J] = P_{\bar z}. \tag{22.81}$$

Now the Jacobi identities imply that Q_+ must be representations of J and in accord with the above discussion we take it to be a spinor with $+$ index implying it satisfies

$$[Q_+, J] = -\tfrac{1}{2}Q_+. \tag{22.82}$$

One consequence of implementing the super-Jacobi identities is the intuitively obvious fact that the tally of Lorentz indices should balance on the left- and right-hand side of each commutator or anticommutator. Bearing this in mind and the Z_2 graded structure of the algebra, we can only have

$$[Q_+, P_z] = 0 = [Q_+, P_{\bar z}]. \tag{22.83}$$

We recall that P_z and $P_{\bar z}$ transform like P_+ and P_- respectively. However, for the anticommutator of two supercharges φ_+. We have

$$\{Q_+, Q_+\} = 2iP_z \tag{22.84}$$

We may always adjust the definition of P_z so as to arrange for a factor 2. The factor i is required as P_z is anti-Hermitian with the usual scalar product:

$$\{Q_+, Q_+\}^* = \{Q_+, Q_+\} = (2iP_z)^* = 2iP_z. \tag{22.85}$$

Next let us consider the $(1, 1)$ superalgebra which contains the real spinors Q_+ and Q_-

$$(Q_+)^* = Q_+, \qquad (Q_-)^* = Q_-.$$

First consider the left-handed generators, for the same arguments as given above they must obey

$$[Q_+, J] = -\tfrac{1}{2}Q_+ \qquad [Q_+, P_z] = 0$$

$$\{Q_+, Q_+\} = 2iP_z. \tag{22.86}$$

Similarly, the right-handed generators satisfy

$$[Q_-, J] = \tfrac{1}{2}Q_- \qquad [Q_-, P_{\bar{z}}] = 0$$

$$\{Q_-, Q_-\} = 2iP_{\bar{z}} \tag{22.87}$$

We now consider the relations between the left and right algebras. The anticommutator $\{Q_+, Q_-\}$ must equal a Lorentz scalar which in principle could be a central charge. We will return to this point later, but for now we will not consider this possibility. The relation $\{Q_+, P_{\bar{z}}\}$ could on grounds of Lorentz character equal Q_- however the super-Jacobi identity

$$[\{Q_-, Q_+\}, P_{\bar{z}}] + \{[P_{\bar{z}}, Q_-], Q_+\} + \{[P_{\bar{z}}, Q_+], Q_-\} = 0 \tag{22.88}$$

implies that $[Q_+, P_{\bar{z}}] = 0$. By a similar argument $[Q_-, P_z] = 0$. Hence the $(1, 1)$ superalgebra is a direct sum of a $(1, 0)$ and a $(0, 1)$ superalgebra and it is given by Eqs. (22.81), (22.86) and (22.87) and the fact that it is a direct sum expressed by the relations

$$\{Q_+, Q_-\} = 0 = [Q_+, P_{\bar{z}}] = [Q_-, P_z] \tag{22.89}$$

We now consider the $(2, 2)$ superalgebra which contains the real Weyl spinors Q_+^i; $i = 1, 2$ and Q_-^i; $i = 1, 2$. The corresponding superalgebra is given by the relations for the left part

$$\{Q_+^i, Q_+^j\} = 2i\delta^{ij}P_z, \qquad [Q_+^i, J] = -\tfrac{1}{2}Q_+^i$$

$$[Q_+^i, P_z] = 0 \tag{22.90}$$

as well as the relations for the right part

$$\{Q_-^i, Q_-^j\} = 2i\delta^{ij}P_{\bar{z}}, \qquad [Q_-^i, J] = \tfrac{1}{2}Q_-^i$$

$$[Q_-^i, P_{\bar{z}}] = 0 \tag{22.91}$$

and the relations between the left and right pieces which in the absence of central charges express the fact that the $(2,2)$ superalgebra is the direct sum of a $(2,0)$ and $(0,2)$ superalgebra:

$$\{Q_+^i, Q_-^j\} = 0, \qquad [Q_+^i, P_{\bar{z}}] = [Q_-^i, P_z] = 0. \tag{22.92}$$

The arguments for the $(1,1)$ superalgebra given above can be carried directly over to the $(2,2)$ superalgebra to yield the above relations. The new feature that arises in the latter case is the existence of internal symmetry generators which rotate the i indices. There exist such independent generators in the left and right sectors, denoted by L and R respectively with the commutators

$$[Q_+^i, L] = \varepsilon^{ij}Q_+^j, \qquad [Q_-^i, R] = \varepsilon^{ij}Q_-^j$$

$$[Q_+^i, R] = 0 \qquad [Q_-^i, L] = 0. \tag{22.93}$$

The reader is encouraged to verify all the super Jacobi identities for the $(2,2)$ algebra.

The $(2,2)$ superalgebra can be written in terms of complex Weyl spinors defined by

$$Q_\pm = \tfrac{1}{2}(Q_\pm^1 - iQ_\pm^2), \qquad \bar{Q}_\pm = \tfrac{1}{2}(Q_\pm^1 + iQ_\pm^2) \tag{22.94}$$

for which $(Q_\pm)^* = \bar{Q}_\pm$.

With these variables the superalgebra becomes for the $(2,0)$ subalgebra

$$\{Q_+, Q_+\} = 0, \qquad \{Q_+, \bar{Q}_+\} = iP_z$$

$$[Q_+, P_z] = 0 = [\bar{Q}_+, P_z]$$

$$[Q_+, J] = -\tfrac{1}{2}Q_+, \qquad [\bar{Q}_+, J] = -\tfrac{1}{2}\bar{Q}_+$$

$$[Q_+, L] = iQ_+, \qquad [\bar{Q}_+, L] = -i\bar{Q}_+ \tag{22.95}$$

and for the $(0,2)$ subalgebra

$$\{Q_-, Q_-\} = 0, \qquad \{\bar{Q}_-, Q_-\} = iP_{\bar{z}}$$

$$[Q_-, P_{\bar{z}}] = 0 = [\bar{Q}_-, P_{\bar{z}}]$$

$$[Q_-, J] = \tfrac{1}{2}Q_- \qquad [\bar{Q}_-, J] = \tfrac{1}{2}\bar{Q}_-$$

$$[Q_-, R] = iQ_-, \qquad [\bar{Q}_-, R] = -i\bar{Q}_- \tag{22.96}$$

while all (anti) commutators between the left and right algebras vanish.

Let us now take the general (N, M) algebra consisting of real Weyl spinors

$$Q_+^i \quad i = 1, 2, \ldots, N$$

and

$$Q_-^l \quad l = 1, 2, \ldots, M.$$

The arguments given above can be carried over and we find the relations

$$\{Q_+^i, Q_+^j\} = 2i\delta^{ij}P_z, \qquad [Q_+^i, P_z] = 0$$

$$[Q_+^i, J] = -\tfrac{1}{2}Q_+^i \qquad\qquad\qquad (22.97)$$

and

$$\{Q_-^l, Q_-^m\} = 2i\delta^{lm}P_{\bar{z}}, \qquad \{Q_-^l, P_{\bar{z}}\} = 0$$

$$\{Q_-^l, J\} = \tfrac{1}{2}Q_-^l \qquad\qquad\qquad (22.98)$$

We note that generators other then P_z in $\{Q_+^i, Q_+^j\}$ would carry i, j indices and be Lorentz tensors with two $+$ indices. Such generators are forbidden if we invoke the type of no-go theorem discussed in Chapter 2.

Consider now internal anti-Hermitian generators L_r which rotate the internal indices

$$[Q_+^i, L_r] = (U_r)^i{}_j Q_+^j \qquad\qquad\qquad (22.99)$$

and have the commutation relations of a Lie algebra

$$[L_r, L_s] = f_{rs}{}^t L_t. \qquad\qquad\qquad (22.100)$$

Taking the adjoint of this equation implies that

$$\{[Q_+^i, L_r]\}^\dagger = [Q_+^i, L_r] = [(U_r)^i{}_j]^* Q_+^j = (U_r)^i{}_j Q_+^j \qquad (22.101)$$

and consequently

$$[(U_r)^i{}_j]^* = (U_r)^i{}_j$$

On the other hand the super-Jacobi identity

$$[\{Q_+^i, Q_+^j\}, L_r] + \{[L_r, Q_+^i], Q_+^j\} + \{[L_r, Q_+^j], Q_+^i\} = 0 \quad (22.103)$$

implies that

$$(U_r)^i{}_k \delta^{kj} + (U_r)^j{}_k \delta^{ki} = 0. \qquad\qquad\qquad (22.104)$$

Consequently, the (U_r) are real antisymmetric matrices and so the L_r generate the group SO(N).

The same logic applies to the right sector and we find

$$[Q_-^l, R_p] = (V_p)^l{}_m Q_-^m$$

$$[R_p, R_q] = f_{pq}{}^r R_r \qquad\qquad\qquad (22.105)$$

where V_p is a real antisymmetric matrix and the R_s generate the group SO(M).

Thus the maximal continuous symmetry admitted by the (N, M) algebra is $SO(N) \times SO(M)$. We could also add the discrete transformations to have $O(N) \times O(M)$.

In the absence of any central charges the (anti) commutators of the left and right algebra vanish.

One might suppose that L_r could also rotate Q^i_- and vice-versa. However, in this case L_r would have to be part of $SO(N)$ and R_s part of $SO(M)$. We may always choose to label our generators so that

$$[Q^i_+, R_p] = 0, \qquad [Q^i_-, L_r] = 0$$

and hence

$$[R_p, L_r] = 0.$$

Consequently, in the absence of central charges, the (N, M) superalgebra is a direct sum of a $(N, 0)$ and $(0, M)$ superalgebra.

However, if central charges are present one can have relations of the form

$$\{Q^i_+, Q^j_-\} = Z^{ij}. \tag{22.106}$$

One readily finds using the $(Q_+{}^i, Q_-{}^j, J)$ Jacobi identity that Z^{ij} commutes with J. Assuming a no-go theorem which states that Z^{ij} commutes with P_z and $P_{\bar{z}}$, we find that Z^{ij} commutes also with $Q^i_\pm$. The super-Jacobi identities involving Q^i_+, Q^j_+ and R_r and L_s imply that Z^{ij} should transform under R_r and L_s as

$$[Z^{ij}, L_r] = (U_r)^i{}_k Z^{kj}$$

$$[Z^{ij}, R_p] = (V_p)^j{}_k Z^{ik}. \tag{22.107}$$

In view of these relations the use of the words central charge above is somewhat misleading. The no-go theorem stated in Chapter 2 states that the internal symmetry group had to be the direct product of an internal group and $U(1)$ factor. If we were to assume this part also for these two-dimensional supersymmetric theories then we would require

$$[Z^{ij}, L_r] = 0, \qquad [Z^{ij}, R_p] = 0 \tag{22.108}$$

and the internal symmetry group would be reduced so as to obey the relations

$$(U_r)^i{}_k Z^{kj} = 0 \qquad (V_p)^j{}_k Z^{ik} = 0 \tag{22.109}$$

From the same theorem we would insist that

$$[Z^{ij}, Z^{kl}] = 0. \tag{22.110}$$

The no-go theorem in four-dimensional Minkowski space relies on certain properties of the S-matrix. Whether one should insist on these in two-dimensions is a moot point, certainly for off-shell quantities it does not hold and equations such as (22.107) can be realized nontrivially. We refer the reader to Ref. 232 for more details concerning central charges.

Let us now turn our attention to superalgebra in Euclidean space. In this case Majorana-Weyl spinors do not exist, although Majorana spinors do. We can still label algebras by the (M, N) notation but we must take $M = N$. Since many of the arguments of the Minkowski space concerning super-Jacobi identities can be carried over to the Euclidean case, we mainly give the results.

Let us consider first the algebra $(1, 1)$ which consists of one Majorana spinor Q_A which has the algebra

$$[P_\gamma, J] = \varepsilon_{\gamma\delta} P_\delta$$

$$[Q_A, J] = \tfrac{1}{2}(\gamma_5)_A{}^B Q_B, \qquad [Q_A, P_z] = [Q_A, P_{\bar{z}}] = 0$$

$$\{Q_A, Q_B\} = 2(\gamma^\delta)_{AB} P_\delta. \tag{22.111}$$

In the basis discussed in the previous section Q_A has components Q_+ and Q_- and the Majorana constraint reads

$$(Q_+)^* = Q_-. \tag{22.112}$$

The algebra becomes

$$[Q_+, J] = +\frac{i}{2} Q_+ \qquad [Q_+, P_z] = 0$$

$$\{Q_+, Q_+\} = 2P_z \qquad [P_z, J] = iP_z \tag{22.113}$$

and

$$[Q_-, J] = -\frac{i}{2} Q_-, \qquad [Q_-, P_{\bar{z}}] = 0$$

$$\{Q_-, Q_-\} = 2P_{\bar{z}} \qquad [P_{\bar{z}}, J] = -iP_{\bar{z}}, \tag{22.114}$$

as well as the fact that the left and right parts of the superalgebra (anti) commute

$$\{Q_+, Q_-\} = 0, \qquad [Q_+, P_{\bar{z}}] = [Q_-, P_z] = 0. \tag{22.115}$$

We observe that the $(1, 1)$ algebra is the direct sum of a left and right algebra, i.e. a $(1, 0)$ and a $(0, 1)$ algebra. Each algebra is consistent by itself in the sense that it obeys the super-Jacobi identities. However, the adjoint operation which arises from the usual scalar product in Euclidean space takes the left algebra into the right algebra and vice-versa. If one asks for an algebra which did not admit this operation of adjoint, then one could take either the left or the right algebra alone.

The $(2, 2)$ algebra contains supercharges Q_A^i; $i = 1, 2$ which we take to be Majorana, i.e.

$$(Q_A^i)^* = Q_A^i.$$

The anticommutator is given by

$$\{Q_A^i, Q_B^j\} = 2\delta^{ij}(\gamma^\delta)_{AB}P_\delta \tag{22.116}$$

while the internal symmetry generators T_1 and T_2 have the commutators

$$[Q_A^i, T_1] = \varepsilon^{ij}Q_A^j$$

$$[Q_A^i, T_2] = \varepsilon^{ij}(\gamma_5)_A{}^B, Q_B^j. \tag{22.117}$$

It is convenient to introduce the complex supercharges

$$Q_A = \frac{1}{2}(Q_A^1 - iQ_A^2), \qquad \bar{Q}_A = \frac{1}{2}(Q_A^1 + iQ_A^2) \tag{22.118}$$

where $(Q_A)^* = \bar{Q}_A$. The above relations become

$$\{Q_A, Q_B\} = 0 = \{\bar{Q}_A, \bar{Q}_B\}$$

$$\{Q_A, \bar{Q}_B\} = (\gamma^\delta)_{AB}P_\delta \tag{22.119}$$

and

$$[Q_A, T_1] = iQ_A, \qquad [\bar{Q}_A, T_1] = -i\bar{Q}_A$$

$$[Q_A, T_2] = i(\gamma_5)_A{}^B Q_B, \qquad [\bar{Q}_A, T_2] = -i(\gamma_5)_A{}^B\bar{Q}_B. \tag{22.120}$$

Finally, we rewrite the above in left-right notation, the components of the supercharge are Q_+ and Q_- and $\bar{Q}_+$ and $\bar{Q}_-$ with the condition

$$\bar{Q}_+ = (Q_-)^*, \qquad \bar{Q}_- = (Q_+)^* \tag{22.121}$$

The algebra becomes the direct product of a left and a right algebra. The left algebra is given by

$$\{Q_+, Q_+\} = \{\bar{Q}_+, \bar{Q}_+\} = 0$$

$$\{Q_+, \bar{Q}_+\} = P_z, \qquad [Q_+, J] = \frac{i}{2}Q_+, \qquad [\bar{Q}_+, J] = \frac{i}{2}\bar{Q}_+$$

$$[Q_+, L] = iQ_+, \qquad [\bar{Q}_+, L] = -iQ_+$$

$$[Q_+, P_z] = [\bar{Q}_+, P_z] = 0 \tag{22.122}$$

while the right-handed algebra is given by

$$\{Q_-, Q_-\} = 0 = \{\bar{Q}_-, \bar{Q}_-\}, \qquad \{Q_-, \bar{Q}_-\} = P_{\bar{z}}$$

$$[Q_-, J] = -\frac{i}{2}Q_-, \qquad [\bar{Q}_-, J] = -\frac{i}{2}\bar{Q}_-$$

$$[Q_-, R] = iQ_-, \qquad [\bar{Q}_-, R] = -i\bar{Q}_-$$

$$[Q_-, P_{\bar{z}}] = 0 = [\bar{Q}_-, P_{\bar{z}}] \tag{22.123}$$

where

$$L = \frac{1}{2}(T_1 - iT_2); \qquad R = \frac{1}{2}(T_1 + iT_2). \qquad (22.124)$$

The fact that the algebras are a direct product implies the relations

$$[R, L] = 0, \qquad [Q_+, R] = 0 = [\bar{Q}_+, R]$$

$$[Q_-, L] = [\bar{Q}_-, L] = 0 \text{ etc.} \qquad (22.125)$$

The superalgebras of (M, M) type obey a similar pattern and apart from the different reality condition of Eq. (22.112) are similar to the Minkowski case.

22.3   Irreducible Representations of Two-Dimensional Supersymmetry

The on-shell states of supersymmetric theories are described by the irreducible representations of supersymmetry. These were discussed for four-dimensions in Chapter 8. In two-dimensions there are a number of different features and in this section we give the analogous discussion for the two-dimensional case.

We recall that the irreducible unitary representations of such groups are constructed using the theory of induced representations. Since the translation generators are anti-Hermitian Abelian operators we may take the representation to be such that these operators are diagonal. Put more simply, we work in terms of momentum eigenstates. A massless particle in an eigenstate of momentum has momentum components of the form $(k, \pm k) = k^\mu$. Unlike four-dimensions, in two-dimensions we have only one Lorentz generator and this does not leave a given momentum inert, excluding the case $k = 0$. Consequently, the little group contains only the translations, P_z and $P_{\bar{z}}$.

Further in two-dimensions a particle can only move to the left or the right and which of the two directions it takes is a Lorentz invariant statement. A left-moving particle of given momentum is of the form $\exp[-ik(x^0 + x^1)]$ while a right-moving particle is of the form $\exp[-ik(x^0 - x^1)]$. Consequently, for such a left-moving particle $P_z = -ik$ and $P_{\bar{z}} = 0$ while for a right-moving particle $P_{\bar{z}} = 0$ and $P_z = -ik$. Excluding the point $P_z = P_{\bar{z}} = 0$, the orbit for left-moving particles are $P_z = -ik$ for all k, $P_{\bar{z}} = 0$ and for right-moving particles $P_{\bar{z}} = -ik$ for all k, $P_z = 0$. Since the supersymmetry generators commute with the momenta, all the states in an irreducible representation of supersymmetry are either all left- or right-moving.

Before classifying the irreducible representations let us examine how many states the casual actions propagate in two dimensions. The usual scalar equation $\partial^2 A = 0$ is equivalent to $\partial_z \partial_{\bar{z}} A = 0$ and so it propagates a left-moving and a right-moving particle. The usual Dirac equation i.e. see Eq. (22.39) implies that spinors of Lorentz type ψ_+ are left-moving (i.e. $\partial_{\bar{z}} \psi_+ = 0$)

while those of ψ_- type are right-moving (i.e. $\partial_z \psi_- = 0$). Setting ψ_+ or ψ_- to zero in the action and we still propagate ψ_- and ψ_+ respectively. We may consider a spin 0 to be left-moving or right-moving only, corresponding to the equations $\partial_{\bar{z}} A = 0$ or $\partial_z A = 0$, however there does not exist any very simple action that describes this possibility (see Ref. 184 in this context). Consider the Maxwell action

$$\int d^2 \xi \left\{ -\frac{1}{4} (F_{\mu\nu})^2 \right\}$$

where $F_{\mu\nu} = \partial_\mu A_\nu - \partial_\nu A_\mu$. Let us suppose for the sake of definiteness that the particle is left-moving. In this case its momentum in a chosen frame is of the form $k_\mu = (k, k)$.

We may write A_μ in as

$$A_\mu = k_\mu \varphi_1 + \bar{k}_\mu \varphi_2$$

where $\bar{k}_\mu = (-k, k)$. Using the original gauge invariance we may choose the gauge condition $\partial_\mu A^\mu = 0 = k_\mu A^\mu$ which implies that $\varphi_2 = 0$. The equation of motion reduces to $\partial^2 A_\mu = 0$ or $\partial^2 \varphi_1 = 0$, but it possesses a residual on-shell invariance namely $\delta A_\mu = \partial_\mu \Lambda$ with $\partial^2 \Lambda = 0$. Using this invariance we may set $\varphi_1 = 0$. Consequently, the two-dimensional Maxwell action propagates no states. The usual Einstein action is in fact a topological invariant and so it also propagates no states, while the gravitino action, $\psi_\mu \gamma^{\mu\nu\rho} \partial_\nu \psi_\rho$ vanishes automatically in two dimensions.

Thus in two dimensions, the only propagating states which possess an action and have good physical properties are spin 0 or spin 1/2. However, other spins can arise in supersymmetric theories such as the ghosts in superstring theory. As is to be expected these states do not satisfy the usual positivity requirements of physical states, but this does not of course stop them forming supermultiplets.

Let us first consider the $(1, 0)$ superalgebra which for a left-moving particle becomes

$$[Q_+, J] = -\frac{1}{2} Q_+, \qquad \{Q_+, Q_+\} = 2k \qquad (22.126)$$

or $Q_+^2 = k$. Given a state $|\psi\rangle$ which has $J|\lambda\rangle = \lambda|\lambda\rangle$ then an irreducible representation is given by

$$|\lambda\rangle, \qquad \left| \lambda + \frac{1}{2} \right\rangle \equiv Q_+ |\lambda\rangle. \qquad (22.127)$$

If $\lambda = -\frac{1}{2}$ we have one real scalar and one Majorana-Weyl spin 1/2.

Consider now a right-moving particle, the superalgebra reads

$$[Q_+, J] = -\frac{1}{2}Q_+ \qquad Q_+^2 = 0. \tag{22.128}$$

Hence, if we begin with the state $|\lambda\rangle$ then the state $Q_+|\lambda\rangle$ has zero norm. Assuming that physical states must have positive norm we conclude that $Q_+|\lambda\rangle = 0$. This leaves us with the one original state $|\lambda\rangle$ which is a singlet under $(1, 0)$ supersymmetry.

For a left-moving particle the $(1, 1)$ superalgebra takes the form

$$[Q_+, J] = -\frac{1}{2}Q_+, \qquad Q_+^2 = k$$

$$[Q_-, J] = +\frac{1}{2}Q_- \qquad Q_-^2 = 0 \tag{22.129}$$

Given a state $|\lambda\rangle$ then $Q_-|\lambda\rangle$ has zero norm and must be zero, hence the only other state is $Q_+|\lambda\rangle$ and the representation structure is the same as the $(1, 0)$ superalgebra. Similarly right-moving particles fall into supersymmetric representations of a $(0, 1)$ superalgebra and are inert under Q_+.

Consider now the $(2, 0)$ superalgebra of Eq. (22.95) for a left-moving particle it becomes

$$Q_+^2 = \bar{Q}_+^2 = 0 \qquad Q_+\bar{Q}_+ + \bar{Q}_+Q_+ = k. \tag{22.130}$$

In accord with the general theory of representation of Clifford algebras, we choose a vacuum state $|\lambda\rangle$ such that $Q_+|\lambda\rangle = 0$. The states

$$|\lambda\rangle \quad \text{and} \quad \bar{Q}_+|\lambda\rangle \tag{22.131}$$

together with their complex conjugates form an irreducible representation. If $\lambda = -\frac{1}{2}$ this means two (real) spin 0's and a complex Weyl spin 1/2.

On the other hand for a right-moving particle

$$Q_+^2 = \bar{Q}_+^2 = Q_+\bar{Q}_+ + \bar{Q}_+Q_+ = 0 \tag{22.132}$$

and we conclude that all physical states must form an irreducible singlet representation of this part of the algebra. It is clear that this conclusion holds for any $(N, 0)$ superalgebra.

Thus the representations of the $(2, 2)$ superalgebra follow the pattern of the $(1, 1)$ superalgebra. The left-moving states are inert under the $(0, 2)$ superalgebra, but form a nontrivial irreducible representations of the $(2, 0)$ superalgebra. Similarly for right-moving particles.

Consider now the $(4, 0)$ superalgebra, in a complex basis we have two complex supercharges $Q_+^i, i = 1, 2$, which obey the relation

$$\{Q_{+i}, \bar{Q}_+{}^j\} = i\delta_i{}^j P_z \tag{22.133}$$

and transform as a complex doublet of SU(2). For left-moving states we can define the vacuum

$$\bar{Q}_+{}^j|\lambda\rangle = 0, \tag{22.134}$$

and then the states

$$|\lambda\rangle, \qquad |\lambda + \tfrac{1}{2}, i\rangle \equiv Q_+{}^i|\lambda\rangle \qquad |\lambda + 1\rangle \equiv Q_+{}^i Q_+{}^j|\lambda\rangle \tag{22.135}$$

form a representation of (4,0) supersymmetry. For the case of $\lambda = -1/2$ we have the complex states

$$|-\tfrac{1}{2}\rangle, \qquad |0, i\rangle, \qquad |+\tfrac{1}{2}\rangle. \tag{22.136}$$

These involve a state of $+\tfrac{1}{2}$ helicity. We recall from our previous discussion that the usual Dirac equation does not propagate a left-moving state with this helicity and so the on-shell states we wish to describe are not of this type. One could try to change the helicity by dividing by k_z, but it seems that an alternative construction must be used to find the on-shell states of interest.

The problem is that starting with a $-\tfrac{1}{2}$ helicity state and creating with two supercharges we will inevitably arrive at a state of $+\tfrac{1}{2}$ helicity. To avoid this possibility we must eliminate the action of three of the supercharges acting on the vacuum state in terms of the remaining one supercharge. To carry out this construction let us consider the (4, 0) superalgebra with real supercharges $Q_+{}^r, r = 1, 2, 3, 4$.

$$\{Q_+{}^r, Q_+{}^s\} = iP_z\delta^{rs}.$$

Consider the vacuum states $|-\tfrac{1}{2}, i\rangle$ $i = 1, 2, 3, 4$ and adopt the constraint

$$Q_+{}^r|-\tfrac{1}{2}, i\rangle = (I^r)^i{}_j \qquad Q_+{}^1|-\tfrac{1}{2}, j\rangle \qquad r = 2, 3, 4$$

where $(I^r)^i{}_j$ are some matrices. Applying $Q_+{}^s$, taking the anticommutator and using the above equation we find that the states vanish unless

$$\{I^r, I^s\} = -\delta^{rs} \qquad r, s = 2, 3, 4.$$

Thus the three matrices I^r have a quaternionic-like structure. The on-shell states are then given by

$$|-\tfrac{1}{2}, i\rangle, \qquad Q_+{}^1|-\tfrac{1}{2}, i\rangle$$

since there is acting on these states, only one effective supercharge. They describe 4 real left-moving spin 0 and 4 left-moving helicity $-\tfrac{1}{2}$ states. The constraint on the Q_+'s acting on the states is reminiscent of the SU(2) Majorana constraints on spinors of Chapter 12.

The discussion for the representations of (N, O) supersymmetry proceed in a similar way. Although somewhat brief, the above discussion will prove adequate to recognize the on-shell states arising in supersymmetric models in terms of irreducible representations of supersymmetry.

Chapter 23

Two-Dimensional Superspace and the Construction of Models

We will now construct the (N, M) superspaces corresponding to the (N, M) superalgebras given in the previous chapter. This construction follows closely that of Chapter 14, where superspaces in four-dimensions were constructed as coset spaces. The coset approach renders the construction a straightforward and somewhat mechanical exercise in the composition of group elements. Not only will we find the symmetry transformations on these superspaces but also invariant vielbein and spin connections which allow us to define covariant derivatures.

Equipped with these tools it is a relatively simple task to construct super-symmetric theories in superspace and find their subsequent component field expressions. We will cover all the above in somewhat laborious detail in view of the importance of the results for string theory. The superspace corresponding to the $(2, 2)$ superalgebra was given in Ref. 215 the $(1, 0)$ and $(2, 0)$ superspaces were given in Ref. 216 and Refs. 217, 218 respectively.

23.1 Minkowski Superspaces

23.1a (1, 0) Superspace

We begin from the $(1, 0)$ superalgebra of Eqs. (22.81)–(22.83) whose general group element may be written in the form

$$g_0 = \exp(aP_z + \varepsilon^+ Q_+)\exp(\bar{a}P_{\bar{z}})\exp(\omega J). \tag{23.1}$$

The $(1, 0)$ superspace is defined to be the coset space

$$\frac{\text{Supergroup } (1, 0)}{\text{Lorentz group}} \tag{23.2}$$

and as such it has two even and one odd Grassmann coordinates. We label our coset representative by the group element

$$g(z, \theta^+; \bar{z}) = \exp(zP_z + \theta^+ Q_+)\exp(\bar{z}P_{\bar{z}}). \tag{23.3}$$

The transformations on superspace are the natural action of the group on the coset space:

$$g_0 g(z, \theta^+; \bar{z}) = \exp\{z'P_z + \theta^{+'}Q_+\}\exp(\bar{z}'P_{\bar{z}})\exp(uJ).\qquad(23.4)$$

Using the equation

$$\exp(\varepsilon A)\exp(B) = \exp\left\{B + \varepsilon A + \varepsilon \sum_{n=1}^{\infty} \frac{1}{(n+1)!}\underbrace{[\ldots[[A,B],B]\ldots B]}_{\text{n times}} + O(\varepsilon^2)\right\}$$

$$(23.5)$$

we find that the induced transformations on superspace are

$$z' = z - i\varepsilon^+ \theta^+ + a, \qquad \bar{z}' = \bar{z} + \bar{a}$$

$$\theta'^+ = \theta^+ + \varepsilon^+ \qquad\qquad\qquad (23.6)$$

and a Lorentz rotation

$$z' = z + \omega z, \qquad \bar{z}' = \bar{z} - \omega\bar{z}, \qquad \theta^{+'} = \theta^+ + \tfrac{1}{2}\omega\theta^+. \qquad (23.7)$$

We will label the coordinates by

$$z^\pi = (z, \theta^+; \bar{z}).$$

A scalar superfield is define by the equation

$$\phi'(z^{\pi'}) = \phi(z^\pi) \qquad\qquad (23.8)$$

and for an infinitesimal transformation

$$\delta\phi \equiv \phi(z^\pi + \delta z^\pi) - \phi(z^\pi) = \delta g^N f_N^\pi \partial_\pi \phi \qquad (23.9)$$

where δg^N is the infinitesimal group element that induces

$$z^\pi \to z^{\pi'} = z^\pi + \delta g^N f_N^\pi.$$

The action of the supercharge on a scalar superfield can be read off to be

$$l_+ = \frac{\partial}{\partial\theta^+} - i\partial_z \theta^+. \qquad\qquad (23.10)$$

We refer the reader to the discussion in Chapter 14 on the difference between active and passive interpretations and the closure of the supersymmetry transformations on the superspace coordinates.

The vielbein and spin connection are defined through Eq. (14.32), which in this case becomes

$$1 + \delta z^\pi E_\pi{}^M K_M + \delta z^\pi \omega_\pi J = g(-z^\pi)g(z^\pi + \delta z^\pi)$$

$$= \exp\{-\bar{z}P_{\bar{z}}\}\exp\{-zP_z - \theta^+ Q_+\}\exp\{(z + \delta z)P_z$$

$$+ (\theta^+ + \delta\theta^+)Q_+\}\exp\{(\bar{z} + \delta\bar{z})P_{\bar{z}}\}. \qquad (23.11)$$

In the above equation the tangent index M runs over the tangent indices z,

$+$ and $\bar{z}$ that is

$$K_M = \{P_z, Q_+; P_{\bar{z}}\}.$$

Evaluating (23.11) we find that the components of the supervielbein are

$$E_\pi{}^M = \begin{pmatrix} 1 & 0 & 0 \\ -i\theta^+ & 1 & 0 \\ 0 & 0 & 1 \end{pmatrix} \qquad (23.12)$$

while the spin connection ω_π vanishes. Defining the inverse supervielbein by Eq. (14.36) we find the result

$$E_M{}^A = \begin{pmatrix} 1 & 0 & 0 \\ i\theta^+ & 1 & 0 \\ 0 & 0 & 1 \end{pmatrix}. \qquad (23.13)$$

The superdeterminant of E_M^A is one.

The covariant derivatives are defined by Eq. (14.38) and they become

$$\partial_z, \ D_+ \equiv \left(\frac{\partial}{\partial \theta^+} + i\partial_z \theta^+ \right) \qquad \text{and} \qquad \partial_{\bar{z}}.$$

As it should D_+ anticommutes with Q_+ and satisfies the anticommutator

$$\{D_+, D_+\} = 2i\partial_z. \qquad (23.14)$$

which implies that $D_+{}^2 = i\partial_z$. We note that a chiral field satisfies $D_+\phi = 0$ which necessarily imply $\partial_z \phi = 0$ and so are right-moving.

Defining the torsion and curvature by Eq. (14.42) we discover that the only nonvanishing component of the torsion is

$$T_{++}{}^{\ddagger} = 2i \qquad (23.16)$$

where $\ddagger$ corresponds to the tangent index associated to the coordinate z, while the curvature satisfies.

$$R_{MN} = 0. \qquad (23.17)$$

A general scalar superfield ϕ contains the component fields

$$\phi = A + i\theta^+ \chi_+. \qquad (23.18)$$

The transformations of the component fields are given by

$$\delta\phi = \delta A + i\theta^+ \delta\chi_+ = \varepsilon^+ l_+ \phi. \qquad (23.19)$$

with the results

$$\delta A = i\varepsilon^+ \chi_+; \qquad \delta\chi_+ = \partial A \varepsilon^+. \qquad (23.20)$$

Let us define superspace integration by

$$\int dz\, d\bar{z}\, d\theta^+ \equiv \int dz\, d\bar{z}\, D_+ = \int dz\, d\bar{z}\, \frac{\partial}{\partial\theta^+}. \qquad (23.21)$$

This completes our discussion of the $(1,0)$ superspace and we now construct the actions of interest. The free action for ϕ must be bilinear in ϕ and Lorentz invariant, the simplest such candidate is

$$-i\int dz\, d\bar{z}\, d\theta^+ \{\phi \partial_{\bar{z}} D_+ \phi\}. \qquad (23.22)$$

In components this action becomes[216]

$$-i\int dz\, d\bar{z}\, D_+ \{\phi \partial_{\bar{z}} D_+ \phi\} = -i\int dz\, d\bar{z}\, \{D_+ \phi \partial_{\bar{z}} D_+ \phi + \phi \partial_{\bar{z}} D_+^2 \phi\}$$

$$= \int dz\, d\bar{z}\, \{-\partial_{\bar{z}} A\, \partial_z A + i\chi_+ \partial_{\bar{z}}\chi_+\}. \qquad (23.23)$$

The equation of motion corresponding to the above action are

$$\partial_{\bar{z}} D_+ \phi = 0 \qquad (23.24)$$

and in components this equation reads

$$\partial_z \partial_{\bar{z}} A = 0, \qquad \partial_{\bar{z}}\chi_+ = 0. \qquad (23.25)$$

Thus we encounter one of the theories considered in Chapter 21; the on-shell states of this theory fall into a singlet of $(1,0)$ supersymmetry $(\partial_z A = 0)$ and the multiplet of $(1,0)$ supersymmetry of Eq. (22.127).

The current multiplet J_{z+} for $(1,0)$ superconformal theories has as components the supercurrent and the T_{zz} part of the energy-momentum tensor

$$J_{z+} = j_{z+} + \theta^+ T_{zz}.$$

It satisfies $\partial_{\bar{z}} J_{z+} = 0$. The $T_{\bar{z}\bar{z}}$ part of the energy momentum tensor is in general nonzero, but it satisfies $\partial_z T_{\bar{z}\bar{z}} = 0$ and is a singlet under $(1,0)$ supersymmetry. In a non-superformal theory, one finds that the above equations are modified by anomalies. In the above classically superconformal theory it is given by

$$J_{z+} = \partial_z \phi D_+ \phi, \qquad T_{\bar{z}\bar{z}} = \partial_{\bar{z}} A \partial_{\bar{z}} A.$$

The reader will observe that the above kinetic terms of Eq. (23.23) do not have their usual normalization. To restore these normalizations and recover the supersymmetry transformations of Chapter 21 we must make the changes

$$A \to \sqrt{2}A, \qquad \chi_+ \to \chi_+$$

and

$$\varepsilon^+ \to \varepsilon^+ \sqrt{2}.$$

To work with fields with the standard normalization, would require putting $\sqrt{2}$'s in many superspace expressions. An analogous discussion will apply to all component field expressions in this chapter.

We can also consider a spinor superfield $\lambda_-(z, \theta^+, \bar{z})$ is Grassmann odd and has the components

$$\lambda_- = \zeta_- + \theta^+ b \tag{23.26}$$

An appropriate superspace action is

$$-\int dz \, d\bar{z} \, d\theta^+ \lambda_- D_+ \lambda_- = \int dz \, d\bar{z} \{+i\zeta_- \partial_z \zeta_- - b^2\}. \tag{23.27}$$

The corresponding equation of motion is given by

$$D_+ \lambda_- = 0 \tag{23.28}$$

and in components it reads

$$\partial_z \zeta_- = 0, \qquad b = 0. \tag{23.29}$$

Thus b is an auxiliary field while ζ_- describes one Majorana-Weyl spinor. The possibility of singlet on-shell states under $(1, 0)$ supersymmetry was discussed in Chapter 21.

Since ϕ is Grassmann even, but the superspace integration $\int dz \, d\bar{z} \, d\theta^+$ is Grassmann odd we cannot have a potential for ϕ alone. We may, however, write one if we include the superfield λ_-. Introducing many ϕ^i, $i = 1, \ldots, n$ and λ_-^p, $p = 1, \ldots, m$ the most general interacting model is provided by the expression[219,259]

$$\int dz \, d\bar{z} \, d\theta^+ \{iG_{ij}(\phi)D_+\phi^i \partial_{\bar{z}}\phi^j - \lambda_-^p \nabla_+ \lambda_-^q H_{pq}(\phi) + \lambda_-^p g_p(\phi)\} \tag{23.30}$$

where

$$\nabla_+ \lambda_-^p = D_+ \lambda_-^p + D_+ \phi^i A_i{}^p{}_q \lambda_-^q$$

and

$$G_{ij}(\phi) = g_{ij}(\phi) + b_{ij}(\phi), \qquad H_{pq}(\phi), \qquad A_i{}^p{}_q(\phi)$$

and g_p are arbitrary functions of ϕ and g_{ij} and b_{ij} can be taken to be symmetric and antisymmetric respectively in i and j. A repeated index implies a summation in the usual way.

We recognize ∇_+ as a covariant derivative in the sense that the second term in the action is invariant under

$$\lambda_-{}^p \to f^p{}_q(\phi)\lambda_-{}^q.$$

$$H_{pq} \to (f^{-1})_p{}^s(f^{-1})_q{}^t H_{st}$$

$$A_i{}^p{}_q \to f^p{}_s \partial_i (f^{-1})^s{}_q + f^p{}_r A_i{}^r{}_s (f^{-1})^s{}_q,$$

where f_q^p is an arbitrary function of ϕ^i. The third term will possess such an invariance if g_p also transforms a $g_p \to (f^{-1})_p{}^q g_q$, that is it is a vector.

The above transformations involving $f^p{}_q$ can be thought of as local $GL(m,R)$ transformations and although they leave the above action invariant they do not in general preserve its form. In particular only a specific subgroup of $GL(m,R)$ will preserve a given metric $H_{pq}(\phi)$.

We can without loss of generality take $H_{pq} = H_{qp}$ by a suitable redefinition of $A_i{}^p{}_q$ and we may use the f_q^p above to diagonalize the remaining real symmetric matrix, i.e. $H_{pq} = \delta_{pq}$. The only surviving part of the connection in the action then satisfies

$$A_i{}^p{}_q = -A_i{}^q{}_p.$$

The connection $A_i{}^p{}_q$ will be automatically antisymmetric in p and q if the metric H_{pq} was covariantly constant. In this case the action will then be left invariant in form by local $SO(N)$ transformations and $A_i{}^p{}_q$ is a $SO(N)$ valued connection with the corresponding transformation law.

In components the above action becomes

$$A = A_1 + A_2 + A_3$$

where

$$A_1 = \int dz \, d\bar{z} \{ -(\partial_z A^i \partial_{\bar{z}} A^j)(g_{ij} + b_{ij}) + ig_{ij}\chi_+{}^i \nabla_{\bar{z}}\chi_+{}^j \} \qquad (23.31a)$$

$$A_2 = \int dz \, d\bar{z} \{ -(b^p)^2 + i\lambda_-^p \nabla_z \lambda_-^p - \tfrac{1}{2}\chi_+{}^i \chi_+{}^j F_{ij}{}^p{}_q(A)\lambda_-^p \lambda_-^q \} \qquad (23.31b)$$

and

$$A_3 = \int dz \, d\bar{z} \{ b^p g_p(A) - i\zeta_-^p \chi_+{}^i g_{p,i}(A) \} \qquad (23.31c)$$

where

$$\nabla_{\bar{z}}\chi_+{}^i \equiv \partial_{\bar{z}}\chi_+{}^i + \hat{\Gamma}^i_{jk} \partial_{\bar{z}} A^j \chi_+{}^k$$

$$\hat{\Gamma}^i_{jk} = \Gamma^i_{jk} + \tfrac{1}{2}H^i{}_{jk}, \qquad H^i{}_{jk} = g^{li} H_{ijk}$$

$$H_{ijk} = \partial_i b_{jk} + \partial_k b_{ij} + \partial_j b_{ki} \qquad (23.32a)$$

and Γ^i_{jk} is the usual Christoffel symbol

$$\Gamma^i_{jk} = \tfrac{1}{2}g^{ih}(\partial_j g_{kh} + \partial_k g_{jh} - \partial_h g_{jk})$$

also we use

$$\nabla_z \lambda^{\underline{p}} = \partial_z \lambda^{\underline{p}} + \partial_z A^i A_i{}^p{}_q(A)\lambda^{\underline{q}}$$

$$g_{p,i}(A) = \partial_i g_p(A).$$

In deriving Eq. (23.31a) we used the fact that $\chi_+{}^i\chi_+{}^k$ is antisymmetric in i and k. The torsion is the j, k antisymmetric part of the connection $\hat{\Gamma}^i_{jk}$ and as such we see that it is the field strength for b_{ij}.

To evaluate A_2 and A_3 we used the following trick due to the σ-model invariance of A_2. We may replace $d\theta^+$ by ∇_+ and then

$$\int dz\, d\bar{z}\, d\theta^+(-\lambda^{\underline{p}}\nabla_+\lambda^{\underline{p}}) = \int dz\, d\bar{z}\nabla_+(-\lambda^{\underline{p}}\nabla_+\lambda^{\underline{p}})$$

$$= \int dz\, d\bar{z}\{-\nabla_+\lambda^{\underline{p}}\nabla_+\lambda^{\underline{p}} + \lambda^{\underline{p}}(\nabla_+)^2\lambda^{\underline{p}}\}. \quad (23.32b)$$

Using the relation

$$\{\nabla_+,\nabla_+\}\lambda^{\underline{p}} \equiv 2i\partial_z\lambda_-{}^p + F_{++}{}^p{}_q\lambda_-{}^q \quad (23.32c)$$

where

$$F_{++}{}^p{}_q = D_+\phi^i D_+\phi^j F_{ij}{}^p{}_q$$

$$F_{ij}{}^p{}_q(\phi) = \partial_i A_j{}^p{}_q - \partial_j A_i{}^p{}_q + A_i{}^p{}_s A_j{}^s{}_q - A_j{}^p{}_s A_i{}^s{}_q \quad (23.33a)$$

and identifying at $\theta = 0$

$$\nabla_+\lambda_-{}^p = b^p \qquad \lambda_-{}^p = \zeta_-{}^p$$

we find the expression for A_2 given above. Tacitly assuming that g_p is a vector we find A_3 by a similar manœuvre

We may also write the heterotic string which has a $(1,0)$ supersymmetry in superspace. We recall from Chapter 21 that it has the field x^μ and $\chi_+{}^\mu$, $\mu = 0, 1, \ldots, 9$ and $\chi_-{}^i$, $i = 1, 2, \ldots, 32$. We may fit these component fields into the superfields

$$\phi^\mu = x^\mu + i\theta^+\chi_+^\mu \quad (23.33b)$$

and

$$H_-{}^i = \chi_-{}^i + \theta^+ b^i. \quad (23.34)$$

The field b^i will be an auxiliary field. The superspace action for the heterotic string is given by the coupling of these fields to $(1,0)$ supergravity; it will be given in the next chapter after we have found a superspace description of $(1,0)$

supergravity. In the usual superconformal gauge (see Chapter 21) these fields obey the equations of motion:

$$\partial_z D_+ \phi^\mu = 0, \qquad D_+ H^i_- = 0 \qquad (23.35)$$

as well as constraint equations that follow from the variation of the supergravity multiplet. These constraints essentially state that the energy-momentum tensor and supercurrent vanish.

23.1b (1, 1) Superspace

We now contruct $(1, 1)$ superspace corresponding to the $(1, 1)$ superalgebra of Eqs. (22.86) and (22.87). This superspace is the coset

$$\frac{(1, 1) \text{ supergroup}}{\text{Lorentz group}} \qquad (23.36)$$

and can be parametrized by z, $\bar{z}$, θ^+ and θ^-. Since the $(1, 1)$ superalgebra is the direct sum of a $(0,1)$ and a $(0, 1)$ superalgebra, the coset space is a direct product space of a $(1,0)$ and $(0,1)$ superspace. This means that the supersymmetry group acts separately on the left and right space and that the vierbein and spin connection have a direct product structure. To made this apparent, we write the group element in the form

$$\exp(zP_z + \theta^+ Q_+)\exp(\bar{z}P_{\bar{z}} + \theta^- Q_-)\exp(\omega J)$$

when carrying out the calculations. The reader can readily extend this argument in such a way that it is coordinate and patch independent.

The supersymmetry transformations are given by

$$z' = z - i\varepsilon^+ \theta^+, \qquad \bar{z}' = \bar{z} - i\varepsilon^- \theta^-$$
$$\theta^{+'} = \theta^+ + \varepsilon^+, \qquad \theta^{-'} = \theta^- + \varepsilon^-. \qquad (23.37)$$

Under a Lorentz rotation we find

$$z' = z + \omega z, \qquad \bar{z}' = \bar{z} - \omega\bar{z}$$
$$\theta^{+'} = \theta^+ + \tfrac{1}{2}\omega\theta^+, \qquad \theta^{-'} = \theta^- - \tfrac{1}{2}\omega\theta^-. \qquad (23.38)$$

Denoting the coordinates by

$$z^\pi = (z, \theta^+; \bar{z}, \theta^-).$$

a scalar superfield is defined by the equation

$$\phi'(z^{\pi'}) = \phi(z^\pi). \qquad (23.39)$$

The action of the two supercharges Q_+ and Q_- are given by the infinitesimal operators

$$l_+ = \frac{\partial}{\partial\theta^+} - i\partial_z\theta^+ \qquad \text{and} \qquad l_- = \frac{\partial}{\partial\theta^-} - i\partial_z\theta^-. \qquad (23.40)$$

The superspin connection vanishes, but the supervielbein is given by

$$E_\pi{}^M = \begin{pmatrix} 1 & 0 & 0 & 0 \\ -i\theta^+ & 1 & 0 & 0 \\ 0 & 0 & 1 & 0 \\ 0 & 0 & -i\theta^- & 1 \end{pmatrix} \qquad (23.41)$$

where M and π run over all the tangent and world indices corresponding to $z, +, \bar{z}, -$ respectively.

The inverse supervierbein is

$$E_M{}^A = \begin{pmatrix} 1 & 0 & 0 & 0 \\ i\theta^+ & 1 & 0 & 0 \\ 0 & 0 & 1 & 0 \\ 0 & 0 & i\theta^+ & 1 \end{pmatrix}. \qquad (23.42)$$

The corresponding covariant derivatures are given by

$$\partial_z, \ D_+ = \frac{\partial}{\partial\theta^+} + i\partial_z\theta^+; \qquad \partial_{\bar{z}}, \qquad D_- = \frac{\partial}{\partial\theta^-} + i\partial_{\bar{z}}\theta^- \qquad (23.43)$$

and obey the relations

$$\{D_+, D_+\} = 2i\partial_z, \qquad \{D_-, D_-\} = 2i\partial_{\bar{z}}$$

$$\{D_+, D_-\} = 0. \qquad (23.44)$$

The curvature of $(1, 1)$ superspace vanishes and the nonzero components of the torsion are

$$T_{++}{}^{\pm} = 2i, \qquad T_{--}{}^{=} = 2i \qquad (23.45)$$

where $\pm$ and $=$ are the tangent indices associated with the coordinates z and $\bar{z}$ respectively.

A general scalar superfield contains the component fields

$$\phi = A + i\theta^+\chi_+ + i\theta^-\chi_- + i\theta^+\theta^- F. \qquad (23.46)$$

Their supersymmetry transformations are derived to be

$$\delta A = i\varepsilon^+\chi_+ + i\varepsilon^-\chi_-$$

$$\delta\chi_+ = \partial_z A\varepsilon^+ + F\varepsilon^-, \qquad \delta\chi_- = -F\varepsilon^+ + \partial_{\bar{z}} A\varepsilon^-$$

$$\delta F = -i\varepsilon^+\partial_z\chi_- + i\varepsilon^-\partial_{\bar{z}}\chi_+. \qquad (23.47)$$

If we take ϕ to be a real superfield then all its components including the spinors are real.

A chiral constraint on ϕ, i.e. $D_+\phi = 0$ implies from Eq. (23.44) that $\partial_z\phi = 0$ and hence the component fields of ϕ are on-shell and right-moving. Similarly, $D_-\phi = 0$ implies $\partial_{\bar z}\phi = 0$.

We define the $(1,1)$ superspace integral by

$$\int dz\, d\bar z\, d\theta^+\, d\theta^- \equiv \int dz\, d\bar z\, D_+D_- = \int dz\, d\bar z\, \frac{\partial}{\partial\theta^+}\frac{\partial}{\partial\theta^-}. \qquad (23.48)$$

The free action for a real scalar superfield ϕ with a potential is

$$\int dz\, d\bar z\, d\theta^+\, d\theta^-\, \{\phi D_+D_-\phi + ig(\phi)\} \qquad (23.49)$$

where g is an arbitrary functiion. The equation of motion is

$$D_+D_-\phi = \frac{i}{2}g'(\phi). \qquad (23.50a)$$

In terms of component fields the action becomes

$$\int dz\, d\bar z\, D_+D_-\{\phi D_+D_-\phi + ig(\phi)\} = \int dz\, d\bar z\, \{D_+D_-\phi D_+D_-\phi - D_-\phi i\partial_z D_-\phi$$

$$- D_+\phi i\partial_{\bar z}D_+\phi - \phi i\partial_z i\partial_{\bar z}\phi + ig'(\phi)D_+D_-\phi$$

$$+ ig''(\phi)D_+\phi D_-\phi\}$$

$$= \int dz\, d\bar z\, \{-\partial_z A\partial_{\bar z}A + \chi_+ i\partial_{\bar z}\chi_+ + \chi_- i\partial_z\chi_-$$

$$- F^2 + Fg'(A) - i\chi_+\chi_- g''(A)\}. \qquad (23.50b)$$

The most general action for a number of general superfields ϕ^i; $i = 1, \ldots, n$ is given by

$$\int dz\, d\bar z\, d\theta^+\, d\theta^-\, \{-(g_{ij} + b_{ij})D_+\phi^i D_-\phi^j + ig(\phi)\} \qquad (23.52)$$

where

$$g_{ij} = g_{ji}, \qquad b_{ij} = -b_{ji}$$

and g are arbitrary functions of ϕ^i. In components it corresponds to a σ-model whose target space has metric g_{ij}, but also torsion associated with the field strength of b_{ij}. The above σ model without torsion was given in Ref. 231 and its extension to realize a central change in Ref. 232. The addition of torsion is to be found in Refs. 233 and 221.

We will evaluate in components the result for $b_{ij} = 0$, $g = 0$. Using the same techniques as in the $(1, 0)$ case the action takes the form

$$\int dz\, d\bar{z}\, \nabla_+ \nabla_- \{ -D_+ \phi^i D_- \phi^j g_{ij} \}$$

$$= -\int dz\, d\bar{z} \{ \nabla_+ \nabla_- \nabla_+ \phi^i \nabla_- \phi^j + \nabla_- \nabla_+ \phi^i \nabla_+ \nabla_- \phi^j$$

$$- \nabla_+^2 \phi^i \nabla_-^2 \phi^j + \nabla_+ \phi^i \nabla_+ \nabla_-^2 \phi^j \} g_{ij}$$

$$= -\int dz\, d\bar{z} \{ R_{+-}{}^i{}_k \nabla_+ \phi^k \nabla_- \phi^j - \nabla_- i \partial_z \phi^i \nabla_- \phi^j - \nabla_- \nabla_+ \phi^i \nabla_- \nabla_+ \phi^j$$

$$+ \partial_z \phi^i \partial_{\bar{z}} \phi^j + \nabla_+ \phi^i \nabla_+ i \partial_{\bar{z}} \phi^j \} g_{ij}$$

$$= \int dz\, d\bar{z} \{ -\partial_z A^i \partial_{\bar{z}} A^j g_{ij} + i g_{ij} \chi_+^i \nabla_{\bar{z}} \chi_+^j + i g_{ij} \chi_-^i \nabla_z \chi_-^j$$

$$- F^i F^j g_{ij} - R_{hljk} \chi_+^h \chi_-^l \chi_+^k \chi_-^j \}.$$

where

$$\nabla_z \chi_{\pm}^i = \partial_z \chi_{\pm}^i + \partial_z A^k \Gamma_{kh}^i \chi_{\pm}^h.$$

In the above we have introduced the covariant derivative which on a tensor T^i acts as

$$\nabla_N T^i = D_N T^i + D_N \phi^k \Gamma_{kh}^i(\phi) T^h.$$

The advantage of using the connection constructed from the Christoffel symbol is that the covariant derivative of the metric g_{ij} vanishes. The x-space fields are the $\theta = 0$ components of

$$\phi^i = A^i, \qquad \nabla_{\pm} \phi^i = i \chi_{\pm}^i, \qquad \nabla_- \nabla_+ \phi^i = i F^i.$$

In x-space the Christoffel symbol is

$$\Gamma_{kh}^j(\phi^i = A^i).$$

and R_{kljk} is the usual Riemann curvature.

The usual open and closed superstring in the Neveu-Schwarz-Ramond formulation has a manifest $(1, 1)$ supersymmetry.[7] The component fields of this formulation are x^μ and χ_A^μ, $\mu = 0, 1, \ldots, 9$; $A = 1, 2$. The spinor χ_A^μ is Majorana, but not Weyl and so has the components χ_+^μ and χ_-^μ. These fields belong to a superfield ϕ^μ

$$\phi^\mu(z, \theta^+; \bar{z}, \theta^-) = x^\mu(z, \bar{z}) + i\theta^+ \chi_+^\mu(z, \bar{z}) + i\theta^- \chi_-^\mu(z, \bar{z}) + i\theta^+ \theta^- N^\mu(z, \bar{z}).$$

The equations of motion in the usual superconformal gauge (see Chapter 21) are

$$D_+ D_- \phi^\mu = 0$$

and we find that N^μ is an auxiliary field. Of course one also has the constraints associated with the variations of the supergravity fields which in this gauge essentially set the supercurrent and energy-momentum tensor to zero. The superspace action will be given in Chapter 24.

For a superconformal $(1,1)$ theory the currents belong to a supermultiplet which is a vector spinor superfield J_{aA}. It has as its components the supercurrent J_{aA} and the energy-momentum tensor T_{ab}. The superfield J_{aA} is subject to the constraints

$$(\gamma^a)_A{}^B J_{aB} = 0, \qquad D^A J_{aA} = 0, \qquad \partial^a J_{aA} = 0.$$

As a result J_{aA} has the θ expansion

$$J_{aA} = j_{aA} + \bar\theta^B (\gamma_b C)_{AB} T_{ab}$$

where

$$(\gamma^a)_A{}^B j_{aB} = 0, \qquad T_{ab} = T_{ba}, \qquad T_a{}^a = 0; \qquad \partial^a j_{aA} = 0 = \partial^a T_{ab}.$$

In the theory above the supermultiplet J_{aA} is realized by the expression

$$J_{aA} = (\gamma^b \gamma^a)_A{}^B \partial_b \phi D_B \phi.$$

For theories which are $(1,1)$ supersymmetric, but not superconformally invariant the above equations are modified by anomaly terms.

23.1c   (2, 0) Superspace

The $(2,0)$ superspace is the coset

$$\frac{(2,0)\ \text{supergroup}}{\text{Lorentz group} \otimes \text{Chiral } U(1)} \tag{23.53}$$

The $(2,0)$ superalgebra was given in Eq. (22.95) and the superspace is parametrized by

$$z^\pi = (z, \theta^+, \bar\theta^+; \bar z). \tag{23.54}$$

The coset representatives can be chosen as

$$g(z^\pi) = \exp\{z P_z + \theta^+ Q_+ + \bar\theta^+ Q_+\} \exp\{\bar z P_{\bar z}\}. \tag{23.55}$$

The supersymmetry transformations are given by

$$z' = z - \frac{i}{2}\varepsilon^+\bar{\theta}^+ - \frac{i}{2}\bar{\varepsilon}^+\theta^+, \qquad \theta^{+'} = \theta^+ + \varepsilon^+$$

$$\bar{\theta}^{+'} = \bar{\theta}^+ + \bar{\varepsilon}^+; \qquad \bar{z}' = \bar{z}, \tag{23.56}$$

while the tangent group transformations are

$$z' = z + \omega z, \qquad \theta^{+'} = \theta^+ + \frac{\omega}{2}\theta^+ - il\theta^+$$

$$\bar{\theta}^{+'} = \bar{\theta}^+ + \frac{\omega}{2}\bar{\theta}^+ + il\bar{\theta}^+, \qquad \bar{z}' = \bar{z} - \omega z, \tag{23.57}$$

where ω and l are the parameters of Lorentz and $U(1)$ transformations (i.e. L transformations) respectively. The infinitesimal operator corresponding to the supersymmetry generators are read off to be given by

$$l_+ = \frac{\partial}{\partial\theta^+} - \frac{i}{2}\partial_z\bar{\theta}^+; \qquad \bar{l}_+ = \frac{\partial}{\partial\bar{\theta}^+} - \frac{i}{2}\partial_z\theta^+. \tag{23.58}$$

The supervielbein is of the form

$$E_\pi{}^M = \begin{pmatrix} 1 & 0 & 0 & 0 \\ -\frac{i}{2}\bar{\theta}^+ & 1 & 0 & 0 \\ -\frac{i}{2}\theta^+ & 0 & 1 & 0 \\ 0 & 0 & 0 & 1 \end{pmatrix} \tag{23.59}$$

where M and π run over all the indices, corresponding to the coordinates z, $\theta^+, \bar{\theta}^+, \bar{z}$ while as usual the spin connection vanishes. The inverse supervielbein is given by

$$E_M{}^A = \begin{pmatrix} 1 & 0 & 0 & 0 \\ \frac{i}{2}\bar{\theta}^+ & 1 & 0 & 0 \\ \frac{i}{2}\theta^+ & 0 & 1 & 0 \\ 0 & 0 & 0 & 1 \end{pmatrix}. \tag{23.60}$$

The covariant derivatives constructed from the inverse supervielbein are of the form

$$\partial_z, D_+ = \frac{\partial}{\partial \theta^+} + \frac{i}{2} \partial_z \bar{\theta}^+, \qquad \bar{D}_+ = \frac{\partial}{\partial \bar{\theta}^+} + \frac{i}{2} \partial_z \theta^+$$

and

$$\partial_{\bar{z}}. \tag{23.61}$$

They obey the relation

$$\{D_+, D_+\} = 0 = \{\bar{D}_+, \bar{D}_+\}$$

$$\{D_+, \bar{D}_+\} = i\partial_z. \tag{23.62}$$

The curvature of $(2, 0)$ superspace vanishes, but the nonzero component of the torsion tensor is

$$T_{+\oplus}{}^+ = i \tag{23.63}$$

where $\oplus$ corresponds to the tangent index associated with $\bar{\theta}^+$. Unfortunately, there would seem no convenient notation to distinguish the two tangent indices associated with θ^+ and $\bar{\theta}^+$.

The general scalar superfield has the component fields

$$\phi = A + i\theta^+ \chi_+ + i\bar{\theta}^+ \bar{\chi}_+ + i\theta^+ \bar{\theta}^+ F_z \tag{23.64}$$

their supersymmetry transformations are given by

$$\delta A = i\varepsilon^+ \chi_+ + i\bar{\varepsilon}^+ \bar{\chi}_+$$

$$\delta \chi_+ = \tfrac{1}{2} \partial_z A \bar{\varepsilon}^+ + F_z \bar{\varepsilon}^+; \qquad \delta \bar{\chi}_+ = \tfrac{1}{2} \partial_z A \varepsilon_+ - F_z \varepsilon^+$$

$$\delta F_z = \frac{i}{2} \varepsilon^+ \partial_z \chi_+ - \frac{i}{2} \bar{\varepsilon}_+ \partial_z \bar{\chi}_+. \tag{23.65}$$

Unlike for the previous superspaces we can impose chiral constraints on ϕ without putting it on-shell. The chiral constraint

$$\bar{D}_+ \phi = 0 \tag{23.67}$$

implies that

$$\bar{\chi}_+ = 0, \qquad F_z = \tfrac{1}{2} \partial_z A. \tag{23.68}$$

We may write such a chiral superfield in a chiral basis:

$$\phi(y, \theta^+; \bar{z}) = A(y, \bar{z}) + i\theta^+ \chi_+(y, \bar{z}) \tag{23.69}$$

where

$$y = z + \frac{i}{2} \theta^+ \bar{\theta}^+. \tag{23.70}$$

Denoting the complex conjugate of ϕ by $\bar{\phi}$. The complex conjugate of equation reads

$$D_+\bar{\phi} = 0 \tag{23.71}$$

and $\bar{\phi}$ contains the fields A^* and $\bar{\chi}_+$.

The corresponding free action is[217,218]

$$2i \int dz \, d\bar{z} \, d\theta^+ \, d\bar{\theta}^+ \{\bar{\phi}\partial_{\bar{z}}\phi\} = \int dz \, d\bar{z} \{-\partial_z A \partial_{\bar{z}} \bar{A} + i\bar{\chi}_+ \partial_z \chi_+\} \tag{23.72}$$

where the integration is defined by

$$\int dz \, d\bar{z} \, d\theta^+ \, d\bar{\theta}^+ \equiv \int dz \, d\bar{z} \, D_+ \bar{D}_+ = \int dz \, d\bar{z} \frac{\partial}{\partial\theta^+} \frac{\partial}{\partial\bar{\theta}^+}. \tag{23.73}$$

The potential term for such an action must involve a chiral subintegral in order to avoid the occurrence of derivatives. However, the chiral subintegrals are given by

$$\int dz \, d\bar{z} \, d\theta^+ = \int dz \, d\bar{z} \, D_+ \tag{23.74}$$

and its complex conjugate

$$\int dz \, d\bar{z} \, d\bar{\theta}^+ = \int dz \, d\bar{z} \, D_+ \tag{23.75}$$

Since these are Grassmann odd and ϕ is Grassmann even we require further odd fields as in the case of $(1,0)$ superspace. Consequently, we introduce the chiral superfield λ_- (i.e. $\bar{D}\lambda_- = 0$) which is odd and write the action

$$\int dz \, d\bar{z} \, d\theta^+ \, d\bar{\theta}^+ \{2i\bar{\phi}\partial_{\bar{z}}\phi - 2\bar{\lambda}_-\lambda_-\} + \left\{ \int dz \, d\bar{z} \, d\theta^+ \lambda_- f(\phi) + \text{h.c.} \right\} \tag{23.76}$$

where $f(\phi)$ is an arbitrary function. The superfield λ_- and its complex conjugate $\bar{\lambda}_-$ contribute one complex Weyl spinor on-shell. We could propagate this field alone by setting $\phi = 0$ in the above action.

The most general $(2,0)$ action involving the fields ϕ^i, $\bar{\phi}^i$ $i = 1, \ldots n$ and λ^p, $\bar{\lambda}^{\bar{p}}$ $p = 1, \ldots m$ is of the form[226,260]

$$\int dz \, d\bar{z} \, d\theta^+ \, d\bar{\theta}^+ \{i(K_i \partial_{\bar{z}} \phi^i - K_{\bar{i}} \partial_{\bar{z}} \bar{\phi}^{\bar{i}}) - 2\bar{\lambda}^{\bar{p}}\lambda^q H_{\bar{p}q}\}$$

$$+ \left\{ \int dz \, d\bar{z} \, d\theta^+ \lambda^p_- g_p(\phi) + \text{h.c.} \right\}$$

where

$$\bar{K}_{\bar{\imath}} = (K_i)^* \qquad \text{and} \qquad g_D$$

are arbitrary functions of ϕ^i. In components the kinetic part of this action reads[219]

$$\int dz\,d\bar{z}\{-g_{i\bar{\jmath}}(\partial_z A^i \partial_{\bar{z}} \bar{A}^{\bar{\jmath}} + \partial_{\bar{z}} A^i \partial_z \bar{A}^{\bar{\jmath}}) - b_{i\bar{\jmath}}(\partial_z A^i \partial_{\bar{z}} \bar{A}^{\bar{\jmath}}$$

$$- \partial_{\bar{z}} A^i \partial_z \bar{A}^{\bar{\jmath}}) + i g_{i\bar{\jmath}} \chi^i_+ \nabla_{\bar{z}} \bar{\chi}^{\bar{\jmath}}_+ + i H_{\bar{p}q} \bar{\zeta}^{\bar{p}}_- \nabla_z \zeta^q_- - 2 b^{\bar{p}} b^q H_{\bar{p}q}\}$$

where

$$g_{i\bar{\jmath}} = K_{i,\bar{\jmath}} + K_{\bar{\jmath},i}, \qquad b_{i\bar{\jmath}} = K_{i,\bar{\jmath}} - K_{\bar{\jmath},i}$$

$$\nabla_z \bar{\chi}^{\bar{\jmath}}_+ = \partial_{\bar{z}} \bar{\chi}^{\bar{\jmath}}_+ + \Gamma^{\bar{\jmath}}_{k\bar{h}} \partial_{\bar{z}} A^k \chi^{\bar{h}}_+ + \Gamma^{\bar{\jmath}}_{k\bar{h}} \partial_{\bar{z}} \bar{A}^{\bar{k}} \bar{\chi}^{\bar{h}}_+$$

$$\Gamma^{\bar{\jmath}}_{k\bar{h}} = \hat{\Gamma}^{\bar{\jmath}}_{k\bar{h}}, \qquad \Gamma^{\bar{\jmath}}_{k\bar{h}} = \hat{\Gamma}^{\bar{\jmath}}_{k\bar{h}} + \frac{1}{2} H^{\bar{\jmath}}_{k\bar{h}}$$

$$H_{jk\bar{h}} = \partial_j b_{k\bar{h}} + \partial_k b_{\bar{h}j}, \qquad \nabla_z \zeta^p_- = \partial_z \zeta^p_- + \Gamma^p_{q\bar{r}} \zeta^q_- \partial_z \bar{A}^{\bar{r}}.$$

and

$$\hat{\Gamma} \qquad \text{and} \qquad \hat{\Gamma}$$

is the usual Christoffel symbol constructed from the matrices $g_{i\bar{\jmath}}$ and $H_{\bar{p}q}$ respectively. We note that the σ-model manifold or target space is not Kähler, but is hermitian.

In a $(2,0)$ superconformal theory the currents belong to a superfield which carries a z index, i.e. J_z which satisfies

$$\partial_{\bar{z}} J_z = 0.$$

It has the component expansions

$$J_z = j_z + \theta^+ j_{z+} + \bar{\theta}^+ \bar{j}_{z+} + \theta^+ \bar{\theta}^+ T_{zz}.$$

where j_z is the current for the $U_L(1)$, $j_{z+}, \bar{j}_{z+}$ the two-supercurrents and T_{zz} a component of the energy momentum tensor. The remaining current is $T_{\bar{z}\bar{z}}$, it obeys $\partial_z T_{\bar{z}\bar{z}} = 0$ and is a singlet under $(2,0)$ supersymmetry.

The action of (23.72) realizes the above multiplet of currents by

$$J_z = D_+ \phi \bar{D}_+ \bar{\phi}.$$

For a $(2,0)$ theory that is not superconformally invariant the above constrains on J_z are modified. In particular one can have

$$\partial_{\bar{z}} J_z = eA$$

where A is an anomaly multiplet that contains $U(1)$, supersymmetry and gravitational anomalies.[216] This type of anomaly is particularly important in string theory.

Given a set of superfields ϕ^i $i = 1, \ldots, n$, we can adopt more complicated chiral constraints than those of Eq. (23.67). In terms of real covariant derivatives $D_+{}^i$ $i = 1, 2$ we can adopt the condition[222]

$$D^2{}_+\phi^i = I^i{}_j D^1{}_+\phi^j$$

where $I^i{}_j$ is itself a function of ϕ^k. Acting on this equation with $D_+{}^2$ we find that

$$(D_+{}^2)^2\phi^i = i\partial_z\phi^i = (\partial_k I^i{}_j)D_+{}^2\phi^k D_+{}^1\phi^j - I^i{}_j D_+{}^1 D_+{}^2\phi^j$$

$$= \partial_k I^i{}_j I^k{}_m D_+{}^1\phi^m D_+{}^1\phi^j - I^i{}_i \partial_j I^l{}_m D_+{}^1\phi^j D_+{}^1\phi^m$$

$$- I^i{}_j I^j{}_{ki}\partial_z\phi^k.$$

Since we do not want to impose a space-time derivative condition on the component fields we require that[222]

$$I^i{}_j I^j{}_k = -1$$

and

$$N_{mj}{}^i \equiv I^k{}_m\partial_k I^i{}_j + I^i{}_i\partial_j I^l{}_m - (j \leftrightarrow m)$$

$$= 0$$

The first condition implies that we have an almost complex structure on the target space and the second equation tells us that the Nijenhuis tensor vanishes and consequently that the structure is actually complex or in other words, one can locally introduce coordinates such that the complex structure is $i \equiv \sqrt{-1}$ and that given charts of complex coordinate the transition functions from chart to chart are analytic functions. Good references to the theory of complex manifolds are *Complex Manifolds Without Potential Theory* by S. S. Chern (Springer Verlag) and *Foundations of Differential Geometry*, Vol. 1 by Kobayashi and Nomizu. Consequently, we recover the chiral conditions for $(2, 0)$ of Eq. (23.67) in complex coordinates. Indeed the above shows we can always find coordinates in which (23.67) holds.

Due to the chiral condition on the superfield in the $(2, 0)$ sigma model the field content is the same as for the $(1, 0)$ models. Consequently, one can view $(2, 0)$ sigma models as $(1, 0)$ models by demanding the existence of an extra supersymmetry which will restrict the couplings of the original $(1, 0)$ model.[222] The final results are the same as those given above as long as the final supersymmetry algebra that is realized is of the standard form of Eq. (22.95). However, using this alternative procedure one can realize a more complicated supersymmetry algebra and find correspondingly more esoteric σ-models. The same discussion applies between $(1, 1)$ and $(2, 1)$ and $(2, 2)$ σ-models.

Let us briefly explore the above line of attack for $(2,0)$ supersymmetric σ-models. Given the $(1,0)$ superfields ϕ^i $i = 1, \ldots, n$ we can write a second supersymmetry variation in the form.

$$\delta\phi^i = \varepsilon I^i{}_j D_+ \phi^j$$

where $I^i{}_j$ is a function of ϕ^k. This transformation must form an invariance of the most general action of Eq. (23.30) with $\lambda^p_- = 0$ and must close. Examining the latter we have

$$[\delta_1, \delta_2]\phi^i = \varepsilon_2 I^i{}_j D_+{}^1 (\varepsilon_1 I^j{}_k D_+{}^1 \phi^k) + \varepsilon_1 \partial_k I^i{}_j \varepsilon_2 I^k{}_l D_+{}^1 \phi^l D_+{}^1 \phi^j - (1 \leftrightarrow 2)$$

$$= -\varepsilon_2 \varepsilon_1 I^i{}_j I^j{}_k i\partial_z \phi^k - \varepsilon_2 \varepsilon_1 \{I^j{}_j \partial_l I^i{}_k + I^j{}_l \partial_j I^i{}_k\} D_+{}^1 \phi^l D_+{}^1 \phi^k.$$

Demanding that we realize the supersymmetry algebra we find that $I^i{}_j I^j{}_k = -1$ and that $N^i{}_{lk} = 0$. In other words we have a complex structure.

For simplicity let us consider the action of Eq. (23.30) with $\lambda_-{}^p = 0$ and $b_{ij} = 0$ which we vary under the above additional supersymmetry transformation:—

$$\delta A = \delta \int dz\, d\bar{z}\, d\theta^+ \{ig_{ij} D_+ \phi^i \partial_z \phi^j\}$$

$$= -i \int dz\, d\bar{z}\, d\theta^+ \{2g_{ij}\delta\phi^i D_+ \partial_z \phi^j + \partial_k g_{ij} D_+ \phi^k \partial_z \phi^j \delta\phi^i$$

$$+ \partial_k g_{ij} D_+ \phi^i \partial_z \phi^k \delta\phi^j - \partial_k g_{ij} D_+ \phi^i \partial_z \phi^j \delta\phi^k\}$$

$$= -2i \int dz\, d\bar{z}\, d\theta^+ \nabla_+ \partial_z \phi^j g_{ij} \delta\phi^i$$

where

$$\nabla_+ \partial_z \phi^j = D_+ \partial_z \phi^j + D_+ \phi^k \partial_z \phi^i \Gamma^j_{ki}$$

and Γ^j_{ki} is the Christoffel symbol which has its usual relation to the metric g_{ij}. Substituting for $\delta\phi^i$ we find that

$$\delta A = -2i \int dz\, d\bar{z}\, d\theta^+ g_{ij} \nabla_+ \partial_z \phi^j \varepsilon I^i{}_k D_+ \phi^k$$

$$= +2i\varepsilon \int dz\, d\bar{z}\, d\theta^+ I_{jk} \nabla_+ \partial_z \phi^j D_+ \phi^k$$

$$= 2i\varepsilon \int dz\, d\bar{z}\, d\theta^+ I_{jk} (\nabla_z D_+ \phi^j) D_+ \phi^k$$

where

$$I_{jk} \equiv g_{ji} I^i{}_k$$

For this to vanish we must demand that

$$I_{ji} + I_{ij} = 0$$

and then the variation of the action becomes

$$\delta A = -i\varepsilon \int dz\, d\bar{z}\, d\theta^+ \, \nabla_l I_{jk} \nabla_z \phi^l D_+ \phi^j D_+ \phi^k$$

from which we conclude that

$$\nabla_l I_{jk} = 0.$$

The first condition can be rewritten as $g_{kj} = I_j^l g_{li} I_k{}^i$ and is the condition of hermiticity of the metric. The second condition is the Kähler condition.[219,222]

It is straightforward to repeat the above calculation with b_{ij} nonzero; one finds that $I_{(ij)} = 0$ and $\nabla_l I_{jk} = 0$ but now the covariant derivative has a connection with torsion and the target manifold is no longer Kähler in agreement with the results found previously.

23.1d (2, 2) Superspace

Our final example of a two-dimensional Minkowski superspace is based on the $(2, 2)$ superalgebra of Eqs. (22.95) and (22.96). For the same reasons as in the $(1, 1)$ case this $(2, 2)$ superspace is the direct product of a $(2, 0)$ and a $(0, 2)$ superspace. The superspace has the coordinates

$$z^\pi = (z, \theta^+, \bar{\theta}^+; z, \theta^-, \bar{\theta}^-) \tag{23.77}$$

with the supersymmetry transformations

$$z' = z - \frac{i}{2}\varepsilon^+ \bar{\theta}^+ - \frac{i}{2}\bar{\varepsilon}^+ \theta^+, \qquad \theta^{+\prime} = \theta^+ + \varepsilon^+, \qquad \bar{\theta}^{+\prime} = \bar{\theta}^+ + \bar{\varepsilon}^+$$

$$\bar{z}' = \bar{z} - \frac{i}{2}\varepsilon^- \bar{\theta}^- - \frac{i}{2}\bar{\varepsilon}^- \theta^-, \qquad \theta^{-\prime} = \theta^- + \varepsilon^-, \qquad \bar{\theta}^{-\prime} = \bar{\theta}^- + \bar{\varepsilon}^-. \tag{23.78}$$

The Lorentz and chiral tangent space transformations are

$$z' = z + \omega z, \qquad \theta^{+\prime} = \theta^+ + \frac{\omega}{2}\theta^+ - il\theta^+, \qquad \bar{\theta}^{+\prime} = \bar{\theta}^+ + \frac{\omega}{2}\bar{\theta}^+ + il\bar{\theta}^+$$

$$\bar{z}' = \bar{z} - \omega\bar{z}, \qquad \theta^{-\prime} = \theta^- - \frac{\omega}{2}\theta^- - ir\theta^-, \qquad \bar{\theta}^- = \bar{\theta}^- - \frac{\omega}{2}\bar{\theta}^- + ir\bar{\theta}^-$$

$$\tag{23.79}$$

where ω is the Lorentz parameter, and l and r are the left and right $U(1)$'s respectively.

The supersymmetry generators are

$$l_+ = \frac{\partial}{\partial\theta^+} - \frac{i}{2}\partial_z\bar{\theta}^+, \qquad \bar{l}_+ = \frac{\partial}{\partial\bar{\theta}^+} - \frac{i}{2}\partial_z\theta^+$$

$$l_- = \frac{\partial}{\partial\theta^-} - \frac{i}{2}\partial_{\bar{z}}\bar{\theta}^-, \qquad \bar{l}_- = \frac{\partial}{\partial\bar{\theta}^-} - \frac{i}{2}\partial_{\bar{z}}\theta^-. \qquad (23.80)$$

The supervielbein and its inverse are given by

$$E_\pi{}^M = \begin{pmatrix} 1 & 0 & 0 & 0 & 0 & 0 \\ -\dfrac{i\bar{\theta}^+}{2} & 1 & 0 & 0 & 0 & 0 \\ -\dfrac{i\theta^+}{2} & 0 & 1 & 0 & 0 & 0 \\ 0 & 0 & 0 & 1 & 0 & 0 \\ 0 & 0 & 0 & -\dfrac{i\bar{\theta}^-}{2} & 1 & 0 \\ 0 & 0 & 0 & -\dfrac{i\theta^-}{2} & 0 & 1 \end{pmatrix} \qquad (23.81)$$

$$E_M{}^A = \begin{pmatrix} 1 & 0 & 0 & 0 & 0 & 0 \\ \dfrac{i\bar{\theta}^+}{2} & 1 & 0 & 0 & 0 & 0 \\ \dfrac{i\theta^+}{2} & 0 & 1 & 0 & 0 & 0 \\ 0 & 0 & 0 & 1 & 0 & 0 \\ 0 & 0 & 0 & \dfrac{i\bar{\theta}^-}{2} & 1 & 0 \\ 0 & 0 & 0 & \dfrac{i\theta^-}{2} & 0 & 1 \end{pmatrix} \qquad (23.82)$$

where the indices correspond to the coordinates in the order z, θ^+, $\bar{\theta}^+$, $\bar{z}$, θ^-, $\bar{\theta}^-$.

The covariant derivatives are

$$\partial_z, D_+ = \frac{\partial}{\partial\theta^+} + \frac{i}{2}\partial_z\bar{\theta}^+, \qquad \bar{D}_+ = \frac{\partial}{\partial\bar{\theta}} + \frac{i}{2}\partial_z\theta^+$$

$$\partial_{\bar{z}}, D_- = \frac{\partial}{\partial\theta^-} + \frac{i}{2}\partial_{\bar{z}}\bar{\theta}^-, \qquad \bar{D}_- = \frac{\partial}{\partial\bar{\theta}^-} + \frac{i}{2}\partial_{\bar{z}}\theta^- \qquad (23.83)$$

and they obey relations for which the only nonzero anticommutators are

$$\{D_+, \bar{D}_+\} = i\partial_z; \qquad \{D_-, \bar{D}_-\} = i\partial_{\bar{z}}. \tag{23.84}$$

The $(2, 2)$ superspace has zero curvature and the nonzero components of the torsion are

$$T_{+\oplus}{}^+ = i, \qquad T_{-\ominus}{}^= = i. \tag{23.85}$$

A general scalar superfield contains the component fields

$$\phi = A + i\theta^+\chi_+ + i\bar{\theta}^+\bar{\chi}_+ + i\theta^+\bar{\theta}^+ F_z + i\theta^-\chi_-$$
$$+ i\theta^-\bar{\chi}_- + i\theta^-\bar{\theta}^- F_{\bar{z}} + i\theta^-\theta^+ H + i\bar{\theta}^-\bar{\theta}^+ K + i\bar{\theta}^-\theta^+ L$$
$$+ i\theta^-\bar{\theta}^+ N + \theta^-\theta^+\bar{\theta}^+\zeta_+ + \theta^+\theta^-\bar{\theta}^-\zeta_- + \bar{\theta}^+\theta^-\bar{\theta}^-\bar{\zeta}_-$$
$$+ \bar{\theta}^-\theta^+\bar{\theta}^+\bar{\zeta}_+ + \theta^+\bar{\theta}^+\theta^-\bar{\theta}^- D. \tag{23.86}$$

One can place chiral constraints on ϕ in two ways; either

$$\bar{D}_+\phi = \bar{D}_-\phi = 0 \tag{23.87}$$

or

$$\bar{D}_+\phi = D_-\phi = 0 \tag{23.88}$$

We call the former conditions simply chiral while the latter are twisted chiral.[220]

Let us first consider the chiral case. A chiral superfield when written in a chiral basis has the form

$$\phi(y, \bar{y}, \theta^+, \theta^-) = A(y, \bar{y}) + i\theta^+\chi_+(y, \bar{y})$$
$$+ i\theta^-\chi_-(y, \bar{y}) + i\theta^+\theta^- F(y, \bar{y}). \tag{23.89}$$

where

$$y = z + \frac{i}{2}\theta^+\bar{\theta}^+$$
$$\bar{y} = z + \frac{i}{2}\theta^-\bar{\theta}^-. \tag{23.90}$$

Since $\phi\bar{\phi}$ is a general superfield, the free action is given by

$$-\int dz\, d\bar{z}\, d^2\theta\, d^2\bar{\theta}\, \phi\bar{\phi}, \tag{23.91}$$

where the full superspace integral is defined by

$$\int dz\, d\bar{z}\, d^2\theta\, d^2\bar{\theta} \equiv \int dz\, d\bar{z}\, D_+\bar{D}_+ D_-\bar{D}_-. \tag{23.92}$$

The most general action[220] for many fields $\phi^i(\bar{\phi}^i)$; $i = 1, \ldots n$ is given by

$$-\int dz\,d\bar{z}\,d^2\theta\,d^2\bar{\theta}\,K(\phi^i, \bar{\phi}^i) + \left\{ \int dz\,d\bar{z}\,d^2\theta\,g(\phi^i) + \text{h.c.} \right\} \qquad (23.93)$$

where K and g are arbitrary functions of ϕ^i and $\bar{\phi}^i$ and ϕ^i alone respectively.

To evaluate this action in components for the case $g_i = 0$ we first write it as

$$-\int dz\,d\bar{z}\,\bar{D}_+ D_- \{g_{i\bar{j}} D_+ \phi^i \bar{D}_- \bar{\phi}^{\bar{j}}\} = -\int dz\,d\bar{z}\,\bar{\nabla}_+ \nabla_- \{g_{i\bar{j}} \nabla_+ \phi^i \bar{\nabla}_- \bar{\phi}^{\bar{j}}\} \qquad (23.94)$$

where

$$g_{i\bar{j}} = \frac{\delta}{\delta\phi^i} \frac{\delta}{\delta\bar{\phi}^{\bar{j}}} K$$

and ∇_N is the connection with the usual Christoffel symbol. On T^i it acts as usual:

$$\nabla_N T^i = D_N T^i + D_N \phi^k \Gamma_{kh}{}^i T^h.$$

Evaluating the action further and using the fact that the covariant derivative of the metric vanishes we have

$$-\int dz\,d\bar{z}\,g_{i\bar{j}}\{\bar{\nabla}_+ \nabla_- \nabla_+ \phi^i \bar{\nabla}_- \bar{\phi}^{\bar{j}} + \nabla_- \nabla_+ \phi^i \nabla_+ \bar{\nabla}_- \bar{\phi}^{\bar{j}}$$

$$- \bar{\nabla}_+ \nabla_+ \phi^i \nabla_- \bar{\nabla}_- \bar{\phi}^{\bar{j}} + \nabla_+ \phi^i \bar{\nabla}_+ \nabla_- \bar{\nabla}_- \bar{\phi}^{\bar{j}}\}$$

$$= \int dz\,d\bar{z}\{-g_{i\bar{j}}\partial z A^i \partial_{\bar{z}} \bar{A}^{\bar{j}} - F^i \bar{F}^{\bar{j}} g_{i\bar{j}} + i\chi_+{}^i \nabla_{\bar{z}} \bar{\chi}_+{}^{\bar{j}} g_{i\bar{j}}$$

$$+ i\chi_-{}^i \nabla_z \bar{\chi}_-{}^{\bar{j}} g_{i\bar{j}} - R_{\bar{h}k\bar{j}l} \bar{\chi}_+{}^{\bar{h}} \chi_-{}^k \chi_+{}^l \bar{\chi}_-{}^{\bar{j}}\} \qquad (23.95)$$

where the covariant derivatives involve the usual Christoffel symbol constructed from the metric $g_{i\bar{j}}$ and $R_{\bar{h}k\bar{j}l}$ is the corresponding curvature.

We recognize it as a σ-model which has as target space a complex manifold whose metric is given by

$$g_{i\bar{j}} = \frac{\partial}{\partial A^i} \frac{\partial}{\partial \bar{A}^{\bar{j}}} K(A, \bar{A}), \qquad (23.96)$$

In other words, it is a Kähler manifold.

The superspace action of Eq. (23.93) has the local symmetry

$$K(\phi, \bar{\phi}) \rightarrow K(\phi, \bar{\phi}) + f(\phi) + \bar{f}(\bar{\phi}) \qquad (23.97)$$

where f and $\bar{f}$ are arbitrary analytic functions of ϕ and $\bar{\phi}$. In terms of components this symmetry implies the well-known symmetry of the Kähler potential

$$K(A, \bar{A}) \rightarrow K(A, \bar{A}) + f(A) + \bar{f}(\bar{A}). \qquad (23.98)$$

The reader who is having a feeling of '*déjà vu*' will realize that the $(2, 2)$ case is the direct dimensional reduction of $N = 1, D = 4$ superspace and the model above is the reduction of the $N = 1$ supersymmetric σ-model.[220]

Since Eq. (22.93) is the most general $(2, 2)$ superspace action constructed from chiral superfields we must conclude that any $(2, 2)$ chiral superfield model describes a σ-model with a Kähler manifold as target space.

One can also contemplate an action involving chiral ϕ and twisted chiral superfields Φ. This will involve the arbitrary function $K(\phi, \bar{\phi}, \Phi, \bar{\Phi})$. However, in this case one finds that the target manifold associated with the corresponding σ-model is in general not a Kähler manifold and in general it has torsion which is associated with the mixed terms involving chiral and twisted chiral fields in K.[221]

Essentially the metric is given by

$$g_{i\bar{j}} = \text{diag} \left\{ \frac{\partial^2 K}{\partial \phi^i \partial \bar{\phi}^j}, \frac{\partial^2 K}{\partial \Phi^i \partial \bar{\Phi}^j} \right\}$$

and the field strength of the field

$$b_{i\bar{j}} = \frac{\partial^2 K}{\partial \phi^i \partial \bar{\Phi}^j}$$

is proportional to the torsion.

The discussion at the end of $(2, 0)$ superspace can be carried over to the $(2, 2)$ case and we refer the reader to the reviews of Refs. 223 and 224 where this is more fully discussed.

Let us consider now a linear superfield which is a real superfield φ subject to the constraints

$$D_+ D_- \varphi = 0, \qquad \bar{D}_+ \bar{D}_- \varphi = 0.$$

We evaluate its component field content by the method explained in Chapter 15 and define the component fields to be the $\theta = 0$ components of the action of the covariant derivatives on φ

$$A \equiv \varphi, \qquad i\chi_+ \equiv D_+ \varphi, \qquad i\chi_- \equiv D_- \varphi, \qquad i\bar{\chi}_+ \equiv \bar{D}_+ \varphi$$

$$i\bar{\chi}_- = \bar{D}_- \varphi, \qquad \tfrac{1}{2}[D_+ \bar{D}_+]\varphi \equiv if_z, \qquad if_{\bar{z}} = \tfrac{1}{2}[D_-, \bar{D}_-]\varphi$$

$$iN \equiv D_+ \bar{D}_- \varphi, \qquad iM \equiv D_- \bar{D}_+ \varphi.$$

The above constraints imply that there are no further components as the action of further covariant derivatives yields space-time derivatives on the component fields we already have. One such example is provided by

$$D_+ \bar{D}_+ D_- \varphi = \frac{i}{2} \partial_z D_- \varphi = -\tfrac{1}{2} \partial_z \chi_-$$

at $\theta = 0$. Systematically exhausting the action of all possible covariant derivatives we find the equation

$$\frac{2}{i}\{\bar{D}_-, D_-\}[D_+, \bar{D}_+]\varphi = \frac{2}{i}(\bar{D}_- D_- D_+ \bar{D}_+ \varphi - D_- \bar{D}_- \bar{D}_- D_+ \varphi)$$

$$= \partial_{\bar{z}} D_+ \bar{D}_+ \varphi - \partial_z D_- \bar{D}_- \varphi$$

which at $\theta = 0$ implies the result

$$\partial_{\bar{z}} f_z + \partial_z f_{\bar{z}} = 0.$$

Locally this equation can be solved by

$$f_z = i\partial_z B, \qquad f_{\bar{z}} = -i\partial_{\bar{z}} B.$$

Thus the field component content of a linear multiplet is given by

$$(A, B; \chi_+, \chi_-, \bar{\chi}_+, \bar{\chi}_-; N, M)$$

where A, B, N and M are real and $(\chi_+)^* = \bar{\chi}_+$, $(\chi_-)^* = \bar{\chi}_-$. By examining the supersymmetry transformations of these fields and taking suitable complex combinations, one realizes that one can identify the linear multiplet with the twisted chiral multiplet provided one disguards the integration constant when determining B from f_z and $f_{\bar{z}}$.[221]

The free action for a linear multiplet is $\int dz\, d\bar{z}\, d^2\theta\, d^2\bar{\theta}\, \varphi^2$.

In a superconformal $(2, 2)$ theory the currents belong to a supermultiplet which in superspace is a vector superfield J_a subject to the constraint

$$(\gamma^a)_A{}^B D_B J_a = 0.$$

It contains a U(1) current, the four supercurrents and the energy-momentum tensor. The way these currents occur in J_a can be obtained by direct dimensional reduction of the $N = 1$, $D = 4$ case given in Chapter 20.

The action of Eq. (23.91) has the supermultiplet of currents

$$J_a = (\bar{D}^A \bar{\phi})(\gamma_a)_A{}^B D_B \phi.$$

23.2 Euclidean Superspace

The construction in the Euclidean case is almost identical to the Minkowski case considered previously. The major difference is that Majorana-Weyl spinors do not exist in Euclidean space and so we can only consider superalgebras with equal numbers of left and right generators, i.e. of (N, N) type. Nevertheless an (N, N) algebra is the direct sum of a (N, O) and a (O, N) superalgebra, the operation of complex conjugations taking us from one algebra to the other. As such, we may consider a (N, O) superalgebra and its associated superspace even in Euclidean space. When using complex conjugation the (N, O) superspace is taken to the (O, N) superspace in accord with the different way complex conjugation acts in the Euclidean and Minkowski cases. Consequently, the two corresponding superalgebras differ by the insertion or removal of various factors of i.

23.2a   (1, 1) Superspace

Consider the $(1, 1)$ superalgebra of Eq. (22.113) and Eq. (22.114), the corresponding $(1, 1)$ superspace has coordinates

$$z^\pi = (z, \theta^+; \bar{z}, \theta^-). \tag{23.99}$$

The superspace transformations on z, θ^+ are

$$z' = z - \varepsilon^+ \theta^+ + a - i\omega z$$

$$\theta'^+ = \theta^+ + \varepsilon^+ - \frac{i}{2}\omega\theta^+ \tag{23.100}$$

while those for $\bar{z}$, θ^- are obtained by taking $i \to -i$, $z \to \bar{z}$, $\theta^+ \to \theta^-$, $\varepsilon^+ \to \varepsilon^-$.
The supervielbein and its inverse are given by

$$E_\pi{}^M = \begin{pmatrix} 1 & 0 & 0 & 0 \\ -\theta^+ & 1 & 0 & 0 \\ 0 & 0 & 1 & 0 \\ 0 & 0 & -\theta^- & 1 \end{pmatrix} \tag{23.101}$$

$$E_M{}^\pi = \begin{pmatrix} 1 & 0 & 0 & 0 \\ \theta^+ & 1 & 0 & 0 \\ 0 & 0 & 1 & 0 \\ 0 & 0 & \theta^- & 1 \end{pmatrix}. \tag{23.102}$$

The supercovariant derivatives are

$$\partial_z, D_+ = \frac{\partial}{\partial\theta^+} + \partial_z\theta^+$$

$$\partial_{\bar{z}}, D_- = \frac{\partial}{\partial\theta^-} + \partial_{\bar{z}}\theta^- \tag{23.103}$$

and satisfy

$$D_+^2 = \partial_z, \qquad D_-^2 = \partial_{\bar{z}}; \qquad \{D_+, D_-\} = 0. \tag{23.104}$$

The general action for the fields ϕ^i is

$$\int dz\, d\bar{z}\, d\theta^+ \, d\theta^- \{(g_{ij} + b_{ij})D_+\phi^i D_-\phi^j + f(\phi)\}$$

$$\equiv \int dz\, d\bar{z}\, D_+ D_- \{(g_{ij} + b_{ij})D_+\phi^i D_-\phi^j + f(\phi)\} \qquad (23.105)$$

where $g_{ij} = g_{ji}$, $b_{ij} = -b_{ji}$ and f are arbitrary functions of ϕ^i.

23.2b (2, 2) Superspace

This superspace is the coset

$$\frac{(2, 2)\ \text{Supergroup}}{\text{Lorentz} \otimes U(1)_C \otimes U_R(1)}$$

We take the $(2, 2)$ supergroup in its complex formulation of Eq. (22.122). The superspace is parametrized by

$$z^\pi = (z, \theta^+, \bar{\theta}^+; \bar{z}, \theta^-, \bar{\theta}^-) \qquad (23.106)$$

and the supersymmetry transformations on the left-handed coordinates are

$$z' = z - \tfrac{1}{2}\varepsilon^+\bar{\theta}^+ - \tfrac{1}{2}\bar{\varepsilon}^+\theta^+ - i\omega z$$

$$\theta^{+\prime} = \theta^+ + \varepsilon^+ - il\theta^+ - \frac{i\omega}{2}\theta^+, \qquad \bar{\theta}^{+\prime} = \bar{\theta}^+ + \bar{\varepsilon}^+ + il\bar{\theta}^+ - \frac{i\omega}{2}\bar{\theta}^+. \qquad (23.107)$$

The transformations of the right-handed coordinates are obtained by taking

$$z \to \bar{z}, \qquad \theta^+ \to \theta^-, \qquad \bar{\theta}^+ \to \bar{\theta}^-, \qquad \varepsilon^+ \to \varepsilon^-,$$

$$\bar{\varepsilon}^+ \to \bar{\varepsilon}^-, \qquad l \to -r \qquad \text{and} \qquad i \to -i. \qquad (23.108)$$

The supervielbein and its inverse are

$$E_\pi{}^M = \begin{pmatrix} \begin{matrix} 1 & 0 & 0 \\ -\tfrac{1}{2}\bar{\theta}^+ & 1 & 0 \\ -\tfrac{1}{2}\theta^+ & 0 & 1 \end{matrix} & \mathbf{0} \\ \mathbf{0} & \begin{matrix} 1 & 0 & 0 \\ -\tfrac{1}{2}\bar{\theta}^- & 1 & 0 \\ -\tfrac{1}{2}\theta^- & 0 & 1 \end{matrix} \end{pmatrix} \qquad (23.109)$$

$$E_\pi{}^M = \begin{pmatrix} \begin{matrix} 1 & 0 & 0 \\ \tfrac{1}{2}\bar{\theta}^+ & 1 & 0 \\ \tfrac{1}{2}\theta^+ & 0 & 1 \end{matrix} & \mathbf{0} \\ \mathbf{0} & \begin{matrix} 1 & 0 & 0 \\ \tfrac{1}{2}\bar{\theta}^- & 1 & 0 \\ \tfrac{1}{2}\theta^- & 0 & 1 \end{matrix} \end{pmatrix}. \qquad (23.110)$$

The corresponding covariant derivatives become

$$\partial_z, D_+ = \frac{\partial}{\partial\theta^+} + \frac{1}{2}\partial_z\bar\theta^+, \qquad \bar D_+ = \frac{\partial}{\partial\bar\theta^+} + \frac{1}{2}\partial_z\theta^+ \qquad (23.111)$$

and

$$\partial_{\bar z}, D_- = \frac{\partial}{\partial\theta^-} + \frac{1}{2}\partial_{\bar z}\bar\theta^-, \qquad \bar D_- = \frac{\partial}{\partial\bar\theta^-} + \frac{1}{2}\partial_{\bar z}\theta^- \qquad (23.112)$$

All anticommutators vanish except for

$$\{D_+, \bar D_+\} = \partial_z, \qquad \{D_-, \bar D_-\} = \partial_{\bar z}. \qquad (23.113)$$

We may introduce chiral superfields

$$\bar D_+ \varphi = 0 = \bar D_- \varphi$$

where φ is a complex field. Taking the complex conjugate yields the conditions on $\bar\varphi$ to be respectively

$$D_- \bar\varphi = 0 = D_+ \bar\varphi.$$

The most general action involving chiral fields φ^i, $\bar\varphi^i$, $i = 1, \ldots, n$ is

$$\int dz\, d\bar z\, d^2\theta\, d^2\bar\theta K(\varphi^i, \bar\varphi^i) + \left\{\int dz\, d\bar z\, d^2\theta\, g(\varphi^i) + \text{h.c.}\right\} \qquad (23.114)$$

where K and g are arbitrary functions of φ^i and $\bar\varphi^i$ and φ^i alone respectively. The target space for this σ-model is Kähler.

The twisted chiral superfields do not generalize to Euclidean superspace if one wishes to also have the usual operation of complex conjugation. If we demand that φ satisfy

$$\bar D_+ \varphi = 0, \qquad D_- \varphi = 0, \qquad (23.115)$$

then complex conjugation implies that

$$D_- \bar\varphi = 0 \qquad \text{and} \qquad \bar D_+ \bar\varphi = 0 \qquad (23.116)$$

respectively. Consequently, φ and $\bar\varphi$ satisfy the same constraints and in particular so does $\varphi\bar\varphi$. As a result, the usual kinetic term

$$\int dz\, d\bar z\, d^2\theta\, d^2\bar\theta\, \varphi\bar\varphi \qquad (23.117)$$

vanishes.

Chapter 24

Superspace Formulations of Two-Dimensional Supergravities

For each (N, M) superalgebra there exists a corresponding (N, M) super-gravity which contains the graviton, Majorana gravitinos, N of which are left-handed and M of which are right-handed, as well as other fields of lower spin. It is important to remember that gravity in two-dimensions does not propagate. Indeed the only coordinate invariant action which is bilinear in the derivatives is Einstein's action and this is a topological invariant, the Euler invariant. By supersymmetry, it follows that the gravitinos and all the other members of the supergravity multiplet also lead to no propagating on-shell states. The free gravitino action in an arbitrary dimension is

$$\bar{\psi}_\mu \gamma^{\mu\nu\rho} \partial_\nu \psi_\rho = \bar{\psi}_\mu \gamma^{[\mu} \gamma^\nu \gamma^{\rho]} \partial_\nu \psi_\rho$$

but this vanishes in two-dimensions.

Despite this the two-dimensional supergravity multiplets are of consider-able importance in view of the role they play in string theory, namely the coordinates of the superstring string are coupled in the action to the super-gravity multiplet in a superconformally invariant manner. Since the super-gravity fields have no kinetic terms their variation leads to a supermultiplet of constraints. These constraints play an important role in string theory for upon quantization they describe the infinite number of space-time particles of the string.

In Chapter 21 we considered $(1, 0)$ and $(1, 1)$ supergravities in components and give their transformation rules and coupling to matter multiplets. In this chapter we give their superspace formulations.

24.1 Geometrical Framework

To each of the (N, M) superalgebras of Chapter 22, there corresponds a rigid superspace which was given in Chapter 23. Our aim here is to construct the corresponding local superspaces. The (N, M) local superspace contains the (N, M) rigid superspace as a particular solution and it has two bosonic coordinates x^δ and $N + M$ anticommuting coordinates. In null plane notation these are the real bosonic coordinates z, $\bar{z}$, and the real anticommuting coordinates $\theta^{+i}, i = 1, \ldots, N$ and $\theta^{-j'}, j' = 1, \ldots, M$. We denote the coordinates by

$$z^\pi = \{z, \bar{z}; \theta^{+i} i = 1, \ldots, N; \theta^{-j'} j' = 1, \ldots, M\}. \qquad (24.1)$$

Our aim is to construct a theory invariant under super-general coordinate transformations

$$z^\pi \rightarrow z^{\pi'} = z^\pi + \xi^\pi \qquad (24.2)$$

whre ξ^π is an arbitrary function of z^π.

The tangent group associated with (N, M) local superspace is

$$SO(1, 1) \cdots \times SO(N) \times SO(M). \qquad (24.3)$$

This choice coincides with the isotropy group of the coset corresponding to rigid (N, M) superspace. The $SO(1, 1)$ is the two-dimensional Lorentz group while the $SO(N)$ and $SO(M)$ are the internal groups associated with the left- and right-parts of (N, M) superalgebra.

A tangent vector V^N has the components

$$V^N = (V^+, V^=, V^{+i}, V^{-j'}) \qquad (24.4)$$

where $+$ and $=$ correspond to the analogue of the $z, \bar{z}$ directions, but in the tangent space. We will systematically use $+$ and $=$ for tangent indices, but z and $\bar{z}$ to indicate the corresponding world indices. Under the Lorentz group V^N transforms as

$$\delta V^N = -V^M \Lambda_M{}^N = (\Lambda \cdot J) V^N$$

$$\delta V_N = \Lambda_N{}^M V_M = (\Lambda \cdot J) V_N \qquad (24.5)$$

where J is the generator of the Lorentz group and Λ is an arbitrary function of z^π. We caution the reader that we have a minus sign up on our analogous definition in Chapter 16. On V^N the generator J has the representation

$$JV^+ = V^+, \qquad JV^= = -V^=$$

$$JV^{+i} = \tfrac{1}{2} V^{+i}, \qquad JV^{-j'} = -\tfrac{1}{2} V^{-j'} \qquad (24.6)$$

or equivalently

$$\Lambda_M{}^N = -\Lambda \begin{pmatrix} 1 & 0 & 0 & 0 \\ 0 & -1 & 0 & 0 \\ \hline 0 & & 1/2 & 0 \\ \hline 0 & & 0 & -1/2 \end{pmatrix} \qquad (24.7)$$

where 1 is the unit matrix. Clearly V^N carries a highly reducible representation of the Lorentz group.

Under $SO(N)$ transformations

$$\delta V^N = (l^r L_r) V^N = l_r V^M (L_r)_M{}^N \tag{24.8}$$

where l^r is an arbitrary function of z^π and L_r is the generator of SO(N) under which all components of V^N are inert except for

$$\delta V^{+i} = (l^r L_r) V^{+i} \equiv V^{+j} (U_r)^i{}_j l_r \tag{24.9}$$

where $(U_r)^i{}_j$ are the real antisymmetric matrices of Eq. (22.99).

Similarly under SO(M) transformations

$$\delta V^N = (r^s R_s) V^N \equiv r^s V^M (R_s)_M{}^N \tag{24.10}$$

where r^s is an arbitrary function of z^π and R_s are the generators of SO(M) under which all components of V_N are inert except for

$$\delta V^{-j'} = (r^s R_s) V^{-j'} = +r^s V^{-j'} (V_s)^{i'}{}_{j'} \tag{24.11}$$

where $(V_s)^{i'}{}_{j'}$ are the real antisymmetric matrices of Chapter 22. The indices associated with SO(N) and SO(M) run, in general over different ranges and are distinct. However, we will take this to be understood and not continue to put primes on the indices associated with SO(M).

The discussion of the geometry local two-dimensional superspace closely parallels that for four-dimensions given in Chapter 16.1. We introduce super-vielbein's E_π^N which transform under supergeneral coordinate transformations as

$$\delta E_\pi{}^N = \xi^A \partial_A E_\pi{}^N + \partial_\pi \xi^A E_A^N. \tag{24.12a}$$

While under the Lorentz group the tangent transformations are

$$\delta E_\pi{}^N = -E_\pi{}^M \Lambda_M{}^N \tag{24.12b}$$

and like V^N in Eqs. (24.9) and (24.11) under SO(N) and SO(M).

The inverse supervielbein is given defined by

$$E_N{}^A E_A{}^M = \delta_N^M. \tag{24.13}$$

Corresponding to each tangent group generator we introduce a gauge field. The more familiar spin connection associated with Lorentz transformations is

$$\Omega_{\pi N}{}^M = \Omega_\pi \cdot J.$$

It acts on tangent indices like a Lorentz parameter, for example,

$$\Omega_\pi \cdot J V^+ = \Omega_\pi V^+, \qquad \Omega_\pi J V^{+i} = \tfrac{1}{2} \Omega_\pi V^{+i}$$

$$\Omega_\pi \cdot J V^{-j} = -\tfrac{1}{2} \Omega_\pi V^{-j}, \qquad (\Omega_\pi \cdot J) V_\pm{}^i = \mp \Omega_\pi V_\pm{}^i \text{ etc.} \tag{24.14}$$

For SO(N) and SO(M) we have the gauge fields A_π^r and B_π^s. On this world index π, Ω_π, A_π^r and B_π^S are world tensors, i.e.

$$\delta\Omega_\pi = \xi^A \partial_A \Omega_\pi + \partial_\pi \xi^A \Omega_A \qquad (24.15)$$

but under tangent rotations they are true gauge fields, i.e.

$$\delta\Omega_\pi = -\partial_\pi \Lambda$$

$$\delta A_\pi{}^r = -\partial_\pi l^r + f_L{}^{rst} l_s A_\pi{}^t$$

$$\delta B_\pi{}^s = -\partial_\pi r^s + f_R{}^{stv} r_t B_\pi{}^v \qquad (24.16)$$

where f_L^{rst} and f_R^{str} are the structure constants of $SO(N)$ and $SO(M)$ repectively.

Using the above connections we can define the covariant derivative

$$\mathcal{D}_A = \partial_A + \Omega_A \cdot J + A_A^t L_t + B_A^s R_s \qquad (24.17)$$

which when referred to the supervielbein frame is of the form

$$\mathcal{D}_N = E_N{}^A \mathcal{D}_A. \qquad (24.18)$$

The covariant derivatives define the torsion, curvature and $SO(N)$ and $SO(M)$ field strengths through the relation

$$[\mathcal{D}_N, \mathcal{D}_M\} = T_{NM}{}^R \mathcal{D}_R + R_{NM} \cdot J + F_{NM}^{(L)\,t} \cdot L_t + F_{NM}^{(R)s} R_s. \qquad (24.19)$$

The identity

$$[\mathcal{D}_M, [\mathcal{D}_N, \mathcal{D}_R\}\} - [[\mathcal{D}_M, \mathcal{D}_N\}, \mathcal{D}_R\} + (-1)^{RN}[[\mathcal{D}_M, \mathcal{D}_R\}, \mathcal{D}_N\} = 0 \qquad (24.20)$$

implies Bianchi identities for the torsion, curvature and $SO(N)$ and $SO(M)$ field strengths. In the absence of the latter these identities are those of Eqs. (16.21) and (16.22).

A convenient and compact way to write the finite transformations of the supervielbein and gauge connections is the expression

$$\mathcal{D}_N' = e^{-K} \mathcal{D}_N e^K$$

where

$$K = \xi^\pi \partial_\pi - \Lambda \cdot J - l^r L_r - r^s R_s.$$

As for the case of four-dimensional supergravity, and unlike Einstein's general relativity, the above geometrical framework is not sufficient to specify the local superspace and one requires in addition superspace constraints. Since the constraints must be supergeneral coordinate invariant, they must be given in terms of the torsion, curvature and $SO(N)$ and $SO(M)$ field strengths. The constraints must also be invariant under tangent group transformations. As the above tensors carry a highly reducible representation of the tangent group, it is possible to impose that only parts of one of the above tensors vanish or be equal to a numerical invariant where such an object exists. However, there is still considerable ambiguity in how to choose a good set of constraints. The ultimate test for a set of constraints is that they lead to a correct set of supergravity fields in x-space. This criterion is rather remote from the initial starting point, but to find a suitable set, we may implement the program of Ref. 57. Therefore we are to impose conventional and representation preserving constraints and, if we wish to describe supergravity rather than conformal supergravity, a further set of conformal choice constraints. We refer the reader to Chapter 16 for a general discussion of this procedure albeit in the context of $N = 1, D = 4$ local superspace.

Having found the required superspace constraints one must then analyze them by, for example, solving the Bianchi identities. This allows us to find the number of independent components of the torsions and curvatures. By an extension of Dragons theorem the supercurvature is always expressible in terms of the supertorsions and so it suffices to consider only the torsions.

The next step is to deduce the x-space component fields contained in the local superspace. We proceed by making a gauge choice on the supervielbein and labelling the remaining x-space fields. In terms of the latter we may calculate the remaining nonzero torsions and curvatures. Finally, we calculate the supersymmetry variations of the x-space component fields from the variation under super-diffeomorphisms of the supervielbein.

The above program parallels closely that given in Chapter 16, where it is explained in more detail. In two-dimensions, however, one has fewer components and therefore the technical complications are much less. Indeed $(1,0)$ local superspace is a kindergarten exercise that can be completed in a few hours and illustrates well the principles of Chapter 16. As such, we will give this $(1,0)$ case in considerable detail and treat the other two-dimensional supergravities in less detail. We will also discuss the coupling of supergravity to matter and give the superspace actions for the heterotic string and older superstring.

The first two-dimensional supergravity to be worked out in superspace was $(1,1)$ local superspace.[225] The discussion of Ref. 225 contains the superspace constraints, the subsequent analysis of the Bianchi identities, and the identification of the x-space component field.

The analogous discussion of the $(1,0)$ case is found from Ref. 225 by demanding that the anticommuting coordinates θ^A be Weyl; θ^A decomposes into θ^+ and θ^- and we therefore set $\theta^- = 0$.

24.2 $(1,0)$ Supergravity

24.2a The constraints

The local $(1,0)$ superspace has the coordinates

$$z^\pi = (z, \theta^+; \bar{z}). \tag{24.21}$$

In what follows we will often drop the $+$ on θ^+ and simply write θ. The tangent group is just the Lorentz group. The supervielbein $E_\pi{}^N$ has the components

$$(E_\pi{}^+, E_\pi{}^=, E_\pi{}^+) \tag{24.22}$$

and we will denote the values of the world index π by $z, \bar{z}$ and θ i.e.

$$E_\pi{}^N = (E_z{}^N, E_{\bar{z}}{}^N, E_\theta{}^N).$$

We now consider what constraints to impose on the torsion and curvature in order that the local superspace describes supergravity. In Chapter 23 we found that, given a general superfield $\phi(z^\pi)$ of rigid $(1,0)$ supersymmetry, there are no covariant constraints we can impose without setting ϕ on-shell. Consequently, there are no representations whose existence in local superspace requires constraints and so there are no representation preserving constraints.

The conventional constraints oblige us to solve for as many components of the supervielbein and spin-connection as possible. The torsion tensor has the form

$$T_{NM}{}^R = E_M{}^A \partial_A E_N{}^\pi E_\pi{}^R + \Omega_{MN}{}^R - (-1)^{MN} (M \leftrightarrow N). \tag{24.23}$$

The simplest to solve for is the spin-connection which has the three components Ω_+, $\Omega_=$ and Ω_+. We may solve for these in terms of the supervielbein by setting respectively

$$T_{+=}{}^+ = 0, \qquad T_{+=}{}^= = 0 \qquad \text{and} \qquad T_{+=}{}^=. \tag{24.24}$$

To discover which components of the supervielbein we may solve for, in terms of the other components, it is instructive to analyze the problem at the linearized level. We define

$$E_\pi{}^N = \mathring{E}_\pi{}^N - H_M{}^N \mathring{E}_\pi{}^M \tag{24.25}$$

where $\mathring{E}_\pi{}^N$ is the supervielbein of rigid $(1,0)$ superspace of Eq. (23.12). To order H^1 the inverse supervielbein is given by

$$E_N{}^\pi = \mathring{E}_N{}^\pi + H_N{}^M \mathring{E}_M{}^\pi \tag{24.26}$$

and hence the covariant derivature at this order becomes

$$D_N = (\delta_N{}^M + H_N{}^M) D_M + \Omega_N \cdot J \tag{24.27}$$

where D is the rigid superspace covariant derivative. Evaluating the (anti) commutator of two covariant derivatives we find that to lowest order in

$$H_M{}^N \qquad \text{and} \qquad \Omega_N = \mathring{E}_N{}^\pi \Omega_\pi$$

the torsion tensor is given by

$$
\begin{aligned}
T_{NM}{}^R = {} & t_{NM}{}^R + D_N H_M{}^R - (-1)^{NM} D_M H_N{}^R \\
& - t_{NM}{}^P H_P{}^R + H_N{}^S t_{SM}{}^R - (-1)^{NM} H_M{}^P t_{PN}{}^R \\
& + \Omega_{NM}{}^R - (-1)^{NM} \Omega_{MN}{}^R
\end{aligned} \tag{24.28}
$$

where $t_{NM}{}^R$ is the torsion of rigid superspace. We recall from Chapter 23 that $t_{NM}{}^R$ has all its components vanishing except for $t_{++}{}^+ = 2i$.

The two torsion tensors of dimension-zero are $T_{++}{}^+$ and $T_{++}{}^=$ for which

$$T_{++}{}^+ = 2i + 2D_+H_+{}^+ - 2iH_+{}^+ + 4iH_+{}^+ \qquad (24.29)$$

$$T_{++}{}^= = 2D_+H_+{}^= - 2iH_+{}^=. \qquad (24.30)$$

Setting

$$T_{++}{}^+ = 2i, \qquad T_{++}{}^= = 0 \qquad (24.31)$$

allows us to solve for $H_+{}^=$ and $H_+{}^+ - 2H_+{}^+$ in terms of $H_+{}^+$ and $H_+{}^=$.

For the dimension-1/2 torsions we have the relations

$$T_{++}{}^+ = D_+H_+{}^+ - D_+H_+{}^+ - H_+{}^+ \cdot 2i - \Omega_+$$

$$T_{++}{}^= = D_+H_+{}^= - D_+H_+{}^=$$

$$T_{+=}{}^+ = D_+H_=^+ - D_=H_+{}^+ - H_=^+ \cdot 2i$$

$$T_{+=}{}^= = D_+H_=^= - D_=H_+{}^= + \phi_+$$

$$T_{++}{}^+ = 2D_+H_+{}^+ - 2iH_=^+ - \tfrac{1}{2}\Omega_+ \qquad (24.32)$$

We have already used $T_{+=}{}^=$ to solve for Ω_+ and $T_{++}{}^=$ cannot be used to solve for any field. Setting

$$T_{+=}{}^+ = 0 \qquad \text{and} \qquad T_{++}{}^+ \qquad (24.33)$$

we can solve for $H_=^+$ and $H_+{}^+$. At dimension-one and higher, we find that the only fields we may solve for are the spin connections Ω_+ and Ω_- by the former two constraints in Eq. (24.24). Clearly, the dimension of the constraint and the field it solves for must coincide, and since these are the only dimension one-fields, we cannot adopt any further constraints.

The superfields we have not solved are

$$H_+{}^+ + 2H_+{}^+, \qquad H_-{}^=, \qquad H_-{}^+ \text{ and } H_=^+, \qquad H_+{}^+. \qquad (24.34)$$

Should we want to make a gauge choice, we have at our disposal four local superparameters, three for supergeneral coordinate transformations, and one for the local Lorentz transformaton. Hence of the above five remaining fields we could expect that it is possible to remove all but one. Should we in addition use super Weyl invariance, we could use its superparameter to remove even this remaining field.

This result is the sypersymmetric extension of the well known fact that locally, using reparametrization invariance, the metric on a two dimensional space can be brought to the form $\eta_{ab}e^{\Omega}$ and further using Weyl symmetry we may set $\Omega = 1$.

Although more complicated, it is evident that one can also, using the above constraints, solve for the same components of the supervielbein and spin-connection at the nonlinear level. We now summarize the constraints we have adopted[225]

$$\text{dim } 0 \quad T_{++}{}^{+} = 2i, \qquad T_{++}{}^{=} = 0$$

$$\text{dim } \tfrac{1}{2} \quad T_{+=}{}^{=} = T_{+=}{}^{+} = 0 = T_{++}{}^{+}$$

$$\text{dim } 1 \quad T_{+=}{}^{=} = 0 = T_{=+}{}^{+}. \tag{24.35}$$

24.2b Solving the Bianchi identities

This course of action involves analyzing the consequences of the constraints of Eq. (24.35) for the Bianchi identities, to find the independent components of the torsions and curvatures. From the above count at the order $(H)^1$ level we may expect that all the torsion and curvature tensors should be expressible in terms of one superfield.

The Bianchi identities are the same as those in Eqs. (16.23)–(16.25). Considering $I_{+++}{}^{+}$ and $I_{+++}{}^{=}$ we find respectively that

$$T_{++}{}^{+} = 0 \quad \text{and} \quad T_{++}{}^{=} = 0. \tag{24.36}$$

Consequently, all dimension-1/2 torsions vanish. The reader may have noticed that, when adopting the dimension-1/2 conventional constraints there was a choice. However, this choice is not in effect there, as no matter what dimension-1/2 torsions we constrain to vanish, the remainder will be set to zero by the Bianchi identities.

The dimension-one Bianchi identities upon substituting the constraints become:

$$T_{+++}{}^{+} = -\frac{3}{2}R_{++} + 6iT_{++}{}^{+} = 0$$

$$I_{+++}{}^{+} = -R_{++} + 2iT_{++}{}^{+} = 0$$

$$I_{++=}{}^{+} = 4iT_{=+}{}^{+} = 0$$

$$I_{++=}{}^{=} = +R_{++} = 0; \qquad I_{+++}{}^{=} = 0 = 0 \tag{24.37}$$

and so

$$R_{++} = 0 = T_{++}{}^{+} = T_{=+}{}^{+}. \tag{24.38}$$

Hence all dimension-one tensors vanish.

Proceeding to the dimension 3/2 Bianchi identities we have

$$I_{+++}{}^+ = -R_{++} = 0$$

$$I_{+=+}{}^+ = -2iT_{+=}{}^+ + R_{=+} = 0$$

$$I_{++=}{}^+ = -R_{=+} + 2iT_{+=}{}^+ = 0$$

$$I_{++=}{}^= = R_{++} = 0 \tag{24.39}$$

and we conclude that

$$R_{++} = 0, \qquad 2iT_{+=}{}^+ = R_{=+}. \tag{24.40}$$

The only nontrivial dimension-two Bianchi is

$$I_{+=+}{}^+ = -D_+T_{+=}{}^+ - \frac{1}{2}R_{+=} = 0. \tag{24.41}$$

The dimension 5/2 Bianchi identity is trivially zero and so we complete our analysis of the Bianchi identity. The result is that all torsions and curvatures vanish except for

$$T_{+=}{}^+, R_{=+} \qquad \text{and} \qquad R_{+=} \tag{24.42}$$

and of these only $T_{+=}{}^+$ is an independent superfield since

$$R_{=+} = 2iT_{+=}{}^+, \qquad R_{+=} = -2D_+T_{+=}{}^+. \tag{24.43}$$

The result is to be expected from a knowledge of the component fields. The supergravity multiplet for $(1,0)$ has only the vielbein e_α^a and the gravitino ψ_α^+ as component fields. Therefore the lowest dimension nontrivial gauge co-variant object is the curl of the gravitino, i.e.

$$\partial_z\psi_{\bar{z}}{}^+ - \partial_{\bar{z}}\psi_z{}^+$$

which has the tangent index structure of $T_{+=}{}^+$ and is of dimension 3/2.

24.2c Recovery of the x-space fields

Given the superspace formalism above, we now show how to recover the x-space component results. This discussion closely parallels that of Chapter 16. The variation of the E_θ^N and Ω_θ begin with the terms

$$E_\theta^N = \partial_\theta\zeta^\pi \hat{E}_\pi{}^N + \cdots$$

$$\Omega_\theta = -\partial_\theta\Lambda + \cdots. \tag{24.44}$$

We may threfore choose the order θ terms of ζ^π and Λ to set the $\theta = 0$ components of E_θ^N and Ω_θ to be their values for rigid superspace:

$$\Omega_\theta(\theta = 0) = 0, \qquad E_\theta^+(\theta = 0) = 1, \qquad E_\theta^+(\theta = 0) = 0 = E_\theta^-(\theta = 0). \tag{24.45}$$

Examining the transformations of

$$E_z{}^N, E_{\bar{z}}{}^N \quad \text{and} \quad \Omega_z{}^N$$

under ξ^π and Λ at $\theta = 0$ suggests that we make the identifications

$$E_z{}^a(\theta = 0) = e_z{}^a, \qquad E_{\bar{z}}{}^a(\theta = 0) = e_{\bar{z}}{}^a$$

$$E_z{}^+(\theta = 0) = \tfrac{1}{2}\psi_z{}^+, \qquad E_{\bar{z}}{}^+(\theta = 0) = \tfrac{1}{2}\psi_{\bar{z}}{}^+ \tag{24.46}$$

where a takes the values $+$ and $=$ and

$$\Omega_z(\theta = 0) = \omega_z, \qquad \Omega_{\bar{z}}(\theta = 0) = \omega_{\bar{z}}.$$

At this stage however this identification can be taken to be just a matter of labelling.

To summarize, the supervielbein contains the component fields

$$E_\Lambda^N(\theta = 0) = \begin{pmatrix} e_z^+ & e_z^= & \tfrac{1}{2}\psi_z{}^+ \\ e_{\bar{z}}^+ & e_{\bar{z}}^= & \tfrac{1}{2}\psi_{\bar{z}}{}^+ \\ 0 & 0 & 1 \end{pmatrix}. \tag{24.47}$$

In principle the supervielbein could contain new component fields at order θ. However as we will now discover, these higher components are determined in terms of the above identification by the torsion constraints. The inverse supervielbein at $\theta = 0$ is of the form

$$E_N{}^\Lambda(\theta = 0) = \begin{pmatrix} e_+^z & e_+^{\bar{z}} & -\tfrac{1}{2}\psi_+{}^+ \\ e_=^z & e_=^{\bar{z}} & -\tfrac{1}{2}\psi_={}^+ \\ 0 & 0 & 1 \end{pmatrix} \tag{24.48}$$

where

$$\psi_+{}^+ = e_+{}^a\psi_a{}^+ = e_+{}^z\psi_z{}^+ + e_+{}^{\bar{z}}\psi_{\bar{z}}{}^+$$

$$\psi_={}^+ = e_=^a\psi_a{}^+ = e_=^z\psi_z{}^+ + e_=^{\bar{z}}\psi_{\bar{z}}{}^+. \tag{24.49}$$

We therefore evaluate at $\theta = 0$ all the torsions which are most conveniently taken to be in the form

$$T_{NM}{}^R = -E_M{}^{\Lambda}E_N{}^\pi\partial_\Lambda E_\pi{}^R(-1)^{(N+\pi)\Lambda} + \Omega_{MN}{}^R - (-1)^{NM}(N \leftrightarrow M). \tag{24.50}$$

We find at $\theta = 0$

dim 0

$$T_{++}{}^= = -2\partial_\theta E_\theta{}^= = 0; \qquad T_{++}{}^+ = 2i = -2\partial_\theta E_\theta{}^+ \tag{24.51}$$

dim $\tfrac{1}{2}$

$$T_{++}{}^+ = -2\partial_\theta E_\theta^+ = 0$$

$$T_{+a}{}^+ = -e_a^z\partial_\theta E_z^+ - e_a^{\bar{z}}\partial_\theta E_{\bar{z}}{}^+ + \tfrac{1}{2}\Omega_a = 0$$

$$T_{+a}{}^{=} = -e_a^z \partial_\theta E_z{}^{=} - e_a^{\bar z} \partial_\theta E_{\bar z}{}^{=} = 0$$

$$T_{+a}{}^{+} = +i\psi_a^{+} - e_a^z \partial_\theta E_z{}^{+} - e_a^{\bar z} \partial_\theta E_{\bar z}{}^{+}. \tag{24.52}$$

where a is $+$ or $=$

Summarizing the above results, the full expressions for the supervielbein are

$$E_\pi{}^N = \begin{pmatrix} e_z{}^+ & e_z{}^= & \frac{1}{2}\psi_z{}^+ \\ e_{\bar z}{}^+ & e_{\bar z}{}^= & \frac{1}{2}\psi_{\bar z}{}^+ \\ 0 & 0 & 1 \end{pmatrix} + \theta^+ \begin{pmatrix} i\psi_z & 0 & \frac{1}{2}\omega_z \\ i\psi_{\bar z} & 0 & \frac{1}{2}\omega_{\bar z} \\ -i & 0 & 0 \end{pmatrix}. \tag{24.53}$$

The spin connections Ω_+, $\Omega_=$, and Ω_+ have been solved for in terms of E_π^N by setting to zero the torsions

$$T_{+=}{}^{=}, T_{=+}{}^{+} \quad \text{and} \quad T_{+=}{}^{=}.$$

Consequently, we can use Eq. (24.53) to compute these spin connections. At $\theta = 0$, Ω_+ was chosen by gauge choice to vanish. One finds that $\Omega_z(\theta = 0) = -\omega_z$ and $\Omega_{\bar z}(\theta = 0) = -\omega_{\bar z}$ are the usual expressions for the spin connection in terms of the vielbein and gravitino (see Chapter 21). The full expressions are

$$\Omega_z = \omega_z + \frac{i}{2}\theta\psi_z^+$$

$$\Omega_{\bar z} = \omega_{\bar z} + \frac{i}{2}\theta\psi_{\bar z}^+, \qquad \Omega_\theta = 0. \tag{24.54}$$

The results is Eqs. (24.53) and (24.54) were given in Ref. 224 for $(1, 1)$ supergravity, $(1,0)$ supergravity is obtained by setting $\theta^- = 0$.

It finally remains to identify $T_{+=}{}^+$ in terms of the component fields. Following Eqs. (16.133)–(16.135) we find that

$$T_{z\bar z}{}^+(\theta = 0) = -\frac{1}{2}\partial_z\psi_{\bar z}{}^+ + \Omega_{z\bar z}{}^+ - (z \leftrightarrow \bar z)$$

$$= \frac{1}{2}(-D_z\psi_{\bar z}{}^+ + D_{\bar z}\psi_z{}^+) \equiv -\frac{1}{2}\psi_{z\bar z}{}^+ \tag{24.55}$$

where

$$D_z\psi_{\bar z} = (\partial_z + \frac{1}{2}\omega_z)\psi_{\bar z}. \tag{24.56}$$

Converting the world indices on $T_{z\bar z}{}^+$ to tangent indices we have

$$T_{+=}{}^+(\theta = 0) = -\frac{1}{2}e_+{}^z e_={}^{\bar z}\psi_{z\bar z}{}^+. \tag{24.57}$$

Repeating the exercise on the curvature we find that

$$R_{z\bar z}(\theta = 0) = \partial_z\omega_{\bar z} - \partial_{\bar z}\omega_z \tag{24.58}$$

and converting to flat indices we establish the result

$$R_{z\bar z}(\theta = 0) = e_z{}^a e_{\bar z}{}^b R_{ab}(\theta = 0) \tag{24.59}$$

or

$$R_{+=}(\theta = 0) = e_+{}^z e_-{}^{\bar z} R_{z\bar z}(\theta = 0).$$

To make complete contact with the component fields of $(1,0)$ supergravity given in Chapter 21, we deduce the transformation rules of $e_z{}^a$, $e_{\bar z}{}^a$, $\psi_z{}^+$ and $\psi_{\bar z}{}^+$ from the above local superspace. The discussion follows closely that given for the case of four-dimensional supergravity at the end of Chapter 16. We first label at $\theta = 0$ the components of the 'flattened' super-differmorphism parameter, $\xi^N = E^N{}_\pi \xi^\pi$

$$\xi^a(\theta = 0) = \zeta^a(z, \bar z); \qquad \xi^+(\theta = 0) = \varepsilon^+(z, \bar z). \tag{24.60}$$

A super-differmorphism of the supervielbein in terms of ξ^N is given by

$$\delta E_\pi{}^M = D_\pi \xi^M - E_\pi{}^R \xi^N T_{NR}{}^M \tag{24.61}$$

where

$$D_\pi \xi^M = \partial_\pi \xi^M + \Omega_\pi{}^M{}_N \xi^N. \tag{24.62}$$

Taking $\theta = 0$ of this equation, using the labelling of Eq. (24.46) and the constraints on the torsion we find

$$\delta e_z{}^+ = i\varepsilon^+ \psi_z{}^+, \qquad \delta e_{\bar z}{}^+ = i\varepsilon^+ \psi_{\bar z}{}^+$$

$$\delta e_z{}^= = 0 = \delta e_{\bar z}{}^=$$

$$\delta \psi_z^+ = 2D_z \varepsilon^+, \qquad \delta \psi_{\bar z}^+ = 2D_{\bar z}\varepsilon^+ \tag{24.63}$$

24.2d Local Matter Multiplets

The matter multiplet of $(1,0)$ supersymmetry is a general superfield ϕ. We may define in analogue with the rigid superspace limit its two-components to be the $\theta = 0$ parts of

$$A \equiv \phi(\theta = 0), \qquad i\chi_+ \equiv \mathcal{D}_+ \phi(\theta = 0). \tag{24.64}$$

Under a supergeneral coordinate transformation

$$\delta\phi = \xi^N \mathcal{D}_N \phi. \tag{24.65}$$

At $\theta = 0$ with the above identifications this equation reads

$$\delta A = i\varepsilon^+ \chi_+. \tag{24.66}$$

While for the spinor we find that

$$\delta\chi_+ = -i\xi^N \mathcal{D}_N \mathcal{D}_+ \phi|_{\theta=0} = \varepsilon^+ \mathcal{D}_+ \phi|_{\theta=0}$$

$$= \left\{ (e_+{}^z \partial_z + e_+{}^{\bar z}\partial_{\bar z}) A - \frac{i}{2}\psi_+^+ \chi_+ \right\} \varepsilon^+. \tag{24.67}$$

Let us now verify that these transformations agree with those given in Chapters 21 and 23. In the rigid limit $\psi_a^+ = 0$ and $e_a^\alpha = \delta_a^\alpha$ and we recover Eq. (23.20). To recover Eqs. (21.67) and (21.68) we take

$$A \to A\sqrt{2}, \qquad \chi \to \chi, \qquad \varepsilon^+ \to \varepsilon^+\sqrt{2} \qquad \text{and} \qquad \psi_a^+ \to K\sqrt{2}\psi_a^+$$

in Eqs. (24.63) and (24.67), and we identify

$$\varepsilon_A = \begin{pmatrix} 0 \\ \varepsilon^+ \end{pmatrix}, \qquad \psi_{\alpha A} = \begin{pmatrix} 0 \\ \psi_\alpha^+ \end{pmatrix} \tag{24.68}$$

in the former equations.

The local coupling of the (A, χ_+) matter multiplet which was given in components in Chapter 21 is found in superspace by simply taking the rigid expression of (23.22), inserting a determinant of E_π^N and covarianting the derivatives:

$$-i \int dz \, d\bar{z} \, d\theta^+ (s \det E_\pi{}^N) \phi \mathscr{D}_= \mathscr{D}_+ \phi \tag{24.69}$$

The other representation of $(1, 0)$ supersymmetry is contained in a superfield λ_-, identifying

$$\lambda_-(\theta = 0) = \zeta_-, \qquad \mathscr{D}_+ \lambda_-(\theta = 0) = b \tag{24.70}$$

the component transformations are

$$\delta \zeta_- = b\varepsilon^+$$
$$\delta b = \varepsilon^+ \mathscr{D}_+ \mathscr{D}_+ \lambda_-|_{\theta=0} = i\varepsilon^+ \mathscr{D}_{\neq} \lambda_-|_{\theta=0} = i\varepsilon^+ \mathscr{D}_{\neq} \zeta_- \tag{24.71}$$

where

$$\mathscr{D}_{\neq} \zeta_- = (\partial_{\neq} + \tfrac{1}{2}\omega_{\neq})\zeta_- - \tfrac{1}{2}b\psi_{\neq}^+ \equiv D_{\neq}\zeta_- - \tfrac{1}{2}b\psi_{\neq}^+. \tag{24.72}$$

In superspace the local coupling of this multiplet is given by

$$-\int dz \, d\bar{z} \, d\theta^+ (s \det E_\pi^N) \lambda_- \mathscr{D}_+ \lambda_-. \tag{24.73}$$

The most general couplings of the above two multiplets given in the rigid case in Chapter 23 generalize in the obvious way to be locally invariant.

24.2e The Heterotic string in superspace

The principle application of $(1, 0)$ supergravity is to the heterotic string. As explained in Eqs. (23.33)–(23.35) the string moves through the superspace-time described by the superfields $\phi^\mu(z^\pi)$, $\mu = 0, 1, \ldots, 9$ and $H_-^i(z^\pi)$, $i = 1, 2, \ldots, 32$. The motion is such as to be an extremum of the action

$$\frac{1}{2\pi\alpha'} \int dz\, d\bar{z}\, d\theta^+ (s \det E_\pi^N) \left\{ -i\phi^\mu \mathscr{D}_= \mathscr{D}_+ \phi^\nu \eta_{\mu\nu} - \sum_{i=1}^{32} H_-^i \mathscr{D}_+ H_-^i \right\}. \quad (24.74)$$

24.2f   Super-Weyl transformations

In this section we wish to express the local super-Weyl transformations in superspace. Since not all components of the supervielbein are independent we must be careful when specifying their transformations. We therefore adopt the transformation

$$E_+^{\pi'} = e^L E_+^{\pi} \quad (24.75)$$

where L is an arbitrary function of z^π and find the transformations of the remaining components of the supervielbein by demanding that the super-Weyl transformations respect the constraints on the torsions. The transformation of E_+^π can only be of the above form since it is the lowest dimension component and a super-Weyl transformation commutes with a local Lorentz transformation.

The brute force method would be to write down the most general variation for the other fields and substitute it in the torsions that are constrained. A better, but equivalent method,[57] is to write down the most general expression for the transformed covariant derivatives and use these to calculate the transformed torsion. Under a super-Weyl transformation we have

$$\mathscr{D}_+' = e^L(\mathscr{D}_+ + f_+ \cdot J)$$

$$\mathscr{D}_a' = f_a{}^b e^{2L}(\mathscr{D}_b + f_b \cdot J + f_b^+ \mathscr{D}_+) \quad (24.77)$$

where $f_a{}^b$, f_b^+ and f_+ are to be determined. We may use a Lorentz transformation to choose $f_=^{=} = 1$. Evaluating the anti-commutator of two-spinor derivatives we have

$$\{\mathscr{D}_+', \mathscr{D}_+'\} = T_{++}{}^{N'}\mathscr{D}_N' + R'_{++}J = T_{++}{}^{+'}e^L(\mathscr{D}_+ + f_+ \cdot J)$$

$$+ T_{++}{}^a f_a{}^b e^{2L}(\mathscr{D}_b + f_b \cdot J + f_b^+ \mathscr{D}_+) + R_{++}{}' \cdot J$$

$$= e^{2L}(T_{++}{}^N \mathscr{D}_N + R_{++} \cdot J) + 2e^{2L}\mathscr{D}_+ L(\mathscr{D}_+ + f_+ \cdot J)$$

$$- e^{2L}f_+ \mathscr{D}_+ + 2e^{2L}\mathscr{D}_+ f_+ \cdot J. \quad (24.78)$$

Using the superspace constraints and equating the coefficients of $\mathscr{D}_+$, $\mathscr{D}_{\neq}$, $\mathscr{D}_=$ and J respectively we find the equations

$$2if_+^+ + f_+ = 2\mathscr{D}_+ L, \qquad f_+^+ = 1, \qquad f_+^= = 0$$

$$if_+ = \mathscr{D}_+ L f_+ + \mathscr{D}_+ f_+. \quad (24.79)$$

Repeating the exercise for $\{\mathscr{D}_a', \mathscr{D}_-'\}$ we find in addition that

$$f_= = 2\mathcal{D}_= L, \qquad f_+ = -2\mathcal{D}_+ L, \qquad f_=^{\;+} = 0$$

$$f_=^{\;+} = 0. \tag{24.80}$$

as well as the transformation of R_{-a}. The final result is then

$$D_+' = e^L(\mathcal{D}_+ - 2\mathcal{D}_+ L \cdot J)$$

$$\mathcal{D}_+' = e^{2L}(\mathcal{D}_+ - 2\mathcal{D}_+ L \cdot J - 2i\mathcal{D}_+ L\mathcal{D}_+)$$

$$\mathcal{D}_-' = e^{2L}(\mathcal{D}_= + 2\mathcal{D}_= L \cdot J) \tag{24.81}$$

or in component superfields

$$E_+{}^{\pi'} = e^L E_+{}^\pi, \qquad \Omega_+' = e^L(\Omega_+ - 2\mathcal{D}_+ L)$$

$$E_+{}^{\pi'} = e^{2L}(E_+{}^\pi - 2i\mathcal{D}_+ L E_+{}^\pi) \qquad E_={}^{\pi'} = e^{2L} E_={}^\pi$$

$$\Omega_+' = e^{2L}(\Omega_+ - 2\mathcal{D}_+ L) \qquad \Omega_=' = e^{2L}(\Omega_= + 2\mathcal{D} = L). \tag{24.82}$$

One may readily verify that in terms of components Eq. (24.82) is equivalent to the results given in Chapter 21:

$$e_a{}^{\alpha'} = e^{-2l} e_a^\alpha, \qquad \psi_{\frac{\theta'}{+}} = e^{-2e}(\psi_{\frac{\theta}{+}} + \eta^\theta), \qquad \psi_=^{\theta'} = e^{-2l}\psi_=^{\theta'}$$

or

$$e_\alpha{}^{a'} = e^{+2l} e_\alpha^a \qquad \psi_z{}^{+'} = e^l \psi_z{}^+ + \eta^+, \qquad \psi_{\bar{z}}^{+'} = e^l \psi_{\bar{z}}{}^+$$

where $l(z, \bar{z})$ and $\eta(z, \bar{z})$ are the x-space components of L.

The superspace form of super-Weyl transformations were first given for $(1, 1)$ supergravity,[225] the $(1, 0)$ case being obtained by a trivial reduction.

24.2g   Solution of the constraints

A count of components of the supervielbein showed that five components were not solved for algebraically. However, we have at our disposal three general super-coordinate transformations, one local Lorentz and one super-Weyl transformation. Consequently, it would seem likely, that one could gauge the supervielbein to its rigid superspace values. One way, therefore, to solve the constraints, is to start form the latter rigid supervielbein and introduce parameters for all the above symmetries as compensating superfields.

In fact this program has not been carried out, but the constraints have been solved directly. This task is easier than in four-dimensions as unlike in the four-dimensional case there are no representation preserving constraints or conformal choice constraints, but only conventional constraints. Thus the solution of the constraints is an algebraic problem. At the linearized level the solution can be read off from Sec. 24.2b. In Refs. 225–227, the constraints of

$(1,0)$ superspace were solved although we quote the results in the form given in Ref. 227.

In terms of two-superfields $H_+{}^{\bar{z}}$ and $H_={}^z$ we define the derivatives

$$\hat{E}_+ \equiv D_+ + H_z{}^{\bar{z}}\partial_{\bar{z}}, \qquad \hat{E}_- = \partial_- + H_-{}^z\partial_z$$

where D_+ is the rigid space-covariant derivative. Although we use the symbol H we will give a nonlinear solution. From these we construct

$$\hat{E}_{\pm} \equiv -\frac{i}{2}\{\hat{E}_+,\hat{E}_+\} \equiv \partial_+ + H_+{}^{\bar{z}}\partial_{\bar{z}} \tag{24.83}$$

and the objects $\hat{C}_.$: by

$$\{\hat{E}_+,\hat{E}_+\} = 2i\hat{E}_{\pm}, \qquad [\hat{E}_+,\hat{E}_{\pm}] = 0$$

$$[\hat{E}_+,\hat{E}_-] = i\hat{C}_{--} + \hat{E}_{++} + \hat{C}_+\hat{E}_{--}$$

$$[\hat{E}_{\pm},\hat{E}_-] = \hat{C}_{--}E_{++} + \hat{C}_{++}E_{--}. \tag{24.84}$$

After defining yet more E's

$$\tilde{E}_+ = \hat{E}_+, \qquad \tilde{E}_- = \hat{E}_- + \tfrac{1}{2}\hat{C}_- + \hat{E}_+$$

$$\tilde{E}_{\pm} = \hat{E}_{\pm} - \frac{i}{2}\hat{C}_+\hat{E}_+ \tag{24.85}$$

we can give the solution. The supervielbein is given in terms of the five superfields

$$H_+{}^{\bar{z}}, H_-{}^z, H^+, L \text{ and } \psi \tag{24.86}$$

by

$$E_+ = e^{(1/2)H^+D_+}(e^{L/2}\psi^{-1/2}\tilde{E}_+)e^{-(1/2)H^+D_+}$$

$$E_- = e^{(1/2)H^+D_+}(e^{-L}\psi^{-1}\tilde{E}_-)e^{-(1/2)H^+D_+}$$

$$E_{\pm} = e^{(1/2)H^+D_+}(e^L\psi^{-1}\tilde{E}_{\pm} + i\tilde{E}_+(\ln\psi)\tilde{E}_+)e^{-(1/2)H^+D_+}. \tag{24.87}$$

Although not very illuminating, we can see that ψ is a super-Weyl compensator, L a local-Lorentz compensator and the remaining fields H^+, $H_{+\bar{z}}$ and $H_-{}^z$ corresponding to the three super-general coordinate transformations. For further details see Ref. 227. Refs. 228, the last reference in 226, and 229 discuss questions such as global gauge fixing and super moduli when applying $(1,0)$ supergravity to string theory.

24.3 $(1,1)$ Supergravity

The construction of $(1,1)$ superspace supergravity[225] closely parallels that of $(1,0)$ superspace supergravity given in the previous section. The superspace contains the coordinates

$$z^\pi = (z, \theta^+; \bar{z}, \theta^-) \tag{24.88}$$

or in the more usual formation

$$z^\pi = (x^\delta, \theta^A) \tag{24.89}$$

where θ^A is a Majorana spinor. The index A refers to a world spinor index. The tangent space group is the Lorentz group. For the construction we will utilize the more standard formalism. In this formalism the tangent Lorentz group rotates a vector as

$$\delta V^N = -V^M V_M{}^N = \Lambda \cdot J V^N,$$

$$\delta V_N = \Lambda_N{}^M V_M = \Lambda \cdot J V_N \tag{24.90}$$

and J the Lorentz generator acts as

$$(\Lambda \cdot J) V_d = \Lambda \varepsilon_d{}^c V_c, \qquad (\Lambda \cdot J) V^d = \Lambda \varepsilon^d{}_c V^c$$

$$(\Lambda \cdot J) V_A = -\tfrac{1}{2} \Lambda (\gamma_5)_A{}^B V_B \tag{24.91}$$

where

$$\varepsilon^{01} = +1, \qquad \varepsilon^{ab} = -\varepsilon^{ba}, \qquad \varepsilon_0{}^1 = -\varepsilon^{01}. \tag{24.92}$$

The corresponding spin connection is valued in the same representation of the Lorentz group namely $\phi_{NM}{}^P = +\phi_N J$, examples are

$$\phi_{Nd}{}^e = \varepsilon_d{}^e \phi_N, \qquad \phi_{NA}{}^c = -\tfrac{1}{2}\phi_N(\gamma_5)_A{}^c. \tag{24.93}$$

The spinorial covariant derivatives of rigid superspace obey

$$\{D_A, D_B\} = i(\gamma^d C)_{AB} P_d \tag{24.94}$$

where our conventions are those given at the beginning of Chapter 21. Therefore, the only nonzero torsion of rigid superspace is

$$t_{AB}{}^d = i(\gamma^d C)_{AB}. \tag{24.95}$$

Recalling the discussion of Ref. 57 given in Chapter 16 we choose a suitable set of constraints. There are no constrained superfields in $(1, 1)$ supersymmetry which are not on-shell, therefore, there are no representation preserving constraints or integrability conditions.

The conventional constraints are most easily chosen by examining the linearized torsions. Using Eqs. (24.25) and (24.28) we have

dim 0

$$T_{AB}{}^d = i(\gamma^d C)_{AB} + D_A H_B{}^d + D_B H_A{}^d - i(\gamma^e C)_{AB} H_e{}^d$$

$$+ i H_A{}^D (\gamma^d C)_{DB} + i H_B{}^D (\gamma^d C)_{DA} \tag{24.96}$$

dim 1/2

$$T_{AB}{}^C = D_A H_B{}^C + D_B H_A{}^C - i(\gamma^d C)_{AB} H_d^C + \phi_{AB}{}^C + \phi_{BA}{}^C$$

$$T_{Ae}{}^d = D_A H_e{}^d - D_e H_A{}^d + \phi_{Ae}{}^d. \tag{24.97}$$

Taking the trace on A, B of $T_{AB}{}^c$ with the symmetric matrices we find that

$$T_{AB}{}^D = 0 \Leftrightarrow (C\gamma_5)^{AB} T_{AB}{}^D = 0 = (C\gamma^e)^{AB} T_{AB}{}^D \tag{24.98}$$

solve for the spin connection ϕ_A and H_d^D respectively. Repeating the exercise for $T_{AB}{}^d$ we find

$$0 = (C\gamma_d)^{AB} T_{AB}{}^e + 2i\delta_d^e = -2H_d^e + 2iH_B{}^D(\gamma^e\gamma_d)_D{}^B + \text{derivative terms}$$

$$0 = (C\gamma_5)^{AB} T_{AB}{}^e = 2iH_B{}^D(\gamma_d\gamma_5)_D{}^B + \text{derivative terms}. \tag{24.99}$$

Hence we may solve for

$$H_d^e \quad \text{and} \quad H_B{}^D(\gamma_d)_D{}^B. \tag{24.100}$$

At dimension one we may use

$$T_{ab}{}^e = 0 \tag{24.101}$$

to solve for the dimension one field $\phi_{ab}{}^e$. Clearly, the dimension of the constraint is equal to the dimension of the field solved for and hence at dimension-one we cannot solve for more fields or impose higher dimensional constraints.

Left unsolved for are

$$H_A^d, H_D{}^D \quad \text{and} \quad H_B{}^D(\gamma_5)_D{}^B \tag{24.102}$$

which constitute 6 superfields in all. In terms of symmetries we have at our disposal four supergeneral coordinate transformations and one local Lorentz transformation; we may use these to remove five of the above 6 superfields leaving only one. Even this may be removed if we use super Weyl transformations.

Summarizing, the constraints we have adopted are the following

$$T_{AB}{}^d = 0 \qquad T_{AB}{}^e = i(\gamma^e C)_{AB}$$

$$T_{AB}{}^D = 0. \tag{24.103}$$

As we shall see the above constraints are super-Weyl invariant essentially by construction since the fields they solve for algebraically, will by definition transform in such a way as to preserve the constraints. Unlike in four-dimensions where (see Chapter 16) to recover supergravity rather than conformal supergravity, we made further 'conformal choice constraints'[57] we cannot impose such constraints in the $(1,0)$ or $(1,1)$ two-dimensional cases. The reason is that the parameter of super-Weyl transformations is a general superfield and we cannot further decompose such a field without putting it on-shell.

The next step is to 'solve the Bianchi identities' although tedious this is identical in principle to the $N = 1, D = 4$ case of Chapter 16 and the $(1, 0)$ case of the previous section. The results[225] are that all dimension-1/2 torsions are zero and the only nonzero tensors are given by

dim 1

$$T_{bA}{}^B = -\tfrac{1}{2}(\gamma_b)_A{}^B S$$

$$R_{AB} = -i(\gamma_5 C)_{AB} S$$

dim 3/2

$$T_{ab}{}^A = -i\varepsilon_{ab}(C\gamma_5)^{AB}\mathcal{D}_B S$$

$$R_{Ab} = -(\gamma_5\gamma_b)_A{}^B\mathcal{D}_B S$$

dim 2

$$R_{ab} = -4i\varepsilon_{ab}\mathcal{D}_A\mathcal{D}^A S - \varepsilon_{ab} S^2. \tag{24.104}$$

Thus all covariant tensors can be expressed in terms of one scalar superfield S.

The next step is to locate the x-space component fields. To this end we first gauge to their flat space values the $\theta = 0$ components of E_A^N

$$E_{\underline{A}}{}^a(\theta = 0) = 0, \qquad E_{\underline{A}}{}^{NB}(\theta = 0) = \delta_{\underline{A}}^B \tag{24.105}$$

as well as the spin connection

$$\Omega_a(\theta = 0) = 0.$$

and then label the $\theta = 0$ components of E_α^N:

$$E_\alpha{}^a(\theta = 0) = e_\alpha^a, \qquad E_\alpha{}^A(\theta = 0) = \tfrac{1}{2}\bar\psi_\alpha^A.$$

Note that we have used both ϕ_π and Ω_π to denote the same spin connection. Evaluating the torsions in terms of E_π^N and comparing them with Eqs. (24.103) and (24.104), we find the all orders in θ expressions of E_π^N and Ω_π to be[224]

$$E_\alpha{}^a = e_\alpha^a + \frac{i}{2}\bar\theta\gamma^a\psi_\alpha + \frac{i}{4}\bar\theta\theta e_\alpha^a A$$

$$E_\alpha{}^A = \frac{1}{2}\psi_\alpha{}^A + \frac{1}{2}\bar\theta^B(\gamma_5)_B{}^A\omega_\alpha - \frac{1}{4}\bar\theta^B(\gamma_a)_B{}^A e_\alpha^a A - \frac{3i}{16}\bar\theta\theta\psi_\alpha{}^A A - \frac{1}{8}\bar\theta\theta(C\gamma_a)^{AB}e_\alpha^a\psi_B$$

$$E_B{}^a = \frac{i}{2}\bar\theta^A(\gamma^a C)_{AB}, \qquad E_A{}^B = \delta_A{}^B - \frac{i}{8}\bar\theta\theta\delta_A^B A$$

and

$$\Omega_\alpha \equiv \phi_\alpha \equiv \omega_\alpha - \frac{i}{2}\bar{\theta}\gamma_5\psi_\alpha A + \frac{i}{4}\bar{\theta}\theta\omega_\alpha A - \frac{i}{4}\bar{\theta}\theta\varepsilon_\alpha{}^\beta\partial_\beta A$$

$$+ \frac{i}{8}\bar{\theta}\theta\bar{\psi}\gamma_5\gamma_a\gamma^b\psi_b e_\alpha^a - \frac{1}{2}\varepsilon_{\alpha\beta}\bar{\theta}\gamma^a\psi e_\alpha^\beta$$

$$\phi_A \equiv \Omega_A = -\frac{i}{2}\bar{\theta}^B(\gamma_5)_{BA} A \tag{24.106}$$

where

$$\psi = +\frac{i}{2}\varepsilon_{\alpha\beta}\gamma_5 D_\alpha\psi_\beta - \frac{i}{4}\gamma^\delta\psi_\delta A \tag{24.107}$$

We recall that we have adopted the convention that a world spinor index is in bold face, i.e. $E_\mathbf{B}{}^\alpha$. The anticommuting coordinates θ^A naturally have their index upstairs and in the above equation we identify these with $\bar{\theta}^A$. We may lower the index with the C matrix in the usual way, i.e. $\bar{\theta} = -\theta^+\gamma^0$.

We see that $(1,1)$ supergravity contains the x-space fields.

$$e_\alpha^a, \psi_{\alpha A} \quad \text{and} \quad A \tag{24.108}$$

The nonzero torsions are expressed in terms of S which has the components

$$S = A + \bar{\theta}\psi + \frac{i}{2}\bar{\theta}\theta C,$$

where

$$C = R - \frac{1}{4}\bar{\psi}_a\gamma^a\psi - \frac{i}{16}\varepsilon^{ab}\bar{\psi}_a\gamma_5\psi_b A - \frac{1}{2}A^2$$

and

$$R = -2\varepsilon^{\alpha\beta}\partial_\alpha\omega_\beta \tag{24.109}$$

To complete the contact with the x-space results, we compute the supersymmetry variatons. Following the $(1,0)$ case or the discussion in Chapter 16 we find

$$\delta e_\alpha^a = i\bar{\varepsilon}\gamma^a\psi_\alpha, \qquad \delta\psi_\alpha = 2D_\alpha\varepsilon - (\gamma_a)e_\alpha^a A\varepsilon$$

$$\delta A = \bar{\varepsilon}\psi \tag{24.110}$$

The field A is auxiliary and setting it to zero we obtain the transformation laws of Chapter 21.

Super-Weyl transformations are obtained in an identical manner to the $(1,0)$ case of the previous section, with the result[225]

$$E_A^{\pi'} = e^{L/2}E_A^\pi$$

$$E_a^{\pi'} = e^L E_a^{\pi} - 2i(C\gamma_a)^{AB}D_B LE_A^{\pi}$$

$$\Omega_A' = e^{L/2}\left(-(\gamma_5)_A^{\ B}D_B L + \Omega_A\right)$$

$$\Omega_a' = e^L\left(\Lambda_a + \varepsilon_a^{\ b}D_b L - 2i(C\gamma_a)^{AB}D_B L\Omega_A\right) \tag{24.111}$$

where the super-Weyl parameter L is a general superfield.

It is possible by using supergeneral coordinate transformations and local Lorentz transformations to bring the supervielbein and spin-connections to be such that they are related to the supervielbein and spin-connection of rigid superspace by a super-Weyl transformation. This result is most easily established at the level of x-space component fields. Using in addition the local super-Weyl transformation, we may therefore bring the supervielbein and spin-connection to their rigid superspace values.

As discussed in Chapter 23 the $(1, 1)$ matter multiplet is a scalar superfield ϕ. We may define its x-space component fields by

$$\phi(\theta = 0) \equiv A, \qquad \mathcal{D}_A\phi(\theta = 0) = i\chi_A$$

$$\mathcal{D}_A\mathcal{D}^A\phi|_{\theta=0} = iF \tag{24.112}$$

and recover their local x-space transformation rules of Chapter 21 using the standard technique.

The superspace formulation of the local coupling of the $(1, 1)$ supermultiplet is found by taking the rigid result of Eq. (23.49) covarianting the derivatives and inserting the superdeterminant of the vielbein:

$$\int dz\, d\bar{z}\, d\theta\, d\bar{\theta}(s \det E_{\pi}^{\ N})\{-\tfrac{1}{2}\mathcal{D}_A\phi\mathcal{D}^A\phi + ig(\phi)\}. \tag{24.113}$$

The generalization of the most general coupling of Eq. (23.52) being obtained in the same way.

The usual superstring[7], which admits a $(1, 1)$ supersymmetry, moves through a superspace with coordinates $\phi^{\mu}(x^{\delta}, \theta^A)$ in such a way as to extremize the action[225]

$$-\frac{1}{4\pi\alpha'}\int dz\, d\bar{z}\, d\theta\, d\bar{\theta}(s \det E_{\pi}^{\ N})\mathcal{D}_A\phi^{\mu}\mathcal{D}^A\phi^{\nu}\eta_{\mu\nu} \tag{24.114}$$

The solution to the superspace constraints of $(1, 1)$ supergravity at the linearized level can be read off from the equations above. The nonlinear solution was given in Ref. 233.

Application of $(1, 1)$ supergravity to superstring theory can be found in Ref. 234. The $(2, 2)$ supergravity in superspace was discussed in Ref. 235 and $(4, 4)$ supergravity in Ref. 236.

Chapter 25

The Superconformal Group

The conformal group contains as a subgroup the dilations (i.e. $x^\mu \rightarrow \lambda x^\mu$) which imply that conformally invariant theories do not in general possess a length scale and in particular are massless. The prototype example of a conformal system is Maxwells equations $\partial^\mu F_{\mu\nu} = 0$ in four dimensions for a spin-1 particle. The fundamental role that massless particles play in nature has motivated physicists for many years to study conformal invariance. However, it is in recent years that the most direct applications have been found.

One of the most important applications has been in the theory of critical phenomena. It has been found that a system in the neighborhood of its critical point is dominated by long range fluctuations of no fixed frequency and thus the system has in effect massless excitations and no fixed scale. The appropriate mathematical description[238] is a theory with an infrared stable fixed point. To find the long range behavior one uses the repeated application of the renormalization group which drives the theory to its fixed point.

However, a system at a fixed point has by definition vanishing β function corresponding to dilation invariance and for most such systems, dilation invariance implies invariance under the full conformal group. An important feature of this framework is that all systems whose couplings are in the domain of attraction of the fixed point have the same critical phenomena which are dominated by the behavior at the fixed point. This mechanism explains the remarkable universality found in the theory of critical phenomena. Namely, the behavior of critical systems are in general the 'same' provided they have the same dimension and symmetries.

Consequently, critical systems correspond to conformally invariant systems, although clearly not every conformally invariant system need, correspond to a system at its critical point. One way to proceed in the study of critical phenomena is to characterize all conformal systems and examine which of these correspond to those found in nature. In two-dimensions, as we shall see, the conformal group is rather powerful and this method of attack[239] has been particularly important in such systems some of which are supersymmetric. Indeed, for a large class of two-dimensional conformal models one can use the conformal symmetry to solve for the Greens functions.[239]

Another important application of two-dimensional (super) conformal invariance is in (super) string theories. Conformal symmetry is part of the group of two-dimensional differmorphisms which arise in the string from its world sheet reparametrization invariance and in fact, it is this (super) conformal invariance which plays the dominant role in (super) string theory. For example, the constraints of the classical string generate the conformal group and it is these constraints which upon quantization lead to the field equations for the infinite number of particles contained in the string. Much of the development of string theory was within context of the bosonic strings in 26-dimensions and the superstrings in 10-dimensions. However, more recently what can constitute an acceptable string theory has been better understood. Essentially, any two-dimensional (super) conformal theory gives rise to a string theory when coupled to conformal (super) gravity provided it satisfies two conditions:
(i) it possesses no conformal anomalies
(ii) it has no space-time anomalies, i.e. is modular invariant.

Although this has lead to string theory in four-dimensions, there are an embarrassing number of such theories.It is therefore, essential for the development of string theory to understand two-dimensional conformal theories and their local couplings.

25.1 The Conformal Group in Arbitrary Dimensions

Given a D dimensional space parametrized by x^μ; a conformal transformation is a diffeomorphism, $x^\mu \to \bar{x}^\mu(x)$ such that the line element changes only by a scale factor, namely

$$ds^2 \equiv d\bar{x}^\mu d\bar{x}^\nu \eta_{\mu\nu} = \Omega \, dx^\mu dx^\nu \eta_{\mu\nu} \equiv \Omega \, ds. \tag{25.1}$$

The metric $\eta_{\mu\nu}$ is diagonal and is of the form

$$\eta_{\mu\nu} = \text{diag}(\underbrace{+1,\ldots,+1}_{r}, \underbrace{-1,-1,\ldots,-1}_{s}). \tag{25.2}$$

Setting $\bar{x}^\mu \equiv f^\mu(x)$, the condition on f^μ of Eq. (25.1) is given by

$$\partial_\rho f^\mu \partial_\kappa f^\nu \eta_{\mu\nu} = \eta_{\rho\kappa}\Omega. \tag{25.3}$$

Taking the trace we find that

$$\Omega = \frac{1}{D}\partial_\rho f^\mu \partial_\kappa f^\nu \eta_{\mu\nu}\eta^{\rho\kappa}. \tag{25.4}$$

Let us consider an infinitesimal conformal transformation

$$f^\mu = x^\mu + \varepsilon^\mu(x)$$

then at lowest order, Eq. (25.4) becomes

$$\partial_\mu \varepsilon_\nu + \partial_\nu \varepsilon_\mu = \frac{2}{D} \eta_{\mu\nu} \partial^\rho \varepsilon_\rho \qquad (25.5)$$

where

$$\varepsilon_\nu = \varepsilon^\mu \eta_{\mu\nu}.$$

Taking the ∂^ν of this equation we find that

$$(-D + 2)\partial_\mu \partial^\kappa \varepsilon_\kappa = D\partial^2 \varepsilon_\mu. \qquad (25.6)$$

Acting on this equation with ∂^μ we conclude that provided $D \neq 1$ then

$$\partial^2 \partial^\mu \varepsilon_\mu = 0. \qquad (25.7)$$

For $D \neq 2$ and using Eq. (25.6) we note that

$$\partial_\rho \partial_\lambda \partial^\kappa \varepsilon_\kappa = \frac{D}{(2 - D)} \partial_\rho \partial^2 \varepsilon_\lambda. \qquad (25.8)$$

However, the left-hand side is ρ, λ symmetric and hence so must be the right-hand side. Consequently, using (25.5) and then (25.7) we conclude that

$$\partial_\rho \partial_\lambda \partial^\kappa \varepsilon_\kappa = \frac{1}{(2 - D)} \eta_{\rho\lambda} \partial^2 \partial^\kappa \varepsilon_\kappa. \qquad (25.9)$$

provided also $D \neq 1$.

As we now show the above equations imply that the third derivative of ε_μ vanishes. We have, using (25.5) and (25.9)

$$\partial_\rho \partial_\lambda \partial_\kappa \varepsilon_\mu = -\partial_\rho \partial_\lambda \partial_\mu \varepsilon_\kappa + \frac{2}{D} \eta_{\kappa\mu} \partial_\rho \partial_\lambda \partial^\tau \varepsilon_\tau$$

$$= -\partial_\rho \partial_\lambda \partial_\mu \varepsilon_\kappa = \partial_\rho \partial_\kappa \partial_\mu \varepsilon_\lambda$$

$$= -\partial_\rho \partial_\kappa \partial_\lambda \varepsilon_\mu = 0. \qquad (25.10)$$

Hence ε^μ, when expressed as a polynomial in x^μ, has no higher terms than quadratic:

$$x^{\mu'} = x^\mu + a^\mu + t^\mu{}_\nu x^\nu + t^\mu{}_{\nu\rho} x^\nu x^\rho. \qquad (25.11)$$

Applying Eq. (25.5) we find that

$$t_{\mu\nu} + t_{\nu\mu} = \frac{2}{D} \eta_{\mu\nu} t^\rho{}_\rho. \qquad (25.12)$$

$$t_{\mu\nu\rho} + t_{\nu\mu\rho} = \frac{2}{D} \eta_{\mu\nu} t^\kappa{}_{\kappa\rho}. \qquad (25.13)$$

We may solve (25.12) by writing

$$t_{\mu\nu} = \omega_{\mu\nu} + \lambda_{\eta\mu\nu}$$

where

$$\omega_{\mu\nu} = -\omega_{\nu\mu}.$$

To solve Eq. (25.13) we consider

$$(t_{\mu\nu\rho} + t_{\nu\mu\rho}) + (t_{\rho\mu\nu} + t_{\mu\rho\nu}) - (t_{\rho\nu\mu} + t_{\nu\rho\mu})$$

$$= 2t_{\mu\nu\rho}$$

$$= \frac{2}{D}(\eta_{\mu\nu}t^{\kappa}{}_{\kappa\rho} + \eta_{\mu\rho}t^{\kappa}{}_{\kappa\nu} - \eta_{\rho\nu}t^{\kappa}{}_{\kappa\mu}).$$

Inserting these solutions in $x^{\mu\prime}$ of Eq. (25.11), it becomes for $D \neq 1$ or 2

$$x^{\mu\prime} = x^{\mu} + a^{\mu} + \omega^{\mu}{}_{\nu}x^{\nu} + \lambda x^{\mu} + \lambda^{\mu}x^{\kappa}x_{\kappa} - 2\lambda_{\kappa}x^{\kappa}x^{\mu} \qquad (25.15)$$

where

$$\lambda^{\mu} = -\frac{2}{D}t^{\kappa}{}_{\kappa}{}^{\mu}.$$

Since the Poincaré group is by definition the group which preserves the line element we find these transformations among the above in the guise of the translation a^{μ} and Lorentz rotation, $\omega^{\mu}{}_{\nu}$. The rescaling of x^{μ} with parameter λ is a dilation and the transformation with parameter λ_{μ} is called a special conformal transformation.

Counting the number of generators, we have

$$D + D\left(\frac{D}{2} - 1\right) + 1 + D = \frac{1}{2}(D + 1)(D + 2).$$

The infinitesimal operators can be read off from Eq. (25.15) and one ready finds that they realize the Poincaré algebra plus the relations

$$[D, \kappa_{\nu}] = \kappa_{\nu} \qquad [D, P_{\mu}] = -P_{\mu}$$

$$[K_{\mu}, J_{\rho\kappa}] = \eta_{\mu\rho}K_{\kappa} - \eta_{\mu\kappa}K_{\rho}, \qquad [K_{\mu}, K_{\nu}] = 0$$

$$[\kappa_{\mu}, P_{\nu}] = -2\eta_{\mu\nu}D + 2J_{\mu\nu}. \qquad (25.16)$$

This algebra, the conformal algebra, is isomorphic to $SO(r + 1, s + 1)$.

The finite transformations are obtained by exponentiating the above infinitesimal transformations. For dilations this is straightforward, we find

$$x^{\mu\prime} = \lambda x^{\mu}. \qquad (25.17)$$

For special conformal transformations the finite transformation is most easily found by realizing that at the infinitesimal level it implies the relation

$$\frac{x^{\mu\prime}}{x^{\prime 2}} = \frac{x^{\mu}}{x^2} + \lambda. \tag{25.18}$$

A finite transformation is, of course, of the same form and solving for $x^{\mu\prime}$ it implies that

$$x^{\mu\prime} = \frac{x^{\mu} + \lambda^{\mu}x^2}{1 + 2x\cdot\lambda + \lambda^2 x^2}. \tag{25.19}$$

Our definition of conformal transformations utilized the flat metric. A more esthetic definition uses the notion of a conformally flat metric. The latter is a metric for which there exists local coordinates such that the line element is of the form

$$ds^2 = \Omega\, dx^{\mu}\, dx^{\nu} \eta_{\mu\nu}. \tag{25.20}$$

A conformal transformation is defined to be one which takes one conformally flat metric to another. Clearly, this leads to the same transformations as our original definition.

One could also not use a conformally flat metric, but any metric and defined conformal transformation to be those general coordinate transformations which scaled the given metric. The resulting conformal group will then change as one takes different metrics.

Finally, we note an important relation between dilation invariance and full conformal invariance. Let us consider a theory in an arbitrary number of dimensions which is Poincaré invariant. For such a theory, there exists an energy-momentum tensor $T_{\mu\nu}$ which we can choose to be symmetric. This tensor is the conserved current, i.e., $\partial^{\mu} T_{\mu\nu} = 0$ whose charge generates translations. The current corresponding to Lorentz rotations is a moment of energy-momentum tensor, namely $x_{\mu} T_{\nu\varrho} - x_{\nu} T_{\mu\varrho}$. It is conserved on its ρ index due to the symmetry of $T_{\mu\nu}$ and its conservation.

If the theory is dilation invariant, it is invariant under $x^{\mu} \to \lambda x^{\mu}$ and the associate current is given by a moment of the energy-momentum tensor

$$J_{\mu} = x^{\nu} T_{\nu\mu}. \tag{25.21}$$

It is conserved provided that $T^{\nu}{}_{\nu} = 0$. However, this condition also allows us to construct further conserved moments of $T_{\mu\nu}$, in fact any current of the form

$$\varepsilon^{\nu} T_{\nu\mu} \tag{25.22}$$

provided

$$\partial^{\mu}\varepsilon^{\nu} + \partial^{\nu}\varepsilon^{\mu} - \phi\eta^{\mu\nu} = 0 \tag{25.23}$$

where ϕ is an arbitrary function of x^{μ}. One example is given by

$$x_\mu x^\rho T_{\rho\nu} - x^2 T_{\mu\nu} \qquad (25.24)$$

which generates the special translations of the conformal group.

These additional conserved currents define corresponding generators which together with the Poincaré and dilation generator have the conformal group as their algebra. This follows from the fact that the charges generate the transformations $x^\mu \rightarrow x^\mu + \varepsilon^\mu$ and these are precisely the complete set of transformations that leave the metric invariant up to a scale factor provided it satisfies Eq. (25.5).

We have shown therefore that any theory that has $T^\nu{}_\nu = 0$ is not only dilation, but also conformally invariant. In fact, the converse can be shown, a conformally invariant theory has a traceless energy-momentum tensor and its currents are the above moments of the energy-momentum tensor (see Chapter 20).

25.2 The Two-Dimensional Conformal Group

In the previous section, we constructed the conformal algebra and transformations, however, an exception that we did not cover was when $D = 2$. In this case, we find Eq. (25.5) becomes in Minkowski space

$$\partial_0 \varepsilon_1 = -\partial_1 \varepsilon_0 \qquad \text{and} \qquad \partial_0 \varepsilon_0 + \partial_1 \varepsilon_1 = 0 \qquad (25.25)$$

and in Euclidean space

$$\partial_1 \varepsilon_2 = -\partial_2 \varepsilon_1 \qquad \text{and} \qquad \partial_1 \varepsilon_1 - \partial_2 \varepsilon_2 = 0. \qquad (25.26)$$

Those in Minkowski space are equivalent to the wave equation for g where $\varepsilon_\mu = (\partial_0 g, -\partial_1 g)$ and those in Euclidean space are the Cauchy-Riemann equations. Unlike the case $D \neq 2$, for $D = 2$ these equations admit an infinite number of conformal transformations which are most easily analyzed by moving to the light cone coordinates of Sec. 22.1. Making this transformation and restricting our attention to the Euclidean case the line element becomes (i.e. Eq. (22.21))

$$ds^2 = dz \, d\bar{z}. \qquad (25.27)$$

Clearly, it is preserved up to a scale factor [i.e., $\Omega = f'(z)g'(z)$] by a transformation of the form

$$z \rightarrow f(z), \qquad \bar{z} \rightarrow g(\bar{z}) \qquad (25.28)$$

for any functions f and g. The only other transformations which preserve the line element up to scale are

$$z \rightarrow h(\bar{z}), \qquad \bar{z} \rightarrow k(z). \qquad (25.29)$$

However, these transformations change the orientation defined on the two-dimensional surface and will not be considered further.

The infinitesimal transformations of Eq. (25.28) are of the form

$$z \to z + a_n z^{n+1}, \qquad \bar{z} \to \bar{z} + \bar{a}_n \bar{z}^{n+1} \qquad (25.30)$$

and are therefore generated by

$$L_n = z^{n+1} \frac{\partial}{\partial z}, \qquad \bar{L}_n = \bar{z}^{n+1} \frac{\partial}{\partial \bar{z}}. \qquad (25.31)$$

The Lie algebra these generators satisfy is given by

$$[L_n, L_m] = (n - m) L_{n+m}$$

$$[L_n, \bar{L}_m] = 0$$

$$[\bar{L}_n, \bar{L}_m] = (n - m) \bar{L}_{n+m}. \qquad (25.32)$$

Naive substitution of Eq. (25.31) in Eq. (25.32) yields the result above with a minus sign. This is an active viewpoint; to arrive at the sign in (25.32), one must take the passive viewpoint and carry out the manœuvre on functions. Thus the conformal group is the direct product of two identical groups and can be denoted by $V \otimes V$.

The infinitesimal conformal transformations of Eq. (25.30) for the left-hand coordinate z generate the vector fields

$$\sigma_n(z) = z^{n+1} \frac{\partial}{\partial z}. \qquad (25.33)$$

on the Riemann sphere $\hat{C}$. At $z = 0$ only those for $n \geqslant -1$ are well-defined. However, $\hat{C}$ requires two-coordinate patches associated with the North and South poles. To move to the other coordinate path we make the coordinate change $u = -1/z$, upon which the above vector fields become

$$v_n(u) = u^{-n+1} \frac{\partial}{\partial u}. \qquad (25.34)$$

At the other pole, $u = 0$ the vector fields are only well-defined for $n \leqslant 1$. Consequently, only the conformal subgroup generated by L_0, $L_{\pm 1}$ and $\bar{L}_0$, $\bar{L}_{\pm 1}$ is globally defined and invertible on $\hat{C}$. Consequently one arrives at the following:

Theorem: The most general conformal transformation which maps the Riemann sphere $\hat{C}$ onto itself in a one-to-one way is a Möbius transformation which is of the form

$$z \to z' = \frac{az + b}{cz + d}; \qquad \bar{z} \to \bar{z}' = \frac{\bar{a}\bar{z} + \bar{b}}{\bar{c}\bar{z} + \bar{d}} \qquad (25.35)$$

where a, b, c, and d are complex parameters. By scaling these parameters, without affecting the transformation, we may set

$$ad - bc = 1 \qquad (25.36)$$

and so the Möbius group is a six real-parameter group. Infinitesimal Möbius transformations are of the form

$$z' = z + \varepsilon_{-1} + \varepsilon_0 z + \varepsilon_1 z^2$$
$$\bar{z}' = \bar{z} + \bar{\varepsilon}_{-1} + \bar{\varepsilon}_0 \bar{z} + \bar{\varepsilon}_1 \bar{z}^2 \qquad (25.37)$$

and so are generated by L_0, $L_{\pm 1}$ and $\bar{L}_0$, $\bar{L}_{\pm 1}$. Indeed one may verify that (25.35) is the most general transformation generated by L_0, $L_{\pm 1}$.

We are also interested (for open strings) in the upper half plane plus the point at infinity, denoted by $\hat{H}$. This space is mapped onto itself in a 1-to-1 manner only by conformal transformations of the form

$$z \to z' = \frac{az + b}{cz + d}, \qquad \bar{z} \to \bar{z}' = \frac{a\bar{z} + b}{c\bar{z} + d} \qquad (25.38)$$

where now a, b, c, and d are real. We may scale the parameters, without affecting the transformation, to obtain either $ad - bc = 1$ or $ad - bc = -1$. This three-parameter group is generated by $L_0 + \bar{L}_0$, $L_{\pm 1} + \bar{L}_{\pm 1}$.

One easily verifies that two Möbius transformations compose to give a third and that the resulting coefficients are the same as one obtains by multiplying the two-by-two matrices. The matrix corresponding to the transformation of Eq. (25.38) being $\begin{pmatrix} a & b \\ c & d \end{pmatrix}$. The correspondence between two-by-two matrices and Möbius transformations is not one-to-one as the matrices $\begin{pmatrix} 1 & 0 \\ 0 & 1 \end{pmatrix}$ and $\begin{pmatrix} -1 & 0 \\ 0 & -1 \end{pmatrix}$ lead to the same Möbius transformation. Apart from this ambiguity, they are in one-to-one correspondence. Consequently, the Möbius group is isomorphic to $SL(2, C)/Z_2$ and the Möbius group with real parameters and $ad - bc = 1$ is isomorphic to $SL(2, R)/Z_2$. To summarize, only the Möbius subgroup of the conformal group is globally defined and invertible on $\hat{C}$. We often refer to these as global transformations.

The further significance of two-dimensional conformal transformations emerges from the following.

Theorem: On any orientated Riemann surface, we may locally choose a coordinate system such that the metric takes the form

$$g_{\alpha\beta}(\xi) = \eta_{\alpha\beta} e^{\sigma(\xi)} \qquad (25.39)$$

or in terms of the line element

$$ds^2 = e^\sigma \eta_{\alpha\beta} \, d\xi^\alpha \, d\xi^\beta = e^\sigma \, dz \, d\bar{z} \tag{25.40}$$

where

$$\eta_{\alpha\beta} = \text{diag}(+1, +1).$$

In other words, on any orientated two-dimensional surface the metric is conformally flat. A proof of this theorem can be found in many text books, for example, see Ref. 240.

We may therefore cover the manifold by patches for which the metric is of the above form and then all our transformations between patches will be conformal transformations. The metric for the above line element is given by

$$g_{\bar{z}\bar{z}} = g_{zz} = 0; \qquad g_{z\bar{z}} = g_{\bar{z}z} = \tfrac{1}{2} e^\sigma \tag{25.41}$$

and its inverse by

$$g^{zz} = g^{\bar{z}\bar{z}} = 0; \qquad g^{z\bar{z}} = g^{\bar{z}z} = 2e^{-\sigma}. \tag{25.42}$$

The Christoffel symbol vanishes except for the components

$$\Gamma^z_{zz} = \partial_z \sigma, \qquad \Gamma^{\bar{z}}_{\bar{z}\bar{z}} = \partial_{\bar{z}} \sigma. \tag{25.43}$$

In the system of patches or coordinate systems above, the tensor calculus of general relativity reduces to a simpler conformal tensor calculus. The most general tensor, as discussed in Chapter 21, is of the form

$$
T^{\overbrace{z\ldots z}^{n} \; \overbrace{\bar{z}\ldots\bar{z}}^{\bar{n}}}_{\underbrace{z\ldots z}_{m} \; \underbrace{\bar{z}\ldots\bar{z}}_{\bar{m}}} = T(z, \bar{z}). \tag{25.44}
$$

and transforms under the coordinate change

$$z \to \omega(z), \qquad \bar{z} \to \bar{\omega}(\bar{z}) \tag{25.45}$$

as

$$T(z, \bar{z}) \to T'(\omega, \bar{\omega}) = \left(\frac{d\omega}{dz}\right)^{n-m} \left(\frac{d\bar{\omega}}{d\bar{z}}\right)^{\bar{n}-\bar{m}} T(z, \bar{z}). \tag{25.46}$$

We call $h = n - m$ and $\bar{h} = \bar{n} - \bar{m}$ the conformal weights or dimensions of the tensor T. A time translation $x^0 \to x^0 + a$ is induced by $z \to z + a$, $\bar{z} \to \bar{z} + a$ for a real and so is generated by $L_{-1} + \bar{L}_{-1}$ which is identified to be the Hamiltonian. A space shift $x^1 \to x^1 + \lambda$ is induced by $z \to z + \lambda$, $\bar{z} \to \bar{z} - \lambda$ and is generated by $L_{-1} - \bar{L}_{-1}$ which is the total momentum. The rotation is generated by $L_0 - \bar{L}_0$ and the dilation $z \to \lambda z$, $\bar{z} \to \lambda \bar{z}$ for λ real is generated by $L_0 + \bar{L}_0$.

We find that under a dilation

$$T \to \lambda^{h+\bar{h}} T$$

while under a rotation

$$T \to e^{i(h-\bar{h})\phi} T.$$

Consequently, we call $h + \bar{h}$ the dilation weight of T and $h - \bar{h}$ its spin.

Any two-dimensional Euclidean invariant theory, possesses a symmetric energy-momentum tensor with components T_{zz}, $T_{\bar{z}\bar{z}}$ and $T_{z\bar{z}} = T_{\bar{z}z}$. If the theory is in addition dilation invariant then

$$T_{z\bar{z}} = 0 \tag{25.47}$$

leaving only T_{zz} and $T_{\bar{z}\bar{z}}$. The conservation condition becomes

$$\partial_{\bar{z}} T_{zz} = 0, \qquad \partial_z T_{\bar{z}\bar{z}} = 0 \tag{25.48}$$

which implies that T_{zz} and $T_{\bar{z}\bar{z}}$ are functions of only z and $\bar{z}$ respectively. Further, for any two analytic functions $f(z)$ and $g(\bar{z})$, we have an infinite set of conserved currents given by

$$f(z) T_{zz} \qquad \text{and} \qquad g(\bar{z}) T_{\bar{z}\bar{z}} \tag{25.49}$$

as clearly $\partial_{\bar{z}}(f(z) T_{zz}) = \partial_z(g(z) T_{\bar{z}\bar{z}}) = 0$ etc. Thus we find an infinite number of generators

$$L_n = \oint \frac{dz}{z} z^{n+2} T_{zz}, \qquad \bar{L}_n = \oint \frac{d\bar{z}}{z} (\bar{z})^{n+2} T_{\bar{z}\bar{z}} \tag{25.50}$$

where the $(2\pi i)^{-1}$ is to be included in the definition of the contour integral. As expected from the general discussion of the previous section we have recovered the infinite dimensional conformal group in a theory for which $T_{z\bar{z}} = 0$.

We have defined the conformal group with respect to the flat metric, however at least locally any metric on a Riemann surface may be brought to be conformally flat; defining the conformal group with respect to this con-formally flat metric will yield the same results as above.

25.3 The (1, 1) Superconformal Group

We will concentrate on the Euclidean superconformal group. The analog-ous Minkowski space results can easily be found by the same procedures and correspond, for the most part, to the insertion of factors of i. The first discus-sion of the (1, 1) superconformal group was given in Ref. 241.

The (1, 1) superspace line element can be taken to be of the form

$$ds^2 = dL \, d\bar{L} \tag{25.51}$$

where

$$dL = dz^\pi E_\pi{}^+ = dz - d\theta^+ \, \theta^+$$

$$d\bar{L} = dz^\pi E_\pi{}^= = d\bar{z} - d\theta^- \theta^- \tag{25.52}$$

and E_π^N is the super vielbein of $(1,1)$ rigid superspace of Eq. (23.101). We could define the $(1,1)$ superconformal group to be a superdiffeomorphism that preserves the above line element up to scale, i.e.

$$ds^2 \to ds'^2 = \Omega \, ds^2. \tag{25.53}$$

However, we will restrict our definition to a superdiffeomorphism for which

$$dL \to dL' = \mu \, dL; \tag{25.54}$$

$$d\bar{L} \to d\bar{L}' = \lambda \, d\bar{L} \tag{25.55}$$

where μ and λ are functions of z, θ^+ and $\bar{z}$, θ^- respectively and $\mu\lambda = \Omega$. This restriction rules out the possibility of orientation nonpreserving transformations. The $(1,1)$ superconformal group will be the direct product of a $(1,0)$ superconformal group obeying Eq. (25.53) and a $(0,1)$ superconformal group obeying Eq. (25.54). Since these groups are mirror images of each other under $z \leftrightarrow \bar{z}$, $\theta^+ \leftrightarrow \theta^-$ we restrict our attention to the left-handed, i.e. $(1,0)$ part and label θ^+ by θ when no ambiguity is possible.

The line element in supersymmetric theories is not unique as a consequence of the reducible representation of the tangent group, the Lorentz group, carried by the superspace. In particular, we could add the supersymmetric invariant $d\theta^+ \, d\theta^-$ times any number to the line element of Eq. (25.51) and considered the resulting conformal group. However, the definition of Eqs. (25.54) and (25.25) which works directly with only parts of the supervielbein will provide us with a superconformal group with the desired properties. This is yet another example of the difference between x-space and superspace geometry.

The most general superdiffeomorphism on z, θ is of the form

$$z \to z' = g + \theta\kappa$$

$$\theta \to \theta' = \mathcal{G} + \theta B. \tag{25.56}$$

where g, K, $\mathcal{G}$ and B are functions of z. Imposing Eqs. (25.54) and (25.55) implies that

$$\kappa = \mathcal{G}B, \qquad B = \sqrt{g_z - \mathcal{G}_z\mathcal{G}} \tag{25.57}$$

where

$$g_z = \partial_z g, \qquad \mathscr{G}_z = \partial_z \mathscr{G}.$$

A complete basis for infinitesimal transformations provided by the choices $g(z) = az^{n-1}$, $\mathscr{G} = 0$ under which

$$z' = z + az^n, \qquad \theta' = \theta + a\left(\frac{n+1}{2}\right)z^n\theta \qquad (25.58)$$

and $g(z) = z$, $\mathscr{G} = z^{r-1/2}\varepsilon$ for ε Grassmann odd under which

$$z' = z + \theta\varepsilon z^{r+1/2}, \qquad \theta' = \theta + z^{r+1/2}\varepsilon. \qquad (25.59)$$

The corresponding infinitesimal operators are

$$l_n = z^n\left\{z\frac{d}{dz} + \frac{1}{2}(n+1)\theta\frac{d}{d\theta}\right\}; \qquad n = 0, \pm 1, \pm 2, \dots.$$

and

$$\mathscr{G}_r = z^{r+1/2}\left(\frac{\partial}{\partial\theta} - \theta\frac{\partial}{\partial z}\right). \qquad (25.60)$$

These operators correspond to an algebra with generators L_n and G_r and this $(1,0)$ superconformal algebra is given by

$$[L_n, L_m] = (n-m)L_{n+m}$$

$$[G_r, L_n] = \left(r - \frac{n}{2}\right)G_{n+r}$$

$$\{G_r, G_s\} = 2L_{r+s}. \qquad (25.61)$$

The reader is referred to Chapter 16 for the discussion of active and passive transformations and the sign of the terms on the right-hand side.

For $r = \pm 1/2, \pm 3/2, \pm 5/2; \dots$ we recognize the Neveu-Schwarz algebra[7] while for $r = 0, \pm 1, \pm 2, \dots$ we have the Ramond algebra.[7] In the latter case, it is usual to replace G_r by F_m; $m = 0, \pm 1, \pm 2$.

For the Neveu-Schwarz case the usual $(1,1)$ supersymmetry algebra is contained in the $(1,1)$ superconformal algebra and generated by the translation L_{-1}, the usual supersymmetry $G_{-1/2}$ and the rotation J. The superextension of the Möbius group generated by L_{-1}, L_0, L_{+1} is given by including the generators $G_{-1/2}$ and $G_{+1/2}$. The $G_{+1/2}$ generate what are usually called special or S-supersymmetry transformations. One may verify that only these transformation lead to vector fields which are globally defined on the super Riemann sphere.

For the Ramond case there is no straightforward extension of the Möbius group since if one includes any F_{+m} generator we must also include L_{+m-1},

L_m, L_{m+1} and L_{2m}, and any generators these require for closure. One easily verifies that one ends up by including the whole of the superconformal algebra.

In Eq. (25.57) we can take either sign of the square root for B and still have a superconformal transformation. Indeed under $z \to e^{2\pi i}z$, θ behaves as

$$\theta \to \sqrt{e^{2\pi i}}\theta = \pm\theta$$

as one might expect from a 'fermionic' object. We will not discuss this sign ambiguity further, but note that it has important consequences in string theory.

Under the transformations of Eq. (25.56) subject to Eq. (25.57) we find

$$dL' = (D\theta')^2 \, dL. \tag{25.62}$$

Our definition of the superconformal group involved $E^+{}_\pi$ part of the super-vielbein; an alternative, but equivalent definition is to demand superdiffeo-morphisms which preserve the $E_-{}^\pi$ and $E_+{}^\pi$ parts of the rigid supervielbein up to Weyl and Lorentz rotation:

$$E_\pm{}^A(z',\theta') = \tau L_\pm{}^\pm E_\pm{}^\pi(z,\theta)\partial_\pi z'^A \tag{25.63}$$

where τ is the Weyl factor and

$$L_\pm{}^\pm = e^{\pm i\omega/2}$$

the Lorentz rotation, both τ and ω being superfields. Recalling the rigid supervielbein of Eq. (23.42), Eq. (25.63) becomes

$$D_+\theta^{-\prime} = 0 = D_-\theta^{+\prime}, \qquad D_+\bar{z}' = 0 = D_-z' \tag{25.64}$$

and

$$D_+z' = \tau^{-1}e^{-i\omega/2}\theta'_+, \qquad D_+\theta^{+\prime} = \tau^{-1}e^{-i\omega/2}$$

$$D_-\bar{z}' = \tau^{-1}e^{i\omega/2}\theta'_-, \qquad D_-\theta^{-\prime} = \tau^{-1}e^{i\omega/2} \tag{25.65}$$

The first equation ensures that left and right coordinates do not mix under superconformal transformations. The content of the second equation is that $\theta^{+\prime}$ and $\theta^{-\prime}$ is unrestricted, but[241]

$$D_+z' = \theta^{+\prime}D_+\theta^{+\prime}$$

$$D_-\bar{z}' = \theta^{-\prime}D_-\theta^{-\prime} \tag{25.66}$$

Substituting the most general superdiffeomorphism of Eq. (25.56) we find the constraints of Eq. (25.57). Thus our two definitions of superconformal trans-formations are equivalent.

The definition of Eq.(25.63) is equivalent to demanding that the spinorial covariant derivatives must transform homeogeneously under a supercon-

formal transformation. Indeed, we find that this provides the quickest derivation of a superconformal transformation. Restricting our attention to the left
coordinates, we find that

$$D = (Dz')\partial_{z'} + (D\theta')\partial_{\theta'}$$

$$= (D\theta')D' + (Dz' - \theta'D\theta')\partial_{z'}. \tag{25.67}$$

Consequently we have the condition of Eq. (25.66) and

$$D = (D\theta')D'. \tag{25.68}$$

The other covariant derivatives do not transform covariantly, indeed

$$\partial_z = (\partial_z z')\partial_{z'} + (\partial_z \theta')\partial_{\theta'}$$

$$= (\partial_z z' - \partial_z \theta'\theta')\partial_{z'} + (\partial_z \theta')D'. \tag{25.69}$$

We define a primary field to have super weight $(H, \bar{H})$, if under a superconformal transformation it transforms as

$$\phi'(Z', \bar{Z}')(dL')^H (d\bar{L}')^{\bar{H}} = \phi(Z, Z')(dL)^H (d\bar{L})^{\bar{H}} \tag{25.70}$$

where we adopt the notation

$$Z = (z, \theta^+), \qquad \bar{Z} = (z, \theta^-). \tag{25.71}$$

Using Eq. (25.62) this implies that

$$\phi'(Z', \bar{Z}')(D_+\theta^{+\prime})^{2H}(D_-\theta^{-\prime})^{2\bar{H}} = \phi(Z, \bar{Z}). \tag{25.72}$$

In what follows below, we will indicate the Z dependence, but suppress the
$\bar{Z}$ dependence and its corresponding weight factors. It is straightforward to
rewrite these factors. For an infinitesimal transformation

$$Z \to Z' = Z - F,$$

Eq. (25.72) becomes

$$\delta\phi(Z) \equiv \phi'(Z') - \phi(Z) = F^\pi \partial_\pi \phi + 2H(DF^\theta)\phi = F^z \partial_z \phi + F^\theta \partial_\theta \phi + 2H(DF^\theta)\phi$$

$$= F^+\partial_z \phi + F^+ D\phi + 2H(DF^\theta)\phi \tag{25.73}$$

where

$$F^+ = F^\pi E_\pi^+ = F^z - F^\theta \theta$$

$$F^+ = F^\pi E_\pi^+ = F^\theta. \tag{25.74}$$

From Eqs. (25.56) and (25.67) and by taking

$$g(z) = z - h(z), \qquad \mathcal{G} = -\varepsilon(z)$$

we find that

$$F^z = h(z) + \theta\varepsilon(z), \qquad F^\theta = \varepsilon(z) + \tfrac{1}{2}\theta\partial_z h(z).$$

Consequently, one has

$$F^+ = h(z) + 2\theta\varepsilon(z), \qquad F^+ = \varepsilon(z) + \tfrac{1}{2}\theta\partial_z h(z).$$

Using Eq. (25.66) or by direct observation of the above, we find that

$$F^+ = \tfrac{1}{2}DF^+, \qquad DF^+ = \tfrac{1}{2}\partial_z F^+.$$

Thus the infinitesimal transformation of a superfield of superweight H is

$$\delta\phi = F\partial_z\phi + \tfrac{1}{2}(DF)D\phi + H(\partial_z F)\phi \qquad (25.75)$$

where $F = F^+$ is an arbitrary superfield.

It is instructive to compute the Jacobian corresponding to a superconformal transformation. The integration measure transforms as

$$dz'\, d\theta' = dz\, d\theta\, J \qquad (25.76)$$

where J is the superdeterminant of the matrix

$$\begin{pmatrix} \dfrac{dz'}{dz} & \dfrac{d\theta'}{dz} \\[2mm] \dfrac{dz'}{d\theta} & \dfrac{d\theta'}{d\theta} \end{pmatrix}. \qquad (25.77)$$

We now summarize some of the most important properties of the super-determinant. For any matrix which is of the form

$$M = \begin{pmatrix} A & B \\ C & D \end{pmatrix}$$

where B and C and Grassmann odd and A and D Grassmann even, we define its superdeterminant or Berezinian[242] by

$$s\det M = \det A[\det(D - CA^{-1}B)]^{-1}. \qquad (25.78)$$

This definition of the superdeterminant satisfies

$$s\det(M_1 M_2) = (s\det M_1)(s\det M_2) \qquad (25.79)$$

and

$$s\det M = \exp s\,\mathrm{Tr}\,\ln M \qquad (25.80)$$

provided we define the $s\,\mathrm{Tr}$ by

$$s \operatorname{Tr} M = \sum_j A_{jj} - \sum_k D_{kk}. \tag{25.81}$$

One can also show, as mentioned above, that it is the superdeterminant that arises as the Jacobian factor when changing the variables of integration.

Applying Eqs. (25.78) and (25.56) we find that J of Eq. (25.77) is given by

$$J = D\theta' = B + \theta \partial_z \mathcal{G} \tag{25.82}$$

which is the square roots of the scale factor induced on dL by a superconformal transformation.

A quicker way of obtaining this result is to realize that

$$J = s \det \begin{vmatrix} \partial_z z' & \partial_z \theta' \\ Dz' & D\theta' \end{vmatrix}$$

since the addition of the bottom row is proportional to the top row. Using Eq. (25.66) we find that

$$J = s \det \begin{vmatrix} (D\theta')^2 - \theta' \partial_z \theta' & \partial_z \theta' \\ \theta' D\theta' & D\theta' \end{vmatrix} = D\theta'.$$

Consequently, for a superfield ϕ of superweight $(1/2, 1/2)$

$$\int dz\, d\bar{z}\, d\theta^+\, d\theta^-\, \phi \tag{25.83}$$

is superconformally invariant.

We recall from Chapter 24 that any supervielbein can be chosen, using a super diffeomorphism, to be superconformally flat, namely it is related to the supervielbein of rigid superspace by a super Weyl transformation. Defining the conformal group with respect to any superconformally flat metric will yield the same superconformal group as defining it with respect to the rigid supervielbein as was done in this section. This fact follows from the observation that only parts of the supervielbein are independent since the torsion constraints are used to solve for certain parts in terms of the certain other components of the supervielbein. Regardless of which of the above definitions we choose, the superconformal transformation of a part of the supervielbein is given and the transformations of the remaining components are solved by the torsion constraints which of course also hold for rigid superspace. Now, the components of the supervielbein used in our definitions also scale uniformly under a super Weyl transformation which by definition preserve the torsion constraints. Hence the relation to these particular components for the rigid supervielbein superspace and those for a general superconformally flat supervielbein are related by a scale factor. Examining the above definition involving the rigid supervielbein we realize that the scale factor would not change the final transformations and hence the superconformal groups are the same.

25.4 The (2, 2) Superconformal Group

The (2, 2) case follows, in principle, the (1, 1) superconformal discussion. Adopting the notation

$$Z^\pi = (z, \theta^+, \bar{\theta}^+), \qquad \bar{Z}^\pi = (\bar{z}, \theta^-, \bar{\theta}^-)$$

for the coordinates of (2, 2) superspace we will subsequently suppress any discussion of the right-handed coordinates $\bar{Z}^\pi$ and give only that for the left-handed coordinates. The former are easily reinstated, the (2, 2) conformal group being a direct product of a (2, 0) and (0, 2) superconformal group which act on the left and right coordinates respectively. We will also drop the $+$ indices, i.e. $Z^\pi = (z, \theta, \bar{\theta})$ where there is no ambiguity.

We can define the (2, 0) superconformal group in two ways:

Definition 1: A superconformal transformation is a superdiffeomorphism for which

$$dL \to dL' = \mu(Z)\, dL$$

where $\mu(Z)$ is an arbitrary scale factor and

$$dL \equiv dZ^\pi E_\pi{}^+ = dz - \tfrac{1}{2} d\theta\, \bar{\theta} - \tfrac{1}{2} d\bar{\theta}\, \theta \tag{25.84}$$

or

Definition 2: A superconformal transformation is a superdiffeomorphism which leaves the components of the rigid supervielbein $E_+{}^\pi$, and $E_\oplus^\pi \equiv \bar{E}_+{}^\pi$ invariant up to a local Weyl and Lorentz transformation

$$E_+{}^A(Z') = \mu L_+{}^+ E_+{}^\pi(Z) \partial_\pi Z^{A'}$$
$$\bar{E}_+{}^A(Z') = \mu L_+{}^+ \bar{E}_+{}^\pi(Z) \partial_\pi Z^{A'}. \tag{25.85}$$

The later definition is equivalent to demanding that the spinorial covariant derivatives transform homogeneously. We find that

$$D = (Dz')\partial_{z'} + (D\theta')\partial_{\theta'} + (D\bar{\theta}')\partial_{\bar{\theta}'}$$
$$= (Dz' - \tfrac{1}{2}\bar{\theta}'D\theta' - \tfrac{1}{2}\theta'D\bar{\theta}')\partial_{z'} + (D\theta')D' + (D\bar{\theta}')\bar{D}'. \tag{25.86}$$

Carrying out a similar calculation for $\bar{D}$, we conclude that a superconformal transformation satisfies

$$Dz' = \tfrac{1}{2}\bar{\theta}'D\theta' \qquad D\bar{\theta}' = 0$$

$$\bar{D}z' = \tfrac{1}{2}\theta'\bar{D}\bar{\theta}' \qquad \bar{D}\theta' = 0 \tag{25.87}$$

and the covariant derivatives transform as

$$D = (D\theta')D'; \qquad \bar{D} = (\bar{D}\bar{\theta}')\bar{D}'. \tag{25.88}$$

It is straightforward to verify that Eqs. (25.87) are also the necessary and sufficient conditions for a superconformal transformation as defined by Definition 1. The scaling of dL being

$$dL \rightarrow dL' = (D\theta')(\bar{D}\bar{\theta}')\,dL. \tag{25.89}$$

For an infinitesimal transformation $Z^{\pi'} = Z^{\pi} - F^{\pi}$ Eqs. (25.87) become

$$DF^{\bar{\theta}} = 0 = \bar{D}F^{\theta} \tag{25.90}$$

$$DF^z = \tfrac{1}{2}F^{\bar{\theta}} + \tfrac{1}{2}\bar{\theta}DF^{\theta}$$

$$\bar{D}F^z = \tfrac{1}{2}F^{\theta} + \tfrac{1}{2}\theta\bar{D}F^{\bar{\theta}}. \tag{25.91}$$

We may rewrite these equations as

$$DF^{+} = F^{\bar{\theta}}$$

$$\bar{D}F^{+} = F^{\theta} \tag{25.92}$$

where

$$F^{+} = F^{\pi}E_{\pi}{}^{+} = F^z - \tfrac{1}{2}F^{\theta}\bar{\theta} - \tfrac{1}{2}F^{\bar{\theta}}\theta. \tag{25.93}$$

Note that applying D and $\bar{D}$ to the first and second of Eq. (25.92) respectively we recover Eq. (25.90).

Consequently, an infinitesimal superconformal transformation is determined by one arbitrary superfield $F \equiv F^{+}$, in terms of which

$$z' = z - (F - \tfrac{1}{2}\bar{\theta}\bar{D}F - \tfrac{1}{2}\theta DF)$$

$$\theta' = \theta - \bar{D}F, \qquad \bar{\theta}' = \bar{\theta} - DF. \tag{25.94}$$

Taking

$$F = g(z) + \theta\bar{\mathscr{G}}(z) + \bar{\theta}\mathscr{G}(z) + \theta\bar{\theta}L(z), \tag{25.95}$$

we find that

$$z' = z - \left(g + \tfrac{1}{2}\theta\bar{\mathscr{G}} + \tfrac{1}{2}\bar{\theta}\mathscr{G}\right)$$

$$\theta' = \theta - \left(\mathscr{G} + \theta\left(\frac{g_z}{2} - L\right) + \frac{1}{2}\theta\bar{\theta}\mathscr{G}_z\right)$$

$$\bar{\theta}' = \bar{\theta} - \left(\bar{\mathcal{G}} + \bar{\theta}\left(\frac{g_z}{2} + L\right) - \frac{1}{2}\theta\bar{\theta}\bar{\mathcal{G}}_z\right) \tag{25.96}$$

where

$$g_z = \partial_z g, \qquad \mathcal{G}_z = \partial_z \mathcal{G}, \qquad \bar{\mathcal{G}}_z = \partial_z \bar{\mathcal{G}}. \tag{25.97}$$

We define a primary field to have super weight H and chiral weight T if it transforms under a superconformal transformation as

$$\phi(z) = (D\theta')^{H-T/2}(\bar{D}\bar{\theta}')^{H+T/2}\phi'(z'). \tag{25.98}$$

Under the infinitesimal transformation of Eq. (25.94), ϕ transforms as

$$\delta\phi \equiv \phi(z') - \phi(z)$$

$$= F\partial_z\phi + (\bar{D}F)D\phi + (DF)\bar{D}\phi + \left(H - \frac{T}{2}\right)(D\bar{D}F)\phi$$

$$+ \left(H + \frac{T}{2}\right)\bar{D}DF\phi \tag{25.99}$$

or

$$\delta\phi = F\partial_z\phi + (\bar{D}F)D\phi + (DF)\bar{D}\phi + H(\partial_z F)\phi$$

$$- \frac{T}{2}([D, \bar{D}]F)\phi. \tag{25.100}$$

A complete basis for infinitesimal transformations are provided by the set

$$F = \{z^{n+1}a, \sqrt{2}z^{r+1/2}\bar{\theta}\bar{\varepsilon}, \sqrt{2}z^{r+1/2}\bar{\theta}\varepsilon, \theta\bar{\theta}z^n l\} \tag{25.101}$$

which induce the corresponding coordinate transformations

$$z' = z - az^{n+1}, \qquad \theta' = \theta - \frac{a}{2}(n+1)z^n\theta$$

$$\bar{\theta}' = \bar{\theta} - \frac{a}{2}z^n(n+1)\bar{\theta} \tag{25.102}$$

and

$$z' = z - \frac{1}{2}\sqrt{2}z^{r+1/2}\theta\bar{\varepsilon} - \frac{\sqrt{2}}{2}z^{r+1/2}\bar{\theta}\varepsilon$$

$$\theta' = \theta - \sqrt{2}z^{r+1/2}\varepsilon - \frac{\sqrt{2}}{2}(r + \tfrac{1}{2})z^{r-1/2}\theta\bar{\theta}\varepsilon$$

$$\bar{\theta}' = \bar{\theta} - \sqrt{2}z^{r+1/2}\bar{\varepsilon} + \frac{\sqrt{2}}{2}(r + \tfrac{1}{2})z^{r-1/2}\theta\bar{\theta}\bar{\varepsilon} \tag{25.103}$$

and

$$z' = z, \qquad \theta' = \theta + z^n l\theta$$

$$\bar{\theta}' = \bar{\theta} - z^n l\bar{\theta}. \tag{25.104}$$

Associated with the above transformations we have infinitesimal operators Δ which are defined by writing Eq. (25.100) in the form

$$\delta\phi = \Delta\phi. \tag{25.105}$$

Corresponding to the set of Eq. (25.101) we have the infinitesimal operators,

$$\Delta = \{l_n, \bar{\mathscr{G}}_r, \mathscr{G}_r, \tau_n\} \tag{25.106}$$

where

$$l_n = z^{n+1}\frac{\partial}{\partial z} + \frac{(n+1)}{2}z^n\theta\frac{d}{d\theta} + \frac{(n+1)}{2}z^n\bar{\theta}\frac{d}{d\bar{\theta}} + (n+1)Hz^n$$

$$+ \frac{T}{4}\theta\bar{\theta}(n+1)nz^{n-1}$$

$$\bar{\mathscr{G}}_r = \sqrt{2}\left\{z^{r+1/2}\frac{\partial}{\partial\bar{\theta}} - \frac{1}{2}z^{r+1/2}\theta\frac{\partial}{\partial z} - \frac{1}{2}\left(r+\frac{1}{2}\right)z^{r-1/2}\theta\bar{\theta}\frac{d}{d\theta}\right\}$$

$$- H\theta\left(r+\frac{1}{2}\right)z^{r-1/2} - \frac{T}{2}\theta\left(r+\frac{1}{2}\right)z^{r-1/2}\Big\}$$

$$\mathscr{G}_r = \sqrt{2}\left\{z^{r+1/2}\frac{\partial}{\partial\theta} - \frac{1}{2}z^{r+1/2}\bar{\theta}\frac{\partial}{\partial z} + \frac{1}{2}\left(r+\frac{1}{2}\right)z^{r-1/2}\theta\bar{\theta}\frac{d}{d\theta}\right.$$

$$- H\left(r+\frac{1}{2}\right)z^{r-1/2}\bar{\theta} + \frac{T}{2}\left(r+\frac{1}{2}\right)z^{r-1/2}\bar{\theta}\Big\}$$

$$\tau_n = z^n\left(-\theta\frac{\partial}{\partial\theta} + \bar{\theta}\frac{\partial}{\partial\bar{\theta}}\right) + nHz^{n-1}\theta\bar{\theta} + Tz^n. \tag{25.107}$$

The $(2,0)$ superconformal algebra associated with these infinitesimal operators has the generators L_n, $\bar{G}_r$, G_r and T_n which obey the algebra

$$[L_n, L_m] = (n-m)L_{n+m}, \qquad [L_n, G_r] = \left(\frac{n}{2} - r\right)G_{n+r}$$

$$[L_n, \bar{G}_s] = \left(\frac{n}{2} - s\right)\bar{G}_{n+s}, \qquad \{G_r, G_s\} = \{\bar{G}_r, \bar{G}_s\} = 0$$

$$\{G_r, \bar{G}_s\} = 2L_{r+s} + (r-s)T_{r+s}$$

$$[T_n, T_m] = 0, \qquad [L_n, T_m] = -mT_{n+m}$$

$$[G_r, T_n] = -G_{n+r}, \qquad [\bar{G}_s, T_n] = +\bar{G}_{n+s} \tag{25.108}$$

This algebra for $r \in Z + 1/2$ is the $N = 2$ Neveu-Schwarz algebra. One subalgebra is that of $(2,0)$ supersymmetry of Chapter 22 which consists of the generators of the translation L_{-1}, the two supersymmetries $G_{-1/2}$, $\bar{G}_{-1/2}$ the chiral $U(1)_L$, T_0 and the Lorentz rotation. The $N = 2$ super-Möbius group is generated by $L_{\pm 1}$, L_0, to $\bar{G}_{\pm 1/2}$, $G_{\pm 1/2}$ and T_0. One can verify that these possess the only globally defined vector fields or infinitesimal operators on the super-Riemann sphere.

One can also consider $r \in Z$ and in this case we have an $N = 2$ Ramond algebra. There is then no analogue of a super-Möbius group.

The above algebras are classical and the corresponding algebra for quantum conformal models have in addition the celebrated central terms which play a crucial role in these theories.

We recall that a chiral superfield satisfies $\bar{D}\phi = 0$, demanding that its superconformal variation of Eq. (25.98) also satisfies this equation, we find that

$$\bar{D}'\phi' = 0 \Rightarrow \bar{D}\phi' = 0 \Rightarrow [\bar{D}(D\theta')^{H-T/2}][\bar{D}\bar{\theta}']^{H+T/2}\phi' = 0 \quad (25.109)$$

or that

$$\left(H - \frac{T}{2}\right)\bar{D}D\theta' = \left(H - \frac{T}{2}\right)\partial_z \theta' = 0 \quad (25.110)$$

which implies that

$$H = \frac{T}{2}. \quad (25.111)$$

Similarly an antichiral superfield satisfies

$$D\bar{\phi} = 0 \quad (25.112)$$

and is a primary field only if

$$H = -T/2. \quad (25.113)$$

Finally it remains to calculate the Jacobian of a superconformal transformation. The measure

$$dz' \, d\theta' \, d\bar{\theta}' = dz \, d\theta \, d\bar{\theta} \, J \quad (25.114)$$

where

$$J = s \det \left\{ \frac{\partial z^{\pi'}}{\partial z^A} \right\}. \quad (25.115)$$

We may write this Jacobian as

$$J = s \det \left\{ \begin{array}{ccc} \partial_z z' & \partial_z \theta' & \partial_z \bar{\theta}' \\ Dz' & D\theta' & D\bar{\theta}' \\ \bar{D}z' & \bar{D}\theta' & \bar{D}\bar{\theta}' \end{array} \right\}$$

$$= s \det \left[\begin{array}{ccc} N & \partial_z \theta & \partial_z \bar{\theta}' \\ \frac{1}{2}\bar{\theta}' D\theta' & D\theta' & 0 \\ \frac{1}{2}\theta' \bar{D}\bar{\theta}' & 0 & \bar{D}\bar{\theta}' \end{array} \right] \tag{12.116}$$

where

$$N = \bar{D}\theta' D\theta' - \tfrac{1}{2}\bar{\theta}'\partial_z\theta' - \tfrac{1}{2}\theta'\partial_z\bar{\theta}'. \tag{12.117}$$

In the last line we used Eq. (25.87). Evaluating the superdeterminant according to Eq. (25.78) we find:

$$J = N \det \left\{ \begin{array}{cc} D\theta' - \dfrac{1}{2N}\bar{\theta}' D\theta'\partial_z\theta' & -\dfrac{1}{2N}\bar{\theta}' D\theta'\partial_z\bar{\theta}' \\[2mm] -\dfrac{1}{2N}\theta' \bar{D}\bar{\theta}'\partial_z\theta' & \bar{D}\bar{\theta}' - \dfrac{1}{2N}\theta' \bar{D}\bar{\theta}'\partial_z\bar{\theta}' \end{array} \right\}$$

$$= 1. \tag{12.118}$$

Consequently, for a primary field ϕ of superweight zero and chiral weight zero (i.e. $H = T = 0$)

$$\int dz \, d\theta \, d\bar{\theta} \, \phi \tag{12.119}$$

is superconformally invariant.

We can also consider the chiral integral

$$\int dz \, d\theta \, \phi = \int dy \, d\theta \, \phi \tag{25.120}$$

where

$$y = z + \tfrac{1}{2}\theta\bar{\theta} \quad \text{and} \quad \bar{D}\phi = 0.$$

Under a conformal transformation the chiral variables (y, θ) satisfies

$$\bar{D}y' = 0, \quad Dy' = \bar{\theta}' D\theta'$$

$$\bar{D}\theta' = 0. \tag{25.121}$$

The Jacobian

$$dy' \, d\theta' = J_c \, dy \, d\theta$$

is given by

$$J_c = s \det \begin{bmatrix} \partial_y y' & \partial_y \theta' \\ \dfrac{\partial y'}{\partial \theta} & \dfrac{\partial \theta'}{\partial \theta} \end{bmatrix} = s \det \begin{bmatrix} \partial_y y' & \partial_y \theta' \\ D y' & D \theta' \end{bmatrix}$$

$$= s \det \begin{bmatrix} \bar{D}\bar{\theta}' D\theta' - \bar{\theta}' \partial_y \theta' & \partial_y \theta' \\ \bar{\theta}' D\theta' & D\theta' \end{bmatrix}$$

$$= \bar{D}\bar{\theta}'. \qquad (25.122)$$

Consequently, for a chiral superfield

$$\int dz \, d\theta \, \phi \qquad (25.123)$$

is invariant provided $H = 1/2 = 1/2T$.

The $(2, 2)$ superconformal algebra was discussed in Refs. 215 and 243. The remarks on the definition of the superconformal group given at the end of the previous section also apply to the $(2, 2)$ case.

Chapter 26

Green Functions and Operator Product Expansions in (2, 2) Superconformal Models

In this chapter we will discuss two applications of supersymmetry in conformal models. We will use superconformal invariance to determine the two and three point Green functions and the chiral Green functions for $N = 2$ superconformal theories. We will also use superspace to formulate, in a compact way, the super operator product expansion associated with the superconformal currents. It is beyond the scope of this book to give an account of conformal models, however we will review some of the necessary material within the context of the ordinary conformal models before tackling the superconformal case. There are a number of reviews on conformal theories; as well as the seminal paper of Ref. 239, the author enjoyed reading those of Ref. 244. The reader may also consult Sec. 4 of Ref. 245 where much of the immediate background material can be found. Our aim in this chapter is to give the reader a feeling for the role supersymmetry plays in conformal models.

26.1 Two and Three Point Green Functions

It has been known for many years[246] that the Möbius subgroup is sufficient to determine the two and three-point function in a conformal model. In particular, it specifies completely the coordinate dependence of the Green functions of three primary fields $\phi_i(z, \bar{z})$, $i = 1, 2, 3$ with conformal weights $(h_i, \bar{h}_i)$:

$$G(z_i, \bar{z}_i) \equiv \langle | T\{\phi_1(z_1, \bar{z}_1)\phi_2(z_2, \bar{z}_2)\phi_3(z_3, \bar{z}_3)\} | \rangle$$

$$\equiv \langle \phi_1(z_1, \bar{z}_1)\phi_2(z_2, \bar{z}_2)\phi_3(z_3, \bar{z}_3) \rangle \tag{26.1}$$

in a conformal model which does not possess supersymmetry. We will as before consider in detail only the z dependence; the $\bar{z}$ dependence being easily reinstated.

We begin with a discussion of conformal theories before giving that for superconformal theories. We may use a Möbius transformation to take any three distinct points on $\hat{C}$ to any other three distinct points. For example, the transformation

$$z \rightarrow z' = \frac{(z - z_1)(z_2 - z_3)}{(z - z_3)(z_2 - z_1)} \qquad (26.2)$$

maps the distinct points (z_1, z_2, z_3) to the points $(0, 1, \infty)$ and it has an inverse which can easily be constructed. Hence we may map (z_1, z_2, z_3) to $(0, 1, \infty)$ and then $(0, 1, \infty)$ to any other three distinct points. Since the Möbius group has only three parameters we cannot map four or more arbitrary but distinct points to any other four or more distinct points. The primary fields have a well-defined transformation rule under conformal transformations

$$\phi(z) \rightarrow \left[\frac{dz'}{dz}\right]^n \phi(z') \qquad (26.3)$$

and as such we can expect to change the arguments of the three-point function to $(0, 1, \infty)$ and acquire a corresponding weight factor.

Inserting the operator U which induces the Möbius transformation $z \rightarrow z'$ for the primary field, i.e. $U\phi(z)U^{-1} = (dz'/dz)^h \phi(z')$, in an N point function, we have

$$G(z_i) = \left\langle \prod_{i=1}^{N} (U\phi_i(z_i)U^{-1}) \right\rangle$$

$$= \left\langle \prod_{i=1}^{N} \left(\frac{dz_i'}{dz_i}\right)^{h_i} \phi_i(z_i') \right\rangle = \left\langle \prod_{i=1}^{N} \phi_i(z_i) \right\rangle. \qquad (26.4)$$

We have used the important property that the elements of a $SU(1, 1)$ i.e. Möbius transformation annihilate the vacuum:

$$L_{\pm 1}| \; \rangle = L_0| \; \rangle = 0 \Leftrightarrow U| \; \rangle = | \; \rangle. \qquad (26.5)$$

We could simply insert the map of Eq. (26.2) into the case where $N = 3$, however it will prove useful to find differential equations for $G(z_i)$ for arbitrary N. Equation (26.4) holds for any Möbius transformation and hence in particular for the infinitesimal transformations

$$z \rightarrow z + a, \qquad z \rightarrow z + az, \qquad z \rightarrow z + az^2 \qquad (26.6)$$

generated by L_{-1}, L_0 and L_{+1} respectively. Making these substitutions Eq. (26.4) then implies the differential equations

$$\sum_{i=1}^{N} \Delta^i G = 0 \qquad (26.7)$$

where

$$\Delta^i = (l_{-1}^i, l_0^i, l_{+1}^i)$$

with

$$l^i_{-1} = \frac{\partial}{\partial z_i}, \qquad l^i_0 = z_i \frac{\partial}{\partial z_i} + h_i$$

$$l^i_{+1} = z_i^2 \frac{\partial}{\partial z_i} + 2h_i z_i. \qquad (26.8)$$

The l_{-1} equation implies $G(z_j)$ is a function of differences $z_{ij} \equiv z_i - z_j$. Taking $G(z_{ij})$ of the form

$$G = e \prod_{i<j} \frac{1}{(z_{ij})^{k_{ij}}} \qquad (26.9)$$

and assuming $e \neq 0$, the l_0 and l_1 equations imply

$$-\sum_{i<j} k_{ij} + \sum_{i=1}^{N} h_i = 0 \qquad (26.10)$$

and

$$-\sum_{i<j} (z_i + z_j)k_{ij} + 2 \sum_{i=1}^{N} h_i z_i = 0. \qquad (26.11)$$

Consider now the two point function $G = e(z_1 - z_2)^{-k}$; Eqs. (26.10) and (26.11) imply that

$$-k + h_1 + h_2 = 0; \qquad -(z_1 + z_2)k + 2h_1 z_1 + 2h_2 z_2 = 0. \qquad (26.12)$$

The unique solution is $h_1 = h_2, k = 2h_1$, and hence for any primary fields $\phi_i(z)$

$$G(z_1, z_2) = \langle \phi_i(z_1)\phi_j(z_2) \rangle = \frac{\delta_{ij}}{(z_1 - z_2)^{2h_i}} \qquad (26.13)$$

We may always scale our primary fields so that $e = 1$ and diagonalize any matrix which appears on the right-hand side for primary fields of the same weight.

Turning to the three-point function, which is of the form

$$G = \frac{e}{(z_{12})^{k_1}(z_{23})^{k_2}(z_{31})^{k_3}} \qquad (26.14)$$

Eqs. (26.10) and (26.11) read

$$-(k_1 + k_2 + k_3) + h_1 + h_2 + h_3 = 0 \qquad (26.15)$$

and

$$-k_1(z_1 + z_2) + 2h_1 z_1 + \text{cyclic} = 0. \qquad (26.16)$$

Taking the latter equation and reading off the coefficient of z_1 we have $k_1 + k_3 = 2h_1$ and the analogous cycled equations. Solving these equations we find

$$k_1 = \tfrac{1}{2}\{(k_1 + k_3) + (k_2 + k_1) - (k_3 + k_2)\}$$
$$= h_1 + h_2 - h_3 \qquad (26.17)$$

and the analogous cycled equations. Hence the three-point function is of the form

$$G(z_i) = \langle \phi_1(z_1)\phi(z_2)\phi(z_3)\rangle$$

$$= \frac{e}{(z_1 - z_2)^{h_1+h_2-h_3}(z_2 - z_3)^{h_2+h_3-h_1}(z_3 - z_1)^{h_3+h_1-h_2}}. \qquad (26.18)$$

We cannot determine the constant e which could vanish, however the arguments above ensure that the dependence on z is unique.

For Green functions with more than three primary fields we cannot, in general, use Möbius or indeed conformal invariance to solve for their functional dependence. Equation (26.7) only provides us with three differential equations, hence we can only hope to determine uniquely the three-point Green's function. Indeed for four z_i, $i = 1, 2, 3, 4$. We can construct the 'cross ratio'

$$u = \frac{(z_1 - z_3)\,(z_2 - z_4)}{(z_1 - z_4)\,(z_2 - z_3)} \qquad (26.19)$$

which is Möbius invariant. A solution of Eq. (26.7) is given by

$$\prod_{i<j} \left(\frac{1}{z_{ij}}\right)^{2/(N-2)[h_i+h_j-h/(N-1)]} F(u_i). \qquad (26.20)$$

where

$$h = \sum_{i=1}^{N} h_i$$

and F is an arbitrary function of the N-3 independent cross ratios u_i, $i = 1, \ldots, N$-3 say.

One cannot directly use the conformal transformations outside the Möbius group since these are generated by $L_{\pm n}$ for $n \geqslant 2$ and $L_{-n}|\,\rangle \neq 0$ for $n \geqslant 2$. However, for certain so-called 'rational' theories there exist null states of the form

$$(L_{-p}|\psi\rangle + \text{higher powers of } L_{-m})|\psi\rangle$$

for some integer p, and these can be used to derive further differential equations which determine the Green's functions.[239]

The construction of the two- and three-point Green's functions for super-conformal models follows the same procedures as for the bosonic case given above. For the $(1,1)$ superconformal models the super Möbius group is generated by L_0, $L_{\pm 1}$ and $G_{\pm 1/2}$. It is straightforward to find the result and we recommend this exercise to the reader.

Let us now consider $(2,2)$ superconformal models and restrict our attention to the left-handed coordinates $Z = (z, \theta, \bar{\theta})$ which carry a nontrivial represen-tation of the $(2,0)$ superconformal group. The super Möbius group is gene-rated by L_0, $L_{\pm 1}$, $G_{\pm 1/2}$, $\bar{G}_{\pm 1/2}$ and T_0. Following the same arguments as in the previous section we conclude that an N-point Green's function con-structed from primary fields $\phi_i(Z_i)$ of superweight H_i and chiral weight T_i

$$G(Z_i) = \left\langle \prod_{i=1}^{N} \phi_i(Z_i) \right\rangle \tag{26.21}$$

satisfies the differential equations

$$\sum_{i=1}^{N} \Delta^i G = 0 \tag{26.22}$$

where

$$\Delta^i = (l_{-1}^i, l_0^i, l_{+1}^i; \bar{\mathcal{G}}_{-1/2}^i, \mathcal{G}_{-1/2}^i, \bar{\mathcal{G}}_{1/2}^i, \mathcal{G}_{1/2}^i, \tau_0). \tag{26.23}$$

These operators can be found in Eq. (25.107). The l_{-1} equation implies that G should be a function of $z_{ij} \equiv z_i - z_j$, θ_i, and $\bar{\theta}^i$. While the $G_{-1/2}$ and $\bar{G}_{-1/2}$ equations imply that G should be a function of the supersymmetric line element

$$s_{ij} \equiv z_i - z_j - \tfrac{1}{2}(\theta_i \bar{\theta}_j + \bar{\theta}_i \theta_j) \tag{26.24}$$

and

$$\theta_{ij} \equiv \theta_i - \theta_j, \qquad \bar{\theta}_{ij} = \bar{\theta}_i - \bar{\theta}_j. \tag{26.25}$$

The line element s_{ij} is invariant under supersymmetry, i.e. L_{-1}, $G_{-1/2}$ and $\bar{G}_{-1/2}$ and having chosen this variable, the remaining variables must also be in-variant under supersymmetry and hence must be θ_{ij} and $\bar{\theta}_{ij}$. Consequently, the Green function is of the form

$$G(s_{ij}, \theta_{ij}, \bar{\theta}_{ij}). \tag{26.26}$$

We must now implement the remaining conditions. The l_0, l_{+1} and τ_0 equations in terms of these new variables become

$$\left[\sum_{i<j} \left\{ s_{ij} \frac{\partial}{\partial s_{ij}} + \frac{1}{2} \theta_{ij} \frac{\partial}{\partial \theta_{ij}} + \frac{1}{2} \bar{\theta}_{ij} \frac{\partial}{\partial \bar{\theta}_{ij}} \right\} + \sum_{i=1}^{N} H_i \right] G = 0 \tag{26.27}$$

$$\left[\sum_{i<j}\left\{(z_i + z_j)s_{ij}\frac{\partial}{\partial s_{ij}} + \left[\frac{1}{2}(z_i + z_j)\theta_{ij} + \frac{1}{2}(z_i - z_j)(\theta_i + \theta_j)\right]\frac{\partial}{\partial \theta_{ij}}\right.\right.$$

$$\left.+ \left[\frac{1}{2}(z_i + z_j)\bar{\theta}_{ij} + \frac{1}{2}(z_i - z_j)(\bar{\theta}_i + \bar{\theta}_j)\right]\frac{\partial}{\partial \bar{\theta}_{ij}}\right\}$$

$$\left.+ \sum_{i=1}^{N}\left(2z_i H_i + \frac{T_i}{2}\theta_i\bar{\theta}_i\right)\right]G = 0 \tag{26.28}$$

and

$$\tau_0 G = \left[\sum_{i<j}\left(-\theta_{ij}\frac{\partial}{\partial \theta_{ij}} + \bar{\theta}_{ij}\frac{\partial}{\partial \bar{\theta}_{ij}}\right) + \sum_{i=1}^{N} T_i\right]G = 0. \tag{26.29}$$

In fact, it suffices to ensure that τ_0, l_0 and l_{+1} are satisfied, as then the commutator of l_{+1} with $\mathscr{G}_{-1/2}$ or $\bar{\mathscr{G}}_{-1/2}$ results in $\mathscr{G}_{+1/2}$ or $\bar{\mathscr{G}}_{+1/2}$ being also satisfied.

Let us first consider the two-point Green's function $G(Z_1, Z_2)$ which must be of the form

$$G(Z_1, Z_2) = \frac{a}{(s_{12})^{\kappa_1}} + \frac{b\theta_{12}}{(s_{12})^{\kappa_2}} + \frac{c\bar{\theta}_{12}}{(s_{12})^{\kappa_3}} + \frac{d\theta_{12}\bar{\theta}_{12}}{(s_{12})^{\kappa_4}}. \tag{26.30}$$

Imposing the differential equation arising from chiral transformations, τ_0 we find that $T_1 + T_2$ can either be 0 or ± 1. If $T_1 + T_2 = 0$ then $b = c = 0$ while if $T_1 + T_2 = +1$, then $a = d = c = 0$ and if $T_1 + T_2 = -1$ then $a = d = b = 0$. Suppose that $T_1 + T_2 = 0$, the dilation equation, l_0 implies that

$$\kappa_1 = H_1 + H_2; \qquad \kappa_4 = H_1 + H_2 + 1. \tag{26.31}$$

The special K transformations, l_{+1} lead to the condition

$$0 = \left\{-\kappa_1(z_1 + z_2)a - (\kappa_4 - 1)(z_1 + z_2)d\frac{\theta_{12}\bar{\theta}_{12}}{s_{12}} + \frac{d}{2}(\theta_1 + \theta_2)\bar{\theta}_{12}\right.$$

$$+ \frac{1}{2}\theta_{12}(\bar{\theta}_1 + \bar{\theta}_2) + \left(2z_1 H_1 + 2z_2 H_2 + \frac{T_1}{2}\theta_1\bar{\theta}_1 + \frac{T_2}{2}\theta_2\bar{\theta}_2\right)$$

$$\left.\times \left(a + d\frac{\theta_{12}\bar{\theta}_{12}}{s_{12}}\right)\right\} \tag{26.32}$$

which is satisfied if and only if

$$H_1 = H_2 \qquad \text{and} \qquad -\frac{aT_1}{2} = d. \tag{26.33}$$

Repeating this analysis we find that there is no solution when $T_1 + T_2 = \pm 1$. Consequently for any two-primary superfields[247]

$$\langle \phi_1(Z_1)\phi_2(Z_2) \rangle = G(Z_1, Z_2)$$

$$= \frac{1}{(s_{12})^{H_1 + H_2}} - \frac{T_1}{2} \frac{\theta_{12}\bar{\theta}_{12}}{(s_{12})^{H_1 + H_2 + 1}} \qquad (26.34)$$

if $H_1 = H_2$ and $T_1 = -T_2$ and vanishes otherwise.

Let us consider now the three-point function which depends on 3 bosonic coordinates z_i, $i = 1, 2, 3$, and 6 Grassmann odd coordinates θ_i, $\bar{\theta}_i$, $i = 1, 2, 3$. On the other hand, the super Möbius group contains the three bosonic generators l_0, $l_{\pm 1}$ and four anticommuting generators $G_{\pm 1/2}$, $\bar{G}_{\pm 1/2}$, Given N-points in $(2, 0)$ superspace $z_j = (z_j, \theta_j, \bar{\theta}_j)$ $j = 1, 2, \ldots, N$ we may use the above transformations to map any three of the z_j's to any three chosen values. Similarly, we may map 2 of the θ_i's and 2 of the $\bar{\theta}_i$'s to fixed values. One can verify these statements by constructing explicitly the corresponding super Möbius map.

An alternative method of arriving at the same conclusion is to examine the number of super Möbius invariants. For any four-points Z_i, $i = 1, 2, 3, 4$ a bosonic super Möbius invariant which generalizes the bosonic cross-ratio is given by

$$u = \frac{s_{13}}{s_{14}} \cdot \frac{s_{24}}{s_{23}}. \qquad (26.35)$$

There are N-3 such cross ratios. However, given any three distinct points Z_i, $i = 1, 2, 3$ there are also two Grassmann odd super Möbius invariants

$$\Theta = \frac{1}{(s_{12}s_{23}s_{31})^{1/2}} \left\{ s_{12}\theta_3 + s_{23}\theta_1 + s_{31}\theta_2 + \frac{1}{6}(\theta_1\bar{\theta}_{23}(\theta_2 + \theta_3) \right.$$

$$\left. + \theta_2\bar{\theta}_{31}(\theta_3 + \theta_1) + \theta_3\bar{\theta}_{12}(\theta_1 + \theta_2)) \right\} \qquad (26.36)$$

and

$$\bar{\Theta} = \frac{1}{(s_{12}s_{23}s_{31})^{1/2}} \left\{ s_{12}\bar{\theta}_3 + s_{23}\bar{\theta}_1 + s_{31}\bar{\theta}_2 + \frac{1}{6}(\bar{\theta}_1\theta_{23}(\bar{\theta}_2 + \bar{\theta}_3) \right.$$

$$\left. + \bar{\theta}_2\theta_{31}(\bar{\theta}_3 + \bar{\theta}_1) + \bar{\theta}_3\theta_{12}(\bar{\theta}_1 + \bar{\theta}_2)) \right\}. \qquad (26.37)$$

We are using the words super Möbius invariant in a loose sense here, what we mean is invariant under all super Möbius transformations except chiral rotations under which it transforms covariantly. The super Möbius invariance of u, Θ and $\bar{\Theta}$ is most easily established by applying the differential operators. They are obviously invariant under l_{-1}. Clearly, u is chiral inert while Θ and $\bar{\Theta}$ transform with invariance $U(1)$ weights $+1$ and -1 respectively. Applying

$\mathcal{G}_{-1/2}$ to Θ yields the following condition for invariance

$$[s_{12} + \tfrac{1}{6}\bar{\theta}_{23}(\theta_2 + \theta_3) + \tfrac{1}{3}\theta_1\bar{\theta}_{23} + \text{cyclic}]\varepsilon = 0.$$

Using the definition of s_{ij} one verifies this equation. The $l_0\Theta = 0$ is readily found to hold provided $H_i = 0$ and a slightly more complicated calculation is required to discover that $l_1\Theta = 0$ provided $H_i = 0$, but $T_i = 1/3$ for $i = 1, 2, 3$. Clearly $\bar{\mathcal{G}}_{-1/2}\Theta$ vanishes.

For N-points there are N-2 such Θ invariants and N-2 such $\bar{\Theta}$ invariants. Consequently, the number of original coordinates minus the number of invariants agrees with the number of coordinates, given above, that we may transform to chosen values.

The above counting does not mean we cannot fix the three-point Green's function. Although in principle it can depend on the invariants Θ and $\bar{\Theta}$ these are anticommuting quantities and so

$$\Theta^2 = 0 = \bar{\Theta}^2.$$

Consequently, the way the Green functions can depend on Θ and $\bar{\Theta}$ is very limited. The dependence on Θ and $\bar{\Theta}$ is further restricted in an obvious way by the chiral generator τ_0.

The three-point function can depend on s_{ij}, θ_{ij} and $\bar{\theta}_{ij}$. As there are only two-independent θ_{ij}'s and $\bar{\theta}_{ij}$'s; the τ_0 equation implies that

$$T = \sum_{i=1}^{3} T_i$$

can only take the values 0, ± 1 and ± 2.

When $T = +2$, the Green function must be of the form

$$\frac{e\theta_{12}\theta_{23}}{(s_{12})^{\kappa_1}(s_{23})^{\kappa_2}(s_{31})^{\kappa_3}}$$

where e is a constant. However, the l_1 equation implies that $e = 0$. Hence there is no nontrivial three-point Green's function with $T = +2$ and similarly for $T = -2$.

Let us consider the case $T = 0$ and begin with the simplest possibility, namely $T_i = 0$, $i = 1, 2, 3$. The Green function is of the form

$$G = \frac{e}{(s_{12})^{\kappa_1}(s_{23})^{\kappa_2}(s_{31})^{\kappa_3}} + O(\theta_{ij}\bar{\theta}_{kl}) \tag{26.38}$$

The l_0 equation implies that

$$\kappa_1 + \kappa_2 + \kappa_3 = H_1 + H_2 + H_3. \tag{26.39}$$

The l_1 equation, when $T_i = 0$, does not mix terms with different numbers of θ and $\bar{\theta}$ and one finds that it is satisfied provided

$$G = \frac{e}{(s_{12})^{H_1+H_2-H_3}(s_{23})^{H_2+H_3-H_1}(s_{31})^{H_3+H_1-H_2}}. \tag{26.40}$$

One can verify that no solution exists which includes $\theta\bar{\theta}$ terms.

Suppose now that T_i takes an arbitrary value subject only to the condition $\sum_i T_i = 0$; Then the l_1 equation mixes orders of $\theta\bar{\theta}$ due to the $1/2 \sum_i T_i\theta_i\bar{\theta}_i$ term. Exploiting this fact, we will solve the l_1 equation order by order in $\theta\bar{\theta}$. The quantity

$$1 + \frac{\theta_{12}\bar{\theta}_{12}}{s_{12}} \tag{26.41}$$

is inert under supersymmetry, satisfies the l_0 equation with $H_i = 0$ and the l_1 equation with $H_i = 0$ and $T_1 = -2$, $T_2 = +2$, $T_3 = 0$. We consider therefore a three-point function of the form

$$G = e \left\{ \frac{1}{(s_{12})^{H_1+H_2-H_3}} \frac{1}{(s_{23})^{H_2+H_3-H_1}} \frac{1}{(s_{31})^{H_3+H_1-H_2}} \right\}$$

$$\times \left\{ 1 + \left(b_1 \frac{\theta_{12}\bar{\theta}_{12}}{s_{12}} + \text{cyclic} \right) \right\} + O[(\theta\bar{\theta})^2]. \tag{26.42}$$

The l_0 equation is still satisfied and as l_{+1} also satisfies the distributive rule, its equation becomes

$$l_1 G = \frac{1}{2} \sum_i T_i\theta_i\bar{\theta}_i \left\{ b_1 \frac{\theta_{12}\bar{\theta}_{12}}{s_{12}} + \text{cyclic} \right\}$$

$$\times \frac{e}{(s_{12})^{H_1+H_2-H_3}} \frac{1}{(s_{23})^{H_2+H_3-H_1}} \frac{1}{(s_{31})^{H_3+H_1-H_2}}$$

$$+ O[(\theta\bar{\theta})^3] \tag{26.43}$$

where the $\theta\bar{\theta}$ terms cancel provided

$$b_1 - b_3 = -\frac{T_1}{2}, \qquad b_2 - b_1 = -\frac{T_2}{2}, \qquad b_3 - b_2 = -\frac{T_3}{2}.$$

The cyclic solution to these equations is

$$b_1 = \tfrac{1}{6}(T_2 - T_1), \qquad b_2 = \tfrac{1}{6}(T_3 - T_2), \qquad b_3 = \tfrac{1}{6}(T_1 - T_3). \tag{26.44}$$

It remains to cancel the $(\theta\bar{\theta})^3$ terms of Eq. (26.43). We therefore repeat the procedure and add to G the term

$$\Delta G \equiv \frac{1}{(s_{12})^{\kappa_1}(s_{23})^{\kappa_2}(s_{31})^{\kappa_3}}\left\{a_3 \frac{\theta_{12}\bar{\theta}_{12}}{s_{12}}\cdot\frac{\theta_{23}\bar{\theta}_{23}}{s_{23}} + \text{cyclic}\right\} \qquad (26.45)$$

The action of l_{+1} will cancel the $(\theta\bar{\theta})^3$ term of Eq. (26.43) provided

$$\frac{T_3 b_1}{2} = a_3 - a_2, \qquad \frac{T_1 b_2}{2} = a_1 - a_3, \qquad \frac{T_2 b_3}{2} = a_2 - a_1 \qquad (26.46)$$

whose cyclic solution is given by

$$a_3 = \frac{1}{6\cdot 6}(T_2 - T_1)(T_3 - T_2) \qquad (26.47)$$

and the analogous cycled equations. Finally one can verify that

$$\sum_i T_i \theta_i \bar{\theta}_i \Delta G = 0 \qquad (26.48)$$

as a result of $\sum_i T_i = 0$. Summarizing the result, any three point Green's function for which $\sum_i T_i = 0$ is of the form[247]

$$\langle \phi_1(z_1)\phi_2(z_2)\phi_3(z_3)\rangle = \frac{1}{(s_{12})^{H_1+H_2-H_3}}\frac{1}{(s_{23})^{H_2+H_3-H_1}}\frac{1}{(s_{31})^{H_3+H_1-H_2}}$$

$$\times \left\{1 + \frac{1}{6}(T_2 - T_1)\theta_{12}\bar{\theta}_{12}\right\}\left\{1 + \frac{(T_3 - T_2)}{6}\theta_{23}\bar{\theta}_{23}\right\}$$

$$\times \left\{1 + \frac{1}{6}(T_1 - T_3)\theta_{31}\bar{\theta}_{31}\right\}. \qquad (26.49)$$

When $T = +1$ the dependence of the Green function on θ and $\bar{\theta}$ implies that it must be Grassmann odd. To find the general such Green functions, we will adopt the same strategy as in the $T = 0$ case. Namely we will find one Green function for a given value of the T_i's such that $T = +1$ and then compute the general case by multiplying by the quantity of Eq. (26.41). A natural candidate to consider is

$$G = \frac{1}{(s_{12})^{\kappa_1}}\frac{1}{(s_{23})^{\kappa_2}}\frac{1}{(s_{31})^{\kappa_3}}\Theta \qquad (26.50)$$

where Θ is the super Möbius invariant of Eq. (26.37) which has $T_i = 1/3$ and so $T = +1$. This indeed satisfies all the differential equations provided

$$\kappa_1 = H_1 + H_2 + H_3, \qquad \kappa_2 = H_2 + H_3 - H_1$$

$$\kappa_3 = H_3 + H_1 - H_2 \qquad (26.51)$$

and G has

$$T_i = \tfrac{1}{3} \qquad i = 1, 2, 3. \tag{26.52}$$

To find a G which has arbitrary T_i, subject to only $T = 1$ we consider

$$G = e\Theta \left(\frac{1}{s_{12}}\right)^{H_1+H_2-H_3} \left(\frac{1}{s_{23}}\right)^{H_2+H_3-H_1} \left(\frac{1}{s_{31}}\right)^{H_3+H_1-H_2}$$

$$\times \left\{1 + a_1 \frac{\theta_{12}\bar{\theta}_{12}}{s_{12}} + a_2 \frac{\theta_{23}\bar{\theta}_{23}}{s_{23}} + a_3 \frac{\theta_{31}\bar{\theta}_{31}}{s_{31}}\right\}. \tag{26.53}$$

One finds that this satisfies the differential equations, provided

$$T_1 = 2(a_1 - a_3) + \tfrac{1}{3}, \qquad T_2 = 2(a_2 - a_1) + \tfrac{1}{3}$$

$$T_3 = 2(a_3 - a_2) + \tfrac{1}{3} \tag{26.54}$$

whose cyclic solution is

$$a_1 = \tfrac{1}{6}(T_2 - T_1), \qquad a_2 = \tfrac{1}{6}(T_3 - T_2)$$

$$a_3 = \tfrac{1}{6}(T_1 - T_3). \tag{26.55}$$

Summarizing a Green function for primary fields with superweights H_i, $i = 1, 2, 3$ and chiral weights T_i, $i = 1, 2, 3$ for which $\sum_i T_i = +1$ has the form

$$G = e\Theta \left(\frac{1}{s_{12}}\right)^{H_1+H_2-H_3} \left(\frac{1}{s_{23}}\right)^{H_2+H_3-H_1} \left(\frac{1}{s_{31}}\right)^{H_3+H_1-H_2}$$

$$\times \left\{1 + \frac{(T_2 - T_1)}{6} \frac{\theta_{12}\bar{\theta}_{12}}{s_{12}} + \frac{(T_3 - T_2)}{6} \frac{\theta_{23}\bar{\theta}_{23}}{s_{23}}\right.$$

$$\left. + (T_1 - T_3)\frac{\theta_{31}\bar{\theta}_{31}}{s_{31}}\right\}. \tag{26.56}$$

26.2 Chiral Correlators in (2, 2) Super Conformal Models

To solve for the Green functions for more than three primary fields is in general much more complicated, even in $N = 2$ rational conformal models where null states exist and provide a specific mechanism for their solution. An exception is when the primary fields are chiral. Before carrying out this calculation, we shall give a brief account, which is not strictly essential for what follows, of $N = 2$ minimal theories and how chiral primary fields occur within them.

It has been found[248] that unitary conformal theories possessing (2, 2) supersymmetry and central charge c less than three must belong to a discreet series with central charge

$$c = 3\left(1 - \frac{2}{n}\right) \qquad n = 3, 4, 5. \tag{26.57}$$

We caution the reader that in the literature it is often standard to rescale c for $N = 2$ theories by a factor of 3. The primary fields $\phi_{H,T}$ are labelled by the commuting generators L_0 and T_0; the indices H and T corresponding to L_0 and T_0 respectively, are their superweight and chiral charge. For a given n in the discreet series above, the only allowed values of H and T which occur in the corresponding model, are

$$H_{l,m} = \frac{l(l+2) - m^2}{4n}, \qquad T_{l,m} = \frac{m}{n}$$

$$l = 0, 1, \ldots, n - 2; \qquad m = 0, \pm1, \pm2, \ldots, \pm l. \tag{26.58}$$

As a result, we will adopt the labelling $\phi_{l,m}$ for the primary fields. The above result corresponds to the Neveu-Schwarz sector and there exists an analogous result for the Ramond sector which we will not discuss.

Given a primary field the corresponding highest weight state is as usual defined by

$$|l, m\rangle = \lim_{z \to 0} \phi_{l,m}(Z)|0\rangle. \tag{26.59}$$

One verifies from the definition of a primary field that it satisfies

$$L_n|l, m\rangle = 0, \qquad n \geqslant 1, \qquad G_r|l, m\rangle = 0$$

$$\bar{G}_r|l, m\rangle = 0 \qquad r \geqslant \tfrac{1}{2}, \qquad T_n|l, m\rangle = 0; \qquad n \geqslant 1 \tag{26.60}$$

and

$$L_0|l, m\rangle = H_{l,m}|l, m\rangle$$

$$T_0|l, m\rangle = \frac{m}{n}|l, m\rangle. \tag{26.61}$$

The descendents of this highest weight state are of the form

$$L_{-n_1} \ldots L_{-n_p} G_{-r_1} \ldots G_{-r_q} \bar{G}_{-s_1} \ldots \bar{G}_{-s_t}|l, m\rangle \tag{26.62}$$

and together with the highest weight state, by construction carry a representation of the superconformal group.

A chiral primary field is one for which the state

$$G_{-1/2}|l, m\rangle = 0 \tag{26.63}$$

is also a highest weight state. As a result this state satisfies the conditions of Eqs. (26.60) and (26.61) and so in particular

$$0 = \langle l, m| \bar{G}_{1/2} G_{-1/2} |l, m\rangle = \langle l, m| \{\bar{G}_{1/2}, G_{-1/2}\} |l, m\rangle$$
$$= \langle l, m| (2L_0 - T_0)|l, m\rangle = (2H_{l,m} - T_{l,m})\langle l, m|l, m\rangle. \qquad (26.64)$$

Hence for a chiral primary field

$$H_{l,m} = \tfrac{1}{2} T_{l,m}$$

which occurs when $l = +m$. By a similar argument, the state $G_{-1/2}|l, m\rangle$ has zero scalar product not only with itself, but with all other states. As the theory is unitary, all states have a norm which satisfies the usual properties of a norm, in particular, if the norm of a given state vanishes so does the state itself. Consequently we conclude that

$$G_{-1/2}|l, m\rangle = 0. \qquad (26.65)$$

By a similar argument, an anti-chiral primary field, in a unitary theory, is one whose highest weight state satisfies

$$\bar{G}_{-1/2}|l, m\rangle = 0. \qquad (26.66)$$

One finds that in this case that

$$H_{l,m} = -T_{l,m/2} \qquad (26.67)$$

which occur when $l = -m$. The (anti) chiral field with the lowest superweight is $(\phi_{1,-1})\phi_{1,1}$.

The above argument works backwards; if

$$H_{l,m} = \tfrac{1}{2} T_{l,m}$$

then in a unitary theory

$$G_{-1/2}|l, m\rangle = 0$$

and as such it is trivially a highest weight state.

In general, descendants which are themselves heighest weight states lead to differential equations for the Green functions that contain the primary fields of which the null states are descendents.[239] For the case of chiral primary fields, one finds that a Green function for N chiral primary fields obeys the constraint

$$\bar{D}_i\langle \phi_1(z_1)\phi_2(z_2)\ldots\phi_N(z_N)\rangle = 0 \qquad (26.68)$$

where $\bar{D}_i$ is the usual covariant derivative of rigid superspace

$$\bar{D}_i = \frac{\partial}{\partial\bar{\theta}_i} + \frac{1}{2}\theta_i\frac{\partial}{\partial z_i}. \qquad (26.69)$$

We will discuss Eq. (26.68) below, but refer the reader to Ref. 247 for a derivation.

An important development in the theory of critical phenomena has been the relation between conformal models and the conformal theories that arise as fixed points of scalar field theories[238] which in this context are called Landau-Ginszburg theories. Recently, in string theory it has been shown[249] in accord with a previous conjecture[250] that the Landau-Ginzburg $(2, 2)$ supersymmetric action

$$ -\int d^2x\, d^2\theta\, d^2\bar{\theta}\, \varphi\bar{\varphi} + \left\{ g \int d^2x\, d^2\theta \frac{\varphi^n}{n!} + \text{c.c} \right\} \tag{26.70} $$

possesses an infrared stable fixed point and at this fixed point the resulting $(2, 2)$ superconformal theory is the nth member of the discreet series discussed above. To be more specific it has been shown[249] that at the fixed point, the conformal weights of chiral fields, their operator product expansions and central charge are in agreement with those of the nth member of the $N = 2$ minimal series above. In particular the chiral field $\phi(\bar{\phi})$ of the Landau-Ginzburg action corresponds to $\phi_{1,1}(\phi_{1,-1})$ of the nth superconformal model. Since the infrared fixed point is outside the range of $g = 0$ perturbation theory, one must use another method of computation and in particular use was made of the ε expansion.

In Refs. 252, 253, and 254 some interesting relations between $N = 2$ super-potentials and singularity theory were observed; in particular, it was noted that the singularities of the simplest such potentials fall into an A, D, and E classification as does a class of $N = 2$ superconformal models. An alternative derivation of the central charge was also given in Refs. 254 and 255, using $N = 2$ superconformal theory, but assuming the existence of an infrared stable fixed point in the Landau-Ginzburg Model and a type of nonrenormalization theorem. These assumptions reduce the problem to a question of conformal group theory, but the disadvantage is that one then uses no properties of the original Landau-Ginzburg Model other than the R weight of the chiral field at its fixed point.

The importance of $N = 2$ conformal theories stems from the observation[256] that four-dimensional string theories with space-time supersymmetry must possess at least $(2, 0)$ or $(2, 2)$ supersymmetry on the world sheet. This means that these string theories must be based on $(2, 0)$ or $(2, 2)$ superconformal theories. It is hoped that Landau-Ginzburg models will provide a tool for distinguishing between the many such conformal theories and the many string theories they give rise to.

Returning to chiral-Green functions, we note that from the point of view of Landau-Ginzburg theories the condition of Eq. (26.68) is intuitively obvious.

Indeed, one can calculate directly from the Landau-Ginzburg action of Eq. (26.76) the Green functions for chiral superfields at the fixed point and so that for the minimal model. We refer the reader to Ref. 247 for this calculation and we now find the same result using the conformal model techniques.

Any N point Green function must satisfy the Möbius constraints of Eq. (26.22) of super Möbius invariance, however chiral correlators also satisfy Eq. (26.68). The latter imply that G is a function of $y_i = z_i + 1/2\theta_i\bar\theta_i$ and θ_i alone. The l_{-1} equation implies that

$$G = G(y_{ij}, \theta_i)$$

where $y_{ij} \equiv y_i - y_j$.

In terms of these variables, Eq. (26.22) becomes

$$\sum_i \tau_0^i G = \sum_i \left(-\theta_i \frac{\partial}{\partial\theta_i} + T_i\right) G = 0$$

$$\sum_i l_0^i G = \left\{\sum_i y_i \frac{\partial}{\partial y_i} + \sum_i \left(\frac{1}{2}\theta_i \frac{\partial}{\partial\theta_i} + H_i\right)\right\} G = 0$$

$$\sum_i l_1^i G = \left\{\sum_i \left((y_i)^2 \frac{\partial}{\partial y_i} + \sum_i y_i\theta_i \frac{\partial}{\partial\theta_i} + 2H_i y_i\right)\right\} G = 0$$

$$\sum_i \mathscr{G}^i_{-1/2} G = \left\{\sum_i \frac{\partial}{\partial\theta_i}\right\} G = 0; \qquad \sum_i \bar{\mathscr{G}}^i_{-1/2} = -\left(\sum_i \theta_i \frac{\partial}{\partial y_i}\right) G = 0$$

$$\sum_i \mathscr{G}^i_{1/2} G = \left(\sum_i y_i \frac{\partial}{\partial\theta_i}\right) G = 0;$$

$$\sum_i \bar{\mathscr{G}}^i_{1/2} G = \left\{-\sum_i \left(y_i\theta_i \frac{\partial}{\partial y_i} + 2H_i\theta_i\right)\right\} G = 0$$

where $\partial/\partial\theta$ is evaluated at constant y. The τ_0 equation implies that $\sum_i T_i$ is an integer. When the total charge is $+1$, G has the form

$$G = \sum_{i=1}^N \theta_i A_i \tag{26.72}$$

where the A_i's are functions of the y_{ij}'s. The l_0 equation then tells us that the A_i's are homogeneous functions of degree -1. The $\mathscr{G}_{-1/2}$ and $\bar{\mathscr{G}}_{-1/2}$ equations then yield

$$\sum_{i=1}^N A_i = 0 \qquad \text{and} \qquad \partial_i A_j - \partial_j A_i = 0 \tag{26.73}$$

so that

$$A_i = \partial_i A \qquad \text{and} \qquad \sum_{i=1}^N \partial_i A = 0. \tag{26.74}$$

The $\mathcal{G}_{+1/2}$ equation tells us that A is a homogeneous function of degree zero while the $\bar{\mathcal{G}}_{+1/2}$ equation is

$$y_{ij}\partial_i\partial_j A = (H_j\partial_i - H_i\partial_j)A. \tag{26.75}$$

The solution to (26.75) is most easily expressed as a Fourier transform from momentum space. It is

$$A = \int \prod_{i=1}^{N} dp_i \, \frac{e^{i(\sum_{j=1}^{N} p_j y_j)}}{\prod_{i=1}^{N} (p_i)^{1-2H_i}} \, \delta\left(\sum_{j=1}^{N} p_j\right). \tag{26.76}$$

It is straightforward to verify that the remaining Eqs. (26.74) and (26.73) are satisfied and that the solution is unique. The final result[247] for the Green function for N chiral fields with superweights H_i, $i = 1, \ldots, N$ and chiral charges $T_i = 2H_i$ such that $T \equiv \sum_i T_i = +1$ is

$$G(z_1,\ldots,z_n) = \int \prod_{i=1}^{N} dp_i \, \frac{\sum_{j=1}^{N} \theta_j p_j}{\prod_{j=1}^{N} (p_i)^{1-2H_j}} \cdot e^{i\sum_{j=1}^{N} p_j y_j} \cdot \delta\left(\sum_{i=1}^{N} p_i\right). \tag{26.77}$$

For the two- and three-point function one can explicitly evaluate this integral and verify that the result agree with that of the previous section.

26.3   Super Operator Product Expansions

An important tool in conformal field theory is Wilson's operator product expansion.[256] This states that given two quantum operators $A(x^\mu)$, $B(y^\mu)$ of a quantum field theory, then their time-ordered product behaves as x^μ approaches y^μ as

$$A(x^\mu)B(y^\mu) \sim \sum_l c_l(x^\mu - y^\mu)O^l(x^\mu + y^\mu) \qquad \text{as } x^\mu - y^\mu \to 0 \quad (26.78)$$

where O^l are a set of local operators and C_l are coefficients depending on $(x - y)^\mu$. This equation is to be understood as being valid when the product $A(x)B(y)$ is inserted in a Green function with other operators, that is

$$\ln_{x\to y} \langle |T\{A(x)B(y) - \sum_l C_l(x - y)O^l\}\phi(\omega_1)\phi(\omega_2)\ldots| \rangle = 0 \quad (26.79)$$

where the ϕ are operators of the theory. It is in this sense that Eq. (26.78) is automatically time-ordered. Generally, Eq. (26.79) is not valid if some of the ϕ's are composite operators, but it always holds if they are the elementary fields of a Lagrangian quantum field theory. The convergence properties of

Eq. (26.79) when valid depend on the fields inserted in the Green function and in general it is an asymptotic series. The coefficients $C_l(x)$ can be shown to be of the form

$$C_l(x) = x^{-d_A - d_B + d_O} \quad \text{(times a polynomial of ln } x) \tag{26.80}$$

where d_A, and d_b and d_O are the dimensions of the operators A, B, and O.

Let us now consider the operators product expansion for two-dimensional conformal field theories. We use $\tau = \ln|z|$ or equivalently $|z|$ as our time-ordering parameter. Given two conformal operators $R(z)$ and $S(z)$ of weights $(d_R, 0)$ and $(d_S, 0)$ respectively, their operator product expansion is

$$S(z)R(z') \sim \sum_Q \frac{Q(z')}{(z - z')^{d_R + d_S - d_Q}} \tag{26.81}$$

where $Q(z')$ are conformal operators of dimension d_Q. The Wilson coefficients are determined by conformal invariance to be of the form shown. It is useful to consider the arguments of the operators Q to be z' rather than $(z + z'/2)$, this corresponds only to a rearrangement of the series.

One way to compute the operator product expansion is to normal order the operators. Given the operators $R(z)$ and $S(z')$ we may normal order them to find

$$S(z)R(z') = \sum_l \frac{P_l(z')}{(z - z')^{-d_l + d_R + d_S}} + :S(z)R(z'): \tag{26.82}$$

which is valid only if $|z| > |z'|$, where P_l are normal-ordered operators. The singularities arise from taking past each other the infinite number of annihilation and creation operators, in order to achieve the normal-ordered product of the two operators. Any normal-ordered product is finite as $z \to z'$ in the sense that it has finite expectation values between any states with finite occupation number. Hence, the singular terms as $z \to z'$ are found in the first terms and as $|z| > |z'|$ corresponds to the correct time ordering for the operator product expansion, we may identify the terms in the operator product expansion with these in the normal-ordered expression.

An important application of the above is in the evaluation of the commutator of S_n and R_m where

$$S_n = \oint_{\Gamma_1} \frac{dz}{z} S(z)z^{n + d_S}; \quad R_n = \oint_{\Gamma_2} \frac{dz}{z} R(z)z^{m + d_R}. \tag{26.83}$$

The closed contour Γ_1, say, must enclose the point $z = 0$, but is otherwise arbitrary. The result for S_n being independent of the particular contour chosen as $S(z)$ is analytic away from $z = 0$ and similarly for R_n

We may write

$$[S_n, R_m] = S_n R_m - R_m S_n$$

$$= \oint_{\Gamma_1} \frac{dz}{z} z^{n+d_S} \oint_{\Gamma_2} \frac{d\zeta}{\zeta} \zeta^{m+d_R} S(z) R(\zeta)$$

$$- \oint_{\Gamma_3} \frac{dz}{z} z^{n+d_S} \oint_{\Gamma_4} \frac{d\zeta}{\zeta} \zeta^{m+d_R} R(\zeta) S(z). \tag{26.84}$$

We choose the contours such that Γ_1 encloses Γ_2 and Γ_4 encloses Γ_3. This coincides with the time ordering used for the operator product expansion. It also coincides with the point where the normal ordering of the two operators converges. We may now use the operator product expansion of Eq. (26.81) but as $T(S(z)R(\zeta)) = S(z)R(\zeta)$ for $|z| > |\zeta|$ and $T(R(\zeta)S(z)) = T(S(z)R(\zeta)) = R(\zeta)S(z)$ for $|\zeta| > |z|$, we use the same expression in the first and second terms. Consequently, we have

$$[S_n, R_m] = \left\{ \oint_{\Gamma_1} \frac{dz}{z} z^{n+d_S} \oint_{\Gamma_2} \frac{d\zeta}{\zeta} \zeta^{m+d_R} - \oint_{\Gamma_3} \frac{dz}{z} z^{n+d_S} \oint_{\Gamma_4} \frac{d\zeta}{\zeta} \zeta^{m+d_R} \right\}$$

$$\times \left\{ \sum_Q \frac{Q(\zeta)}{(z-\zeta)^{d_S + d_R - d_Q}} \right\}. \tag{26.85}$$

Here we have taken S and R to be such that one of them is Grassmann odd. The same conclusion may be achieved by thinking about the normal ordering of the two operators. Let us choose to identify the contours Γ_2 and Γ_4. The contour Γ_1 encloses Γ_2 which in turn encloses Γ_3 (see Fig. 26.1) and we consider

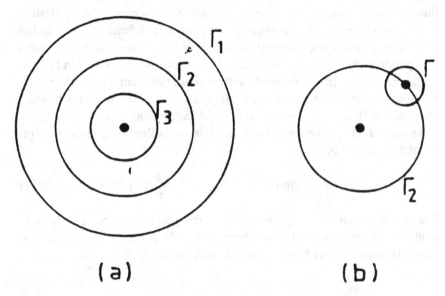

(a)   (b)

Fig. 26.1 Contours occurring in the evaluation of a commutator.

carrying out the z integration. Since the integrand is analytic everywhere except at $z = \zeta$. We have the result

$$[S_n, R_m] = \oint_{\Gamma_2} \frac{d\zeta}{\zeta} \zeta^{m+d_R} \oint_\Gamma \frac{dz}{z} z^{n+d_S} \sum_Q \frac{Q(\zeta)}{(z-\zeta)^{d_S+d_R-d_Q}} \qquad (26.86)$$

where Γ is a contour which surrounds the point $z = \zeta$ (see Fig. 26.1). By standard techniques in complex analysis, in particular Cauchy's theorem, we may then evaluate the integral. By using the same argument we may show that

$$[S_n, R(z)] = \oint_\Gamma \frac{dz}{z} z^{n+d_S} \sum_Q \frac{Q(\zeta)}{(z-\zeta)^{d_S+d_R-d_Q}}. \qquad (26.87)$$

What the above demonstrates is the close relation between the commutator of two operators and their operator product expansion. To illustrate this fact, consider $S(z) \equiv T_{zz} = T$ and $R(z)$ to be any conformal operator of dimension d which must therefore satisfy

$$[L_n, R(z)] = z^n \left\{ z \frac{d}{dz} + (n+1)d \right\} R(z). \qquad (26.88)$$

The operator product expansion to recover this result upon using Eq. (26.87) is read off to be

$$T(z)R(\zeta) = \frac{dR(\zeta)}{(z-\zeta)^2} + \frac{\frac{d}{d\zeta}R(\zeta)}{(z-\zeta)}. \qquad (26.89)$$

Similarly in order to recover the Virasoro algebra using Eq. (26.81), the operator product of two energy-momentum tensors must be of the form

$$T(z)T(\zeta) = \frac{1}{2} \frac{c}{(z-\zeta)^4} + \frac{2T(\zeta)}{(z-\zeta)^2} + \frac{\frac{d}{d\zeta}T(\zeta)}{(z-\zeta)} \qquad (26.90)$$

where c is a number. Indeed, using Eq. (26.86) in Eq. (26.90) we find that

$$[L_n, L_m] = (n-m)L_{n+m} + \frac{c}{12} n(n^2-1)\delta_{n+m,0} \qquad (26.91)$$

which is the quantum version of our classical algebra given in Chapter 25. The last term contains the object c which is called the central charge. We refer the reader to Refs. 244 and 245 for further discussions, which include worked examples, of the operator product expansion.

Our subsequent discussion of operator product expansions in (2, 2) super-conformal models is taken from Ref. 257. Before plunging into this exposition we will acquire some supercontour integral formulae. Given any analytic function F of $(2,0)$ superspace that is of $Z = (z, \theta, \bar{\theta})$ the usual Cauchy integral can be generalized to superspace as

$$\oint_z dz' \, F(z') \frac{(\bar{\theta}' - \bar{\theta})(\theta' - \theta)}{(Z' - Z)} = F(Z) \qquad (26.92)$$

where

$$\oint dZ \equiv \oint_z dz \, d\theta \, d\bar{\theta}$$

$$\oint dz \equiv \frac{1}{2\pi i} \oint dz \qquad (26.93)$$

and $Z' - Z$ is the supersymmetric distance, also called s in the previous section

$$Z' - Z = z' - z - \tfrac{1}{2}(\theta'\bar{\theta} + \bar{\theta}'\theta). \qquad (26.94)$$

Other useful formulae which can be derived from the above or by direct evaluation are

$$DF(Z) = \oint_z dZ' \, F(Z') \frac{(\bar{\theta}' - \bar{\theta})}{Z' - Z}, \qquad \bar{D}F(Z) = -\oint_z dZ' \frac{F(Z')(\theta' - \theta)}{(Z' - Z)} \qquad (26.95)$$

and

$$[D, \bar{D}]F(Z) = 2 \oint_z dZ' \frac{F(Z')}{(Z' - Z)^2}. \qquad (26.96)$$

An operator product expansion of two $(2,0)$ superfield operators at points Z, Z' can be written in terms of other $(2,0)$ superfield operators with Wilson coefficients which due to supersymmetry depend only on $Z - Z'$, $\theta - \theta'$ and $\bar{\theta} - \bar{\theta}'$. Following the same arguments as in the nonsupersymmetric case above, we find that Eqs. (26.86) and (26.87) generalize in the obvious way.

The $(2,0)$ superconformal algebra arises from the energy-momentum tensor $T(z) = \sum_n L_{+n} z^{-n-2}$, the two supercurrents denoted $\bar{G}(z) = \sum_r \bar{G}_r z^{-r-3/2}$ and $G(z) = \sum_r G_r z^{-r-3/2}$ and the U(1) chiral current $I(z) = \sum_n T_n z^{-n-1}$. These may be incorporated into a superfield

$$J(z) = -I(z) + \frac{i}{\sqrt{2}}(\theta \bar{G}(z) + \bar{\theta} G(z)) - \theta \bar{\theta} T(z). \qquad (26.97)$$

The corresponding generators are of the form

$$J_F = \oint dZ' \, F(Z') J(Z'). \qquad (26.98)$$

Taking the basis of functions

$$F = (z^{n+1}, +i\sqrt{2}\bar{\theta}z^{r+1/2}\bar{\eta}, -i\sqrt{2}\theta z^{r+1/2}\eta, +\theta\bar{\theta}z^n) \qquad (26.99)$$

where η is a constant anticommuting parameter. We recover from Eq. (26.98) the generators

$$L_n, \bar{\eta}\bar{G}_r, \eta G_r, \text{ and } T_n. \qquad (26.100)$$

Given a primary field ϕ of superweight H and chiral weight T we take its super-operator product expansion with the supercurrent multiplet J to be

$$J(Z')\phi(Z) = \frac{(\bar{\theta}' - \bar{\theta})(\theta' - \theta)}{(Z' - Z)}\partial_z\phi(Z) + \frac{(\bar{\theta}' - \bar{\theta})}{(Z' - Z)}\bar{D}\phi(Z) - \frac{(\theta' - \theta)}{(Z' - Z)}D\phi(Z)$$

$$+ H\frac{(\bar{\theta}' - \bar{\theta})(\theta' - \theta)}{(Z' - Z)^2}\phi(Z) - \frac{T\phi(Z)}{(Z' - Z)^2}. \qquad (26.101)$$

The superconformal variation of ϕ is given by

$$\delta_F\phi(Z) = [J_F, \phi(Z)] = \oint_Z dZ'\, F(Z')J(Z')\phi(Z) \qquad (26.102)$$

upon using the superspace analogue of Eq. (26.87). Inserting the operator product expansion above and using Eqs. (26.92), (26.95) and (26.96) we discover that

$$\delta\phi = F\partial_z\phi + (DF)\bar{D}\phi + (\bar{D}F)D\phi + \left\{H\partial_z F - \frac{T}{2}[D, \bar{D}]F\right\}\phi. \qquad (26.103)$$

Thus we recover our definition of the transformation for a primary field given in the previous chapter.

From the operator product expansion of $J(z)$ with itself we can recover the superconformal algebra. To ensure this we take

$$J(Z')J(Z) = \frac{(\bar{\theta}' - \bar{\theta})(\theta' - \theta)}{(Z' - Z)}\partial_z J(Z) + \frac{(\bar{\theta}' - \bar{\theta})}{(Z' - Z)}\bar{D}J(Z) - \frac{(\theta' - \theta)}{(Z' - Z)}DJ(Z)$$

$$+ \frac{(\bar{\theta}' - \bar{\theta})(\theta' - \theta)}{(Z' - Z)^2}J(Z) + \frac{c_2}{(Z' - Z)^2}. \qquad (26.104)$$

We note that apart from the last term this is the operator product expansion that would result if we took $\phi(z) = J(z)$ and $H = 1$, $T = 0$ in Eq. (26.101). Using the superspace analogue of Eq. (26.86), we find that

$$[J_{F_1}, J_{F_2}] = \left[\oint dZ'\, F_1(Z')J(Z'), \oint dZ\, F_2(Z)J(Z)\right]$$

$$= J_{F_3} + \frac{c_2}{2}\oint dZ\, F_2(Z)[D, \bar{D}]F_1(Z) \qquad (26.105)$$

where

$$F_3 = F_2 \partial_z F_1 - \partial_z F_2 F_1 + DF_2 \bar{D} F_1 + \bar{D} F_2 DF_1 \qquad (26.106)$$

Introducing the basis of Eq. (26.99), we find the $N = 2$ superconformal algebra with a central charge

$$[L_n, L_m] = (n - m)L_{n+m} + \frac{c_2}{4} n(n^2 - 1)\delta_{n+m,0}$$

$$\{G_r, \bar{G}_s\} = 2L_{r+s} + (r - s)T_{r+s} + \tfrac{1}{4}c_2(4r^2 - 1)\delta_{r+s,0}$$

$$\{G_r, G_s\} = 0 = \{\bar{G}_r, \bar{G}_s\}, \qquad [L_n, G_s] = \left(\frac{n}{2} - s\right)G_{n+s}$$

$$[L_n, \bar{G}_s] = \left(\frac{n}{2} - s\right)\bar{G}_{n+s}, \qquad [T_n, T_m] = \frac{c_2}{16} n\delta_{n+m,0}$$

$$[T_n, G_r] = \tfrac{1}{4}G_{n+r}, \qquad [T_n, \bar{G}_r] = -\tfrac{1}{4}\bar{G}_{n+r}, \qquad [L_n, T_m] = -nT_{n+m}. \quad (26.107)$$

Upon setting the central charge $c_2 = 0$, this algebra agrees with the classical superconformal algebra found in the previous chapter.

Chapter 27

Gauge Covariant Formulation of Strings

Before beginning to covariantly quantize an extended object, namely the string, it will be instructive to consider the corresponding path from the classical to the second quantized point particle.

27.1 The Point Particle

The trajectory of a classical point particle in spacetime is parametrized by its proper time and is given by $x^\mu(\tau)$. The path it takes is such as to minimize the action

$$A = -m \int d\tau \sqrt{-\dot{x}^\mu \dot{x}^\nu \eta_{\mu\nu}} = \frac{1}{2} \int d\tau \left\{ V^{-1} \dot{x}^\mu \dot{x}^\nu \eta_{\mu\nu} - m^2 V \right\} \qquad (27.1)$$

where $\dot{x}^\mu = \dfrac{dx^\mu}{d\tau}$. These actions are invariant under reparametrizations of the proper time: $\tau \to f(\tau)$ with the transformation of x^μ being $\delta x^\mu = f(\tau) \dot{x}^\mu$.

Let us now give a Hamiltonian treatment of the first action of Eq. (27.1), although the same results can be found from the second action. The canonical momentum is given by

$$p_\mu = \frac{\delta A}{\delta \dot{x}^\mu(\tau)} = \frac{m \dot{x}^\mu}{\sqrt{-\dot{x}^\mu \dot{x}^\nu \eta_{\mu\nu}}} \qquad (27.2)$$

and the equation of motion is of the form

$$\partial_\tau p_\mu = 0 \qquad (27.3)$$

Due to the reparametrization invariance, the system is constrained by

$$\phi \equiv p_\mu^2 + m^2 = 0 \qquad (27.4)$$

and the Hamiltonian vanishes

$$H = p^\mu \dot{x}_\mu - L = 0 \qquad (27.5)$$

The method of dealing with such a system was given in Ref. 188 and we now apply this in outline for the point particle. The reader who wishes to read further details is encouraged to consult Ref. 189. We take the Hamiltonian to be proportional to the constraint, i.e.,

$$H = v(\tau)(p^2 + m^2) \tag{27.6}$$

where $v(\tau)$ is an arbitrary function of τ. One may verify that in this case there are no further constraints and that H generates time translations or reparametrizations in the sense that

$$\frac{dx^\mu}{d\tau} = \{x^\mu(\tau), H\} = 2v(\tau)p^\mu \tag{27.7}$$

The fundamental Poisson Brackets vanish except for

$$\{x^\mu, p^\nu\} = \eta^{\mu\nu} \tag{27.8}$$

To quantize the theory we make the usual transition from Poisson Brackets to commutators, which are represented by the replacements

$$x^\mu \to x^\mu, \qquad p^\mu \to -i\hbar \frac{\partial}{\partial x^\mu} \tag{27.9}$$

The constraint then becomes

$$\hat{\phi} = (-\partial^2 + m^2) \tag{27.10}$$

To find the second quantized field theory we consider the state to be described by a field $\psi(x^\mu, \tau)$ and we impose the constraint

$$\hat{\phi}\psi = 0 = (-\partial^2 + m^2)\psi \tag{27.11}$$

We also impose the Schrödinger equation

$$i\hbar \frac{\partial \psi}{\partial \tau} = H\psi \tag{27.12}$$

The right-hand side of this equation vanishes and we find that ψ is independent of τ. In the second quantized theory there is in any case more than one particle and so the concept of a single proper time becomes problematical.

The action that leads to the above Klein-Gordon equation is

$$A = \int d^4x \psi(-\partial^2 + m^2)\psi \tag{27.13}$$

and we may use it to weight the Feynman path integral that can then be used to find the Green's functions of the second quantized theory.

We note that the original reparametrization invariance of the proper time which was so important for determining the form of the classical action is absent in the second quantized theory, its only remnant being the field equation itself. Since we performed a Hamiltonian quantization with respect to the proper time this is only to be expected, but one might wonder if one could second quantize in such a way as to maintain this invariance. We now wish to repeat the above steps for the string.

27.2 The Bosonic String

The bosonic string whose length is parametrized by σ sweeps out in time τ, a 2-dimensional surface parametrized by $\xi^\alpha = (\tau, \sigma)$ in a space-time x^μ according to the function $x^\mu(\xi)$. This trajectory is in such away as to sweep out an extremal area, and so its action, is given by[190]

$$A = -\frac{1}{2\pi\alpha'} \int d^2\xi \sqrt{-\det \partial_\alpha x^\mu \partial_\beta x^\nu \eta_{\mu\nu}} \tag{27.14}$$

where α' has the dimensions of $(\text{mass})^{-2}$. It is invariant under arbitrary reparametrizations of the 2-dimensional surface

$$\xi^\alpha \to \xi^\alpha + f^\alpha(\xi) \qquad x^\mu \to x^\mu + f^\alpha \partial_\alpha x^\mu \tag{27.15}$$

as well as invariant under the Poincaré group transformations acting on the space-time x^μ.

The canonical momentum is given by

$$P_\mu = \frac{\delta A}{\delta \left(\frac{\partial x^\mu}{\partial \tau} \right)} = \frac{\partial_\tau x_\mu (x'^\nu)^2 - x'_\mu ((\partial_\tau x^\nu) x'_\nu)}{2\pi\alpha' \sqrt{-\det(\partial_\alpha x^\mu \partial_\beta x^\nu \eta_{\mu\nu})}} \tag{27.16}$$

where $x'^\mu = \frac{\partial x^\mu}{\partial \sigma}$.

Due to the invariance mentioned above, we have the constraints

$$P_\mu^2 + \frac{1}{(2\pi\alpha')^2} (x'^\mu)^2 = 0 \tag{27.17}$$

$$x'^\mu P_\mu = 0 \tag{27.18}$$

It is convenient mathematically to extend the range of σ to be between $-\pi$ to π by requiring

$$x^\mu(\sigma) = \begin{cases} x^\mu(\sigma) & 0 \leqslant \sigma \leqslant \pi \\ x^\mu(-\sigma) & -\pi \leqslant \sigma \leqslant 0 \end{cases} \tag{27.19}$$

i.e., $x^\mu(\sigma) = x^\mu(-\sigma)$. Using this extension the above constraints can then be written as

$$(\wp^\mu)^2 = 0$$

where

$$\wp^\mu \equiv \pi\sqrt{2\alpha'}\left(P^\mu + \frac{x^{\mu\prime}}{2\pi\alpha'}\right) \tag{27.20}$$

It will be advantageous to take the Fourier transform of these constraints, and so we define

$$L_n \equiv \frac{1}{4\pi}\int_{-\pi}^{\pi} d\sigma (\wp^\mu)^2 e^{in\sigma} \tag{27.21}$$

One finds that the L_n obey the algebra[191]

$$\{L_n, L_m\} = -i(n-m)L_{n+m} \tag{27.22}$$

Since the Hamiltonian vanishes, we take it to be proportional to the constraints, i.e.,

$$H = \sum_{n=-\infty}^{\infty} c_n L_n \tag{27.23}$$

One may verify that the L_n are the generators of 2-dimensional conformal transformations in the world sheet of the string. Generally, the conformal group has only a finite number of generators, but for two dimensions only it is an infinite dimensional algebra which corresponds to making an arbitrary analytic transformation in $z = \tau + i\sigma$. Clearly, the 2-dimensional flat metric which can be written as $dz\,d\bar{z}$ scales under $z \to f(z)$. The emergence of the 2-dimensional conformal group rather than the original 2-dimensional general coordinate group is presumably related to the choice of τ as the time to be used in the Hamiltonian approach. As we shall see, the Virasoro algebra, and hence the 2-dimensional conformal group, plays an important role in the second quantized gauge covariant theory.

The fundamental Poisson Brackets of the theory are

$$\{x^\mu(\sigma), P^\nu(\sigma')\} = \delta(\sigma - \sigma')\eta^{\mu\nu}$$

$$\{x^\mu(\sigma), x^\nu(\sigma')\} = 0 = \{P^\mu(\sigma), P^\nu(\sigma')\} \tag{27.24}$$

To quantize the theory we replace these relations by the commutators, and insert a factor of $i\hbar$. The commutators are represented by the changes

$$x^\mu(\sigma) \to x^\mu(\sigma) \qquad P^\mu(\sigma) \to -i\hbar\frac{\delta}{\partial x^\mu(\sigma)} \tag{27.25}$$

Making these replacements in the generators of the constraints, we find that

$$L_n = \frac{1}{4\pi} \int_{-\pi}^{\pi} d\sigma e^{+in\sigma} \wp^\mu(\sigma) \wp^\nu(\sigma) \eta_{\mu\nu} \qquad (27.26)$$

where

$$\wp^\mu(\sigma) = -i\hbar\pi\sqrt{2\alpha'} \frac{\delta}{\partial x^\mu(\sigma)} + \frac{x^{\mu'}(\sigma)}{\sqrt{2\alpha'}}. \qquad (27.27)$$

We now impose the following constraints on the functional $\psi[x^\mu(\sigma)]$, which, like the particle, is τ independent:

$$(L_0 - 1)\psi = 0 \qquad L_n\psi = 0 \qquad n \geqslant 1 \qquad (27.28)$$

These are not entirely what one might naively expect. The -1 in the first equation corresponds to the possibility of there being, due to L_0 not being uniquely defined by the classical theory, a normal ordering constant.

We shall see that this is fixed to be -1 by requiring a ghost free spectrum of on-shell states. In the second equation we do not require all the L_n's to vanish on ψ, since this would imply that ψ itself would vanish due to the central term in the Virasoro algebra, which we will discuss shortly. However, the latter equation implies that

$$(\psi, L_n\psi) = (L_{-n}\psi, \psi) = 0 \qquad \forall n, n \neq 0 \qquad (27.29)$$

as $L_n^\dagger = L_{-n}$. In this way, one recovers in the classical limit that all the L_n vanish in accord with Eq. (27.20). The procedure is the same as for the Gupta-Bleuler formulation of quantum electrodynamics.

The necessity of the constraints of Eq. (27.28) is guaranteed by the following.

Theorem[192]: Equation (27.28) describes a ghost-free set of on-shell states provided the dimension, D, of space-time is less than or equal to 26.

In fact for $D < 26$ there are other problems, and for the remainder of this contribution we will take $D = 26$. It may be helpful to recall the distinction between a ghost and a tachyon, a scalar with action

$$A = \int d^4x(-c(\partial_\mu A)^2 - dA^2) \qquad (27.30)$$

is a ghost if $c < 0$ and a tachyon if $d < 0$. We will see that the open bosonic string does indeed possess a tachyon.

We are now in a position to specify what requirements a second quantized gauge covariant formulation of strings must satisfy. We must demand that there be an action whose equations of motion imply Eq. (22.28). Of course, we may have to make some gauge choices, and we expect $L_n = 0 \, (n \geqslant 1)$ to be the result of these gauge choices, while $(L_0 - 1)\psi = 0$ is the remaining equation of motion. We further expect the action to be local in that the free theory should contain no more than two space-time derivatives. The actions we will obtain will contain the fields of Yang-Mills and gravity for the open and closed bosonic strings respectively, and thus possess their corresponding gauge invariances. Thus we must find that the gauge symmetries of the string will contain these particular symmetries.

To quantize this system we will use this action, after appropriate gauge fixing and introduction of ghosts, to weight a Feynman path integral. The vacuum-to-vacuum amplitude is given by

$$\int \mathscr{D}\psi \exp\left(\frac{i}{\hbar}S\right) \tag{27.31}$$

where the action S is of the generic form

$$S = \int (\mathscr{D}x^{\mu})f(\psi[x^{\mu}(\sigma)]) + \text{gauge fixing} + \text{ghost} + \text{source terms} \tag{27.32}$$

Up until recently, the second quantization of strings has either been performed with constraints being present, or carried out in a given gauge such as the light-cone gauge[193], where the constraints have in effect been solved. Quantization using BRS techniques of the linearized theory in a given gauge has been discussed in Ref. 194. It is possible that one can, in principle, discover all the properties of a theory by quantizing in a given gauge. However, without the ability to use all the wisdom acquired with second quantization, this may be difficult. An example is the computation of anomalies which are absent in the light cone gauge; however, a careful search of Lorentz invariance would show that it is violated when gauge anomalies are present. Also, the whole subject of non-perturbative semi-classical phenomena and spontaneous symmetry breaking has up to now only been developed in the gauge invariant framework. It is also possible that the finiteness properties of strings may become particularly apparent in a covariant formulation, as they did in the case of supersymmetric theories.

Another advantage of obtaining a second quantized field theory of strings is that it will help us to understand what strings are. One of the remarkable developments of modern physics is that the theories relevant to nature are almost entirely determined by symmetries. For example, a theory possessing local gauge invariance realized on a vector potential A_{μ}, and with action having no more than second order derivatives, can only be based on the Yang-Mills action. Similarly, Einstein's action is determined by general coordinate transformations realized on the metric, and supergravity is controlled by local supersymmetry. The string theories are unique up to distinctions about being open and closed, supersymmetric or bosonic. They also contain the local symmetries mentioned above. It is natural to suppose that the string is also completely determined by a symmetry. A knowledge of this symmetry would explain the many wondrous concellations found in string theory as well as, hopefully, lead to many more. Clearly, possessing an invariant action under a set of transforming fields whose algebra is known would make it much easier to guess the principle which underlies this symmetry and hence string theory itself.

27.3 Oscillator Formalism

In order to analyse the Virasoro conditions of Eq. (27.28) it is useful to re-express the quantities discussed above in terms of creation and annihilation operators. We may write

$$x^\mu(\sigma) = \sum_{n=-\infty}^{\infty} x_n^\mu \cos n\sigma = x_0^\mu + \sum_{\substack{n=-\infty \\ n\neq 0}}^{\infty} x_n^\mu \frac{e^{in\sigma}}{2} \tag{27.33}$$

where $x_0^\mu = x^\mu$ and $x_{-n}^\mu = (x_n^\mu)^* = x_n^\mu$ from the reality of $x^\mu(\sigma)$ and Eq. (27.19) The form of the expansion of $x^\mu(\sigma)$ is such as to obey the boundary conditions for the open string:

$$\frac{\partial x^\mu}{\partial \sigma}\Big|_{\sigma=0} = \frac{\partial x^\mu}{\partial \sigma}\Big|_{\sigma=\pi} = 0 \tag{27.34}$$

We may use the chain rule to rewrite the functional derivatives

$$\frac{\delta}{\delta x^\mu(\sigma)} = \sum_{n=-\infty}^{\infty} \frac{\partial x_n^\nu}{\partial x^\mu(\sigma)} \frac{\partial}{\partial x_n^\nu} = \frac{1}{\pi} \sum_{n=-\infty}^{\infty} e^{in\sigma} \frac{\partial}{\partial x_n^\mu} \tag{27.35}$$

Let us define

$$\wp^\mu(\sigma) = \sum_{n=-\infty}^{+\infty} \alpha_n^\mu e^{-in\sigma} \tag{27.36}$$

Using Eq. (27.33) we find that

$$\alpha_n^\mu = -i\left(\sqrt{2\alpha'} \frac{\partial}{\partial x_n^\mu} + \frac{nx_n^\mu}{\sqrt{2\alpha' \cdot 2}}\right) \tag{27.37}$$

From the reality of $\hat{P}^\mu(\sigma)$ we find that $\alpha_{-n}^\mu = \alpha_n^{\mu\dagger}$. The α's commutation relations are

$$[\alpha_n^\mu, \alpha_m^\nu] = 0, \qquad [\alpha_n^\mu, \alpha_m^{\nu\dagger}] = n\delta_{n,m}\eta^{\mu\nu} \tag{27.38}$$

for $n, m \geq 1$. The Virasoro operators can be expressed in terms of the α_n by

$$L_n = \frac{1}{2}: \sum_{m=-\infty}^{+\infty} \alpha_m^\mu \alpha_{n-m}^\nu \eta_{\mu\nu}: \tag{27.39}$$

where L_0 is understood to be normal ordered. They obey the modified algebra

$$[L_n, L_m] = (n - m)L_{n+m} + \frac{26}{12}n(n^2 - 1)\delta_{n,-m} \qquad (27.40)$$

The extra term arises due to the normal ordering and is called the central term. Its value can most easily be found by taking the vacuum expectation value of this relation. We find in particular that

$$L_0 = \frac{1}{2}\alpha_0^\mu \alpha_{0,\mu} + \sum_{m=1}^{\infty} \alpha_m^{\mu\dagger}\alpha_{m,\mu} \qquad (27.41)$$

$$L_1 = L_{-1}^\dagger = \alpha_0^\mu \alpha_{1,\mu} + \sum_{m=1}^{\infty} \alpha_m^{\mu\dagger}\alpha_{m+1,\mu} \qquad (27.42)$$

where

$$\alpha_0^\mu = -i\sqrt{2\alpha'}\,\frac{\partial}{\partial x_0^\mu}$$

We may write the state of the string in occupation number basis in terms of the creation operators $\alpha_n^{\mu\dagger}$ by

$$\psi[x^\mu(\sigma)] = \{\phi(x) + iA_\mu^1\alpha_1^{\mu\dagger} + h_{\mu\nu}\alpha_1^{\mu\dagger}\alpha_1^{\nu\dagger} + A_\mu^2 i\alpha_2^{\mu\dagger} + \cdots\}\langle x^\mu(\sigma)|0\rangle \quad (27.43)$$

The vacuum satisfies the equation

$$\alpha_n^\mu\langle x^\mu(\sigma)|0\rangle = 0, \qquad n \geqslant 1 \qquad (27.44)$$

The vacuum of Eq. (27.44) is of the form

$$\langle x^\mu(\sigma)|0\rangle = \prod_{n=1}^{\infty} c_n \exp\left(-\frac{n}{2\alpha'}x_n^\mu x_{n,\mu}\right) \qquad (27.45)$$

The action of the $\alpha_n^{\mu\dagger}$ on $\langle x^\mu(\sigma)|0\rangle$ produces the well-known complete set of Hermite polynomials. In terms of component fields we find that Eq. (27.28) has as consequences

$$(\alpha'\partial^2 + 1)\phi(x) = 0 = (\alpha'\partial^2 - (l - 1))A_\mu^l = (\alpha'\partial^2 - 1)h_{\mu\nu} \ldots \text{etc.} \quad (27.46)$$

and

$$\partial^\mu A_\mu^1 = 0 = \sqrt{2\alpha'}\,\partial^\mu A_\mu^{(2)} + 2h_\nu^\nu = -\sqrt{2\alpha'}\,\partial^\mu h_{\mu\nu} + A_\nu^2 \ldots \text{etc.} \quad (27.47)$$

The theorem of Ref. 194 is remarkable, in the sense that the presence of $a_n^{0\dagger}$ could lead to many ghost states. However the Virasoro constraints of Eq. (27.28) ensure that the physical states have positive definite norms.

27.4 The Gauge Covariant Theory at Low Levels

We now wish to find an action in terms of a ψ which does not have the constraints of Eq. (27.28) but instead possesses gauge invariances that allow these constraints to arise as a gauge choice upon the equation of motion. Consequently, we expect an infinite number of gauge invariances. This can be achieved mass level by mass level by successively releasing the constraints on ψ.

At the first level, we release the constraint $L_1 \psi = 0$, but ψ is still subject to

$$L_2 \psi = L_1^2 \psi = L_3 \psi = L_2 L_1 \psi \ldots = 0 \qquad (27.48)$$

Consider now the gauge transformation of ψ

$$\delta \psi = L_{-1} \Lambda_1 \qquad (27.49)$$

corresponding to constraints on ψ of Eq. (27.48). Λ_1 is subject to

$$L_1 \Lambda_1 = L_2 \Lambda_1 = 0 = \ldots . \qquad (27.50)$$

Using the form of L_{-1} given in Eq. (27.42), we find that this invariance contains the transformation $\delta A_\mu^1 = \partial_\mu \Lambda_1(x)$ which is the Abelian transformation expected for a linearized Yang-Mills theory.

An action invariant under $\delta \psi = L_{-1} \Lambda_1$ is given by

$$\frac{1}{2}\left(\psi, \left(L_0 - 1 - \frac{1}{2}L_{-1}L_1\right)\psi\right) \qquad (27.51)$$

The equation of motion is given by

$$\left(L_0 - 1 - \frac{1}{2}L_{-1}L_1\right)\psi = 0 \qquad (27.52)$$

This equation curectly reproduces the linearized Yang-Mills equation; namely $\partial^2 A_\mu - \partial_\mu \partial^\nu A_\mu = 0$. Explicitly testing the invariance, we find that

$$\left(L_0 - 1 - \frac{1}{2}L_{-1}L_1\right)L_{-1}\Lambda_1 = (L_0 L_{-1} - L_{-1} - L_{-1}L_0)\Lambda_1 = 0 \quad (27.53)$$

since $L_1 \Lambda_1 = 0$. Equation (27.52) was probably known to a few people in the old heyday of string theory, but has been rediscovered more recently. See Refs. 195 and 196.

The projector P of a string field onto the physical states of Eq. (27.28) was given in Ref. 197. The projector is an object that has the property $PL_{-n} = 0$, and at lowest order it is given by

$$P = 1 - \frac{1}{2}L_{-1}\frac{1}{L_0}L_1 + \cdots \qquad (27.54)$$

We note that *at this level* the equation of motion is given by

$$(L_0 - 1)P\psi = 0 \tag{27.55}$$

It has been proposed recently, that Eq. (27.55) is the correct equation of motion of all levels of the string, and this speculation has been encouraged by the above coincidence for the spin one at the first level. However, P has been formally computed for all levels, and Eq. (27.55) is explicitly non-local at the second level, and more and more so for higher levels. One's suspicions are further aroused by the fact that P can be constructed in an arbitrary space-time dimension, and that $D = 26$ is not particularly favoured. The clearest way to show that Eq. (27.55) is not the right generalization is to consider the first level of the closed bosonic oriented string. At this level, the covariant degrees of freedom are described by a single symmetric field $h_{\mu\nu}$. The Virasoro condition implies that it satisfies

$$\partial^2 h_{\mu\nu} = 0 \qquad \partial^\mu h_{\mu\nu} = 0 \tag{27.56}$$

These equations tell us that at this level the closed string contains only a spin two and a spin zero. The generalization of Eq. (27.55) for this level is

$$\partial^2 R_\mu{}^\rho R_\nu{}^\lambda h_{\rho\lambda} = 0 \tag{27.57}$$

where

$$R_\mu{}^\rho = \left(\delta_\mu{}^\rho - \frac{\partial_\mu \partial^\rho}{\partial^2} \right) \tag{27.58}$$

The reader immediately recognizes that this is a non-local equation, which does not admit a Hamiltonian formulation. Making it local by multiplication by ∂^2 leads to additional states.

In fact, it is known, (see Ref. 198) that there is no Lorentz invariant, gauge invariant, local action constructed from $h_{\mu\nu}$ alone, which describes both spin two and spin zero. The only way to achieve this is to introduce another, scalar, field to describe the spin zero; this is the well-known Einstein + massless scalar action. As we shall see, rather than the spin one, the first level of the closed string illustrates the general pattern; namely, for a local formulation the theory naturally requires supplementary fields, as we shall now demonstrate. In fact only for spins 0, 1 and $\frac{1}{2}$ does multiplication by ∂^2 lead to a local field equation. Higher spins must be treated differently and the relation between their projectors and field equations is more subtle.

Let us examine whether any supplementary fields are required at the second level of the bosonic open string (see Ref. 199). To count the number of on-shell states at this level we could examine the Virasoro constraints at this level. The Virasoro constraints, however, possess on-shell gauge invariances that must be chosen before the on-shell states become apparent. This is clear even at the first level, for which the conditions are $\partial^2 A^1{}_\mu = 0$ and $\partial^\mu A_\mu{}^1 = 0$. As is well-known, it is the additional on-shell invariance $\delta A_\mu = \partial_\mu \Lambda$ where $\partial^2 \Lambda = 0$ which allows the reduction to two rather than three on-shell states. A much faster method is to use the fact that the constraints are solved in the light cone gauge and so one only has the on-shell states corresponding to oscillators $\alpha^i{}_n$, $i = 1$ to 24. In this case ψ has the expansion:

$$\{\phi + i(\alpha_1{}^i)^\dagger A_i{}^1 + h_{ij}(\alpha_1{}^i)^\dagger (\alpha_1{}^j)^\dagger + i(\alpha_2{}^i)^\dagger (A_i{}^2)...\} \langle x^i(\sigma)|0\rangle \quad (27.59)$$

At the first level we have $(D-2)$ states corresponding to the $(D-2)$ states contained in the vector representation of $SO(D-2)$, which is the little group for massless particles in D dimensions. Since the massive states must belong to representations of the little group $SO(D-1)$, this demonstrates that the normal ordering constant in Eq. (27.28) was chosen correctly. Any other choice would violate Lorentz invariance in D dimensions, since $(D-2)$ states cannot carry a representation of $SO(D-1)$.

At the second level we have

$$\frac{(D-2)(D-1)}{2} + (D-2) = \frac{D(D-1)}{2} - 1 \quad (27.60)$$

on-shell states. These we can only identify with the second rank traceless symmetric representation of $SO(D-1)$. We shall refer to this as "pure spin 2". "Pure spin 2" is described on-shell by the field $h_{\mu\nu} = h_{\nu\mu}$ subject to the equations

$$\partial^\mu h_{\mu\nu} = h_\mu{}^\mu = (\partial^2 - m^2)h_{\mu\nu} = 0 \quad (27.61)$$

The projector is well known (see Ref. 202). It involves terms of the form $\partial_\mu \partial_\lambda \partial_\rho \partial_\sigma/(\partial^2)^2$, and hence multiplication by (∂^2) does not lead to a local field equation. In fact, there is no way to describe in a Lorentz covariant way only a massive spin two particle in terms of only $h_{\mu\nu} = h_{\nu\mu}$ subject to $h_\mu^\mu = 0$. The correct equations of motion involve the introduction of a supplementary field ϕ and are given by

$$(-\partial^2 + m^2)h_{\mu\nu} + (\partial_\mu \partial^\rho h_{\rho\nu} + \partial_\nu \partial^\rho h_{\rho\mu}) - 2\frac{\eta_{\mu\nu}}{D}\partial^\rho \partial^\lambda h_{\rho\lambda}$$

$$= \frac{D-2}{D-1}\left(\partial_\mu \partial_\nu \phi - \frac{\eta_{\mu\nu}}{D}\partial^2 \phi\right) \quad (27.62)$$

$$\partial^\mu \partial^\nu h_{\mu\nu} = \left(\partial^2 - \frac{D}{D-2}m^2\right)\phi$$

Indeed, these coupled equations lead to the desired result, namely

$$0 = \partial^\mu h_{\mu\nu} = (\partial^2 - m^2)h_{\mu\nu} \tag{27.63}$$

The above equations illustrate a more general method (see Ref. 200) of introducing additional fields in order to propagate higher spin fields. At the second level the string contains the fields $h_{\mu\nu}$ and $A_\mu{}^2$; however, $A_\mu{}^2$ is gauged away leaving $h_{\mu\nu}$ traceless and so we require one extra supplementary field ϕ to implement the massive spin-2 field equation. This field ϕ will be the lowest component of a supplementary string field $\chi^{(2)}[x^\mu(\sigma)]$

$$\chi^{(2)}[x^\mu(\sigma)] = \{\phi(x) + \ldots\}\langle x^\mu(\sigma)|0\rangle \tag{27.64}$$

We will now find the gauge covariant action at the next level. Starting from the action of Eq. (27.51) and releasing the constraints of Eq. (27.48) on the first level, we subject ψ and $\chi^{(2)}$ to

$$L_3\psi = L_2 L_1{}^2\psi = L_1{}^3\psi = \ldots = 0$$

$$L_1\chi^{(2)} = L_2\chi^{(2)} = \ldots = 0 \tag{27.65}$$

The most general expression of the correct order is

$$\left(L_0 - 1 - \frac{1}{2}L_{-1}L_1 - \frac{1}{4}\gamma L_{-2}L_2\right)\psi + \left(L_{-1}^2 + \frac{3}{2}\beta L_{-2}\right)\chi^{(2)} = 0 \tag{27.66}$$

$$\left(L_1{}^2 + \frac{3}{2}\beta L_2\right)\psi = (aL_0 + b)\chi^{(2)} \tag{27.67}$$

The use of the same β in (27.66) and (27.67) is required by demanding that the equations of motion follow from an action. An alternative way of searching for a gauge invariance is to demand that L_1 on (27.66) should vanish when we use Eq. (27.67). Carrying this out and using the constraints of (27.65), we find

$$-\frac{1}{2}L_{-1}\left(L_1{}^2 + \frac{3}{2}\gamma L_2\right)\psi + (4L_{-1}L_0 + 2L_{-1})\chi^{(2)} + \frac{9}{2}\beta L_{-1}\chi^{(2)} = 0 \tag{27.68}$$

To eliminate ψ by (27.67) requires $\gamma = \beta$ and we find

$$-\frac{1}{2}L_{-1}(aL_0 + b)\chi^{(2)} + L_{-1}\left(4L_0 + 2 + \frac{9}{2}\beta\right)\chi^{(2)} = 0 \tag{27.69}$$

Hence we conclude that $a = 8$ and $b = 4 + 9\beta$. Applying L_2 in a similar way fixes $\beta = 1$, and we find the equations of motion

$$\left(L_0 - 1 - \frac{1}{2}\sum_{n=1}^{2}\frac{1}{n}L_{-n}L_n\right)\psi + \left(L_{-1}^2 + \frac{3}{2}L_{-2}\right)\chi^{(2)} = 0$$

$$\left(L_1{}^2 + \frac{3}{2}L_2\right)\psi = (8L_0 + 13)\chi^{(2)} \tag{27.70}$$

This system of equations is in fact invariant under the gauge transformations

$$\delta_1\psi = L_{-1}\Lambda_1, \quad \delta_1\chi^{(2)} = \frac{1}{2}L_1\Lambda_1, \quad \delta_2\psi = L_{-2}\Lambda_2, \quad \delta_2\chi^{(2)} = \frac{3}{2}\Lambda_2 \tag{27.71}$$

with

$$L_2\Lambda_1 = L_1{}^2\Lambda_1 = \ldots = 0, \quad L_1\Lambda_2 = L_2\dot{\Lambda}_2 = \ldots = 0 \tag{27.72}$$

We stress that Eqs. (27.70) are Λ_2 invariant only for $D = 26$. It may be possible, with the introduction of further supplementary fields, to relax this condition. The corresponding action is given by[199]

$$\frac{1}{2}\left(\psi,\left(L_0 - 1 - \frac{1}{2}L_{-1}L_1 - \frac{1}{4}L_{-2}L_2\right)\psi\right)$$

$$+ \left(\psi,\left(L^2_{-1} + \frac{3}{2}L_{-2}\right)\chi^{(2)}\right) - \frac{1}{2}(\chi^{(2)},(8L_0 + 13)\chi^{(2)}) \tag{27.73}$$

This completes the second level.

Before constructing the action to all orders it will be instructive to rewrite the second level result in a kind a first order form. Completing the squares in the $L_1\psi$ and $L_2\psi$ terms in the action we may rewrite (27.73) as

$$\frac{1}{2}(\psi,(L_0 - 1)\psi) - \frac{1}{4}(L_1\psi - 2L_{-1}\chi^{(2)}, L_1\psi - 2L_{-1}\chi^{(2)})$$

$$- \frac{1}{8}(L_2\psi - 6\chi^{(2)}, L_2\psi - 6\chi^{(2)}) - 2(\chi^{(2)},(L_0 + 1)\chi^{(2)}) \tag{27.74}$$

where we have used that at this level $L_1\chi^{(2)} = 0$. We now introduce the auxiliary fields $\phi^{(1)}$ and $\phi^{(2)}$ to rewrite the action as

$$\frac{1}{2}(\psi,(L_0 - 1)\psi) + (\phi^{(1)}, L_1\psi + L_{-1}\zeta^1{}_1)$$

$$+ (\phi^{(2)}, L_2\psi + 3\zeta^1{}_1) + (\phi^{(1)}, \phi^{(1)})$$

$$+ 2(\phi^{(2)}, \phi^{(2)}) - \frac{1}{2}(\zeta^1{}_1,(L_0 + 1)\zeta^1{}_1) \tag{27.75}$$

where

$$\zeta^1{}_1 = -2\chi^{(2)}$$

This completes the analysis of the second level. The construction up to the 6th level was given in Ref. 199.

27.5   The Finite Set

We now find an action which is gauge invariant to all levels. We require the gauge invariance

$$\delta\psi = L_{-1}\Lambda^1 + L_{-2}\Lambda^2 \tag{27.76}$$

This was seen to be as a consequence of our lowest order construction, but it can also be seen from the on-shell gauge invariances, which now become off-shell, of the Virasoro constraints or from the associated projector. We note that since $L_{-3} = -[L_{-2}, L_{-1}]$ we will also have the gauge invariance $\delta\psi = L_{-3}\Lambda_3$ and by an extension $\delta\psi = L_{-n}\Lambda_n$ for any $n > 1$.

Given an action A invariant under (27.75) we can deduce that

$$0 = \left(\Lambda^1, L_1 \frac{\delta A}{\delta\psi}\right) + \sum_i \left(\delta\phi^i, \frac{\delta A}{\delta\phi^i}\right) \tag{27.77}$$

The ϕ^i represent all the other string functionals that occur in A. If we enforce all the field equations except that of ψ we find that

$$L_1 \frac{\delta A}{\delta\psi} = 0 \tag{27.78}$$

and by a similar argument $L_2 \dfrac{\delta A}{\delta\psi} = 0$. The ψ equation must be of the form $(L_0 - 1)\psi + \cdots = 0$. Applying L_1 and L_2 we require a knowledge of $L_1\psi$ and $L_2\psi$ in order to get zero. This requires the introduction of two fields ϕ^1 and ϕ^2 in order to specify $L_1\psi$ and $L_2\psi$. As such, we require a term $(\phi^1, L_1\psi) + (\phi^2, L_2\psi)$ in the action. Consequently ϕ^1 and ϕ^2 must occur in the ψ equation as

$$(L_0 - 1)\psi + L_{-1}\phi^1 + L_{-2}\phi^2 = 0 \tag{27.79}$$

Having introduced these new terms we now must specify $L_1\phi^1$, $L_2\phi^1$ and $L_1\phi^2$ and $L_2\phi^2$ which are determined by the field equations, of the new fields $\zeta^1{}_1$, $\zeta^2{}_1$, $\zeta^1{}_2$ and $\zeta^2{}_2$, respectively. We now write these field equations down and add in, with arbitrary coefficients, any other possible terms of the correct level. For example in the ϕ^2 equation which begins $L_2\psi$ we may add a term $c\zeta^1{}_1$, which is also of level 2. We then determine the arbitrary coefficients by applying L_1 and L_2 on $\dfrac{\delta A}{\delta\psi}$ and demanding zero. We find the equations of Ref. 201. These field equations are:

field	field equation
ψ	$(L_0 - 1)\psi + L_{-1}\phi^1 + L_{-2}\phi^2 = 0$

$$\phi^1 \qquad L_1\psi = -2\phi^1 - L_{-1}\zeta^1{}_1 - L_{-2}\zeta^2{}_1$$

$$\phi^2 \qquad L_2\psi = -4\phi^2 - L_{-1}\zeta^1{}_2 - L_{-2}\zeta^2{}_2 - 3\zeta^1{}_1$$

$$\zeta^1{}_1 \qquad L_1\phi^1 = (L_0 + 1)\zeta^1{}_1 - 3\phi^2$$

$$\zeta^2{}_1 \qquad L_2\phi^1 = (L_0 + 2)\zeta^1{}_2$$

$$\zeta^1{}_2 \qquad L_1\phi^2 = (L_0 + 2)\zeta^2{}_1$$

$$\zeta^2{}_2 \qquad L_2\phi^2 = (L_0 + 3)\zeta^2{}_2 \qquad\qquad (27.80)$$

We leave the application of L_1 to the reader and consider the application of L_2 in more detail: we find that

$$(L_0 + 1)L_2\psi + 3L_1\phi^1 + L_{-1}L_2\phi^1 + \left(4L_0 + \frac{D}{2}\right)\phi^2 + L_{-2}L_2\phi^2 = 0$$

$$(27.81)$$

Substituting for $L_2\psi$, $L_1\phi^{(1)}$, $L_2\phi^{(1)}$ and $L_2\phi^{(2)}$, we find that

$$(D - 26)\phi^2 = 0 \qquad\qquad (27.82)$$

Consequently, we discover that this system only exists in the critical dimension $D = 26$.

Given the field equations, we can search for the full gauge invariance. These are entirely determined from $\delta\psi = L_{-1}\Lambda^1 + L_{-2}\Lambda^2$ and one may easily check that they are given by[201]

$$\delta\psi = \sum_{n=1}^{2} L_{-n}\Lambda^n \qquad \delta\phi^n = -(L_0 + n - 1)\Lambda^n \qquad (27.83)$$

$$\delta\zeta^n{}_m = -L_m\Lambda^n - (2m + n)\Lambda^{n+m} \quad \text{for } n, m = 1, 2. \qquad (27.84)$$

It is straightforward to write down an action from which the above equation follows:

$$\frac{1}{2}(\psi, (L_0 - 1)\psi) + \sum_{n=1}^{2}(\phi^n, L_n\psi) + \sum_{n,m=1}^{2}(L_n\phi^m, \zeta^n{}_m)$$

$$+ \sum_{n=1}^{2} n(\phi^n, \phi^n) - \frac{1}{2}\sum_{n,m=1}^{2}(\zeta^n{}_m, (L_0 + n + m - 1)\zeta^m{}_n)$$

$$+ \sum_{n,m=1}^{2}(2n + m)(\phi^{n+m}, \zeta^n{}_m) \qquad\qquad (27.85)$$

We will refer to this system as the "finite set."

In fact, one can find free-gauge covariant formulation of all known strings in this way; the open and closed bosonic string and the open and closed superstring theories and the heterotic string. We refer the reader to Ref. 201 for these other formulations.

27.6   The Infinite Set

One can extend the result for the open bosonic string given above so that it contains an infinite number of supplementary fields.[201,202] This will have the advantage that the generators of the Virasoro algebra will appear on a more equal footing. This is achieved by introducing the fields

$$\psi, \phi^n, \zeta^n_m \quad n, m = 1, 2, \ldots, \infty \tag{27.86}$$

and the action is the same as that of Eq. (27.85) except that now all the sums run from 1 to ∞. Remarkably this action is invariant under the transformations of Eq. (27.83) except that the sums also run from 1 to ∞. This requires the identity

$$-\sum_{m=1}^{n-1} (2n - m)(n + m) + \frac{D}{12} n(n^2 - 1) - 2n(n - 1) = 0 \tag{27.87}$$

which only works of $D = 26$.

It will be useful to write out the equations of motion of this set for future use; they are

$$(L_0 - 1)\psi + \sum_{n=1}^{\infty} L_{-n}\phi^n = 0$$

$$L_n\psi + L_{-n}\zeta^n_m + \sum_{p,m=1}^{\infty} \delta(m + p - n)(2p + m)\zeta^p_m + 2n\phi^n = 0$$

$$L_n\phi^m - (L_0 + n + m - 1)\zeta^m_n = 0 \tag{27.88}$$

Of course, one must not only find a gauge invariant action; one must also find an action which gives the correct count of on-shell states. One way to demonstrate this would be to show that after appropriate gauge choices we recover the Virasoro constraints. It is known that the "finite set" and the ∞ set lead to the correct count of states up to the tenth and sixth levels respectively.[204] Recently,[205] some incomplete arguments have been advanced to suggest that they fail at these levels. Whether this is the case or not will emerge when the count of states is explicitly carried out at these levels.

Even at the third and sixth levels in the infinite and finite sets, respectively, one finds[204] the existence of complicated additional symmetries that are required for the counting of states and, for example, building the interacting theory.

We would, of course, like a formulation in which all the symmetries are manifest. This formulation is easily found by extending the ∞ set by the same process by which it was found. We require essentially that L_n vanish on all equations of motion. For the ϕ^n equation this requires a knowledge of $L_n\zeta^m{}_p$ and so we require a new field $\phi^{mn}{}_p$ which in turn requires a field $\psi^{mn}{}_{pq}$. Repeating the process indefinitely we find the so called master theory which has the field content

$$\psi^k_k \quad \phi^{k+1}_k$$

where the index k indicates the number of indices. We will find that indices on a given level are completely symmetrized.

In fact, with some technology, the transition from the ∞ to the master set is very straightforward, and we will see how this is achieved later. At this point we will discuss some of the technology which enables a simple derivation of the master set.

One might hope to find the finite and infinite sets by a gauge choice from the master set. In this process one would find compensating transformations to preserve the choice of gauge. A preliminary examination of the complicated symmetries of the smaller set does indeed indicate that they are of this type. Later, we will show that the master set does have the correct on-shell count. Consequently, if one could recover the smaller sets, through a gauge choice this would show that the smaller sets also give the correct on-shell count.

27.7 The Master Set

The master set was found in Ref. 204 using the techniques outlined in this chapter. It was also, independently found in Ref. 206 by extending the gauge covariant free theory given in Refs. 203, 207 and 208 and it was mentioned in Ref. 209.

It will prove useful to introduce an extension of space-time to include anticommuting coordinates. We consider the space which is parametrized by the coordinates $x^u(\sigma)$, $c(\sigma)$, $\bar{c}(\sigma)$. The coordinate $x^u(\sigma)$ are the usual 26 bosonic coordinates while $c(\sigma)$ and $\bar{c}(\sigma)$ are fermionic coordinates. We take these coordinates to satisfy the boundary conditions

$$\partial_\sigma x^u(\sigma) = \partial_\sigma c(\sigma) = \bar{c}(\sigma). \tag{27.89}$$

all vanish at $\sigma = 0$ and π. In fact, this 28-dimensional space emerges naturally in the context of string theory when one BRST quantizes the Nambu action for the string.[210] The fields $c(\sigma)$ and $\bar{c}(\sigma)$ correspond to the ghosts one must introduce when one fixes the two-dimensional reparametrization symmetry of the world sheet of the string. The reader should be clear that we are going to find a gauge invarinat second quantized field theory of strings and *not* a

BRST formulation. The latter can be found from the former by the usual method of gauge fixing and the introduction of corresponding ghosts.

However, it has emerged that many of the tools introduced in the *first quantized* string action have a natural role in the *gauge invariant* second quantized theory.

While $x^\mu(\sigma)$ has the usual Fourier expansion, the Fourier expansions for c and $\bar{c}$ corresponding to the above boundary conditions are

$$c(\sigma) = c^0 + 2 \sum_{n=1}^{\infty} c_n \cos n\sigma$$

$$\bar{c}(\sigma) = 2 \sum_{n=1}^{\infty} \bar{c}_n \sin n\sigma \qquad (27.90)$$

In analogy with $x^\mu(\sigma)$ we introduce fermionic annihilation and creation operators:

$$\bar{\beta}(\sigma) = \frac{\delta}{\delta c(\sigma)} - \frac{1}{2\pi}\bar{c}(\sigma) = \frac{1}{\pi}\frac{1}{\sqrt{2}} \sum_{n=-\infty}^{\infty} \bar{\beta}_n e^{in\sigma}$$

$$\beta(\sigma) = -\frac{\delta}{\delta\bar{c}(\sigma)} + \frac{1}{2\pi}c(\sigma) = \frac{1}{\pi}\frac{1}{\sqrt{2}} \sum_{n=-\infty}^{\infty} \beta_n e^{in\sigma} \qquad (27.91)$$

Equipped with the obvious scalar product we find that

$$\bar{\beta}(\sigma)^\dagger = \bar{\beta}(\sigma) \qquad \beta(\sigma)^\dagger = \beta(\sigma) \qquad (27.92)$$

Note that unlike for $\dfrac{\delta}{\delta x^\mu(\sigma)}$ we do not require an i for hemiticity since we are dealing with anticommuting quantities and

$$\left(\frac{\delta}{\delta c(\sigma)}\right)^\dagger = \frac{\delta}{\delta c(\sigma)} \qquad (27.93)$$

as a result we find that

$$\bar{\beta}_n^\dagger = \bar{\beta}_{-n}, \qquad \beta_n^\dagger = \beta_{-n} \qquad (27.94)$$

and, in particular,

$$\bar{\beta}_0^\dagger = \bar{\beta}_0 \qquad \beta_0^\dagger = \beta_0 \qquad (27.95)$$

It is easily seen that these fermionic oscillators obey the relations

$$\{\bar{\beta}_n^\dagger, \beta_m\} = \{\bar{\beta}_n, \beta_m^\dagger\} = \delta_{n,m}$$

$$\{\beta_n, \beta_m\} = 0 = \{\bar{\beta}_n, \bar{\beta}_m\} \quad n, m \geq 0$$

$$\{\beta_n, \beta_m^\dagger\} = 0 = \{\bar{\beta}_n, \bar{\beta}_m^\dagger\} \qquad (27.96)$$

We now define a vacuum with respect to these oscillators. Clearly we can take

$$\beta_n| \ \rangle = \beta_n| \ \rangle = 0 \quad n \geqslant 1 \tag{27.97}$$

as well as the usual condition for the bosonic α_n^μ oscillators. The action of the zero modes on the vacuum, however, require more care.[210] We can define a vacuum $|+\rangle$ by

$$\beta_0|+\rangle = 0 \tag{27.98}$$

then under $\bar{\beta}_0$ we find a new vacuum

$$\bar{\beta}_0|+\rangle = |-\rangle \tag{27.99}$$

since $\bar{\beta}_0^2 = 0$, we find that $\bar{\beta}_0|-\rangle = 0$. From the relation $\{\beta_0, \bar{\beta}_0\} = 1$ we also find $\beta_0|-\rangle = |+\rangle$. We note that

$$\langle +|+\rangle = \langle -|\beta_0\beta_0|-\rangle = 0 \tag{27.100}$$

and similarly for $\langle -|-\rangle = 0$. However, we have the relations

$$\langle +|-\rangle = \langle -|\beta_0\bar{\beta}_0|+\rangle = \langle -|+\rangle \tag{27.101}$$

and we choose $\langle +|-\rangle = 1$. We take the $|-\rangle$ vacuum to be odd and so the $|+\rangle$ vacuum is even as β_0 is an odd object.

Let us consider the most general functional χ of $x^\mu(\sigma)$, $c(\sigma)$, $\bar{c}(\sigma)$. In the oscillator basis it may be written as

$$
\begin{aligned}
|\chi\rangle &\equiv \psi|-\rangle + \varphi|+\rangle \\
&= \sum_{\{n\}\{m\}} \beta^{n_1\dagger} \dots \beta^{n_b\dagger} \bar{\beta}_{m_1}^\dagger \dots \bar{\beta}_{m_a}^\dagger \psi_{n_1 \dots n_b}{}^{m_1 \dots m_a}[x^\mu(\sigma)]|-\rangle \\
&\quad + \sum_{\{n\}\{m\}} \beta^{n_1\dagger} \dots \beta^{n_b\dagger} \bar{\beta}_{m_1}^\dagger \dots \bar{\beta}_{m_{a+1}}^\dagger \varphi_{n_1 \dots n_b}{}^{m_1 \dots m_{a+1}}[x^\mu(\sigma)]|+\rangle \quad (27.102)
\end{aligned}
$$

In fact, these are the string functionals that occur in covariant string field theory. The ghost coordinates, as we shall see, correctly encode the supplementary fields. This is similar to the superspace of sypersymmetric theories which enables one to encode in a systematic way the component fields of a supermultiplet.

We note that $\psi_{n_1 \dots n_b}{}^{m_1 \dots m_a} = \psi_{[n_1 \dots n_b]}{}^{[m_1 \dots m_a]}$ and that if $a + b$ is an odd integer then ψ_b^a is an anticommuting field. Since we want to describe a gauge invariant string field theory and not a BRST invariant theory which has ghost fields, we require χ to be subject to a constraint. Note that the previously discussed master theory had a field content ψ_k^k, ϕ_k^{k+1}. This will be the content of $|\chi\rangle$ if we impose the constraint

$$\sum_{m=1}^\infty (\beta^{m\dagger}\bar{\beta}_m - \bar{\beta}_m^\dagger\beta^m + \beta_0\bar{\beta}_0)|\chi\rangle = 0 \tag{27.103}$$

or

$$N|\chi\rangle = 0$$

where N is defined to be the operator in the first equation. We now adopt this equation. Note that due to the fermionic assignments of the vacuum if $|\chi\rangle$ is odd then the component fields ψ_k^k and ϕ_k^{k+1} are even. An extremely useful operator is the Virasoro charge which is constructed as follows. The Virasoro generators L_n classically obey the algebra $\{L_n, L_m\} = -i(n-m)L_{n+m} = -if_{nm}{}^p L_p$. The corresponding BRST charge is

$$Q' = \sum_{n=-\infty}^{\infty} \beta_{-n} L_n - \frac{1}{2} \sum_{n,m,p=-\infty}^{\infty} \bar{\beta}_p f_{nm}{}^p \beta_{-n} \beta_{-m} \qquad (27.104)$$

This type of construction applied to first quantized Hamiltonian theories can be found in Ref. 211. The above object is not well defined unless it is normal ordered; in which case one may find a normal ordering constant. Hence we consider the object

$$Q = : \sum_{n=-\infty}^{\infty} \beta_{-n} L_n - \frac{1}{2} \sum_{n,m,p=-\infty}^{\infty} \bar{\beta}_p f_{nm}{}^p \beta_{-n} \beta_{-m} - a\beta_0 : \qquad (27.105)$$

Note that for anticommuting quantities we must assign minus signs when normal ordering; for example

$$: \beta_n \bar{\beta}_{-m} : = -\bar{\beta}_{-m} \beta_n \quad \text{for } n, m \geqslant 1 \qquad (27.106)$$

From its definition almost, one easily finds that $Q'^2 = 0$. However, the well defined charge Q only satisfies $Q^2 = 0$ provided $D = 26$ and the intercept is 1 (i.e., $a = 1$).[210]

We may rewrite Q in various ways that we will shortly need

$$Q = \beta_0 K - 2\bar{\beta}_0 M + d + D \qquad (27.107)$$

where K, M, and D do not involve any zero modes, i.e., any β_0 and $\bar{\beta}_0$. We find that

$$K = L_0 - 1 + \sum_{n=1}^{\infty} (n\beta_n^\dagger \bar{\beta}_n + n\bar{\beta}_n^\dagger \beta_n)$$

$$M = + \sum_{n=1}^{\infty} n\beta_n^\dagger \beta_n$$

$$d = \beta^{n\dagger}\left(L_n + f_{m-n}{}^p \bar{\beta}_p^\dagger \beta^m + \frac{1}{2}\beta^{m\dagger} f_{mn}{}^p \bar{\beta}_p \right)$$

$$D = \beta^n\left(L_{-n} + f_{m-n}{}^p \beta^{m\dagger} \bar{\beta}_p + \frac{1}{2}\bar{\beta}_p^\dagger f_{mn}{}^p \beta^m \right) \quad m, n, p \geqslant 1 \qquad (27.108)$$

We note that $D = d^\dagger$ and they satisfy the relations

$$d^2 = D^2 = 0 \quad [K, d] = 0 = [K, D] = [K, M] \tag{27.109}$$

$$[M, d] = [M, D] = 0 \quad \{d, D\} - 2MK = 0 \tag{27.110}$$

As a consequence of these

$$Q^2 = \{d, D\} - 2MK = 0 \tag{27.111}$$

Let us now study the action of Q on an arbitrary functional χ. If $|\chi\rangle$ is odd we find that

$$Q|\chi\rangle = (2M\phi + (d + D)\psi)|-\rangle + (K\psi + (d + D)\phi)|+\rangle \tag{27.112}$$

while if $|\chi\rangle$ is even

$$Q|\chi\rangle = (-2M\phi + (d + D)\psi)|-\rangle + (-K\psi + (d + D)\phi)|+\rangle \tag{27.113}$$

where K, M, d and D are now defined by acting on $\psi_{n_1 \dots n_b}{}^{m_1 \dots m_a}$.

$$(K\psi)_{n_1 \dots n_b}{}^{m_1 \dots m_a} = (L_0 - 1 + m + n)\psi_{n_1 \dots n_b}{}^{m_1 \dots m_a}$$

$$(M\psi)_{n_1 \dots n_{b+1}}{}^{m_1 \dots m_{a-1}} = (-1)^b a(\eta_{[n_1 p}\psi_{n_2 \dots n_{b+1}]}^{pm_1 \dots m_{a-1}})$$

$$(d\psi)_{n_1 \dots n_{b+1}}{}^{m_1 \dots m_a} = L_{[n_1}\psi_{n_2 \dots n_{b+1}]}^{m_1 \dots m_a} + af_{p-n_1}{}^{m_1}\psi_{n_2 \dots n_{b+1}}^{pm_2 \dots m_a} - \frac{1}{2}bf_{n_1 n_2}{}^p\psi_{pn_3 \dots n_{b+1}}^{m_1 \dots m_a}$$

$$(D\psi)_{n_1 \dots n_b}^{m_1 \dots m_{a-1}} = (-1)^b a \left[L_{-p}\psi_{n_1 \dots n_b}^{pm_1 \dots m_{a-1}} + bf_{n-p}{}^q\psi_{qn_2 \dots n_b}^{pm_1 \dots m_{a-1}} \right.$$
$$\left. - \frac{1}{2}(a - 1)f_{ke}{}^{m_1}\psi_{n_1 \dots n_b}^{ke m_2 \dots m_{a-1}} \right]$$

where $m = \sum_{i=1}^{a} m_i$, $n = \sum_{i=1}^{b} n_j$. These definitions differ from the definitions given in the literature by trivial factors. The difference in signs between $|\chi\rangle$ being even and odd comes from pushing the zero modes through the oscillators and fields onto the vacuum.

For example, we find that if

$$\psi_0^0 = \psi \quad \phi_0^1 = \phi^n$$

$$d\psi_0^0 = L_n\psi \quad (D\phi_0^1) = L_{-n}\phi^n$$

$$(d\phi_0^1) = L_m\phi^n + (2m + n)\phi^{n+m}$$

$$(M\phi_0^1) = n\phi^n \quad K\phi^n = (L_0 - 1 + n)\phi^n \tag{27.114}$$

We now realize that the above operators are particularly suited to a discussion of the equations of motion (27.88) of the infinite set. These equations may be rewritten in the form

$$K\psi_0^0 + D\phi_0^1 = 0$$

$$d\psi_0^0 + D\psi_1^1 + 2M\phi_0^1 = 0$$

$$K\psi_1^1 + d\phi_0^1 = 0 \tag{27.115}$$

where $\psi_1^1 = \{\zeta^n{}_m : n, m = 1, 2, \ldots, \infty\}$ and the gauge invariances are

$$\delta\psi_0^0 = D\Lambda_0^1 \quad \delta\phi_0^1 = -K\Lambda_0^1$$

$$\delta\psi_1^1 = d\Lambda_0^1 \tag{27.116}$$

The invariance of equations (27.115) under the gauge invariances of equation (27.116) is now easily shown. The generalization of the infinite set to the master set is obvious. The general equations are[204, 206]

$$K\psi_a^a + D\phi_a^{a+1} + d\phi_{a-1}^a = 0$$

$$d\psi_a^a + D\psi_{a+1}^{a+1} + 2M\phi_a^{a+1} = 0 \tag{27.117}$$

while the gauge invariances are

$$\delta\psi_a^a = d\Lambda_{a-1}^a + D\Lambda_a^{a+1} - 2M\Omega_{a-1}^{a+1}$$

$$\delta\phi_a^{a+1} = -K\Lambda_a^{a+1} + d\Omega_{a-1}^{a+1} + D\Omega_a^{a+2} \tag{27.118}$$

where Ω is a new symmetry. Some traces of this symmetry can be found in the infinite set, although due to gauge compensation, it occurs in a rather complicated form.

The full use of the formalism given above gives an extremely simple description of the master set. Before giving this, however let us momentarily return to the point particle.[194] Here, we only have one constraint and so

$$Q = \beta_0(-\partial^2 + m^2) \tag{27.119}$$

since the classical algebra of the constraints is Abelian. The well known Klein-Gordon action can be written as

$$\int d^4x \psi(-\partial^2 + m^2)\psi = \langle \chi | Q | \chi \rangle \tag{27.120}$$

where $|\chi\rangle = \psi|-\rangle + \phi|+\rangle$ is the most general functional of x^μ and c and $\bar{c}$. Clearly, $Q^2 = 0$ and so the action is invariant under

$$\delta|\chi\rangle = Q|\Lambda\rangle \tag{27.121}$$

In components this reads

$$\delta\psi = 0 \quad \delta\phi = (-\partial^2 + m^2)\Lambda \tag{27.122}$$

where $|\Lambda\rangle = \Lambda|-\rangle + \Omega'|+\rangle$ and so the symmetry is trivially realized.

Returning to the string the action for the master set is in fact none other than[204,206,209]

$$\frac{1}{2}\langle\chi|Q|\chi\rangle \tag{27.123}$$

and the gauge invariance is

$$\delta|\chi\rangle = Q|\Lambda\rangle \tag{27.124}$$

we recall that $|\chi\rangle$ is subject to the algebraic constraint $N|\chi\rangle = 0$ and this reflects itself, using $[\phi, N] = -N$ to give the constraint

$$(N + 1)|\Lambda\rangle = 0 \tag{27.125}$$

on $|\Lambda\rangle$. The functional $|\Lambda\rangle = \Lambda|-\rangle + \Omega|+\rangle$ then contains the following functionals of $x^u(\sigma)$

$$\Lambda_k^{k+1} \quad \Omega_k^{k+2} \quad \text{for all } K. \tag{27.126}$$

The action may be expressed in functional form:

$$\frac{1}{2}\int \mathscr{D}x^u(\sigma)\mathscr{D}c(\sigma)\mathscr{D}\bar{c}(\sigma)\chi[x^u(\sigma), c(\sigma), \bar{c}(\sigma)]Q\chi[x^u(\sigma), c(\sigma), \bar{c}(\sigma)] \tag{27.127}$$

where Q, in functional form, is given by

$$Q = \frac{i\pi}{2} : \oint \frac{dz}{z}\left[\sqrt{2}\pi\hat{\wp}_\mu^2 - \frac{1}{\pi^2} + z\frac{d}{dz}(\beta\bar{\beta}) + \left(z\frac{d}{dz}\beta\right)\bar{\beta}\right]\beta : \tag{27.128}$$

where $z = e^{-i\sigma}$. Note that x and Q are odd, but so is $\mathscr{D}c(\sigma)\mathscr{D}\bar{c}(\sigma) = \prod_{n=0}^{\infty} dc_n \prod_{n=1}^{\infty} d\bar{c}_n$ due to the fact that $c(\sigma)$ has one more zero mode than $\bar{c}(\sigma)$.

The equation of motion is given by

$$Q|\chi\rangle = 0 \tag{27.129}$$

which, using Eq. (27.112), can be written in terms of component fields ψ_k^k and ϕ_k^{k+1} as

$$K\psi + d\phi + D\phi = 0$$

$$d\psi + D\psi + 2M\phi = 0 \tag{27.130}$$

The invariance in terms of the component fields Λ_k^{k+1}, Ω_k^{k+2} is given by

$$\delta\psi = (d + D)\Lambda - 2M\Omega$$

$$\delta\phi = -K\Lambda + (d + D)\Omega \tag{27.131}$$

These are of course identical to Eq. (27.117) where the index structure is explicitly shown.

The action $\langle\chi|Q|\chi\rangle$ becomes, in component fields,

$$\frac{1}{2}(\psi, K\psi) + (\psi, d\phi) + (D\phi, \psi) + (\phi, M\phi) \qquad (27.132)$$

In deriving this result we have taken account of the vacuum properties of Eqs. (27.98) and (27.99) and used the fact that $(\psi, d\phi) = (D\psi, \phi)$ for any ψ and ϕ.

We note that $|\Lambda\rangle = Q|\Lambda'\rangle$ leaves $|\chi\rangle$ inert and is a so-called hidden invariance. This automatically leads to the phenomenon of "ghost for ghosts" since the gauge fixing term must involve $|\chi\rangle$ alone and so the corresponding ghost action will inevitably be invariant under this invariance. We will discuss this shortly.

Unlike the finite and infinite sets it is straightforward to show that the master set leads to the correct on-shell states for the string.

27.8   The On-shell Spectrum of the Master Set[204]

The string in $D = 26$ has the same number of on-shell degrees of freedom as there are fields in a functional of $x^i(\sigma)$, $i = 1, 2, \ldots, 24$ and x^-. That is because in the light-cone gauge, where only $x^i(\sigma)$ and the x^- centre-of-mass coordinate remain, all degrees of freedom are physical. Hence, the number of on-shell degrees of freedom at each level N is $p(N)$, where $p(N)$ is given by

$$\sum_{n=0}^{\infty} p(N)x^N = \prod_{n=1}^{\infty} (1 - x^n)^{-24}$$

$$= 1 + 24x + \cdots \qquad (27.133)$$

Note that the partition function $\prod_{n=1}^{\infty} (1 - x^n)^{-D}$ just records the number of degrees of freedom in a D-dimensional functional.

We must show that this is the number of on-shell degrees of freedom predicted by $\langle\chi|Q|\chi\rangle$. We could do this in three ways.

(i) By counting the degrees of freedom classically, but Lorentz covariantly. This is non-trivial even for the photon.

(ii) Show we can go to the light-cone formalism by a gauge choice.

(iii) Quantize the system, that is, gauge fix and add ghosts and then show that the bosonic minus the fermionic degrees is $p(N)$ at level N.

We now carry out route (iii). We begin with the classical theory whose component field content is ψ_k^k, ϕ_k^{k+1}. Examining Eq. (27.118), let us fix Λ_k^{k+1} by setting $\phi_k^{k+1} = 0$. Hence we insert $\delta(\phi_k^{k+1})$ in the functional integral and incur the following ghost term $\bar{\Lambda}_{k+1}^k(-K\Lambda_k^{k+1} + D\Omega_k^{k+2} + d\Omega_{k-1}^{k+1})$. We are using the same notation for ghost fields as the symmetry from which they arise. This ghost action, however, possesses an invariance corresponding to the "hidden" invariance $|\Lambda\rangle = Q|\Lambda'\rangle$. Since $[Q, N] = -N$ we find that $N|\chi\rangle = 0$ implies that $(N + 2)|\Lambda'\rangle = 0$ and so it has the field content $\{\Lambda_{k-1}'^{k+1}, \Omega_{k-1}'^{k+2}\}$. We fix this invariance by setting $\Omega_{k-1}^{k+1} = 0$, that is, we insert $\delta(\Omega_{k-1}^{k+1})$ in the functional integral and add the ghost term

$$\bar{A}_{k+1}^{\prime k-1}(K A_{k-1}^{\prime k+1} + d A_{k-2}^{\prime k+1} + D A_{k-1}^{\prime k+2}) \tag{27.134}$$

This term again has an invariance corresponding to $|\Lambda'\rangle = Q|\Lambda''\rangle$ requiring ghost for ghost for ghosts. Clearly this process continues indefinitely.

The net result is the field content

$$\psi_k^k \quad A_k^{k+1} \quad A_{k-1}^{\prime k+1} \quad A_{k-1}^{\prime\prime k+2} \dots$$

and

$$\bar{A}_{k+1}^k \quad \bar{A}_{k+1}^{\prime k-1} \quad \bar{A}_{k+2}^{\prime\prime k-1} \dots \tag{27.135}$$

and they occur in the action with the kinetic operator K. We observe that in this set there is one and only one tensor of the type $A^{m_1 \cdots m_a}{}_{n_1 \dots n_b}$ for each a and b. These fields may be neatly fitted into a functional χ which is completely general except it has no $|+\rangle$ vacuum, i.e., it satisfies

$$\bar{\beta}_0 |\chi\rangle = 0$$

$$|\chi\rangle = \sum_{\{m\}\{n\}} \beta^{n_1\dagger} \dots \beta^{n_b\dagger} \bar{\beta}_{m_1}^\dagger \dots \bar{\beta}_{m_a}^\dagger A^{m_1 \cdots m_a}{}_{n_1 \dots n_b} |-\rangle. \tag{27.136}$$

The free BRST action is given by

$$\langle \chi | \beta_0 K | \chi \rangle = \langle \chi | [\beta_0 \bar{\beta}_0, Q] | \chi \rangle \tag{27.137}$$

Of course, it is no longer gauge invariant but is BRST invariant, namely,

$$\delta |\chi\rangle = \lambda Q |\chi\rangle. \tag{27.138}$$

where λ is the anticommuting BRST parameter.

We will now demonstrate that this gauge fixed action, originally found by Siegel[194] by another approach, has the correct number of degrees of freedom. Let us write the tensor A_{b+c}^{a+c} in the form

$$A_{n_1 \dots n_b; p_1 \dots p_c}{}^{m_1 \cdots m_a; p_1 \cdots p_c} \tag{27.139}$$

indicating that it has c indices in common.

This tensor first contributes at level $M = \sum_{i=1}^a m_i + \sum_{i=1}^b n_i + 2 \sum_{i=1}^c p_i$ and is commuting or anticommuting according to whether $\sum m_i + \sum n_i$ is even or odd. The net (bose-fermi) number of tensors at level M is $c(m)$ where

$$\sum_{m=0}^\infty c(m) x^m = \prod_{n=1}^\infty (1 - 2x^n + x^{2n})$$

$$= \prod_{n=1}^\infty (1 - x^n)^2 \tag{27.140}$$

Since each such tensor contributes $T^{26}(N-M)$, the net total number of degrees of freedom at level N is given by

$$p(N) = \sum_{m=0}^{N} c(m) T^{26}(N-m)$$

and so

$$\sum_{N=0}^{\infty} p(N) x^N = \sum_{N=0}^{\infty} \sum_{m=0}^{N} c(m) T^{26}(N-m) x^{N-m} x^m$$

$$= \prod_{n=1}^{\infty} (1-x^n)^2 \prod_{n=1}^{\infty} \frac{1}{(1-x^n)^{26}}$$

$$= \prod_{n=1}^{\infty} (1-x^n)^{-24} = \sum_{n=0}^{\infty} T^{24}(N) x^N \qquad (27.141)$$

Hence we recover the light-cone count and so confirm that $\langle \chi | Q | \chi \rangle$ does describe the spectrum of string theory.

A much shorter proof is as follows[204]:

$$\sum_{N} \left(n_B(N) - n_F(N) \right) x^N = s \, \mathrm{Tr} \, x^{\sum_{n=1}^{\infty} (\alpha_n^\mu \alpha_n + \beta_n^\dagger \bar{\beta}_n + \bar{\beta}_n^\dagger \beta_n)}$$

$$= \prod_{n=1}^{\infty} (1-x^n)^2 \prod_{n=1}^{\infty} \frac{1}{(1-x)^{26}}$$

$$= \prod_{n=1}^{\infty} \frac{1}{(1-x^n)^{24}} \qquad (27.142)$$

where we have made use of the fermionic partition function.

This concludes our discussion of the open bosonic string. The systematic method of construction given for this case can be easily used to find all the other free gauge covariant string theories. One first finds the finite set, then generalizes this to the infinite set and then to the master set. This one can put in convenient form using extra coordinates. The count of states is given in the same way as above. The closed bosonic string and the open superstring are constructed in the same way as the open bosonic string above. For a discussion of these free gauge covariant theories along the same lines as that given here the reader is refered to the author's Trieste lectures (1986) and references therein. This source also certains a discussion of progress in interacting gauge covariant string theory.

Appendix A
An Explanation of our Choice of Conventions

These conventions are taken verbatim from Ref. 6.

Of course, conventions are conventions, and one set is not better than another set. However, some conventions lead to simpler manipulations, and hence to fewer algebraic errors, than other conventions. Let us now discuss our conventions, which we have chosen with this simplicity in mind.

Our metric in Minkowski space-time is $\eta_{mn} = (-, +, +, +)$. The ε-symbol is defined by

$$\varepsilon^{0123} = -\varepsilon_{0123} = +1 \tag{A.1}$$

Our index convention distinguishes between flat and curved indices. Spinors can have 2 components or 4 components. If we consider spinor and vector indices together, we call them superindices. The table below summarizes these conventions

	Vector	2-Comp Spinor	4-Comp Spinor	Super Indices
Flat	m, n	$A, \dot{A}$	α, β	M, N
Curved	$\underline{\mu}, \underline{\nu}$	$\underline{A}, \underline{\dot{A}}$	$\underline{\alpha}, \underline{\beta}$	$\underline{\Delta}, \underline{\Pi}$

Superindices are contracted in north-westerly fashion: from left-upper to right-lower. For example

$$\chi^M \zeta_M, \quad \varepsilon^M T_{MN}{}^R, \quad E_M{}^{\underline{A}} \xi_{\underline{A}} \tag{A.2}$$

In terms of 2-component spinors and vectors, these contractions decompose per definition as follows

$$\chi^M \zeta_M = \chi^m \zeta_m + \chi^A \zeta_A + \chi^{\dot{A}} \zeta_{\dot{A}} = (-1)^M \chi_M \zeta^M$$

$$\varepsilon^M T_{MN}{}^R = \varepsilon^m T_{mN}{}^R + \varepsilon^A T_{AN}{}^R + \varepsilon^{\dot{A}} T_{\dot{A}N}{}^R$$

$$E_M{}^{\underline{A}} \xi_{\underline{A}} = E_M{}^{\underline{\mu}} \xi_{\underline{\mu}} + E_M{}^{\underline{A}} \xi_{\underline{A}} + E_M{}^{\underline{\dot{A}}} \xi_{\underline{\dot{A}}} \tag{A.3}$$

The symbol $(-)^M$ is $+1$ for $M = m$ and (-1) for $M = A$ or $M = \dot{A}$.

Two component spinors χ^A, $\zeta_{\dot{A}}$ belong to the $(\frac{1}{2}, 0)$ and $(0, \frac{1}{2})$ representations of the Lorentz group. Their indices can be raised and lowered with the invariant tensors

$$\varepsilon_{AB} = \varepsilon^{AB} = -\varepsilon_{\dot{A}\dot{B}} = -\varepsilon^{\dot{A}\dot{B}} \qquad \varepsilon_{12} = +1 \qquad (A.4)$$

Note that $\varepsilon^{CB}\varepsilon_{BA} = -\varepsilon_{AB}\varepsilon^{CB} = -\delta_A^C$. The reason for the minus signs will become clear when we require that the 4-component and 2-component contractions all follow the north-west rule. The convention we choose for raising and lowering spinor indices is the same as for super indices. Hence

$$\chi_A = \chi^B \varepsilon_{BA} , \qquad \chi^A = \varepsilon^{AB}\chi_B$$

$$\zeta_{\dot{A}} = \zeta^{\dot{B}}\varepsilon_{\dot{B}\dot{A}} \qquad \zeta^{\dot{A}} = \varepsilon^{\dot{A}\dot{B}}\zeta_{\dot{B}} \qquad (A.5)$$

As a result $\theta^A\theta_A = -\theta_A\theta^A \equiv \theta^2$ is a Lorentz invariant; as is $+\theta^{\dot{A}}\theta_{\dot{A}} = -\theta_{\dot{A}}\theta^{\dot{A}} \equiv \bar{\theta}^2$. Hence

$$\theta_A\theta_B = -\tfrac{1}{2}\varepsilon_{AB}\theta^2 \qquad (A.6)$$

The matrices $(\sigma^m)^{A\dot{B}} = (1, \boldsymbol{\sigma})^{A\dot{B}}$ where $\boldsymbol{\sigma}$ are the Pauli matrices

$$\boldsymbol{\sigma} = \left(\begin{pmatrix} 0 & 1 \\ 1 & 0 \end{pmatrix}, \begin{pmatrix} 0 & -i \\ i & 0 \end{pmatrix}, \begin{pmatrix} 1 & 0 \\ 0 & -1 \end{pmatrix} \right) \qquad (A.7)$$

are invariant tensors under the Lorentz group. Lowering the indices on $(\sigma^m)^{A\dot{B}}$ we find that

$$(\sigma^m)_{A\dot{B}} = +(\bar{\sigma}^m)_{\dot{B}A}$$

where $(\bar{\sigma}^m)_{\dot{B}A} = (-1, +\boldsymbol{\sigma})_{\dot{B}A}$. Note that

$$\sigma_m = (-1, \boldsymbol{\sigma}) \text{ and } \bar{\sigma}_m = (+1, \boldsymbol{\sigma})$$

both σ^m and $\bar{\sigma}^m$ have the same space-component $\boldsymbol{\sigma}$; this is natural given our metric $(-, +, +, +)$, and it leads to relations between σ^m and $\bar{\sigma}_n$ without any minus sign as we now show.

These matrices satisfy the following relations

$$(\sigma^m)^{A\dot{B}}(\bar{\sigma}_n)_{\dot{B}C} = +\delta_n^m\delta_C^A + (\sigma^m{}_n)^A{}_C$$

$$(\bar{\sigma}^m)_{\dot{B}A}(\sigma_n)^{A\dot{C}} = +\delta_n^m\delta_{\dot{B}}^{\dot{C}} + (\bar{\sigma}^m{}_n)_{\dot{B}}{}^{\dot{C}}$$

$$(\sigma^m)^{A\dot{B}}(\bar{\sigma}^n)_{\dot{B}C} = +\eta^{mn}\delta_C^A + (\sigma^{mn})^A{}_C$$

$$\tfrac{1}{2}\varepsilon_{mnrs}(\sigma^{rs})^A{}_B = i(\sigma_{mn})^A{}_B \qquad (A.8)$$

where

$$(\sigma^m{}_n)^A{}_C = \tfrac{1}{2}(\sigma^m\bar{\sigma}_n - \sigma_n\bar{\sigma}^m)^A{}_C$$

$$(\bar{\sigma}^m{}_n)_{\dot{B}}{}^{\dot{C}} = \tfrac{1}{2}(\bar{\sigma}^m\sigma_n - \bar{\sigma}_n\sigma^m)_{\dot{B}}{}^{\dot{C}} \qquad (A.9)$$

Note that σ^{mn} and $\bar{\sigma}^{mn}$ with both indices down or up are symmetric matrices. Further useful indentities are:

$$(\sigma^m)^{A\dot{B}}(\bar{\sigma}_m)_{\dot{C}D} = 2\delta^A_C \delta^{\dot{B}}_{\dot{C}}$$

$$(\sigma^m)^{A\dot{B}}(\sigma_m)^{C\dot{D}} = +2\varepsilon^{AC}\varepsilon^{\dot{B}\dot{D}}$$

$$(\sigma^m)^{A\dot{B}}(\bar{\sigma}^n)_{\dot{B}D}(\sigma^r)^{D\dot{E}} = +[\eta^{mn}\eta^{rs} + \eta^{ms}\eta^{nr} - \eta^{mr}\eta^{ns}](\sigma_s)^{A\dot{E}} + i\varepsilon^{mnrs}(\sigma_s)^{A\dot{E}}$$

$$\tag{A.10}$$

To establish the connection between four- and two-component spinors, we now consider four-component spinors. A representation useful for the transition to 2-component spinors is

$$\gamma^m = \begin{pmatrix} 0 & -i\sigma^m \\ i\bar{\sigma}^m & 0 \end{pmatrix} \quad \text{where} \quad \begin{aligned} \sigma^m &= (1, \boldsymbol{\sigma}) \\ \bar{\sigma}^m &= (-1, \boldsymbol{\sigma}) \end{aligned}$$

We also define

$$\gamma_5 = i\gamma^1\gamma^2\gamma^3\gamma^0 = \begin{pmatrix} 1 & 0 \\ 0 & -1 \end{pmatrix} \quad \text{and so} \quad \gamma_5^2 = +1 \tag{A.11}$$

A Majorana spinor is one whose Majorana conjugate is equal to its Dirac conjugate. The Majorana conjugate of a four-component spinor χ_α is defined by

$$\bar{\chi}_m = \chi^T C \quad \text{or} \quad (\bar{\chi}_m)^\alpha = \chi_\beta C^{\beta\alpha} \tag{A.12}$$

The Dirac conjugate is defined by

$$\bar{\chi}_D = \chi^\dagger(i\gamma^0) \quad \text{or} \quad (\bar{\chi}_D)^\alpha = (\chi_\beta)^*(i\gamma^0)_\beta{}^\alpha \tag{A.13}$$

Hence a Majorana spinor satisfies

$$\chi^T C = \chi^\dagger \gamma^0 i \quad \text{or} \quad \chi_\beta C^{\beta\alpha} = (\chi_\beta)^*(i\gamma^0)_\beta{}^\alpha \tag{A.14}$$

The four-dimensional charge conjugation matrix $C^{\alpha\beta}$ is antisymmetric while $C\gamma^m$ is symmetric, viz.

$$C^T = -C \qquad C\gamma^m C^{-1} = -(\gamma^m)^T \tag{A.15}$$

In the above representation we have

$$C = i\gamma^0\gamma^2 = \begin{pmatrix} \begin{matrix} 0 & 1 \\ -1 & 0 \end{matrix} & \mathbf{0} \\ \mathbf{0} & \begin{matrix} 0 & -1 \\ 1 & 0 \end{matrix} \end{pmatrix} \tag{A.16}$$

We are now in a position to establish the connection between two- and four-component spinors. Let us decompose a 4-component spinor into two

2-component spinors as follows

$$\chi_\alpha = \begin{pmatrix} \chi^A \\ \zeta_{\dot{A}} \end{pmatrix} \tag{A.17}$$

Hence the 2-component spinors are related to 4-component spinors by

$$\chi^A = \tfrac{1}{2}(1 + \gamma_5)\chi \qquad \bar\zeta_{\dot{A}} = \tfrac{1}{2}(1 - \gamma_5)\chi \tag{A.18}$$

At this point χ^A and $\zeta_{\dot{A}}$ only serve to label the first two and last two components of χ^α. The charge conjugation matrix reads then

$$C = \begin{pmatrix} \varepsilon_{AB} & 0 \\ 0 & +\varepsilon^{\dot{A}\dot{B}} \end{pmatrix} \tag{A.19}$$

The Majorana condition, $\bar\chi_M = \bar\chi_D$, becomes

$$(\chi_A)^* = \zeta_{\dot{A}} \qquad (\chi^A)^* = -\zeta^{\dot{A}} \tag{A.20}$$

where, in agreement with our north-west rule

$$\chi_A = \chi^B \varepsilon_{BA} \qquad \zeta^{\dot{A}} = \varepsilon^{\dot{A}\dot{B}} \zeta_{\dot{B}} \tag{A.21}$$

Hence for a Majorana spinor

$$\chi_\alpha = \begin{pmatrix} \chi \\ \bar\chi_{\dot{A}} \end{pmatrix}$$

and

$$\bar\chi^\alpha = (\chi_A, -\bar\chi^{\dot{B}}) \tag{A.22}$$

where by definition $(\chi_A)^* = \bar\chi_{\dot{A}}$ and consequently $(\chi^A)^* = -\bar\chi^{\dot{A}}$. The bar indicates that $\bar\chi_{\dot{A}}$ is obtained from χ_A by complex conjugation. In the main text we will often omit these bars for notational simplicity.

We now turn to the transformation properties of two- and four-component spinors under the Lorentz group.

We normalize the generators of the Lorentz group by their action on a covariant vector

$$(\tfrac{1}{2}w^{mn} J_{mn}) V_r = w_r{}^s V_s \tag{A.23}$$

Hence

$$(J_{mn})_r{}^s = (\eta_{mr}\delta_n{}^s - \eta_{nr}\delta_m{}^s) \tag{A.24}$$

The commutation relations of the generators J_{mn} are thus

$$[J_{mn}, J_{pq}] = \eta_{np}J_{mq} + 3 \text{ terms} \tag{A.25}$$

In the 4-component spinor representation the generators are given by

$$J_{mn} = \tfrac{1}{2}\gamma_{mn} \qquad \gamma_{mn} = \tfrac{1}{2}(\gamma_m\gamma_n - \gamma_n\gamma_m) \tag{A.26}$$

Hence four-component spinors transform under Lorentz transformation as

$$\delta(w)\chi_\alpha = (\tfrac{1}{2}w^{mn}J_{mn})\chi_\alpha = \tfrac{1}{4}w^{mn}(\gamma_{mn})_\alpha{}^\beta\chi_\beta \tag{A.27}$$

Using the representation

$$\gamma^m = \begin{pmatrix} 0 & -i\sigma^m \\ +i\bar{\sigma}^m & 0 \end{pmatrix} \tag{A.28}$$

the transformation rules of 2-component spinors $\chi^\alpha = (\chi^A, \chi_{\dot{A}})$ are

$$\delta\chi^A = \tfrac{1}{4}w^{mn}(\sigma_{mn})^A{}_B\chi^B$$

$$\delta\chi_{\dot{A}} = \tfrac{1}{4}w^{mn}(\bar{\sigma}_{mn})_{\dot{A}}{}^{\dot{B}}\chi_{\dot{B}} \tag{A.29}$$

where we recall

$$\sigma_{mn} = \tfrac{1}{2}(\sigma_m\bar{\sigma}_n - \sigma_n\bar{\sigma}_m) \tag{A.30}$$

The last result follows immediately from $\lambda^B\chi_B = -\lambda_B\chi^B$. Note that we have omitted the bar on $\chi_{\dot{A}}$, for reasons discussed above. Since σ_{mn} is symmetric in its spinor indices one also has

$$\delta\chi_A = -\chi_B\tfrac{1}{4}w^{mn}(\sigma_{mn})^B{}_A = -\tfrac{1}{4}w^{mn}(\sigma_{mn})_A{}^B\chi_B \tag{A.31}$$

We now give the definition of complex conjugation $*$. For an x-space function $*$ is the usual complex conjugation, hence $[(\partial/\partial x^\mu)f(x)]^* = (\partial/\partial x^\mu)f^*(x)$. But for scalars constructed from two anticommuting objects

$$S = \theta^A\varepsilon_{AB}\theta^B \tag{A.33}$$

we require that complex conjugation and hermitian conjugation are the same. If we view the θ's as matrices, hermitian conjugation means

$$S^\dagger = (\theta^B)^\dagger(\varepsilon_{AB})^*(\theta^A)^\dagger = \theta^{\dot{B}}(\varepsilon_{AB})^*\theta^{\dot{A}} \tag{A.34}$$

Hence, in order that $S^* = S^\dagger$, we require that the order of θ's is reversed under $*$. For example,

$$(\theta_A\theta_B)^* = \theta_{\dot{B}}\theta_{\dot{A}} \qquad (\theta^A\theta^B\theta^C)^* = -\theta^{\dot{C}}\theta^{\dot{B}}\theta^{\dot{A}} \tag{A.35}$$

because $(\theta^A)^* = -\theta^{\dot{A}}$ (omitting the bar). Thus $(\theta^A\theta_A)^* = \theta^{\dot{A}}\theta_{\dot{A}}$.

We next define the fermionic derivatives. They are left-derivatives:

$$\frac{\partial}{\partial\theta^A}\theta^B = \delta_A{}^B \qquad \frac{\partial}{\partial\theta^{\dot{A}}}\theta^{\dot{B}} = \delta_{\dot{A}}{}^{\dot{B}}$$

Hence

$$\frac{\partial}{\partial\theta_A}\theta_B = \delta_B{}^A \qquad \frac{\partial}{\partial\theta_{\dot A}}\theta_{\dot B} = \delta_{\dot B}{}^{\dot A} \tag{A.36}$$

We can now define the complex conjugate of the derivative by demanding that

$$\delta_A^B = \delta_A{}^{\dot B} = \left(\frac{\partial}{\partial\theta^A}\theta^B\right)^* = (\theta^B)^*\left(\frac{\overleftarrow{\partial}}{\partial\theta^A}\right)^* = -\theta^{\dot B}\left(\frac{\overleftarrow{\partial}}{\partial\theta^A}\right)^* = \left(\frac{\overrightarrow{\partial}}{\partial\theta^A}\right)^*\theta^{\dot B} \tag{A.37}$$

Hence

$$\left(\frac{\partial}{\partial\theta^A}\right)^* = \frac{\partial}{\partial\theta^{\dot A}}, \qquad \left(\frac{\partial}{\partial\theta^{\dot A}}\right)^* = \frac{\partial}{\partial\theta^A}$$

but

$$\left(\frac{\partial}{\partial\theta_A}\right)^* = -\frac{\partial}{\partial\theta_{\dot A}} \quad \text{and} \quad \left(\frac{\partial}{\partial\theta_{\dot A}}\right)^* = -\frac{\partial}{\partial\theta_A} \tag{A.38}$$

These results show that the fermionic derivatives of a bosonic scalar form a Majorana spinor. In carrying out complex conjugation with spinor derivatives one must interchange the order of the derivative as well as the θ's. The reader may check that the above definition works on all functions of θ and not only on one θ.

We now give the definitions of the torsions and curvatures. We first define a covariant derivative by

$$\mathbf{D}_M = E_M{}^A(\partial_A + \tfrac{1}{2}\Omega_A{}^{mn}J_{mn}) \tag{A.39}$$

where J_{mn} are the Lorentz generators, $E_M{}^A$ is the inverse supervierbein and $\Omega_A{}^{mn}$ is the spin connection. The torsions and curvatures are given by

$$[\mathbf{D}_M, \mathbf{D}_N] = T_{MN}{}^R\mathbf{D}_R + \tfrac{1}{2}R_{MN}{}^{mn}J_{mn} \tag{A.40}$$

Explicit calculations shows that

$$T_{MN}{}^R = E_M{}^A\partial_A E_N{}^\pi E_\pi^R + \Omega_{MN}{}^R - (-1)^{MN}(M \leftrightarrow N)$$

$$R_{MN}{}^{rs} = E_M{}^A E_N^\pi(-1)^{A(N+\pi)}\{\partial_A\Omega_\pi{}^{rs} + \Omega_A{}^{rk}\Omega_{\pi k}{}^s - (-1)^{A\pi}(A \leftrightarrow \pi)\} \tag{A.41}$$

The symbol $\Omega_{MN}{}^R$ is the Lorentz valued spin connection and so takes the form

$$\Omega_{MN}{}^R = \begin{pmatrix} \Omega_{Mm}{}^n & & \\ & -\tfrac{1}{4}\Omega_M{}^{mn}(\sigma_{mn})_A{}^B & \\ & & \tfrac{1}{4}\Omega_M{}^{mn}(\bar\sigma_{mn})_{\dot A}{}^{\dot B} \end{pmatrix} \tag{A.42}$$

The symbol $(\sigma_{mn})_A{}^B$ is defined according to the rules in (A.5)

$$(\sigma_{mn})_A{}^B = (\sigma_{mn})^C{}_D\varepsilon_{CA}\varepsilon^{BD}$$

The reason for the extra minus sign is that $(\sigma_{mn})_{AB}\chi^B = -(\sigma_{mn})_A{}^B\chi_B$. It is useful to define the Lorentz-valued curvature $\frac{1}{2}R_{MNP}{}^{Q}$ as follows

$$R_{MNP}{}^{Q} = \begin{pmatrix} R_{MNr}{}^{s} & & \\ & -R_{MN}{}^{rs}\frac{1}{4}(\sigma_{rs})_{A}{}^{B} & \\ & & R_{MN}{}^{rs}\frac{1}{4}(\bar{\sigma}_{rs})_{\dot{A}}{}^{\dot{B}} \end{pmatrix} \qquad (A.43)$$

We now discuss the reality properties of $E_{M}{}^{\pi}$ and its inverse $E_{\pi}{}^{M}$. Since $dz^{M}E_{M}{}^{\pi}$ can be considered as a transformation from inertial (flat) to curved coordinates; we require that both flat and curved coordinates have the same reality properties under *; namely,

$$(dz^{m})^{*} = dz^{m}(dz^{A})^{*} = -dz^{\dot{A}}(dz_{\dot{A}})^{*} = dz_{A}$$

$$(dz^{\mu})^{*} = dz^{\mu}(dz^{\underline{A}})^{*} = -dz^{\underline{\dot{A}}}(dz_{\underline{\dot{A}}})^{*} = dz_{\underline{A}}. \qquad (A.44)$$

From the identity $dz^{\pi} = dz^{M}E_{M}{}^{\pi}$ one can deduce the following reality properties

$$(E_{m}{}^{\mu})^{*} = E_{m}{}^{\mu}(E_{m}{}^{A})^{*} = -E_{m}{}^{\dot{A}}$$

$$(E_{A}{}^{\mu})^{*} = E_{\dot{A}}{}^{\mu}(E_{A}{}^{\underline{B}})^{*} = E_{\dot{A}}{}^{\dot{\underline{B}}}(E_{A}{}^{\dot{\underline{B}}})^{*} = E_{\dot{A}}{}^{\underline{B}} \qquad (A.45)$$

These rules are easy to understand $(E_{A}{}^{A})^{*}$ becomes equal to $E^{A^{*}}{}_{A^{*}}$ and commuting A^{*} past A^{*} yields the final result.

The reality properties of the supertorsion tensor can now be deduced, since we know how $E_{M}{}^{A}$ and ∂_{A} behave under the * operation. As an example

$$(T_{AB}{}^{C})^{*} = (E_{A}{}^{A}\partial_{A}E_{B}{}^{\pi}E_{\pi}{}^{C} + \cdots)^{*} \qquad (A.46)$$

It is easiest to take the A and π to be bosonic; in that case

$$(T_{AB}{}^{C})^{*} = (E_{\nu}{}^{C})^{*}(\partial_{\mu}E_{B}{}^{\nu})^{*}(E_{A}{}^{\mu})^{*} + \cdots$$

$$= -E_{\nu}{}^{\dot{C}}\partial_{\mu}E_{\dot{B}}{}^{\nu}E_{\dot{A}}{}^{\mu} + \cdots = E_{\dot{A}}{}^{\mu}\partial_{\mu}E_{\dot{B}}{}^{\nu}E_{\nu}{}^{\dot{C}} + \cdots \qquad (A.47)$$

Hence $(T_{AB}{}^{C})^{*} = T_{\dot{A}\dot{B}}{}^{\dot{C}}$. All rules thus derived can be most easily understood as follows.

$$(T_{MN}{}^{P})^{*} = (-1)^{p(m+n)+mn+p}T_{M^{*}N^{*}}{}^{P^{*}} \qquad (A.48)$$

where the sign is due to pulling P^{*} past $N^{*}M^{*}$, interchanging M^{*} with N^{*} and accounting for the $(-)$ sign in $(\theta^{A})^{*} = -\theta^{\dot{A}}$. An even easier mnemonic is to take for $T_{MN}{}^{P}$ the particular case $Z_{M}Z_{N}Z^{P}$ and using the rules for $(Z_{M})^{*}$. To give two more examples

$$(T_{AB}{}^{m})^{*} = -T_{\dot{A}\dot{B}}{}^{m} \qquad (T_{Am}{}^{B})^{*} = T_{\dot{A}m}{}^{\dot{B}} \qquad (A.49)$$

To determine the reality properties of curvatures we must first determine the reality properties of the connections. From the requirement that both terms in $\mathbf{D}_M = E_M{}^{\Lambda} \partial_{\Lambda} + \frac{1}{2} E_M{}^{\Lambda} \Omega_{\Lambda}{}^{mn} J_{mn}$ have the same reality properties, we first determine how $\Omega_{\Lambda}{}^{mn}$ behaves under the * operation. Taking, for example, a real vector v^k one finds from

$$(\mathbf{D}_M v^k)^* = (E_M{}^{\mu} \partial_{\mu} v^k + E_M{}^{\Lambda} \Omega_{\Lambda}{}^k{}_l v^l + \cdots)^* = E_{\dot{M}}{}^{\mu} \partial_{\mu} v^k + v^l (\Omega_{\Lambda}{}^k{}_l)^* (E_M{}^{\Lambda})^*$$

$$(A.50)$$

that $(\mathbf{D}_m v^k)^*$ must equal $(\mathbf{D}_m v^k)$. Hence

$$(\Omega_{\Lambda}{}^{kl})^* = \Omega_{\dot{\Lambda}}{}^{kl} \tag{A.51}$$

where per definition $\dot{\Lambda} = (m, \dot{A}, A)$ if $\Lambda = (m, A, \dot{A})$.

Next we determine the reality properties of $\Omega_{\Lambda P}{}^Q$. Since the Lorentz generators satisfy

$$[(\sigma_{mn})^A{}_B]^* = -(\bar{\sigma}_{mn})_{\dot{B}}{}^{\dot{A}}$$

i.e.,

$$[(\sigma_{mn})^{AB}]^* = (\bar{\sigma}_{mn})^{\dot{A}\dot{B}} \tag{A.52}$$

one finds

$$(\Omega_{\Lambda P}{}^Q)^* = \Omega_{\dot{\Lambda}\dot{P}}{}^{\dot{Q}} \tag{A.53}$$

(Note the minus sign due to complex conjugation of σ^{mn} is cancelled by the minus sign in the definition of $\Omega_{\Lambda P}{}^Q$.) From this result we can now read off the reality properties of the curvatures. Since

$$R_{\Lambda \pi P}{}^Q = \partial_{\Lambda} \Omega_{\pi P}{}^Q + \cdots$$

one deduces

$$(R_{n\pi P}{}^Q)^* = (-1)^{\Lambda \pi} R_{\dot{\Lambda}\dot{\pi}\dot{P}}{}^{\dot{Q}} \tag{A.55}$$

where we have used $(\partial/\partial \Lambda)^* = \partial/\partial \dot{\Lambda}$. Since we made sure that dz^{Λ} and dz^m have the same reality properties, also the curvatures with flat indices have the reality properties

$$(R_{MNP}{}^Q)^* = (-1)^{MN} R_{\dot{M}\dot{N}\dot{P}}{}^{\dot{Q}} \tag{A.56}$$

To illustrate the above discussion we now apply it to the variation of the supervielbein. From the transformation properties of the supervierbein under general coordinate transformations

$$\delta E_{\Lambda}{}^M = \partial_{\Lambda} \Xi^{\pi} E_{\pi}{}^M + \Xi^{\pi} \partial_{\pi} E_{\Lambda}{}^M \tag{A.57}$$

we deduce that

$$\delta E_{\Lambda}{}^M = -E_{\Lambda}{}^R \varepsilon^N T_{NR}{}^M + \mathbf{D}_{\Lambda} \varepsilon^M \tag{A.58}$$

where

$$\varepsilon^N E_N{}^{\pi} = \Xi^{\pi}$$

Consequently, taking $T_{A\dot{B}}{}^m = -2i(\sigma^m)_{A\dot{B}} = T_{\dot{B}A}{}^m$ we find from

$$E_\mu{}^m(\theta = 0) = e_\mu{}^m, \qquad \varepsilon^A(\theta = 0) = \varepsilon^A$$

and

$$E_\mu{}^A(\theta = 0) = \psi_\mu{}^A \tag{A.59}$$

that

$$\delta_\varepsilon e_\mu{}^m = +\varepsilon^B T_{\dot{B}A}{}^m \psi_\mu{}^A + \varepsilon^{\dot{B}} T_{\dot{B}A}{}^m \psi_\mu{}^A = -2i\varepsilon^B \sigma^m{}_{B\dot{A}} \psi_\mu{}^{\dot{A}} - 2i\varepsilon^{\dot{B}}(\sigma^m)_{A\dot{B}} \psi_\mu{}^A$$

The reader may check that the variation of $e_\mu{}^m$ is indeed real. We have used that the only non-vanishing component of $T_{BM}{}^m$ is $T_{B\dot{A}}{}^m$ (see Chapter 11). In four components this corresponds to

$$T_{\alpha\beta}{}^m = 2(C\gamma^m)_{\alpha\beta} \quad \text{and} \quad \delta e_\mu{}^m = 2\bar{\varepsilon}\gamma^m \psi_\mu \tag{A.61}$$

Care must be taken in going from two to four components, in that the 2-component contractions are

$$\chi^M \zeta_M = \chi^m \zeta_m + \chi^A \zeta_A + \chi^{\dot{A}} \zeta_{\dot{A}} \tag{A.62}$$

while 4-component spinor contractions are

$$\bar{\chi}_\alpha \zeta^\alpha = \chi_A \zeta^A - \chi^{\dot{B}} \zeta_{\dot{B}} = -(\chi^A \zeta_A + \chi^{\dot{B}} \zeta_{\dot{B}}) \tag{A.63}$$

The 2-component and 4-component results can be seen to be in agreement since

$$\delta e_\mu{}^m = \varepsilon_\alpha T_{\alpha\beta}{}^m \psi_{\mu\beta} = 2\bar{\varepsilon}\gamma^m \psi_\mu$$

$$= 2\bar{\varepsilon}^\alpha(\gamma^m)_\alpha{}^\beta \psi_{\mu\beta}$$

$$= 2(\varepsilon_B, -\bar{\varepsilon}^{\dot{B}})\begin{pmatrix} 0 & -i(\sigma^m)^{B\dot{A}} \\ +i(\sigma^m)_{\dot{B}A} & 0 \end{pmatrix}\begin{pmatrix} \psi_\mu{}^A \\ \psi_{\mu\dot{A}} \end{pmatrix}$$

$$= -2i\varepsilon_B(\sigma^m)^{B\dot{A}}\psi_{\mu\dot{A}} + \text{h.c.} \tag{A.64}$$

We conclude our set of conventions by recording the $N = 1$ super-Poincaré algebra.

$$[J_{mn}, J_{pq}] = (\eta_{np}J_{mq} - \eta_{mp}J_{nq} + \eta_{mq}J_{np} - \eta_{nq}J_{mp})$$

$$[P_r, J_{mn}] = \eta_{mr}P_n - \eta_{nr}P_m$$

$$[Q_\alpha, J_{mn}] = \tfrac{1}{2}(\gamma_{mn})_\alpha{}^\beta Q_\beta$$

$$[Q^\alpha, P_r] = 0$$

$$\{Q_\alpha, Q_\beta\} = -2(\gamma^m C^{-1})_{\alpha\beta} P_m \tag{A.65}$$

In 2-component notation one has,

$$[Q^A, J_{mn}] = \tfrac{1}{2}(\sigma_{mn})^A{}_B Q^B$$

$$[Q_{\dot{A}}, J_{mn}] = \tfrac{1}{2}(\bar{\sigma}_{mn})_{\dot{A}}{}^{\dot{B}} Q_{\dot{B}}$$

$$\{Q^A, Q^B\} = 0$$

$$\{Q^A, Q^{\dot{B}}\} = -2i(\sigma^m)^{A\dot{B}} P_m$$

$$[Q^A, P_n] = 0, \qquad [Q_{\dot{A}}, P_n] = 0 \qquad\qquad (A.66)$$

plus the other relations which are unchanged.

We now illustrate how this algebra is realized in the Wess-Zumino model. From $\delta A = \bar{\varepsilon} Q A$ where Q are the abstract generators, one has

$$[\delta_1, \delta_2] A = [\bar{\varepsilon}_1 Q, \bar{\varepsilon}_2 Q] A \qquad\qquad (A.67)$$

On the other hand since $\delta A = \bar{\varepsilon} \chi$, and $\delta \chi = \gamma^\mu \partial_\mu A \varepsilon$ one has

$$[\delta_1, \delta_2] A = \bar{\varepsilon}_2 \delta_1 \chi - (1 \leftrightarrow 2)$$

$$= 2\bar{\varepsilon}_2 \gamma^\mu \varepsilon_1 \partial_\mu A \qquad\qquad (A.68)$$

With $P_m A = \partial_m A$ we find

$$= -2\bar{\varepsilon}_2{}^\beta (\gamma^\mu C^{-1})_{\beta\delta} \bar{\varepsilon}_1{}^\delta P_\mu A$$

$$= -\bar{\varepsilon}_1{}^\beta \bar{\varepsilon}_2{}^\delta \{Q_\beta, Q_\delta\} A \qquad\qquad (A.69)$$

Hence, the abstract generators satisfy the algebra

$$\{Q_\beta, Q_\delta\} = -2(\gamma^\mu C^{-1})_{\beta\delta} P_\mu, \qquad\qquad (A.70)$$

which agrees with the result (A.65).

Finally, we give the Fierz identity which is used extensively when closing the supersymmetry algebra, demonstrating invariance and many other places. We begin with $\delta_\alpha{}^\beta \delta_\gamma{}^\delta$ and we would like to express it in terms of matrices which link α to δ and β to γ. Since the 16 matrices

$$(\gamma_R)_\alpha{}^\beta = \{\delta_\alpha^\beta, (\gamma_5)_\alpha{}^\beta, (\gamma_\mu)_\alpha{}^\beta, (\gamma_\mu \gamma_5)_\alpha{}^\beta, (\gamma_{\mu\nu})_\alpha{}^\beta\} \qquad\qquad (A.71)$$

form a complete set, any four by four matrix may be expressed in terms of them. Applying this to α and δ and to β and γ, we may write

$$\delta_\alpha{}^\beta \delta_\gamma{}^\delta = \frac{1}{4} \sum_{RS} (\gamma_R)_\alpha{}^\delta (\gamma_S)_\gamma{}^\beta c_{RS} \qquad\qquad (A.72)$$

for some coefficients c_{RS}.

To determine the c_{RS} we take the trace of this equation with $(\gamma_T)_\beta^\gamma$ to find

$$(\gamma_T)_\alpha{}^\delta = \frac{1}{4} \sum_{R,S} \mathrm{Tr}(\gamma_T \gamma_R) c_{RS} (\gamma_S)_\alpha{}^\delta. \qquad\qquad (A.73)$$

We evaluate $\mathrm{Tr}(\gamma_T \gamma_R)$ by explicit computations and we observe that it vanishes unless T equals R and in that case, we find

$$\tfrac{1}{4}\mathrm{Tr}(\mathbb{1}\ \ \mathbb{1}) = +1, \quad \tfrac{1}{4}\mathrm{Tr}(\gamma_5 \gamma_5) = +1, \quad \tfrac{1}{4}\mathrm{Tr}\,\gamma_\mu \gamma_\nu = \eta_{\mu\nu},$$

$$\tfrac{1}{4}\mathrm{Tr}(\gamma_\mu \gamma_5 \gamma_\nu \gamma_5) = -\eta_{\mu\nu}, \quad \tfrac{1}{4}\mathrm{Tr}(\gamma_{\mu\nu}\gamma_{\rho\kappa}) = -(\eta_{\mu\rho}\eta_{\nu\kappa} - \eta_{\nu\rho}\eta_{\mu\kappa}) \quad (A.74)$$

where $(\mathbb{1})_\alpha^\beta = \delta_\alpha^\beta$. The last equation is most easily evaluated, as is often the case, by writing down what it can be in terms of a tensor with the correct symmetries and Lorentz character and then evaluating the coefficient by taking a specific case. Here, for example, we can only have

$$\tfrac{1}{4}\mathrm{Tr}\,\gamma_{\mu\nu}\gamma_{\rho\kappa} = e(\eta_{\mu\rho}\eta_{\nu\kappa} - \eta_{\nu\rho}\eta_{\mu\kappa}). \quad (A.75)$$

Taking $\mu = \rho = 1$, $\nu = \kappa = 2$ it becomes

$$\tfrac{1}{4}\mathrm{Tr}(\gamma_1 \gamma_2 \gamma_1 \gamma_2) = e(+1) = -1 \quad (A.76)$$

from which we conclude that $e = -1$. Returning to Eq. (A.73), we have

$$(\gamma_T)_\alpha{}^\delta = \frac{1}{4}\sum_S \mathrm{Tr}(\gamma_T \gamma_T) c_{TS}(\gamma_S)_\alpha{}^\delta \quad (A.77)$$

and as the (γ_R) form an independent set

$$c_{TS} = \delta_{TS} c_S. \quad (A.78)$$

Consequently, we have

$$1 = \tfrac{1}{4}\mathrm{Tr}(\gamma_T \gamma_T) c_T \,\forall T. \quad (A.79)$$

Using Eq. (A.74), we conclude the final result

$$\delta_\alpha^\beta \delta_\gamma^\delta = \frac{1}{4}\left\{ \delta_\alpha^\delta \delta_\gamma^\beta + (\gamma_5)_\alpha^\delta (\gamma_5)_\gamma^\beta + (\gamma_\mu)_\alpha{}^\delta (\gamma^\mu)_\gamma{}^\beta - (\gamma_\mu \gamma_5)_\alpha{}^\delta (\gamma^\mu \gamma_5)_\gamma{}^\beta - \frac{1}{2}(\gamma_{\mu\nu})_\alpha{}^\delta (\gamma^{\mu\nu})_\gamma{}^\beta \right\}$$

$$= \frac{1}{4}\sum_R c_R (\gamma_R)_\alpha^\delta (\gamma_R)_\gamma^\beta \quad (A.80)$$

where

$$c_R = \{1, 1, \eta_{\mu\nu}, -\eta_{\mu\nu}, -\tfrac{1}{4}(\eta_{\mu\rho}\eta_{\nu\kappa} - \eta_{\nu\rho}\eta_{\mu\kappa})\}. \quad (A.81)$$

When applying the Fierz identity, we are often required to contract indices on three spinors χ_α, λ_α and ψ_β in a different order. Consider, for example

$$\chi_\alpha \bar{\lambda}\psi = \delta_\alpha^\beta \chi_\beta \bar{\lambda}^\gamma \psi_\delta \delta_\gamma^\delta$$

$$= \frac{1}{4}\chi_\beta \bar{\lambda}^\gamma \psi_\delta \sum_R (\gamma_R)_\alpha^\delta (\gamma_R)_\gamma^\beta c_R$$

$$= -\frac{1}{4}\sum_R c_R (\gamma_R \psi)_\alpha \bar{\lambda}\gamma_R \chi$$

$$= -\frac{1}{4}\left\{\psi_\alpha\bar{\lambda}\chi + (\gamma_5\psi)_\alpha\bar{\lambda}\gamma_5\chi + (\gamma_\mu\psi)_\alpha\bar{\lambda}\gamma^\mu\chi - (\gamma_\mu\gamma_5\psi)_\alpha\bar{\lambda}\gamma^\mu\gamma_5\psi\right.$$

$$\left. -\frac{1}{2}(\gamma_{\mu\nu}\psi)_\alpha\bar{\lambda}\gamma^{\mu\nu}\psi\right\}. \tag{A.82}$$

The minus sign arises as the spinors anticommute with each other.

We could equally well have defined a Majorana conjugate (see Appendix A) through the equation

$$\chi_\alpha = C_{\alpha\beta}\bar{\chi}^\beta. \tag{A.83}$$

Using Eq. (A.12) we find that

$$C_{\alpha\beta}C^{\alpha\beta} = \delta_\alpha^\gamma \quad \text{or} \quad C_{\alpha\beta}C^{\beta\gamma} = -\delta_\alpha^\gamma. \tag{A.84}$$

Since $C^2 = -1$ from Eq. (A.16), we may conclude that

$$C_{\alpha\beta} = C^{\alpha\beta} = C. \tag{A.85}$$

It is often very useful to change the position of two Majorana spinors χ_α and λ_α. In particular, we have

$$\bar{\chi}\gamma_R\lambda = \bar{\chi}^\alpha(\gamma_R)_\alpha{}^\delta\lambda_\delta = \bar{\chi}^\alpha(\gamma_R)_\alpha{}^\delta C_{\delta\gamma}\bar{\lambda}^\gamma$$

$$= \bar{\chi}^\alpha(\gamma_R C)_{\alpha\gamma}\bar{\lambda}^\gamma = -\bar{\lambda}^\gamma(\gamma_R C)_{\alpha\gamma}\bar{\chi}^\alpha = -\bar{\lambda}^\gamma(\gamma_R C)_{\gamma\alpha}^T\bar{\chi}^\alpha$$

$$= -d_R\bar{\lambda}^\gamma(\gamma_R C)_{\gamma\alpha}\bar{\chi}^\alpha = -d_R\bar{\lambda}\gamma_R\chi. \tag{A.86}$$

In the above

$$d_R = +1 \quad \text{if} \quad \gamma_R = \gamma_\mu, \gamma_{\mu\nu} \tag{A.87}$$

and

$$d_R = -1 \quad \text{if} \quad \gamma_R = \mathbb{1}, \gamma_5, \gamma_\mu\gamma_5, \tag{A.88}$$

as $(\gamma_R C)^T, = d_R(\gamma_R C)$ using Eq. (A.15). Written in a more transparent way, we find that

$$\bar{\chi}\lambda = \bar{\lambda}\chi, \quad \bar{\chi}\gamma_5\lambda = \bar{\lambda}\gamma_5\chi, \quad \bar{\chi}\gamma_\mu\gamma_5\lambda = \bar{\lambda}\gamma_\mu\gamma_5\chi \tag{A.89}$$

but

$$\bar{\chi}\gamma_\mu\lambda = -\bar{\lambda}\gamma_\mu\chi, \qquad \bar{\chi}\gamma_{\mu\nu}\lambda = -\bar{\lambda}\gamma_{\mu\nu}\chi. \tag{A.90}$$

In particular, taking $\lambda = \chi$ we observe that

$$\bar{\chi}\gamma_\mu\chi = 0 \quad \text{and} \quad \bar{\chi}\gamma_{\mu\nu}\chi = 0 \tag{A.91}$$

Appendix B
A List of Reviews and Books

Supersymmetry and Supergravity

P. Fayet and S. Ferrara, *Phys. Rep.* **C32**, 249 (1977).

P. van Nieuwenhuizen, *Phys. Rep.* **C68**, 189 (1981).

S. J. Gates, M. T. Grisaru, M. Rocek and W. Siegel, *Superspace or One Thousand-and-One Lessons in Supersymmetry*, Frontiers in Physics (Benjamin/Cummings, Reading, Mass. 1983).

J. Wess and J. Bagger, *Supersymmetry and Supergravity* (Princeton University Press, Princeton, 1983).

B. de Witt, *Supermanifolds* (Cambridge University Press, 1984).

P. van Nieuwenhuizen and P. West, *Principles of Supersymmetry and Supergravity* (Cambridge University Press, in preparation).

M. Sohnius, *Phys. Rep.* **C128**, 39 (1985).

J. G. Taylor, in *Progress in Particle and Nuclear Physics* 12, ed. Sir Denys Wilkinson.

E. S. Fradkin and A. A. Tseytlin, *Phys. Rep.* **C119**, 235 (1985).

F. Gieres, "Geometry of Supersymmetric Gauge Theories" *Leature Notes in Physics* **302** (Springer, 1988).

R. Arnowitt, A. Chamseddine and P. Nath, *Applied N = 1 Supergravity* (World Scientific, Singapore).

See also lectures contained in:

Supergravity (Stony Brook 1979) eds. D. Z. Freedman and P. van Nieuwenhuizen.

Superspace and Supergravity, eds. S. W. Hawking and M. Rocek (Cambridge University Press, Cambridge, 1981).

Quantum Gravity 2, eds. C. Isham, R. Penrose and D. Sciama (Oxford University Press, Oxford, 1982)

Unification of the Fundamental Particle Interactions (Erice) eds. S. Ferrara, J. Ellis and P. van Nieuwenhuizen (Plenum Press, New York, 1983).

Quantum Structure of Space and Time, eds. C. J. Isham and M. Duff (Cambridge University Press, Cambridge, 1980).

Supersymmetry and Supergravity '81 (Trieste), eds. S. Ferrara, J. G. Taylor and P. van Nieuwenhuizen (Cambridge University Press, Cambridge, 1982).

Supersymmetry and Supergravity '82 (Trieste), eds. S. Ferrara, J. G. Taylor and P. van Nieuwenhuizen (World Scientific Publishing Co., Singapore, 1983).

Recent Developments in Gravitation (Cargese 1978), eds. M. Levy and S. Deser (Plenum Press, New York, 1978).

Supersymmetry and Supergravity '83, eds. S. Ferrara, P. van Nieuwenhuizen, and B. de Wit (World Scientific, Singapore, 1983).

Introduction to Supersymmetry in Particle and Nuclear Physics, eds. O. Castanos, A. Frank, and L. Urrutia (Plenum Press, New York, 1982).

Frontiers in Particle Physics '83, eds. D. Sijacki, N. Bilic, B. Dragovic and D. Popovic (World Scientific, Singapore, 1984).

Problems in Unification and Supergravity, eds. G. Farrar and Frank Henyey, American Institute of Physics No. 116.

Phenomenological Supersymmetry

N. P. Nilles, *Phys. Rep.* **C110**, 1 (1984).

H. E. Haber and G. L. Kane, *Phys. Rep.* **C117**, 75 (1985).

R. Arnowitt, A. Chamseddine and P. Nath, Northeastern University preprint.

R. Barbieri and S. Ferrara, *Surveys in High Energy Physics* **4**, 33 (1983).

Kaluza-Klein

M. Duff, B. Nilsson and C. Pope, *Phys. Rep.* **C130** (1986) 1.

Unified Field Theories of more than 4 Dimensions, eds. V. de Sabbata and E. Schmutzer (World Scientific, Singapore, 1983).

Strings

V. Alessandrini, D. Amati, M. LeBellac and D. I. Olive, *Phys. Rep.* **1C**, 170 (1971).

P. Frampton, *Dual Resonance Models* (Benjamin, Reading, 1974).

S. Mandelstam, *Phys. Rep.* **13C**, 259 (1974).

S. Mandelstam, in *Structural Analysis of Collision Amplitudes* (Les Houches, 1975, June Institute) p. 593.

C. Rebbi, *Phys. Rep.* **12C**, 1 (1974).

J. Scherk, *Rev. Mod. Phys.* **47**, 123 (1975).

J. H. Schwarz, *Phys. Rep.* **8C**, 269 (1973).

G. Veneziano, *Phys. Rep.* **9C**, 199 (1974).

J. H. Schwarz, *Phys. Rep.* **89**, 223 (1982).

M. Green, *Surveys of High Energy Physics* **3**, 127 (1983).

Appendix C
Problems

In this section we give a number of problems on which the reader may clarify his understanding. The problems vary considerably in length and difficulty. However, they introduce a number of subjects that could equally well have been developed further in the text.

Chapter 2

2.1 The De Sitter group $SO(4, 1)$ has an algebra with the generators P_a and J_{ab} which obey the relations

$$[P_a, P_b] = m^2 J_{ab}, \qquad [P_a, J_{bc}] = \eta_{ab} P_c - \eta_{ac} P_b$$

$$[J_{ab}, J_{cd}] = -(\eta_{ac} J_{bd} + \eta_{bd} J_{ac} - \eta_{ad} J_{bc} - \eta_{bc} J_{ad})$$

where

$$\eta_{ab} = \operatorname{diag}(-1, +1, +1, +1, +1).$$

Following the same arguments for the Poincaré case find its superextension. You should for simplicity, introduce only one Majorana supercharge Q_a and assume a no-go theorem as given in Eqs. (2.2) and (2.3). In particular you should find what replaces (2.18) and what becomes of $\{Q_a, Q_\beta\}$. Verify that all the super Jacobi identities hold for the final super algebra.

2.2 Repeat the problem 2.1 for the case of many supercharges. What is the maximal internal symmetry group?

2.3 Beginning with the conformal group generated by P_n, J_{mn}, D and K_n find its superextension assuming a no-go theorem and using the same arguments as for the Poincaré group. This calculation is rather lengthy but the result is given in Eq. (2.41). For simplicity just consider the extension to include only the supercharges Q_a and S_a.

Chapter 3

3.1 The action for the simplest supersymmetric model in four dimensions, the Wess-Zumino model is given by

$$A^{\omega-z} = \int d^4x \left\{ -\frac{1}{2}(\partial_\mu A)^2 - \frac{1}{2}(\partial_\mu B)^2 - \frac{1}{2}\bar{\chi}\partial\!\!\!/\chi - \frac{m^2}{2}(A^2 + B^2) - \frac{m}{2}\bar{\chi}\chi \right.$$

$$\left. - m\lambda A(A^2 + B^2) - \lambda^2(A^2 + B^2) - \lambda\bar{\chi}(A - i\gamma_5 B)\chi \right\}$$

where χ_α is a Majorana spinor and λ the coupling constant. Calculate, at one loop, the wave-function, mass and coupling constant renormalizations; show that the wave-function renormalization of A, B and χ_α are the same and that the mass and coupling constant renormalization are the same no matter what vertex is used to calculate them. What relation do the mass and coupling renormalizations have to the wave-function renormalization?

Note that one-loop infinities will conform to any symmetry relations in the original action regardless of whether the regulator respects the symmetry or not.

Chapter 4

4.1 Show using the extended supersymmetry algebra of Eq. (2.23) that if one supersymmetry is preserved, that is $Q_\alpha^1|0\rangle = 0$ say for the vacuum state, then all supersymmetries are preserved. (This is proved in Chapter 19).

4.2 The superspace of $N = 2$ supersymmetry of Eq. (14.115) carries a representation of supersymmetry in which the Fermi-Bose degrees of freedom do not match; why not?

Chapter 5

5.1 Show that the off-shell transformations of Eq. (5.14) really do close off-shell and verify that they leave the action of Eq. (5.15) invariant.

5.2 Consider the theory of Eq. (5.16) with the transformations of Eq. (5.17) and find the new symmetries that appear when evaluating the commutation of two supersymmetries. Verify that they are indeed symmetries of the action of Eq. (5.16) but consider their implications for interacting theories, in outline.

5.3 Show that the mass term given by

$$m \int d^4x \left\{ FA + BG - \frac{1}{2}\bar{\chi}\chi \right\}$$

is invariant under supersymmetry. Also find the cubic interaction term. (Hint, start by varying $\bar{\chi}A\chi$).

5.4 In the simplest model of supersymmetry we have one spin $1/2$ (Majorana) and two spin 0. Let us represent the spin $1/2$ by the Majorana spinor χ_α

subject to $\not{\partial}\chi = 0$ and the parity odd spin 0 by B subject to $\partial^2 B = 0$. However, the remaining spin 0 is represented by the field $A_{\mu\nu} = -A_{\nu\mu}$ which has a gauge invariance

$$\delta A_{\mu\nu} = \partial_\mu \Lambda_\nu - \partial_\nu \Lambda_\mu$$

and so field strength

$$F_{\mu\nu\rho} = \partial_\mu A_{\nu\rho} + \partial_\rho A_{\mu\nu} + \partial_\nu A_{\rho\mu}$$

with equation of motion

$$\partial^\mu F_{\mu\nu\rho} = 0$$

Following Chapter 5 find:
  (a) the on-shell supersymmetry transformations and determine any un-determined coefficients by demanding closure;
  (b) count the number of Fermi-Bose degrees of freedom and guess the auxil-iary field structure (hint: only on gauge invariant states is P_μ a one-to-one operator);
  (c) find the off-shell supersymmetry transformations. Having tested closure, taking into account the gauge invariance, find the corresponding action.
  The results can be found, if in somewhat hidden form in Chapter 11. In recognizing the result we utilize the fact that at least locally the solution of $\partial_\mu v^\mu = 0$ is given by

$$v^\mu = \varepsilon^{\mu\nu\rho\kappa}\partial_\nu A_{\rho\kappa}.$$

5.5 Now represent both spin 0's by antisymmetric tensors and repeat step (a).

Chapter 6

6.1 Verify that the transformations of Eq. (6.10) close off-shell if $a = +1$ and that they are an invariance of the action of Eq. (6.11).

6.2 Using the Fierz relations in Appendix A show that
(a) if χ is a Majorana spinor that

$$[(1 + \gamma_5)\chi]_\alpha \bar{\chi}(1 + \gamma_5)\chi = 0.$$

(b) if ψ_μ is Majorana spinor that

$$\varepsilon^{\mu\nu\rho\kappa}\gamma^a \psi_\nu \bar{\psi}_\rho \gamma_a \psi_\kappa = 0$$

Chapter 7

7.1 Verify that the supersymmetry transformations of Eq. (7.20) close off-shell and that they leave the action of Eq. (7.21) invariant.

7.2 Writing the metric as

$$g_{\mu\nu} = \eta_{\mu\nu} + \kappa h_{\mu\nu}$$

the lowest order contribution to Einstein's theory is

$$-\frac{1}{2}\int d^4x\, h^{\mu\nu}E_{\mu\nu}$$

where

$$E_{\mu\nu} = R^L_{\mu\nu} - \frac{1}{2}\eta_{\mu\nu}R^L, \qquad R^L = \eta^{\mu\nu}R^L_{\mu\nu}$$

$$R^L_{\mu\nu} = -\partial_\mu\partial_\nu h^\rho_\rho - \partial^2 h_{\mu\nu} + \partial_\mu\partial_\rho h\rho_\nu + \partial_\nu\partial_\rho h^\rho{}_\mu.$$

It is invariant under rigid translations, i.e.

$$\delta h_{\mu\nu} = \xi^\kappa\partial_\kappa h_{\mu\nu}$$

and local Abelian gauge invariances

$$\delta h_{\mu\nu} = \partial_\mu\zeta_\nu + \partial_\nu\zeta_\mu.$$

Using the Noether technique find the lowest order interacting term such that the action possesses a local invariance to order κ^0. What is the new transformation law of $h_{\mu\nu}$?

As a check verify that it agrees with that given by Einstein's theory.

Chapter 8

8.1 Construct the states of $N = 8$ supergravity in terms of the supercharge and verify that the result agrees with the number of states of each spin given in Table 8.2.

8.2 Repeat the analysis of irreducible representations when one central charge is present for: (a) a general value of the central charge and (b) the particular value $|Z| = m$ and verify that one recovers the states given in Table 8.4.

8.3 Massless supersymmetric representations can acquire mass by a super-symmetric Higgs mechanism. Verify, in several examples, by examining Tables 8.1, 8.2, and 8.3 that the count of states works. For example 'massive $N = 1$ supergravity' has one spin-2, two spin-3/2 and one spin-1 and could be obtained from the usual massless $N = 1$ supergravity and one Yang-Mills multiplet given in Tables 8.1 and 8.2 respectively.

Chapter 9

9.1 Show that

$$E^{\mu\nu} = 0, \qquad R^\mu = 0$$

respectively describe a massless spin-2 and a massless spin-3/2 each with two degrees of freedom by:

(a) going to light cone gauge
(b) fixing the gauge, BRST quantize and counting the net degrees of freedom assuming that a ghost degree of freedom cancels one degree of freedom of the original fields.

9.2 Verify the closure of the linearized supersymmetry transformations of Eq. (9.12) and show that they leave the action of Eq. (9.15) invariant.

9.3 Calculate, using the Noether technique, the first interacting term in the supergravity action that is: find an action which possess local supersymmetry to order κ^0.

9.4 Compute that closure on e_μ^a in the full supergravity theory and verify if it agrees with Eq. (9.26).

9.5 There exists another off-shell formulation of $N = 1$ supergravity called 'new minimal' which has the auxiliary fields A_μ and $a_{\kappa\lambda}$ and possess the gauge invariances

$$\delta A_\mu = \partial_\mu \Lambda, \qquad \delta a_{\kappa\lambda} = \partial_\kappa \Lambda_\lambda - \partial_\lambda \Lambda_\kappa.$$

Verify that:

(a) the off-shell Fermi-Bose count holds;
(b) find a gauge invariant term for the above auxiliary fields which leads to no on-shell degrees of freedom;
(c) find the linearized off-shell supersymmetry transformations and their action for this new minimal formulation of supergravity;
(d) from the point of view of additional symmetries, how does new minimal differ from old minimal?
(e) What does the full noninteraction theory look like?

Chapter 10

10.1 Show that (10.11) cancels (10.13) without resorting to inertial coordinates and perform the Freiz transformation required to obtain the latter equation.

Chapter 11

11.1 Carry out the computation suggested below Eq. (11.2). Namely starting from the field C and by systematically carrying out the closure and demanding that it gives the supersymmetry algebra, find the full multiplet $(C; \zeta_\alpha; H, K, A_\mu; \lambda_\alpha; D)$ and their transformation laws. The first step in this calculation is Eq. (11.3).

11.2 Verify Eq. (11.8) by beginning with $C_1 C_2$ and repeatedly carrying out supersymmetry variations.

11.3 Verify that TA really is a chiral multiplet.

11.4 Compute $[A \cdot A \cdot A \cdot A]_F$ and verify it really is invariant by explicit computation.

Chapter 12

12.1 Consider the representation of $N = 2$ matter provided by the fields $L^{ij} = L^{ji}$, $\lambda_{i\alpha}$ (Majorana), V_μ subject to $\partial^\mu V_\mu = 0$ and the two auxiliary fields S and P and find their supersymmetry transformations.

12.2 Work out the action and transformation laws of the $N = 4$ Yang-Mills theory discussed in Sec. 12.4 using the same techniques as we used for the $N = 2$ theories. You could do this for the O(4) formulation or the SU(4) formulation or both.

Chapter 13

13.1 Verify the closure on C and ζ whose transformation laws are given in Eq. (13.8).

13.2 Perform the exercise partially carried out on p. 75; that is take variations of $(C, \zeta; H, K, A_a; \lambda, D)$ given in Eq. (11.2) and using the Noether method on the algebra, find the transformations that satisfy a local supersymmetry algebra. Note that, if one wishes, one can determine both the full transformations of the supergravity fields, the matter fields and the local supersymmetry algebra by this method; however, it is quicker to assume the former.

13.3 Verify the invariance of the D or F density formulae of Eqs. (13.82) and (13.29) respectively.

13.4 Deduce the F density formula of Eq. (13.29) by the Noether coupling method and at the same time find the local transformation laws of the multiplet $(A, B; \chi; F, G)$.

Chapter 14

14.1 Using the $N = 1$ super anti-De Sitter algebra given in Eq. (2.42) construct the anti-De Sitter superspace. Namely find on the coset space $N = 1$ super anti-De Sitter/Lorentz
(a) the transformations
(b) the supervierbein

(c) the covariant derivatives

(d) the superdeterminant of the supervierbein

(e) the torsion and curvature

(f) the analogue of chiral representations.

14.2 Repeat Problem 14.1 for the $N = 1$ superconformal group whose super algebra is given in Eq. (2.41). The coset is given by

$$\frac{N = 1 \text{ superconformal group}}{\{J_{mn}, D, K_n, S_\alpha, A\}}$$

14.3 Interpret the constructions of Eqs. (14.61)–(14.63) on p. 96 in terms of a coset construction.

14.4 Consider a vierbein as a set of coordinate frames at every point of the coset. Show that if we take the components of the frames at the origin to be $e_\pi^n = \delta_\pi^n$ then the frames at a general coset point ξ^n can be obtained by sweeping out the frames at the origin using the group element $\exp \xi^n K_n = L(\xi)$ and that these coincide with the vierbein's given by the equation:

$$L^{-1}(\xi)\partial_\pi L(\xi) = e_\pi^n K_n + \omega_\pi^i H_i.$$

Interpret ω_π^i as the spin connection. This point of view is extensively discussed in Ref. 6.

14.5 Consider the $N = 1$ superconformal transformations to be those that leave the supersymmetric line element

$$ds^2 = (dx^\mu - d\theta \gamma^\mu \theta)^2$$

invariant up to a scale factor. Find all such transformations and verify that they coincide with those in Problem 14.2.

Chapter 15

15.1 The relaxed hypermultiplet formulation of $N = 2$ matter discussed on p. 127 contains the superfileds L^{jk}, L^{ijkl} and G which satisfy the constraints of Eqs. (15.121) and (15.122). Deduce its x-space component field content and its action in x-space and superspace.

Chapter 16

16.1 Take the supervierbein to be given in the form

$$E_N^\pi = \mathring{E}_N^\pi + \kappa H_N{}^\pi$$

where $\mathring{E}_N^\pi$ is the supervierbein of flat superspace given in Eq. (14.37). To lowest nontrivial order in $H_N{}^\pi$ and $\Omega_\pi{}^{bc}$ work out the torsions and curvatures and

(a) solve the Bianchi identities
(b) solve the constraints of $N = 1$ supergravity.

16.2 Solve the constraints

$$T_{\hat A \hat B}{}^c = T_{\hat A \hat B}{}^\pi = 0.$$

References

1. Y. A. Golfand and E. S. Likhtman, *JETP Lett.* **13**, 323 (1971).
2. D. V. Volkov and V. P. Akulov, *Pis'ma Zh. Eksp. Teor. Fiz.* **16**, 621 (1972); *Phys. Lett.* **46B**, 109 (1973).
3. J. Wess and B. Zumino, *Nucl. Phys.* **B70**, 139 (1974).
4. S. Coleman and J. Mandula, *Phys. Rev.* **159**, 1251 (1967).
5. R. Hagg, J. Lopuszanski and M. Sohnius, *Nucl. Phys.* **B88**, 61 (1975).
6. P. van Nieuwenhuizen and P. West, *Principles of Supersymmetry and Supergravity*, forthcoming book to be published by Cambridge University Press.
7. P. Ramond, *Phys. Rev.* **D3**, 2415 (1971); A. Neveu and J. H. Schwarz, *Nucl. Phys.* **B31**, 86 (1971); *Phys. Rev.* **D4**, 1109 (1971); J.-L. Gervais and B. Sakita, *Nucl. Phys.* **B34**, 477, 632 (1971); F. Gliozzi, J. Scherk and D. I. Olive, *Nucl. Phys.* **B122**, 253 (1977).
8. J. Wess and B. Zumino, *Nucl. Phys.* **B78**, 1 (1974).
9. For a discussion of the Noether procedure in the context of supergravity, see: S. Ferrara, D. Z. Freedman and P. van Nieuwenhuizen, *Phys. Rev.* **D13**, 3214 (1976).
10. S. Ferrara and B. Zumino, *Nucl. Phys.* **B79**, 413 (1974); A. Salam and J. Strathdee, *Phys. Rev.* **D11**, 1521 (1975).
11. A. Salam and J. Strathdee, *Nucl. Phys.* **B80**, 499 (1974); M. Gell-Mann and Y. Neeman, (1974) unpublished; W. Nahm, *Nucl. Phys.* **B135**, 149 (1978). For a review, see: D. Z. Freedman in *Recent Developments in Gravitation*, Cargèse (1978), eds. M. Levy and S. Deser (Gordon and Breach, New York, 1979); S. Ferrara and C. Savoy in *Supergravity '81*, eds. S. Ferrara and J. Taylor (Cambridge University Press, Cambridge, 1982).
12. E. P. Wigner, *Ann. of Math.* **40**, 149 (1939).
13. P. van Nieuwenhuizen, *Phys. Rep.* **68**, 189 (1981).
14. D. Freedman, P. van Nieuwenhuizen and S. Ferrara, *Phys. Rev.* **D13**, 3214 (1976); *Phys. Rev.* **D14**, 912 (1976).
15. S. Deser and B. Zumino, *Phys. Lett.* **62B**, 335 (1976).
16. K. Stelle and P. West, *Phys. Lett.* **B74**, 330 (1978).
17. S. Ferrara and P. van Nieuwenhuizen, *Phys. Lett.* **B74**, 333 (1978).
18. A. Chamseddine and P. West, *Nucl. Phys.* **B129**, 39 (1977).
19. P. Townsend and P. van Nieuwenhuizen, *Phys. Lett.* **B67**, 439 (1977).
20. J. Wess and B. Zumino, *Nucl. Phys.* **B78**, 1 (1974).
21. S. Ferrara in *Proceedings of the 9th International Conference on General Relativity and Gravitation* (1980), ed. Ernst Schmutzer.
22. M. Sohnius, K. Stelle and P. West, in *Superspace and Supergravity*, eds. S. W. Hawking and M. Rocek (Cambridge University Press, Cambridge, 1981).
23. A. Salam and J. Strathdee, *Phys. Lett.* **51B**, 353 (1974); P. Fayet, *Nucl. Phys.* **B113**, 135 (1976).

24. M. Sohnius, K. Stelle and P. West, *Nucl. Phys.* **B17**, 727 (1980); *Phys. Lett.* **92B**, 123 (1980).
25. P. Fayet, *Nucl. Phys.* **B113**, 135 (1976).
26. P. Breitenlohner and M. Sohnius, *Nucl. Phys.* **B178**, 151 (1981); M. Sohnius, K. Stelle and P. West, in *Superspace and Supergravity*, eds. S. W. Hawking and M. Rocek (Cambridge University Press, Cambridge, 1981).
27. P. Howe and P. West, "$N = 1$, $d = 6$ Harmonic Superspace", in preparation.
28. G. Sierra and P. K. Townsend, *Nucl. Phys.* **B233**, 289 (1984); L. Mezincescu and Y. P. Yao, *Nucl. Phys.* **B241**, 605 (1984).
29. F. Gliozzi, J. Scherk and D. Olive, *Nucl. Phys.* **B122**, 253 (1977); L. Brink, J. Schwarz and J. Scherk, *Nucl. Phys.* **B121**, 77 (1977).
30. In this context, see: M. Rocek and W. Siegel, *Phys. Lett.* **105B**, 275 (1981); V. O. Rivelles and J. G. Taylor, *J. Phys. A. Math. Gen.* **15**, 163 (1982).
31. S. Ferrara, J. Scherk and P. van Nieuwenhuizen, *Phys. Rev. Lett.* **37**, 1035 (1976); S. Ferrara, F. Gliozzi, J. Scherk and P. van Nieuwenhuizen, *Nucl. Phys.* **B117**, 333 (1976); P. Breitenlohner, S. Ferrara, D. Z. Freedman, F. Gliozzi, J. Scherk and P. van Nieuwenhuizen, *Phys. Rev.* **D15**, 1013 (1977); D. Z. Freedman, *Phys. Rev.* **D15**, 1173 (1977).
32. S. Ferrara and P. van Nieuwenhuizen, *Phys. Lett.* **76B**, 404 (1978).
33. K. S. Stelle and P. West, *Phys. Lett.* **77B**, 376 (1978).
34. S. Ferrara and P. van Nieuwenhuizen, *Phys. Lett.* **78B**, 573 (1978).
35. K. S. Stelle and P. West, *Nucl. Phys.* **B145**, 175 (1978).
36. M. Sohnius and P. West, *Nucl. Phys.* **B203**, 179 (1982).
37. R. Barbieri, S. Ferrara, D. Nanopoulos and K. Stelle, *Phys. Lett.* **113B**, 219 (1982).
38. E. Cremmer, S. Ferrara, B. Julia, J. Scherk and L. Girardello, *Phys. Lett.* **76B**, 231 (1978).
39. E. Cremmer, B. Julia, J. Scherk, S. Ferrara, L. Girardello and P. van Nieuwenhuizen, *Nucl. Phys.* **B147**, 105 (1979).
40. E. Cremmer, S. Ferrara, L. Girardello and A. Van Proeyen, *Nucl. Phys.* **B212**, 413 (1983); *Phys. Lett.* **116B**, 231 (1982).
41. S. Deser, J. Kay and K. Stelle, *Phys. Rev. Lett.* **38**, 527 (1977); S. Ferrara and B. Zumino, *Nucl. Phys.* **B134**, 301 (1978).
42. M. Sohnius and P. West, *Nucl. Phys.* **B198**, 493 (1982).
43. S. Ferrara, L. Girardello, T. Kugo and A. Van Proeyen, *Nucl. Phys.* **B223**, 191 (1983).
44. S. Ferrara, M. Grisaru and P. van Nieuwenhuizen, *Nucl. Phys.* **B138**, 430 (1978).
45. B. de Wit, J. W. van Holten and A. Van Proeyen, *Nucl. Phys.* **B184**, 77 (1981); *Phys. Lett.* **95B**, 51 (1980); *Nucl. Phys.* **B167**, 186 (1980).
46. A. Salam and J. Strathdee, *Phys. Rev.* **D11**, 1521 (1975); *Nucl. Phys.* **B86**, 142 (1975).
47. W. Siegel, *Phys. Lett.* **85B**, 333 (1979).
48. R. Arnowitt and P. Nath, *Phys. Lett.* **56B**, 117 (1975); L. Brink, M. Gell-Mann, P. Ramond and J. Schwarz, *Phys. Lett.* **74B**, 336 (1978); **76B**, 417 (1978); S. Ferrara and P. van Nieuwenhuizen, *Ann. Phys.* **126**, 111 (1980); P. van Nieuwenhuizen and P. West, *Nucl. Phys.* **B169**, 501 (1980).
49. M. Sohnius, *Nucl. Phys.* **B165**, 483 (1980).
50. P. Howe, K. Stelle and P. Townsend, *Nucl. Phys.* **B214**, 519 (1983).
51. M. Grisaru, M. Rocek and W. Siegel, *Nucl. Phys.* **B159**, 429 (1979).

52. E. Berezin, *The Method of Second Quantization* (Academic Press, New York, 1960).
53. A. Salam and J. Strathdee, *Nucl. Phys.* **B76**, 477 (1974); S. Ferrara, J. Wess and B. Zumino, *Phys. Lett.* **51B**, 239 (1974).
54. M. Grisaru, M. Rocek and W. Siegel, *Nucl. Phys.* **B159**, 429 (1979).
55. J. Wess, Lecture Notes in Physics 77 (Springer, Berlin, 1978).
56. J. Gates and W. Siegel, *Nucl. Phys.* **B147**, 77 (1979).
57. J. Gates, K. Stelle and P. West, *Nucl. Phys.* **B169**, 347 (1980).
58. R. Grimm, M. Sohnius and J. Wess, *Nucl. Phys.* **B133**, 275 (1978).
59. P. Breitenlohner and M. Sohnius, *Nucl. Phys.* **B178**, 151 (1981).
60. P. Howe, K. Stelle and P. West, *Phys. Lett.* **124B**, 55 (1983).
61. P. Howe, K. Stelle and P. West, "$N = 1$, $d = 6$ Harmonic Superspace", Kings College preprint.
62. M. Sohnius, K. Stelle and P. West, in *Superspace and Supergravity*, eds. S. W. Hawking and M. Rocek (Cambridge University Press, Cambridge, 1981).
63. A. Galperin, E. Ivanov, S. Kalitzin, V. Ogievetsky and E. Sokatchev, Trieste preprint.
64. L. Mezincescu, JINR report P2-12572 (1979).
65. J. Koller, *Nucl. Phys.* **B222**, 319 (1983); *Phys. Lett.* **124B**, 324 (1983).
66. P. Howe, K. Stelle and P. K. Townsend, *Nucl. Phys.* **B236**, 125 (1984).
67. A. Salam and J. Strathdee, *Nucl. Phys.* **B80**, 499 (1974).
68. S. Ferrara, J. Wess and B. Zumino, *Phys. Lett.* **51B**, 239 (1974).
69. J. Wess and B. Zumino, *Phys. Lett.* **66B**, 361 (1977); V. P. Akulov, D. V. Volkov and V. A. Soroka, *JETP Lett.* **22**, 187 (1975).
70. R. Arnowitt, P. Nath and B. Zumino, *Phys. Lett.* **56**, 81 (1975); P. Nath and R. Arnowitt, *Phys. Lett.* **56B**, 177 (1975); **78B**, 581 (1978).
71. N. Dragon, *Z. Phys.* **C2**, 62 (1979).
72. E. A. Ivanov and A. S. Sorin, *J. Phys. A. Math. Gen* **13**, 1159 (1980).
73. That some representations do not generalize to supergravity was noticed in: M. Fischler, *Phys. Rev.* **D20**, 1842 (1979).
74. P. Howe and R. Tucker, *Phys. Lett.* **80B**, 138 (1978).
75. P. Breitenlohner, *Phys. Lett.* **76B**, 49 (1977); **80B**, 217 (1979).
76. W. Siegel, *Phys. Lett.* **80B**, 224 (1979).
77. W. Siegel, "Supergravity Superfields Without a Supermetric", Harvard preprint HUTP-771 A068, *Nucl. Phys.* **B142**, 301 (1978); S. J. Gates Jr. and W. Siegel, *Nucl. Phys.* **B147**, 77 (1979).
78. See also in this context: V. Ogievetsky and E. Sokatchev, *Phys. Lett.* **79B**, 222 (1978).
79. R. Grimm, J. Wess and B. Zumino, *Nucl. Phys.* **B152**, 1255 (1979).
80. These constraints were first given by: J. Wess and B. Zumino, *Phys. Lett.* **66B**, 361 (1977).
81. J. Wess and B. Zumino, *Phys. Lett.* **79B**, 394 (1978).
82. P. Howe and P. West, *Nucl. Phys.* **B238**, 81 (1983).
83. P. Howe, *Nucl. Phys.* **B199**, 309 (1982).
84. A. Salam and J. Strathdee, *Phys. Rev.* **D11**, 1521 (1975).
85. S. Ferrara and O. Piguet, *Nucl. Phys.* **B93**, 261 (1975).
86. J. Wess and B. Zumino, *Phys. Lett.* **49B**, 52 (1974).
87. J. Iliopoulos and B. Zumino, *Nucl. Phys.* **B76**, 310 (1974).
88. S. Ferrara, J. Iliopoulos and B. Zumino, *Nucl. Phys.* **B77**, 41 (1974).

89. D. M. Capper, *Nuovo Cim.* **25A**, 259 (1975); R. Delbourgo, *Nuovo Cim.* **25A**, 646 (1975).
90. P. West, *Nucl. Phys.* **B106**, 219 (1976); D. Capper and M. Ramon Medrano, *J. Phys.* **62**, 269 (1976); S. Weinberg, *Phys. Lett.* **62B**, 111 (1976).
91. M. Grisaru, M. Rocek and W. Siegel, *Nucl. Phys.* **B159**, 429 (1979).
92. B. W. Lee in *Methods in Field Theory*, Les Houches 1975, eds. R. Balian and J. Zinn-Justin (North Holland, Amsterdam and World Scientific, Singapore, 1981).
93. W. Siegel, *Phys. Lett.* **84B**, 193 (1979); **94B**, 37 (1980).
94. L. V. Avdeev, G. V. Ghochia and A. A. Vladiminov, *Phys. Lett.* **105B**, 272 (1981); L. V. Avdeev and A. A. Vladiminov, *Nucl. Phys.* **B219**, 262 (1983).
95. D. M. Capper, D. R. T. Jones and P. van Nieuwenhuizen, *Nucl. Phys.* **B167**, 479 (1980).
96. G. 't Hooft and M. Veltman, *Nucl. Phys.* **B44**, 189 (1972); C. Bollini and J. Giambiagi, *Nuovo Cim.* **12B**, 20 (1972); J. Ashmore, *Nuovo Cim. Lett.* **4**, 37 (1972).
97. P. Howe, A. Parkes and P. West, *Phys. Lett.* **147B**, 409 (1984); *Phys. Lett.* **150B**, 149 (1985).
98. J. W. Juer and D. Storey, *Nucl. Phys.* **B216**, 185 (1983); O. Piguet and K. Sibold, *Nucl. Phys.* **B248**, 301 (1984).
99. E. Witten, *Nucl. Phys.* **B188**, 52 (1981).
100. E. Witten, in "Unity of the Fundamental Interactions", Proceedings of the 19th Erice School 1981, ed. A Zichichi (Plenum Press, 1983).
101. A. Parkes and P. West, *Phys. Lett.* **138B**, 99 (1984).
102. L. Mezincescu and D. T. R. Jones, *Phys. Lett.* **136B**, 242, 293 (1984).
103. L. Mezincescu and D. T. R. Jones, *Phys. Lett.* **138B**, 293 (1984).
104. P. West, *Phys. Lett.* **136B**, 371 (1984).
105. A. Parkes and P. West, "Three-Loop Results in Two-Loop Finite Supersymmetric Gauge Theories", *Nucl. Phys.* **B256**, 340 (1985); M. T. Grisaru, B. Milewski and D. Zanon, *Phys. Lett.* **155B**, 357 (1985).
106. B. S. deWitt, *Dynamical Theory of Groups and Fields* (Gordon and Breach, New York, 1978). R. Kallosh, *Nucl. Phys.* **B78**, 293 (1974); M. Grisaru, P. van Nieuwenhuizen and C. C. Wu, *Phys. Rev.* **D12**, 3202 (1975); J. Honerkamp, *Nucl. Phys.* **B36**, 130 (1971); **B48**, 269 (1972); G. 't Hooft in *Proceedings of XII Winter School of Theoretical Physics* in Karpacz; B. S. deWitt in *Quantum Gravity*, Vol. 2, eds. C. J. Isham, R. Penrose and D. W. Sciama (Oxford University Press; London, 1980). L. Abbott, *Nucl. Phys.* **B185**, 189 (1981); D. Boulware, "Gauge Dependence of the Effective Action", University of Washington preprint RLO 1388-822 (1980).
107. K. Schaeffer, L. Abbot and M. Grisaru, Brandeis preprint.
108. M. Grisaru and W. Siegel, *Nucl. Phys.* **B201**, 292 (1982); M. Rocek and W. Siegel, *Superspace* (Benjamin/Cummings, Reading, Mass., 1983).
109. S. Ferrara and B. Zumino, *Nucl. Phys.* **B79**, 413 (1974).
110. D. R. T. Jones, *Phys. Lett.* **72B**, 199 (1977); E. Poggio and H. Pendleton, *Phys. Lett.* **72B**, 200 (1977).
111. O. Tarasov, A. Vladimirov, A. Yu, *Phys. Lett.* **93B**, 429 (1980); M. T. Grisaru, M. Rocek and W. Siegel, *Phys. Rev. Lett.* **45**, 1063 (1980); W. E. Caswell and D. Zanon, *Nucl. Phys.* **B182**, 125 (1981).
112. M. Sohnius and P. West, *Phys. Lett.* **100B**, 45 (1981).

113. S. Ferrara and B. Zumino, unpublished.
114. M. Grisaru and W. Siegel, *Nucl. Phys.* **B201**, 292 (1982).
115. P. Howe, K. Stelle and P. Townsend, *Nucl. Phys.* **B236**, 125 (1984).
116. S. Mandelstam, Proc. 21st Int. Conf. on High Energy Physics, eds. P. Petiau and J. Pomeuf, *J. Phys.* *12*, 331 (1982).
117. L. Brink, O. Lindgren and B. Nilsson, *Nucl. Phys.* **B212**, 401 (1983); *Phys. Lett.* **123B**, 328 (1983).
118. D. Freedman, private communication.
119. D. R. T. Jones, *Nucl. Phys.* **B87**, 127 (1975).
120. P. Howe, K. Stelle and P. West, *Phys. Lett.* **124B**, 55 (1983).
121. P. West in *Proceedings of the Shelter Island II Conference on Quantum Field Theory and Fundamental Problems of Physics*, eds. R. Jackiw, N. Khuri, S. Weinberg and E. Witten, M.I.T. Press (June 1983). This work contains a general review of the finiteness properties of supersymmetric theories and in particular an account of the anomalies argument applicable to $N = 2$ theories. This latter work was performed in collaboration with P. Howe.
122. P. Howe and P. West, "The Two-Loop β-Function in Models in Extended Rigid Supersymmetry", King's preprint, *Nucl. Phys.* **B242**, 364 (1984).
123. A. A. Slavnov, *Teor. Mat. Fig.* **13**, 1064 (1972); B. W. Lee and J. Zinn-Justin, *Phys. Rev.* **D5**, 3121 (1972).
124. D. J. Gross and F. Wilczek, *Phys. Rev.* **D8**, 3633 (1973).
125. T. Curtright, *Phys. Lett.* **102B**, 17 (1981); P. West, Higher Derivative Regulation of Supersymmetric Theories, CALTEC preprint, CALT-68-1226.
126. S. L. Adler and W. A. Bardeen, *Phys. Rev.* **182**, 1517 (1969).
127. A. Zee, *Phys. Rev. Lett.* **29**, 1198 (1972); J. H. Lowenstein and B. Schroer, *Phys. Rev.* **D6**, 1553 (1972); **D7**, 1929 (1973).
128. M. Grisaru and P. West, "Supersymmetry and the Adler-Bardeen Theorem", *Nucl. Phys.* **B254**, 249 (1985).
129. P. Howe and P. West, unpublished.
130. S. Ferrara and B. Zumino, *Nucl. Phys.* **B79**, 413 (1974).
131. R. Slansky, *Phys. Rep.* **79**, 1 (1981); D. Olive, "Relation Between Grand Unified and Monopole Theories", Erice (1981); F. Gürsey in *1st Workshop on Grand Unification*, eds. P. H. Frampton, S. H. Glashow and A. Yildiz, (Math. Sci. Press, Brooklyn, Massachusetts, 1980); p. 39; Mehmet Koca, *Phys. Rev.* **D42**, 2636, (1981); D. Olive and P. West, *Nucl. Phys.* **B217**, 248 (1983).
132. A. Parkes and P. West, *Phys. Lett.* **127B**, 353 (1983).
133. R. Barbieri, S. Ferrara, L. Maiani, F. Palumbo and A. Savoy, *Phys. Lett.* **115B**, 212 (1982).
134. L. Susskind, private communication.
135. S. J. Gates, M. T. Grisaru, M. Rocek and W. Siegel, *Superspace* (Benjamin/Cummins, Reading, 1983).
136. L. Alvarez-Gaumé and D. Z. Freedman, *Phys. Lett.* **94B**, 171 (1980); *Comm. Math. Phys.* **80**, 443 (1981).
137. R. Kallosh in *Supergravity '81*, eds. S. Ferrara and J. G. Taylor (Cambridge University Press, Cambridge, 1982).
138. N. Marcus, and A. Sagnotti, *Phys. Lett.* **135B**, 85 (1984).
139. A. Parkes and P. West, *Phys. Lett.* **122B**, 365 (1983).
140. A. Parkes and P. West, *Nucl. Phys.* **B222**, 269 (1983).
141. A. Namazie, A. Salam and J. Strathdee, *Phys. Rev.* **D28**, 1481 (1983).

142. J. J. Van der Bij and Y.-P. Yao, *Phys. Lett.* **125B**, 171 (1983).
143. S. Rajpoot, J. G. Taylor and M. Zaimi, *Phys. Lett.* **127B**, 347 (1983).
144. A. Parkes and P. West, *Phys. Lett.* **127B**, 353 (1983).
145. J.-M. Frère, L. Mezincescu and Y.-P. Yao, *Phys. Rev.* **D29**, 1196 (1984).
146. L. Girardello and M. Grisaru, *Nucl. Phys.* **B194**, 55 (1982); O. Piquet, K. Sibold and M. Schweda, *Nucl. Phys.* **B174**, 183 (1980).
147. L. O'Raifeartaigh, *Nucl. Phys.* **B96**, 331 (1975); P. Fayet, *Phys. Lett.* **58B**, 67 (1975).
148. S. Dimopoulos and S. Raby, *Nucl. Phys.* **B192**, 353 (1981).
149. I. Ibañez and G. G. Ross, *Phys. Lett.* **105B**, 439 (1981); *Phys. Lett.* **110B**, 215 (1982); J. Ellis, I. Ibañez and G. G. Ross, *Phys. Lett.* **113B**, 283 (1982).
150. P. Fayet and J. Iliopoulos, *Phys. Lett.* **51B**, 461 (1974).
151. S. Ferrara, L. Girardello and F. Palumbo, *Phys. Rev.* **D20**, 403 (1979).
152. W. Fischler, N. P. Nilles, J. Polchinski, S. Raby and L. Susskind, *Phys. Rev. Lett.* **47**, 757 (1981).
153. E. Gildener and S. Weinberg, *Phys. Rev.* **D13**, 3333 (1976); M. Veltman, "The Infra-red Ultra-violet Connection", Univ. of Michigan preprint; L. Maiani, in Proc. of the Ecole d'Eté de Physique des Particules, Gif-sur-Yvette, p. 3 (1979).
154. E. Witten, *Nucl. Phys.* **B186**, 513 (1981) 150; S. Dimopoulos and H. Georgi *Nucl. Phys.* **B193**, 150 (1981); N. Sakai, *Z. Physik* **C11**, 153 (1982); R. Kaul, *Phys. Lett.* **B109**, 19 (1982).
155. The above discussion owes much to C. H. Llewellyn-Smith and G. G. Ross, *Phys. Lett.* **105B**, 38 (1981).
156. S. Deser and B. Zumino, *Phys. Lett.* **62B**, 335 (1976).
157. J. Polony, Budapest preprint KFK/1977/83 (1977).
158. E. Cremmer, S. Ferrara, L. Girardello and A. Van Proeyen, *Nucl. Phys.* **B212**, 413 (1983); *Phys. Lett.* **116B**, 231 (1982).
159. H. P. Nilles, *Phys. Lett.* **115B**, 193 (1982); L. Ibañez, *Phys. Lett.* **118B**, 73 (1982); J. Ellis, D. V. Nanopoulos and K. Tamvakis, *Phys. Lett.* **121B**, 123 (1983); L. Ibañez and C. Lopez, *Phys. Lett.* **126B**, 54 (1983). L. Alvarez-Gaumé, J. Polchinski and M. B. Wise, *Nucl. Phys.* **B221**, 495 (1983); J. Ellis, J. S. Hagelin, D. V. Nanopoulos and K. Tamvakis, *Phys. Lett.* **125B**, 275 (1983).
160. H. P. Nilles, *Phys. Rep.* **110C**, 1 (1984); H. E. Aber and G. L. Kane, *Phys. Rep.* **C**, to appear; R. Arnowitt, A. Chamseddine and P. Nath, *Phys. Rev. Lett.* **49**, 970 (1982); **50**, 232 (1983); R. Arnowitt *et al.*, *Applied* $N = 1$ *Supergravity* (World Scientific, Singapore).
161. J. P. Derendinger, S. Ferrara and A. Masiero, *Phys. Lett.* **143B**, 133 (1984).
162. S. Weinberg, *Phys. Rev.* **D8**, 3497 (1973).
163. D. R. T. Jones, L. Mezincescu and Y. P. Yao, *Phys. Lett.* **148B**, 317 (1984).
164. F. del Aguila, M. Dugan, B. Crinstein, L. Hall, G. G. Ross and P. West, *Nucl. Phys.* **B250**, 225 (1985).
165. S. Ferrara and B. Zumino, *Nucl. Phys.* **B87**, 2074 (1975).
166. M. Sohnius and P. West, *Phys. Lett.* **105B**, 353 (1981).
167. M. Sohnius and P. West, Lectures given at Trieste in Supergravity School, (1981), preprint in preparation.
168. G. Mack and A. Salam, *Ann. Phys.* **53**, 174 (1969).
169. T. Clark, O. Pignet and K. Sibold, *Nucl. Phys.* **B143**, 445 (1981).
170. M. Sohnius, *Phys. Lett.* **81B**, 8 (1979).

171. E. Bergshoeff, M. de Roo, B. De Wit and J. van Holten, *Nucl. Phys.* **B182**, 173 (1981).

172. J. Wess and B. Zumino, *Nucl. Phys.* **B70**, 139 (1974); M. Sohnius, Univ. of Karlsruhe, Thesis (1976) unpublished; see also P. van Nieuwenhuizen, *Phys. Rep.* **C68**, 189 (1981).

173. S. Adler, J. Collins and A. Duncan, *Phys. Rev.* **D15**, 1712 (1977); J. Collins, A. Duncan and S. Joglekar, *Phys. Rev.* **D16**, 438 (1977); N. K. Nielsen, *Nucl. Phys.* **B120**, 212 (1977).

174. T. Clark, O. Piguet and K. Sibold, *Ann. Phys.* **109**, 418 (1977); *Nucl. Phys.* **B143**, 445 (1978); *Nucl. Phys.* **B159**, 1 (1979). J. Wess and B. Zumino, unpublished.

175. T. Clark in *Supergravity*, eds. P. van Nieuwenhuizen, D. Freedman (North Holland, Amsterdam, 1985).

176. O. Piguet and M. Schweda, *Nucl. Phys.* **B92**, 344 (1975).

177. M. Grisaru in *Recent Developments in Gravitation*, eds. S. Deser and M. Levy (Gordon and Breach, New York, 1979). M. Grisaru and P. West, *Nucl. Phys.* **B268**, 335 (1985).

178. V. Ogievetski and E. Sokatchev, *Nucl. Phys.* **B124**, 309 (1977).

179. M. Kaku, P. K. Townsend and P. van Nieuwenhuizen, *Phys. Rev.* **D17**, 3179 (1978).

180. S. Ferrara and P. van Nieuwenhuizen, *Phys. Lett.* **74B**, 333 (1978); K. Stelle and P. West, *Phys. Lett.* **74B**, 330 (1978).

181. M. Sohnius and P. West, *Phys. Lett.* **105B**, 353 (1981); V. P. Akulov and V. A. Soroka, *Theor. Mat. Fiz.* **3**, 112 (1977).

182. D. Breitenlohner, *Nucl. Phys.* **B124**, 500 (1977); *Phys. Lett.* **80B**, 217 (1979).

183. M. Sohnius and P. West, *Nucl. Phys.* **B216**, 100 (1983); **B198**, 493 (1982); V. Rivelles and J. G. Taylor, *Phys. Lett.* **113B**, 467 (1982).

184. W. Siegel, U. C. Berkeley preprint PTH-83/22 (1983).

185. L. Brink, P. Di Vecchia and P. Howe, *Phys. Lett.* **65B**, 471 (1976).

186. S. Deser and B. Zumino, *Phys. Lett.* **65B**, 369 (1976).

187. D. Gross, J. Harvey, E. Martinec and R. Rohm, *Phys. Rev. Lett.* **54**, 502 (1985), *Nucl. Phys.* **B256**, 253 (1985) and "The Interacting Heterotic String," Princeton preprint (1985).

188. P. A. M. Dirac, *Lecture in Quantum Mechanics* (Belfer Graduate School of Science, Yeshiva University, New York, 1964).

189. P. Goddard, J. Goldstone, C. Rebbi and C. Thorn. *Nucl. Phys.* **B56**, 109 (1973).

190. V. Nambu, *Proc. Int. Conf. on Symmetries and Quark Modes*, Detroit 1969 (Gordon and Breach, New York, 1970).

191. M. Virasoro, *Phys. Rev.* **D1**, 2933 (1970).

192. R. C. Brower, *Phys. Rev.* **D6**, 1655 (1972); P. Goddard and C. B. Thorn, *Phys. Lett.* **40B**, 235 (1972).

193. E. Cremmer and J.-L. Gervais, *Nucl. Phys.* **B90**, 410 (1975); M. Kaku and K. Kikkawa, *Phys. Rev.* **D10**, 1110, 1823 (1974).

194. W. Siegel, *Phys. Lett.* **148B**, 556 (1984); **149B**, 157 (1984).

195. S. Raby and P. C. West, unpublished.

196. T. Banks and M. Peskin, in *Symposium on Anomalies, Geometry and Topology*, eds. W. A. Bardeen and A. R. White (World Scientific, Singapore, 1985); M. Kaku and J. Lykken, *ibid.*

197. C. Brower and C. B. Thorn, *Nucl. Phys.* **B31**, 163 (1971).

198. P. van Nieuwenhuizen, *Nucl. Phys.* **B60**, 478 (1973).
199. A. Neveu and P. West, *Nucl. Phys.* **B268**, 125 (1986).
200. L. P. S. Singh and C. R. Hagen, *Phys. Rev.* **D9**, 898 (1974).
201. A. Neveu, H. Nicolai and P. West, *Nucl. Phys.* **B264**, 573 (1986).
202. A. Neveu, J. Schwarz and P. West, *Phys. Lett.* **164B**, 51 (1985).
203. T. Banks and M. Peskin, SLAC preprint 3740, 1985.
204. A. Neveu, H. Nicolai and P. West, *Phys. Lett.* **167B**, 307 (1986).
205. Y. Meurice, CERN. preprints.
206. E. Witten, "Non-commutative Geometry and String Field Theory, Princeton preprint.
207. W. Siegel and B. Zweibach, Berkeley preprint UC12-PTH-85130, 1985.
208. K. Itoh, T. Kugo, H. Kunitomo and H. Oogun.
209. A. Restuccia and J. Taylor, unpublished.
210. K. Kato and K. Ogawa, *Nucl. Phys.* **B212**, 443 (1983). These coordinates were also utilized in Ref. 194.
211. E. Fradkin and G. Vilkovisky, *Phys. Lett.* **55B**, 224 (1975); I. A. Batalin and G. Vilkovisky, *Phys. Lett.* **69B**, 309 (1977); E. S. Fradkin and T. E. Fradkina, *Phys. Lett.* **72B**, 343 (1978).
212. B. Zumino, *Nucl. Phys.* **B89**, 535 (1975).
213. A. Parkes, *Phys. Lett.* **156B**, 73 (1985); D. R. T. Jones and A. Parkes, *Phys. Lett.* **160B**, 267 (1985).
214. D. I. Kazakov, *Phys. Lett.* **179B**, 952 (1986); A. V. Ermushev, D. Kazakov and O. Tarasov, *Nucl. Phys.* **B281**, 72 (1987); C. Luccesi, O. Pignet and K. Sibold, *Helv. Phys. Acta.* **61**, 321 (1988).
215. M. Ademollo, L. Brink, A. D'Adda, R. D'Auria, E. Napolitano, S. Sciuto, E. Del Giudice, P. Di Vecchia, S. Ferrara, F. Gliozzi, R. Musto and R. Pettorino, *Phys. Lett.* **62B**, 105 (1976); *Nucl. Phys.* **B111**, 77 (1976).
216. M. Sakamoto, *Phys. Lett.* **151B**, 115 (1985).
217. P. Howe and P. West, *Phys. Lett.* **156B**, 335 (1985).
218. Y. Tanii, *Phys. Lett.* **165B**, 275 (1985).
219. C. Hull and E. Witten, *Phys. Lett.* **160B**, 398 (1985); C. Hull, *Nucl. Phys.* **B267**, 266 (1986).
220. B. Zumino, *Phys. Lett.* **87B** 205 (1979).
221. J. Gates, C. Hull and M. Rocek, *Nucl. Phys.* **B248**, 157 (1984).
222. P. Howe and G. Papadopoulos, *Class. Quant. Grav.* **5**, 1647 (1988).
223. P. Howe, "Two-Dimensional Non-Linear σ Models", *Frontiers in High Energy Physics*, ed. I. G. Halliday (Adam Hilger).
224. C. Hull, "Superfield Theory", *Proceedings of 1986 Vancouver Workshop*, eds. G. Kunstatter and H. C. Lee (Plenum Press, New York).
225. P. Howe, *J. Phys.* **A12**, 393 (1979).
226. R. Brooks, F. Muhammad and S. Gates, *Nucl. Phys.* **B268**, 599 (1986); R. Brooks and S. Gates, *Nucl. Phys.* **B287**, 669 (1987).
227. M. Evans, J. Louis and B. Ovrut, *Phys. Rev.* **D10**, 3045 (1987); S. Randjbar-Daemi, A. Salam and J. Strathdee, preprint IC/88/180.
228. S. Gates and F. Gieres, "Unidexterous Supergravity, Beltrami Parametrization and BRST Quantization", BUTP 88/23.
229. F. Gieres, "Beltrami Differentials and 2d (Super)Gravity in the Chiral Gauge", Bern BUTP-88/28.

230. G. Moore, P. Nelson and J. Polchinsky, *Phys. Lett.* **169B**, 47 (1986).

231. S. Ferrara and P. Di Vecchia, *Nucl. Phys.* **B130**, 93 (1977); E. Witten, *Phys. Rev.* **D16**, 2991 (1977); D. Z. Freedman and P. K. Townsend, *Nucl. Phys.* **B177**, 282 (1981).

232. L. Alvarez-Gaumé and D. Z. Freedman, *Commun. Math. Phys.* **91**, 87 (1983); S. J. Gates, *Nucl. Phys.* **B238**, 349 (1984); A. C. Davis, M. D. Freeman and A. J. Macfarlane, *Nucl. Phys.* **B256**, 299 (1985).

233. T. Curtright and C. Zachos, *Phys. Rev. Lett.* **53**, 53 (1984); P. Howe and G. Sierra, *Phys. Lett.* **148B**, 451 (1984).

234. S. J. Gates and H. Nishino, *Class. Quant. Grav.* **3**, 391 (1986); M. Rocek, P. van Nieuwenhuizen and S. C. Zhang, *Ann. Phys.* **172**, 348 (1986).

235. E. Martinec, *Phys. Rev.* **D28**, 2604 (1983); L. Baulieu, M. Bellon and R. Grimm, *Phys. Lett.* **B198**, 343 (1987).

236. P. Howe and G. Papadopoulos, *Class. Quant. Grav.* **4**, 11 (1987); *Nucl. Phys.* **B289**, 264 (1987).

237. S. Gates, L. Lu and R. Oerter, *Phys. Lett.* **218**, 33 (1989).

238. K. G. Wilson and M. E. Fisher, *Phys. Rev. Lett.* **28**, 240 (1972); K. G. Wilson, *Phys. Rev. Lett.* **28**, 548 (1972); C. di Castro, *Lett. Nuovo Cimento* **5**, 69 (1972); G. Mack, in "Lecture Notes in Physics", International Summer Institute on Theoretical Physics, Kaiserslautern, eds. W. Ruhl and A. Vancura, Vol. 17 (Springer Verlag, Berlin, 1972); E. Brezin, J. C. Le Guillou and J. Zinn-Justin, *Phys. Rev.* **D8**, 434 (1973); *Phys. Rev.* **D8**, 2418 (1973); P. K. Mitter, *Phys. Rev.* **D7**, 2927 (1973); G. Parisi, in "Field Theory Approach to Second-Order Phase Transitions in Three- and Two-Dimensional Systems", lectures given at the 1973 Cargese Summer School, eds. E. Brezin and J. M. Charap (Gordon and Breach, New York, 1974); B. Schroer, *Phys. Rev.* **B8**, 4200 (1973); K. Wilson and J. Kogut, *Phys. Reports* **12C**, 75 (1974); K. Wilson, *Rev. Mod. Phys.* **55**, 583 (1983) and references therein.

239. A. Belavin, A. Polyakov and A. Zamolodchikov, *Nucl. Phys.* **B241**, 333 (1984).

240. M. Schiffer and D. Spencer, *Functionals of Finite Riemann Surfaces*, (Princeton University Press, 1954).

241. R. Horsley, *Nucl. Phys.* **B138**, 493 (1978).

242. A. Berezin, *The Method of Second Quantization* (Academic Press, 1966); R. Arnowitt, P. Nath and B. Zumino, *Phys. Lett.* **56B**, 81 (1975).

243. J. Cohen, *Nucl. Phys.* **B284**, 349 (1987).

244. J. Cardy, *Conformal Invariance, Phase Transitions*, eds. Domb and Lebowitz, Vol. 11 (Academic Press, 1987); P. Ginsparg, *Applied Conformal Field Theory*, Les Houches (1988), to appear, and references therein; P. Goddard, "Meromorphic Conformal Field Theory", preprint DAMTP 89-1, to be published in the Proceedings of C.I.R.M. Conference on Infinite Dimensional Lie Algebras, July 1988.

245. P. West, *Introduction to String Theory*, in 1988 Zakopone Summer School; *Acta Phys. Polonica* **B20**, 471 (1989); Trieste Summer School 1988; *Superstrings, Unified Theories and Cosmology 1988*, eds. G. Ellis, J. C. Pati, S. Randjbar-Daemi, E. Sezgin and Q. Skafi, (World Scientific, 1989); Séminaire de Mathématiques Supérieures, Montreal 1988, Proceedings to appear.

246. A. M. Polyakov, *JETP Lett.* **12**, 381 (1970); *Zh. JETP* **39**, 10 (1974).

247. P. Howe and P. West, *Phys. Lett.* **B227** (1989) 397.

248. V. A. Fateev and A. B. Zamolodchikov, *Sov. Phys. JETP* **62**, 215 (1985); P. Di

Vecchia, J. L. Petersen and H. B. Zheng, *Phys. Lett.* **162B**, 327 (1965); P. Di Vecchia, J. L. Petersen and M. Yu, *Phys. Lett.* **172B**, 211 (1986); P. Di Vecchia, J. L. Petersen, M. Yu and H. B. Zheng, *Phys. Lett.* **174B**, 280 (1986); W. Boucher, D. Friedan and A. Kent, *Phys. Lett.* **172B**, 316 (1986).

249. P. Howe and P. West, *Phys. Lett.* **223B** (1989) 377.

250. A. Zamolodchikov, *Sov. J. Nucl. Phys.* **44**, 529 (1986); D. Kastor, E. Martinec and S. Shenker, preprint EFI 88-31.

251. T. Banks, L. Dixon, D. Friedan and E. Martinec, *Nucl. Phys.* **B299**, 613 (1988); C. Hull and E. Witten, *Phys. Lett.* **160B**, 398 (1985); W. Boucher, D. Friedan and A. Kent, *Phys. Lett.* **B172**, 613 (1986).

252. E. Martinec, "Algebraic Geometry and Effective Lagrangians" EFI preprint 88-76 (1988).

253. C. Vafa and N. Warner, "Catastrophes and the Classification of Conformal Theories", Harvard preprint HUTP-88/A047 (1988).

254. W. Lerche, C. Vafa and N. Warner, "Chiral Rings in $N = 2$ Superconformal Theories", Harvard preprint HUTP-89/A065 (1989).

255. E. Martinec, Seminar at CERN (February 1989).

256. K. Wilson, *Phys. Rev.* **179**, 1499 (1969); W. Zimmerman, *Lectures on Elementary Particles and Quantum Field Theory*, Brandeis (1970), (MIT Press, Cambridge, 1970); C. Callan, Jr., *Methods in Field Theory*, Les Houches (North Holland, 1975).

257. P. Di Vecchia, J. L. Petersen and H. B. Zheng, *Phys. Lett.* **162B**, 327 (1985).

258. "Gauge Covariant String Field Theory", in *Supersymmetry, Supergravity and Superstrings '86*, eds. B. de Wit, P. Fayet and M. Grisaru (World Scientific) "Gauge Covariant and Dual Model Vertices", in *Frontiers of High Energy Physics*, ed. I. Halliday (Adam Hilger, 1987); "A Review of Duality String Vertices, Overlap Identities and the Group Theoretic Approach to String Theory", in *Particle Physics*, Cargèse (1987), eds. M. Lévy, J. L. Basdevant, M. Jacob, D. Speiser, J. Weyers and R. Gastmans (Plenum Press, 1988); *Frontiers in Particle Theory*, Proceedings of the Johns Hopkins Workshop 1987, eds. Y. Duan, G. Domokos and S. Kovesi-Domokos (World Scientific); "A Brief Review of the Group Theoretic Approach to String Theory", in *Conformal Field Theories and Related Topics*, Proceedings of Third Annecy Meeting on Theoretical Physics, LAPP, Annecy le Vieux, France, *Nucl. Phys. B* (Proc. Suppl.) **5B**, 217 (1988), eds. P. Binétruy, P. Sorba and R. Stora (North Holland, 1988); *Ninth Workshop on Grand Unification*, ed. R. Barloutaud (World Scientific, 1988); *1988 Electroweak Interactions and Unified Theories*, Proceedings of the XXIIIrd Rencontre de Moriond, ed. J. Tran Thanh Van (Editions Frontières, 1988).

259. A. Sen, *Phys. Rev.* **D32**, 2102 (1985).

260. M. Dine and N. Sieberg, *Phys. Lett.* **180B**, 104 (1986).

Subject Index

Abelian gauge group 25
Action formulas 58
Adler-Bardeen theorem 189, 194
Akulov and Volkov 2
Algebraically on-shell 21, 24
Anomalies 208, 216, 228, 231
Anomalies argument 186, 187
Anomaly multiplet 224, 285
Anomolous dimension 179
Anti-commutation relations 6
Anti-de Sitter superalgebra 14
Anti-de Sitter superspace 138
Antichiral 165
Auxiliary fermions 72
Auxiliary field 22, 23, 26, 46, 49, 52, 63, 71, 76, 84, 162, 177, 206, 207, 215, 228, 232, 242, 244, 246, 274

B.R.S. invariance 181
BRST quantize 407
Background superfields 182
Background-field formalism 190
Background-field method 180, 187
β-function 179, 186, 188, 189, 191, 193, 318
Bianchi identities 120, 121, 122, 134, 144, 145, 300, 304
Bose-Einstein 15, 17
Breaking of supersymmetry 204

Calculus 3
Callan-Symanzik-function 229
Casimir operator 17
Cauchy-Riemann equations 323
Central charge 11, 40, 72, 109, 110, 263, 351, 406
Charge conjugation 7
Chiral 142, 165, 225, 290
Chiral charge 352
Chiral correlators 351
Chiral current 187, 216, 225, 236, 360
Chiral density 147
Chiral fields 296
Chiral integrals 166

Chiral invariance 232
Chiral matter field 147
Chiral multiplet 55, 118, 159, 198, 206, 214, 228, 231
Chiral rotation 235
Chiral subintegral 284
Chiral superfield 123, 126, 183, 290
Chiral supermultiplet 78, 212
Chiral transformations 226
Christoffel symbol 280, 285, 291
Clifford algebra 37, 38, 268
Clifford vacuum 39, 65
Coleman-Mandula theorem 6, 10
Combinatoric factor 165, 170
Compensator 312
Complex manifold 291
Conformal 53
Conformal algebra 321
Conformal anomalies 319
Conformal chiral transformations R 227
Conformal group 219, 324
Conformal operators 357
Conformal supercurrent 225
Conformal supergravity 142
Conformal symmetry 218
Conjugate spinors 234
Conjugation 234
Conjugation matrix 233
Connection 93, 276
Constraints 311
Conventions 233, 248
Coset 270
Coset space 85, 87, 100, 108, 154, 277
Cosmological constant 81, 214
Counterterms 83
Covariant derivative 88, 93, 107, 134, 272
CPT self conjugates 35, 74, 187
Critical phenomena 318
Cross ratios 347
Curl multiplet 54
Current multiplet 273
Currents 230
Curvature 134, 150, 272, 300

δ-functions 100
D-density formula 83
De Sitter group 13, 403
Degree of divergence 174, 192
Descendents 352
Determinant 309
Dilatation 219, 226, 227, 235, 318, 321
Dilatation generator 219
Dilatation invariant 223
Dilatation weight 327
Dimensional reduction 170
Dimensional regularization 170, 230
Dirac conjugate 234, 253
Dirac operator 253
Dragon's theorem 135

ε-scalars 170
E_4 193
E_5 193
E_6 193
E_7 193
E_8 193
Effective action 190
Effective potential 176, 208
Einstein multiplet 79
Einstein's action 141
Einstein's equations 230
Energy-momentum 216, 360
Energy-momentum tensor 187, 217, 226, 236, 281
Euclidean space 165, 167
Euclidean spinors 254
Euclidean superspace 293
Extended rigid supersymmetry 64
Extended superspace 103

(4,0) superalgebra 268
F density formula 80
Faddeev-Popov determinant 168, 184
Faddeev-Popov ghosts 118, 180
Fayet-Iliopoulos mechanism 207
Fayet-Iliopoulous term 83
Fayet-O'Raifeartaigh mechanism 206, 213
Fermi-Bose balance 24, 223
Fermi-Bose symmetry 3
Fermi-Dirac statistics 15, 17
Fierz identity 170, 234, 398
Fierzing 241
Fierz rearrangement 20
Finiteness 179, 186, 189, 193
First-order formalism 50

Gauge completion 114, 126
Gauge group 65
Gauge prepotential 192
Gauge singlets 198
Gauge-fixing 166, 169, 171, 180, 192, 199
Gauge-hierarchy problem 209, 214
Gaugino 202
Gelfand and Likhtman 2, 6
General relativity 3
Ghosts 166, 168, 171, 190, 192, 199, 267
Goldstone spinor 204, 205
Gravitational anomalies 285
Gravitino 52, 242, 244
Graviton 297
Green's functions 159, 164, 318, 341

Heterotic string 246, 276, 309
Higgs field 210, 212
Higgs mass 212
Higgs mechanism 406
Higher derivatives 188
Hypermultiplet 72, 73, 110, 127, 128, 129, 194

Induced representations 32
Infinitesimal operators 89, 105
Infrared stable fixed point 318
Internal currents 187
Internal symmetry group 5
Invariant tensor 150
Irreducible representation 32, 35, 266

Jacobian 257, 332
Jacobi identities 7, 218, 259

Kähler manifold 291, 296
Killing vectors 89
Kinetic multiplet 58

Lagrange multipliers 72, 128
Landau-Ginzburg theories 354
Leptons 204
Light-cone coordinates 186, 248
Linear anomalies 225
Linear multiplet 55, 59, 72, 79, 99, 224
Linear superfield 142, 292
Line element 319
Local Abelian 27, 42
Local matter multiplets 308
Local tensor calculus 75
Logarithmic divergences 174, 211

Lorentz group 17, 85, 132, 133, 217, 244, 252
Lorentz invariance 15, 25, 194, 218, 238
Lorentz representation 10

Majorana condition 65, 66, 236, 238, 391
Majorana spinor 7, 19, 20, 25, 85, 240, 241, 255
Majorana supersymmetry 236, 240
Majorana vector spinor 42
Mass spectrum 214
Mass splitting 213
Massive QED 75
Massive representation 39
Massless QCD 229
Mass term 194, 202, 227
Maxwell action 82
Maxwell multiplet 77, 78
Maxwell equations 318
Metric 133, 233, 280, 291, 319
Minimal auxiliary field formulation 231
Minkowski line element 249
Minkowski metric 53
Minkowski space 4, 219
Mirror particles 65, 194
Möbius group 325, 342
Möbius transformation 324
Modular invariant 319
Momentum cutoff 209

(N, M) superalgebra 263
(N, M) supergravity 297
$N = 1$ conformal supergravity 231
$N = 1$ super-Feynman rules 157
$N = 1$ supergravity 46, 132, 156, 407
$N = 1$ supersymmetric Yang-Mills 36
$N = 1$ Yang-Mills theory 15, 19, 27, 115, 159
$N = 2$ Yang-Mills theory 66, 110, 127, 186, 191
$N = 2$ linear multiplet 112
$N = 2$ matter 65, 70
$N = 2$ minimal theories 351
$N = 2$ rigid supersymmetric theory 186
$N = 2$ supergravity 84
$N = 2$ supersymmetry 189
$N = 2, D = 10$ chiral supergravity theory 154
$N = 4$ Yang-Mills theory 74, 187, 191, 195, 201, 408
$N = 4$ multiplet 64
$N = 8$ supergravity 406

Neveu-Schwarz algebra 329
New minimal formulation 232, 407
New minimal matter couplings 84
Nielson-Kallash 169, 184
Noether coupling 240
Noether currents 227
Noether method 3, 27, 30, 46, 75, 226, 241, 243, 407
No-go theorems 6, 262
Non-renormalization theorem 175, 196, 203

(1, 0) multiplet 245
(1, 0) supergravity 301
(1, 0) superspace 270
(1, 0) supersymmetry 246, 309
(1, 1) superalgebra 259, 260
(1, 1) supergravity 312
(1, 1) superspace 294
1.5 order formalism 51
Observed sector 213
Off-shell formulation 154, 407
Off-shell infrared divergences 194
Old minimal formulation of supergravity 75, 84, 143, 148, 407
One-particle irreducible 160, 181
On-shell 21, 149
On-shell representation 21
Operator product expansions 341, 356
OSP(4,1) 133

Pauli matrices 233
Poincaré group 4, 5, 8, 32, 53, 88, 104, 217, 219
Poincaré superspace 138
Polonyi superpotential 215
Primary field 339, 341
Projectors 140, 165, 166, 192, 209

Q supersymmetries 242
Quadratic divergences 211
Quantum anomalies 228
Quantum correction 177
Quantum effective potential 154, 177
Quarks 204

R invariant 83, 84, 224, 232
Ramond algebra 329
Ramond-Neveu-Schwarz formulation 245
Rarita-Schwinger field 43, 44, 151, 155
Realistic models 204, 213
Relaxed hypermultiplet 130, 188, 409

Renormalization group 318
Representation preserving constraints 120, 143
Riemannian manifold 95
Riemann sphere 324
Riemann tensor 42
Rigid supersymmetry 3

σ-model 195, 276, 279, 291, 296
S supersymmetries 235, 242
SO(4N) 38
SO(M) field strengths 300
SO(N) 300
SO($r + 1, s + 1$) 321
SP(N) 11
SU(4) 74
SU(5) 210
Scalar multiplet 53, 79
Scalar supermultiplet 96
Second-order formalism 150
Self-dual tensors 74
Singlet 274
Singletons 238
Soft terms 195
Space-time anomalies 319
Special conformal supersymmetry 226, 227, 321
Special translations 226, 227
Spin connection 88, 105, 133, 137, 148, 243, 244, 270, 272, 302
Spin-3/2 42, 46
Spin-5/2 42
Spinor superfield 91
Spurion technique 195, 196, 197, 199
Standard model 2, 65, 214
String 319
Structure constants 27, 199
Superalgebras in two dimensions 259
Supercharge 7, 11, 37, 65, 86, 107, 204, 234, 236, 277
Superconformal algebra 13, 84
Superconformal anomalies 187
Super-conformal choice 143
Superconformal currents 216, 223, 226
Superconformal group 227, 242, 318
Superconformal theories 273, 341
Superconformal transformation 141
Supercovariantizing 76
Supercurrent 187, 189, 216, 226, 360
Superdeterminant 272, 332
Superdiffeomorphism 334
Super-Fermi-Feynman gauge 167

Super-Feynman rules 164, 169, 194, 196
Superfield 4, 86, 132, 157, 196, 198, 272
Supergravity 50, 75, 79, 136, 204, 230, 231, 232
Super integration 100
Supermultiplet 4, 281
Superparticle 239
Superpartners 212
Super-Poincaré group 234
Superpotential 62
Super QED 29, 42, 59, 207
Superspace 4, 85, 113, 132, 133, 194, 209, 271
Superspace constraints 136, 143
Superspace integral 157
Superstring 233, 244, 245, 246, 259, 317
Superstring theory 267
Supersymmetric Yang-Mills 119
Supersymmetry 227, 244, 285
Supersymmetry algebra 5, 17, 234
Supersymmetry-breaking 209
Supersymmetry current 236
Supersymmetry parameter 65
Supervielbein 105, 132, 147, 272, 289, 300, 334
Superweight 352
Super-Weyl transformations 310
Super Yang-Mills theory 54, 166, 228

10-dimensional Poincaré group 244
10-dimensions 319
(2, 0) superalgebra 268
(2, 2) super conformal models 351
(2, 2) superspace 288
(20+20) formulations of supergravity 232
2-component notation 397
2-dimensional Dirac algebra 233
2-dimensional models 233
2-dimensional supergravity 239
26-dimensions 319
Tangent space group 132, 133
Technical gauge hierarchy problem 210, 212
Tensor calculus 4, 53, 214, 215
Three-point function 343
Torsion 134, 150, 272, 300
Trace 229
Tree-level potential 205
Twisted chiral superfield 290, 292
Two auxiliary fields S 408
Two-point functions 162
Two-point Green's function 346

U(N) 11
USp($2N$) 39
Unitary 64, 168
Upper half plane 325

Vacuum amplitude 167, 168, 184
Vacuum energy 16, 176, 205
Vacuum expectation 204, 205, 206
Vector fields 27
Vector-like 65
Vertex 164, 165
Vielbein 93, 270, 305
Vierbein 88, 132, 148, 244

Ward identity 182
Wave-function infinites 185
Wave-function renormalizations 209
Wess-Zumino 15, 16, 59
Wess-Zumino gauge 68, 124, 196

Wess-Zumino model 2, 19, 35, 42, 97, 113, 157, 204, 226, 403
Wess-Zumino multiplet 160, 198, 230
Weyl constraint 237, 238
Weyl invariance 242
Weyl spinor 238, 245, 255, 260
Weyl supersymmetry 237
Weyl tensor 83
Weyl transformation 242, 244
Wick rotation 248
Wigner method 32
Wilson coefficients 360
World sheet 244

Yang-Mills 27, 29, 46, 99, 123, 137, 148, 184, 198, 204
Yukawa couplings 70, 72, 212, 214

Z_2 graded 6, 11, 17, 259